Student Solutions Manual
for Tussy and Gustafson's

# *Developmental Mathematics for College Students*

Kimberly Shockey
Tod Shockey

**BROOKS/COLE**

**THOMSON LEARNING**

Australia • Canada • Mexico • Singapore • Spain • United Kingdom • United States

Assistant Editor: *Rachael Sturgeon*
Marketing Manager: *Leah Thomson*
Marketing Communications: *Samantha Cabaluna*
Marketing Assistant: *Maria Salinas*
Editorial Assistant: *Jonathan Wegner*

Production Coordinator: *Dorothy Bell*
Cover Design: *Roy R. Neuhaus*
Cover Illustration: *George Abe*
Print Buyer: *Christopher Burnham*
Printing and Binding: *Victor Graphics*



*For more information about this or any other Brooks/Cole product, contact:*
BROOKS/COLE
511 Forest Lodge Road
Pacific Grove, CA 93950 USA
www.brookscole.com
1-800-423-0563 (Thomson Learning Academic Resource Center)



Printed in the United States of America

10  9  8  7  6  5  4  3  2  1

ISBN 0-534-38355-6

## Section 1.1 An Introduction to the Whole Numbers

### Vocabulary

1. A **set** is a collection of objects.
3. When 297 is written as 2 hundreds + 9 tens + 7 ones, it is written in **expanded** notation.
5. Using a process know as graphing, whole numbers can be represented on a **number** line.

### Concepts

7. In the numeral 57,634, the **3** is in the tens column.
9. In the numeral 57,634, the **6** is in the hundreds column.
11. whole numbers
13. Graph 1, 3, 5, and 7

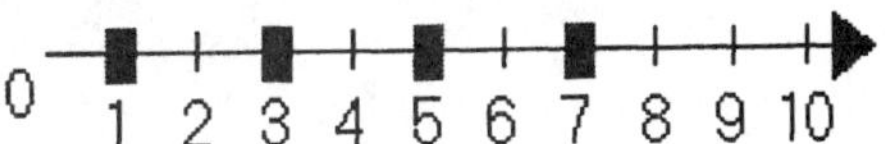

15. Graph the whole numbers less than 6

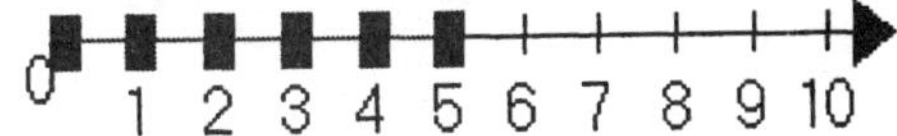

17. 47 > 41
19. 309 > 300
21. 2052 < 2502
23. Since 4 < 7, it is also true that 7 > 4.

### Notation

25. The symbols { }, called **braces**, are used when writing a set.

### Practice

27. 245; 2 hundreds + 4 tens + 5 ones; two hundred forty-five
29. 3609; 3 thousands + 6 hundreds + 9 ones; three thousand six hundred nine
31. 32,500; 3 ten thousands + 2 thousands + 5 hundreds; thirty-two thousand five hundred
33. 104,401; 1 hundred thousand + 4 thousands + 4 hundreds + 1 one; one hundred four thousand four hundred one
35. 425
37. 2736
39. 456
41. 27,598
43. 9,113
45. 10,700,506
47. 79,500
49. 80,000
51. 5,926,000
53. 5,900,000
55. $419,160
57. $419,000

**Applications**

59. EATING HABITS

| Country | Per Person Consumption |
|---|---|
| United States | 261 lb |
| New Zealand | 259 lb |
| Australia | 239 lb |
| Cyprus | 236 lb |
| Uruguay | 230 lb |
| Austria | 229 lb |
| Saint Lucia | 222 lb |
| Denmark | 219 lb |
| Canada | 211 lb |
| Spain | 211 lb |

61. MISSIONS TO MARS
    a.    The 70's had the greatest number of successful missions.
    b.    The 60's had the greatest number of unsuccessful missions.

63. ENERGY RESERVES

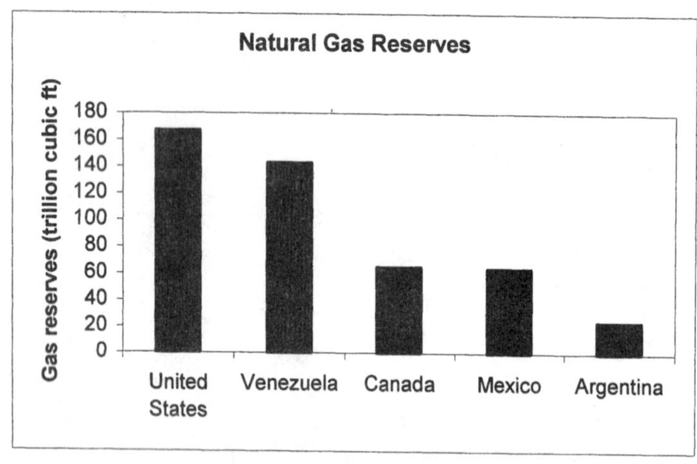


65. COFFEE

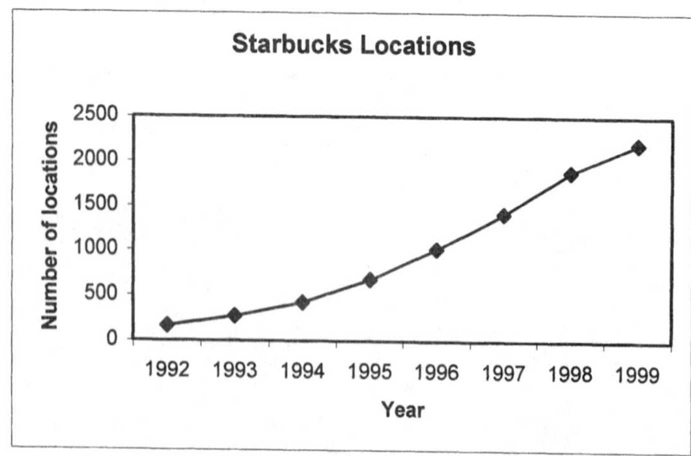

67. CHECKS

a. $15,601$\dfrac{00}{100}$ ; Fifteen thousand six hundred one and $\dfrac{00}{100}$ .

b. $3,433$\dfrac{46}{100}$ ; Three thousand four hundred thirty three and $\dfrac{46}{100}$ .

69. EDITING

one million, eight hundred sixty-five thousand, five hundred ninety-three votes; 1,865,593
four hundred eighty-two thousand, eight hundred eighty; 482,880
one thousand five hundred three; 1,503
two hundred sixty-nine; 269
forty-three thousand four hundred forty-nine;43,449

71. SPEED OF LIGHT

a. Rounded to the nearest hundred thousand meters per second is 299,800,000 m/s.
b. Rounded to the nearest million meters per second is 300,000,000 m/s.

**Writing**

73. Answers will vary.
75. Answers will vary.

## Section 1.2 Adding and Subtracting Whole Numbers

### Vocabulary

1. When two numbers are added, the result is called a **sum**. The numbers that are added are called **addends**.
3. A **rectangle** is a four-sided figure (like a dollar bill) whose opposite sides are of equal length.
5. When two numbers are subtracted, the result is called a **difference**. In a subtraction problem, the **subtrahend** is subtracted from the **minuend**.
7. The property that allows us to group numbers in an addition in any way we want is called the **associative** property of addition.

### Concepts

9. Commutative property of addition
11. Associative property of addition
13. Use $x$ and $y$,
    a. Commutative property of addition: $x + y = y + x$
    b. Associative property of addition: $(x + y) + z = x + (y + z)$

15. Any number added to **0** remains the same.
17. The addition fact illustrated is $4 + 3 = 7$.

### Notation

19. The grouping symbols ( ) are called **parentheses**.

21. $(36 + 11) + 5 = 47 + 5$
$$= 52$$

### Practice

23. $25 + 13 = 38$

25. $156 + 305 = 461$

27. $19 + 39 + 53 = 111$

29. $(95 + 16) + 39 = (111) + 39 = 150$

31. $25 + (321 + 17) = 25 + (338) = 363$

33. $\begin{array}{r} 632 \\ +347 \\ \hline 979 \end{array}$

35. $\begin{array}{r} 1{,}372 \\ +\ 613 \\ \hline 1{,}985 \end{array}$

37. $\begin{array}{r} 6{,}427 \\ +3{,}573 \\ \hline 10{,}000 \end{array}$

39. $\begin{array}{r} 8,539 \\ +7,368 \\ \hline 15,907 \end{array}$

41. $\begin{array}{r} 1,246 \\ 578 \\ +\ \ 37 \\ \hline 1,861 \end{array}$

43. $\begin{array}{r} 3,156 \\ 1,578 \\ +\ 578 \\ \hline 5,312 \end{array}$

45. Perimeter $= 2l + 2w$

    $2(32) + 2(12) = 88$ ft

47. Perimeter $= 4s$

    $4(17) = 68$ in

49. $17 - 14 = 3$

51. $39 - 14 = 25$

53. $174 - 71 = 103$

55. $633 - (598 - 30) = 633 - (568) = 65$

57. $160 - 15 - 4 = 160 - 19 = 141$

59. $29 - 17 - 12 = 29 - 29 = 0$

61. $\begin{array}{r} 367 \\ -343 \\ \hline 24 \end{array}$

63. $\begin{array}{r} 423 \\ -305 \\ \hline 118 \end{array}$

65. $\begin{array}{r} 1,537 \\ -\ 579 \\ \hline 958 \end{array}$

67. $\begin{array}{r} 4,267 \\ -2,578 \\ \hline 1,689 \end{array}$

69. $\begin{array}{r} 17,246 \\ -\ 6,789 \\ \hline 10,457 \end{array}$

71.  $\begin{array}{r} 15,700 \\ -15,397 \\ \hline 303 \end{array}$

73. $43 - 12 + 9 = 43 - 3 = 40$

75. $120 + 30 - 40 = 120 - 10 = 110$

**Applications**

77. TAXIS
$\$23 - \$5 = \$18$
$\$18$ was the fare.

79. MAGAZINE CIRCULATION
Circulation in 1996 was 1,803,566
Circulation in 1997 grew by 15,865
Circulation in 1998 decreased by 69,404

Total circulation in 1997 was $1,803,566 + 15,865 = 1,819,431$
Total circulation in 1998 was $1,819,431 - 69,404 = 1,750,027$

The total circulation in 1998 was 1,750,027.

81. BANKING
Initial balance was \$370.
Deposit of \$40
Withdrawal of \$197
Therefore, after the deposit and withdrawal the account balance
is $\$370 + \$40 - \$197 = \$410 - \$197 = \$213$.

83. TAX DEDUCTION
The total number of miles driven is the sum of the miles driven for each of the first six months which is
$2,345 + 1,712 + 1,778 + 445 + 1,003 + 2,774$
$= 2,345 + 1,712 + 1,778 + 445 + 3,777$
$= 2,345 + 1,712 + 1,778 + 4,222$
$= 2,345 + 1,712 + 6,000$
$= 2,345 + 7,712$
$= 10,057$ miles

85. INCOME
   a.   Step 1/Column 1
      Each year is a Step so sum the salaries for the first five steps from Column 1,
      $\$26,785 + \$28,107 + \$29,429 + \$30,751 + \$32,073 = \$147,145$.
   b.   Step 1/Column 3
      Each year is a Step so sum the salaries for the first five steps from Column 3,
      $\$29,701 + \$31,023 + \$32,345 + \$33,667 + \$34,989 = \$161,725$.

87. BLUEPRINT

The length of the house is the sum of the parts, $24 + 35 + 16 + 16 = 91$ ft.
The house is 91 ft long

89. CAR EMISSIONS

The number of tons that have been removed daily is the sum of the tons removed daily for each step taken. $215 + 85 + 117 + 35 + 70 + 150 + 120 = 792$ tons
792 tons have been removed daily because of the legislation.

91. CITY FLAG

The number of inches of fringe needed is equivalent to the perimeter of the flag.
$P = 2l + 2w = 2(64) + 2(34) = 128 + 68 = 196$ in.
The amount of fringe needed is 196 inches.

**Writing**

93. Answers will vary.

**Review**

95. 3,125; 3 thousands + 1 hundred + 2 tens + 5 ones
97. Rounded to the nearest ten is 6,354,780.
99. Rounded to the nearest ten thousand is 6,350,000

**Section 1.3 Multiplying and Dividing Whole Numbers**

**Vocabulary**

1.  **Multiplication** is repeated addition.
3.  The statement $ab = ba$ expresses the **commutative** property of multiplication.
5.  If a square measures one inch on each side, its area is 1 **square inch**.

**Concepts**

7.  $8 + 8 + 8 + 8 = 4 \cdot 8$

9.  The amount of surface enclosed by a rectangle is found by multiplying its length by its width.

11. Do each multiplication
    a.    $1 \cdot 25 = 25$
    b.    $62(1) = 62$
    c.    $10 \cdot 0 = 0$
    d.    $0(4) = 0$

13. The number of squares is $5 \cdot 12$.

**Notation**

15. Three symbols used for
    a.    Multiplication are $\times$, $\cdot$, $(\ )$.

    b.    Division are $\div$, $-$, $\overline{)\ \ \ }$ .

17. $ft^2$ means square feet.

**Practice**

19. $12 \cdot 7 = 84$

21. $27(12) = 324$

23. $9 \cdot (4 \cdot 5) = 9 \cdot 20 = 180$

25. $5 \cdot 7 \cdot 3 = 35 \cdot 3 = 105$

27.
$$
\begin{array}{r}
99 \\
\times\ \underline{77} \\
693 \\
\underline{693\phantom{0}} \\
7{,}623
\end{array}
$$

29.  $\begin{array}{r} 20 \\ \times\,53 \\ \hline 60 \\ 100 \\ \hline 1{,}060 \end{array}$

31.  $\begin{array}{r} 112 \\ \times\,23 \\ \hline 336 \\ 224 \\ \hline 2{,}576 \end{array}$

33.  $\begin{array}{r} 207 \\ \times\,97 \\ \hline 1449 \\ 1863 \\ \hline 20{,}079 \end{array}$

35.  $13{,}456 \cdot 217 = 2{,}919{,}952$

37.  $3{,}302 \cdot 358 = 1{,}182{,}116$

39.  $\begin{aligned} Area &= lw \\ &= 14(6) \\ &= 84\,\text{in}^2 \end{aligned}$

41.  $\begin{aligned} Area &= lw \\ &= 12(12) \\ &= 144\,\text{in}^2 \end{aligned}$

43.  $40 \div 5 = 8$

45.  $42 \div 14 = 3$

47.  $132 \div 11 = 12$

49.  $\dfrac{221}{17} = 13$

51.  $\begin{array}{r} 73 \\[-2pt] 13\overline{)949} \\ 91 \\ \hline 39 \\ 39 \\ \hline 0 \end{array}$

Answer is 73.

$$\begin{array}{r} 41 \\ 33\overline{)1{,}353} \\ 132 \\ \hline 33 \\ 33 \\ \hline 0 \end{array}$$

53.

Answer is 41.

$$\begin{array}{r} 205 \\ 39\overline{)7{,}995} \\ 78 \\ \hline 19 \\ 0 \\ \hline 195 \\ 195 \\ \hline 0 \end{array}$$

55.

Answer is 205.

$$\begin{array}{r} 210 \\ 29\overline{)6{,}090} \\ 58 \\ \hline 29 \\ 29 \\ \hline 00 \\ 0 \\ \hline 0 \end{array}$$

57.

Answer is 210.

$$\begin{array}{r} 8 \\ 31\overline{)273} \\ 248 \\ \hline 25 \end{array}$$

59.

The quotient is 8 and the remainder is 25.

$$\begin{array}{r} 20 \\ 37\overline{)743} \\ 74 \\ \hline 3 \\ 0 \\ \hline 3 \end{array}$$

61.

The quotient is 20 and the remainder is 3.

$$\begin{array}{r} 30 \\ 42\overline{)1{,}273} \\ \underline{126} \\ 13 \\ \underline{0} \\ 13 \end{array}$$

63. The quotient is 30 and the remainder is 13.

$$\begin{array}{r} 31 \\ 57\overline{)1{,}795} \\ \underline{171} \\ 85 \\ \underline{57} \\ 28 \end{array}$$

65. The quotient is 31 and the remainder is 28.

**Applications**

67. FIGURING WAGES

The cook worked 12 hours at $11 per hour, therefore she earned $12(11) = \$132$.

69. FINDING DISTANCE

The car can hold 14 gallons and goes 29 miles on one gallon; therefore on a full tank (14 gallons) the car can go $14(29) = 406$ miles.

71. CONCERT ATTENDANCE

The total number of concerts is $2(37) = 74$. Since 1700 fans attended each concert the total number of people to hear the group was $1700(74) = 125{,}800$.

73. ORANGES IN JUICE

If it takes 13 oranges to make one can of juice then $24(13) = 312$ oranges are need to make a case of 24 cans.

75. CAPACITY OF AN ELEVATOR

If there are 14 people each averaging 150 pounds then the weight in the elevator is $14(150) = 2100$ pounds. The elevator is overloaded.

77. WORD PROCESSING

If there are 8 columns and 9 rows then there will be $8(9) = 72$ entries in the table.

79. DISTRIBUTING MILK

$$\begin{array}{r} 3 \\ 23\overline{)73} \\ \underline{69} \\ 4 \end{array}$$

If the milk is distributed evenly then there will be **4 cartons** left over.

81. MILEAGE

$$140\overline{)700}$$ with quotient $5$, $700$ subtracted giving $0$.

The bus can travel **5 miles** on one gallon.

83. CONVERSION

$$5,280\overline{)11,000}$$ with quotient $2$, $10,560$ subtracted giving $440$.

11,000 is **440 feet** more than two miles.

85. PRICE OF A TEXTBOOK

$954,193 \div 23,273 = 41$

The cost of each book is $41.

87. VOLLEYBALL LEAGUE

Since a reasonable team size is 7 to 10 girls, divide the total number of girls by each team size to determine which meets all the requirements listed.

$216 \div 7 = 30$ teams with a remainder of 6 girls

$216 \div 8 = 27$ teams

$216 \div 9 = 24$ teams

$216 \div 10 = 10$ teams with a remainder of 6 girls

Since all teams must have the same number of players and the number of teams must be even, there should be **24 teams with 9 girls on each** to meet all the requirements.

89. COMPARING ROOMS

$$Area_{rectangle} = lw \qquad\qquad Area_{square} = lw$$
$$= 14(17) \qquad\qquad\qquad = 16(16)$$
$$= 238 \text{ feet}^2 \qquad\qquad\quad = 256 \text{ feet}^2$$

The **square room** has the greater area.

$$Perimeter_{rectangle} = 2l + 2w \qquad Perimeter_{square} = 4l$$
$$= 2(14) + 2(17) \qquad\qquad = 4(16)$$
$$= 28 + 34 \qquad\qquad\qquad = 64 \text{ feet}$$
$$= 62 \text{ feet}$$

The **square room** has the greater perimeter.

91. GARDENING

$$Area_{garden} = lw$$
$$= 27(19)$$
$$= 513 \text{ ft}^2$$

Since the path is 125 ft$^2$ there will be $513 \text{ ft}^2 - 125 \text{ ft}^2 = 388 \text{ ft}^2$ left for planting.

**Writing**

93. Answers will vary.
95. Answers will vary.

**Review**

97. The 8 is in the hundreds column.
99.    357
        39
      +476
      ‾‾‾‾‾
       872

**Estimation**

1. $\begin{array}{r} 30,000 \\ 10,000 \\ 9,000 \\ 1,000 \\ 10,000 \\ +\,30,000 \\ \hline 90,000 \end{array}$

   The estimate does not seem reasonable; no.

3. $\begin{array}{r} 500 \\ \times\quad 70 \\ \hline 35,000 \end{array}$

   The estimate does not seem reasonable; no.

5. $60,000 \div 30 = 2,000$
   The answer given of 200 is not reasonable.

7. CAMPAIGNING
   She flew approximately 8,900 miles.
   $4,000 + 600 + 1,000 + 300 + 3,000 = 8,900$ mi

9. GOLF COURSE
   The number of bags needed is approximately $90,000 \div 3,000 = 30$.

11. CURRENCY
    The number of \$5 bills in circulation is approximately $8,000,000,000 \div 5 = 1,600,000,000$.

**Section 1.4 Prime Factors and Exponents**

**Vocabulary**

1. Numbers that are multiplied together are called **factors**.
3. To **factor** a whole number means to express it as the product of other whole numbers.
5. Whole numbers, greater than 1, that are not prime numbers are called **composite** numbers.
7. To prime factor a number means to write it as a product of only **prime** numbers.
9. In the exponential expression $6^4$, 6 is called the **base**, and 4 is called the **exponent**.

**Concepts**

11. $1 \cdot 27$ and $3 \cdot 9$

13. Find the number
    a.    44
    b.    100

15. Find the factors
    a.    1 and 11
    b.    1 and 23
    c.    1 and 37
    d.    All the numbers are prime.

17. If 4 is a factor of a whole number then 4 also divides that number exactly.

19. $2 \cdot 3 \cdot 3 \cdot 5 = 90$

21. $11^2 \cdot 5 = 11 \cdot 11 \cdot 5 = 605$

23. The order of the base and exponent in an exponential expression **cannot** be changed.

$3^2 \neq 2^3$ since $3^2 = 9$ and $2^3 = 8$

25. Prime factorization of 30 is $2 \cdot 3 \cdot 5$

        30

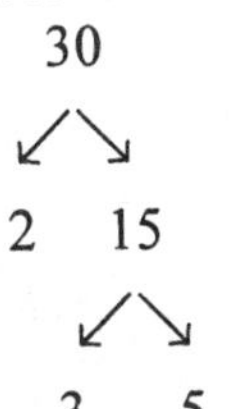

    2   15

        3   5

Prime factorization of 242 is $2 \cdot 11 \cdot 11$

        242

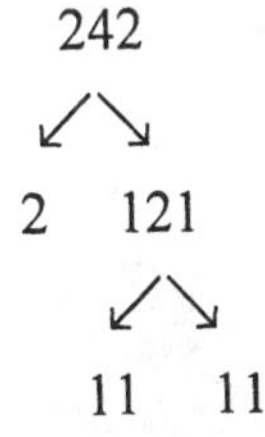

    2   121

        11  11

The prime factor in common for 30 and 242 is 2.

27. Prime factorization of 20 is $2 \cdot 2 \cdot 5$

        20

    2   10

        2   5

Prime factorization of 50 is $2 \cdot 5^2$

        50

    2   25

        5   5

The prime factors in common for 20 and 50 are 2 and 5.

29. 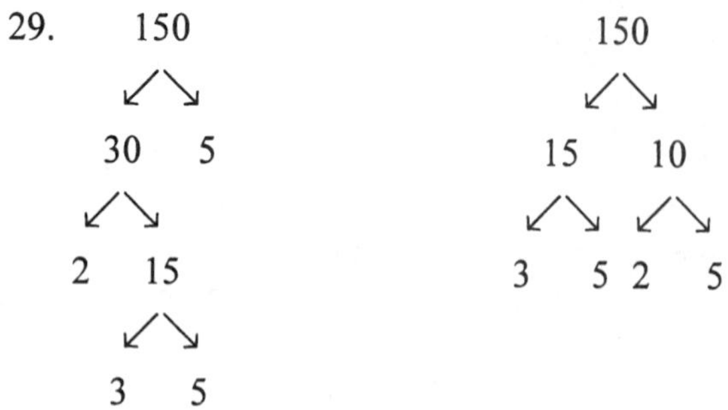

The results are the same.

31. Complete the table

| Product of factors of 12 | Sum of factors of 12 |
| --- | --- |
| $1 \cdot 12$ | $1 + 12 = 13$ |
| $2 \cdot 6$ | $2 + 6 = 8$ |
| $3 \cdot 4$ | $3 + 4 = 7$ |

33. If the number is even the first choice when doing a prime factorization should be 2.

**Notation**

35. $7^3 = 7 \cdot 7 \cdot 7$

37. $3^5 = 3 \cdot 3 \cdot 3 \cdot 3 \cdot 3$

39. $5^2(11) = 5 \cdot 5 \cdot 11$

41. $10^1 = 10$

43. $2 \cdot 2 \cdot 2 \cdot 2 \cdot 2 = 2^5$

45. $5 \cdot 5 \cdot 5 \cdot 5 = 5^4$

47. $4(4)(5)(5) = 4^2(5^2)$

**Practice**

49. The factors of 10 are 1, 2, 5 and 10.
51. The factors of 40 are 1, 2, 4, 5, 8, 10, 20 and 40.
53. The factors of 18 are 1, 2, 3, 6, 9 and 18.
55. The factors of 44 are 1, 2, 4, 11, 22 and 44.
57. The factors of 77 are 1, 7, 11 and 77.
59. The factors of 100 are 1, 2, 4, 5, 10, 20, 25, 50 and 100.
61. The prime-factored form of 39 is $3 \cdot 13$.
63. The prime-factored form of 99 is $3^2 \cdot 11$.
65. The prime-factored form of 162 is $2 \cdot 3^4$.
67. The prime-factored form of 220 is $2^2 \cdot 5 \cdot 11$.
69. The prime-factored form of 64 is $2^6$.
71. The prime-factored form of 147 is $3 \cdot 7^2$.
73. $3^4 = 3 \cdot 3 \cdot 3 \cdot 3 = 81$
75. $2^5 = 2 \cdot 2 \cdot 2 \cdot 2 \cdot 2 = 32$
77. $12^2 = 144$

79. $8^4 = 8 \cdot 8 \cdot 8 \cdot 8 = 4,096$

81. $3^2(2^3) = 3 \cdot 3 \cdot 2 \cdot 2 \cdot 2 = 72$

83. $2^3 \cdot 3^3 \cdot 4^2 = 2 \cdot 2 \cdot 2 \cdot 3 \cdot 3 \cdot 3 \cdot 4 \cdot 4 = 3,456$

85. $234^3 = 234 \cdot 234 \cdot 234 = 12,812,904$

87. $23^2 \cdot 13^3 = 23 \cdot 23 \cdot 13 \cdot 13 \cdot 13 = 1,162,213$

**Applications**

89. PERFECT NUMBERS
   The factors of 28 are 1, 2, 4, 7, 14, and 28
   $1 + 2 + 4 + 7 + 14 = 28$ therefore 28 is a perfect number.

91. LIGHT
   From the picture, 1 yd from the bulb the light energy passes through 1 square unit of area, 2 yds from the bulb the light energy passes through $4 = 2^2$ square units of area, 3 yds from the bulb the light energy passes through $9 = 3^2$ square units of area and 4 yds from the bulb the light energy passes through $16 = 4^2$ square units of area.

**Writing**

93. Answers will vary.
95. Answers will vary

**Review**

97. 230,999 rounded to the nearest thousand is 231,000.
99. $0 \div 15 = 0$
101. Area of a rectangle is $A = lw$.

**Section 1.5 Order of Operations**

**Vocabulary**

1.  The grouping symbols ( ) are called **parentheses**, and the symbols [ ] are called **brackets**.
3.  To **evaluate** $2+5\cdot4$ means to find its value.

**Concepts**

5.  Three operations need to be performed; square, multiply and subtract.
7.  In the numerator, multiplications should be done first.  In the denominator, subtraction should be done first.
9.  $2\cdot3^2 = 2\cdot3\cdot3 = 2\cdot9 = 18$ and $(2\cdot3)^2 = (6)^2 = 36$

**Notation**

11. $\begin{aligned}28 - 5(2)^2 &= 28 - 5(4)\\ &= 28 - 20\\ &= 8\end{aligned}$

13. $\begin{aligned}[4(2+7)] - 6 &= [4(9)] - 6\\ &= 36 - 6\\ &= 30\end{aligned}$

**Practice**

15. $7 + 4\cdot5 = 7 + 20 = 27$

17. $2 + 3(0) = 2 + 0 = 2$

19. $20 - 10 + 5 = 10 + 5 = 15$

21. $25 \div 5\cdot5 = 5\cdot5 = 25$

23. $7(5) - 5(6) = 35 - 30 = 5$

25. $4^2 + 3^2 = 16 + 9 = 25$

27. $2\cdot3^2 = 2\cdot9 = 18$

29. $3 + 2\cdot3^4\cdot5 = 3 + 2\cdot81\cdot5 = 3 + 810 = 813$

31. $\begin{aligned}5\cdot10^3 + 2\cdot10^2 + 3\cdot10^1 + 9\\ = 5\cdot1000 + 2\cdot100 + 3\cdot10 + 9\\ = 5000 + 200 + 30 + 9\\ = 5{,}239\end{aligned}$

33. $3(2)^2 - 4(2) + 12 = 3(4) - 8 + 12 = 12 - 8 + 12 = 4 + 12 = 16$

35. $(8-6)^2 + (4-3)^2 = (2)^2 + (1)^2 = 4 + 1 = 5$

37. $60 - (6 + \dfrac{40}{8}) = 60 - (6 + 5) = 60 - (11) = 49$

39. $6 + 2(5 + 4) == 6 + 2(9) = 6 + 18 = 24$

41. $3 + 5(6 - 4) = 3 + 5(2) = 3 + 10 = 13$

43. $(7 - 4)^2 + 1 = (3)^2 + 1 = 9 + 1 = 10$

45. $6^3 - (10 + 8) = 216 - (18) = 198$

47. $50 - 2(4)^2 = 50 - 2(16) = 50 - 32 = 18$

49. $16^2 - 4(2)(5) = 256 - 40 = 216$

51. $39 - 5(6) + 9 - 1 = 39 - 30 + 9 - 1 = 17$

53. $(18 - 12)^3 - 5^2 = (6)^3 - 25 = 216 - 25 = 191$

55. $2(10 - 3^2) + 1 = 2(10 - 9) + 1 = 2(1) + 1 = 2 + 1 = 3$

57. $6 + \dfrac{25}{5} + 6(3) = 6 + 5 + 18 = 29$

59. $3\left(\dfrac{18}{3}\right) - 2(2) = 3(6) - 4 = 18 - 4 = 14$

61. $(2 \cdot 6 - 4)^2 = (12 - 4)^2 = (8)^2 = 64$

63. $4[50 - (3^3 - 5^2)] = 4[50 - (27 - 25)] = 4[50 - (2)] = 4[48] = 192$

65. $80 - 2[12 - (5 + 4)] = 80 - 2[12 - (9)] = 80 - 2[3] = 80 - 6 = 74$

67. $2[100 - (5 + 4)] - 45 = 2[100 - (9)] - 45 = 2[91] - 45 = 182 - 45 = 137$

69. $\dfrac{10 + 5}{6 - 1} = \dfrac{15}{5} = 3$

71. $\dfrac{5^2 + 17}{6 - 2^2} = \dfrac{25 + 17}{6 - 4} = \dfrac{42}{2} = 21$

73. $\dfrac{(3 + 5)^2 + 2}{2(8 - 5)} = \dfrac{(8)^2 + 2}{2(3)} = \dfrac{64 + 2}{6} = \dfrac{66}{6} = 11$

75. $\dfrac{(5 - 3)^2 + 2}{4^2 - (8 + 2)} = \dfrac{(2)^2 + 2}{16 - (10)} = \dfrac{4 + 2}{6} = \dfrac{6}{6} = 1$

77. $12{,}985 - (1{,}800 + 689) = 12{,}985 - (2{,}489) = 10{,}496$

79. $3{,}245 - 25(16 - 12)^2 = 3{,}245 - 25(4)^2 = 3{,}245 - 25(16) = 3{,}245 - 400 = 2{,}845$

**Applications**

81. BUYING GROCERIES
$2(6) + 4(2) + 2(1) = 12 + 8 + 2 = 22$
The total cost of the groceries is \$22.

83. BANKING

The total amount of cash being deposited is

$24(1) + 0(2) + 6(5) + 10(10) + 12(20) + 2(50) + 1(100)$

$= 24 + 0 + 30 + 100 + 240 + 100 + 100$

$= \$594$

85. SCRABBLE

Brick: $3(3) + 1 + 1 + 3 + 3(5) = 9 + 1 + 1 + 3 + 15 = 29$

Aphid: $3[1 + 3 + 4 + 1 + 2] = 3[11] = 33$

87. CLIMATE

The average temperature is $\dfrac{75 + 80 + 83 + 80 + 77 + 72 + 86}{7} = \dfrac{553}{7} = 79°$.

89. NATURAL NUMBERS

The average of the first nine natural numbers is

$\dfrac{1 + 2 + 3 + 4 + 5 + 6 + 7 + 8 + 9}{9} = \dfrac{45}{9} = 5$

91. FAST FOOD

The average number of calories for the group of sandwiches is

$\dfrac{237 + 289 + 295 + 302 + 303 + 312 + 349}{7} = \dfrac{2087}{7} = 298$

**Writing**

93. Answers will vary.

95. Answers will vary.

**Review**

97.   $4{,}029$

  $+ 3{,}271$

  $7{,}300$

99.   $417$

  $\times\ 23$

  $1251$

  $834$

  $9{,}591$

**Section 1.6 Solving Equations by Addition and Subtraction**

**Vocabulary**

1. An equation is a statement that two expressions are **equal**. An equation contains an $=$ sign.
3. The answer to an equation is called a **solution** or a **root**.
5. **Equivalent** equations have exactly the same solutions.

**Concepts**

7. If $x = y$ and $c$ is any number, then $x + c = y + c$.
9. In $x + 6 = 10$, the addition of 6 is performed on the variable and subtraction of 6 from both sides would undo the operation and isolate the variable.

**Notation**

11. $$x + 8 = 24$$
$$x + 8 - 8 = 24 - 8$$
$$x = 16$$
Check: $x + 8 = 24$
$$16 + 8 \overset{?}{=} 24$$
$$24 = 24$$
So 16 is a solution.

**Practice**

13. $x = 2$ is an equation.
15. $7x < 8$ is not an equation.
17. $x + y = 0$ is an equation.
19. $1 + 1 = 3$ is an equation.
21. 1 is a solution; $1 + 2 = 3$
23. 7 is a solution; $7 - 7 = 0$
25. 5 is not a solution; $8 - 5 \neq 5$
27. 16 is not a solution; $16 + 32 \neq 0$
29. 7 is not a solution; $7 + 7 \neq 7$
31. 0 is a solution; $0 = 0$

33. $$x - 7 = 3$$
$$x - 7 + 7 = 3 + 7$$
$$x = 10$$

35. $$a - 2 = 5$$
$$a - 2 + 2 = 5 + 2$$
$$a = 7$$

37. $$1 = b - 2$$
$$1 + 2 = b - 2 + 2$$
$$3 = b$$

39. $x - 4 = 0$

$x - 4 + 4 = 0 + 4$

$x = 4$

41. $y - 7 = 6$

$y - 7 + 7 = 6 + 7$

$y = 13$

43. $70 = x - 5$

$70 + 5 = x - 5 + 5$

$75 = x$

45. $312 = x - 428$

$312 + 428 = x - 428 + 428$

$740 = x$

47. $x - 117 = 222$

$x - 117 + 117 = 222 + 117$

$x = 339$

49. $x + 9 = 12$

$x + 9 - 9 = 12 - 9$

$x = 3$

51. $y + 7 = 12$

$y + 7 - 7 = 12 - 7$

$y = 5$

53. $t + 19 = 28$

$t + 19 - 19 = 28 - 19$

$t = 9$

55. $23 + x = 33$

$23 - 23 + x = 33 - 23$

$x = 10$

57. $5 = 4 + c$

$5 - 4 = 4 - 4 + c$

$1 = c$

59. $99 = r + 43$

$99 - 43 = r + 43 - 43$

$56 = r$

61. $512 = x + 428$

$512 - 428 = x + 428 - 428$

$84 = x$

63. $$x + 117 = 222$$
$$x + 117 - 117 = 222 - 117$$
$$x = 105$$

65. $$3 + x = 7$$
$$3 - 3 + x = 7 - 3$$
$$x = 4$$

67. $$y - 5 = 7$$
$$y - 5 + 5 = 7 + 5$$
$$y = 12$$

69. $$4 + a = 12$$
$$4 - 4 + a = 12 - 4$$
$$a = 8$$

71. $$x - 13 = 34$$
$$x - 13 + 13 = 34 + 13$$
$$x = 47$$

**Applications**

73. ARCHAEOLOGY

**Analyze the problem**

- The manuscript is <u>1,700 years</u> old.
- The manuscript is <u>425 years</u> older than the jar.
- We are asked to find <u>the age of the jar</u>.

**Form an equation**  Since we want to find the age of the jar, we can let $x = $ <u>the age of the jar</u>.  Now we look for a key word or phrase in the problem.

**Key phrase:**  <u>older than</u>

**Translation:**  <u>add</u>

We can write the age of the manuscript in two ways.

**The age of the manuscript** is 425 plus the age of the jar.

$$1,700 = 425 + x$$

**Solve the equation**

$$1,700 = 425 + x$$
$$1,700 - 425 = 425 + x - 425$$
$$1,275 = x$$

**State the conclusion**  <u>The jar is 1,275 years old.</u>

**Check the result**  If the jar is 1,275 years old, then the manuscript is $1,275 + 425 = 1,700$ years old. The answer checks.

75. ELECTIONS

Let $t = $ the total number of votes received by the three major candidates.

The total number of votes cast is the sum of the number of votes received by the three major candidates.

$$t = 47,401,185 + 39,197,469 + 8,085,294$$

$$t = 94,683,948 \text{ votes}$$

## 77. PARTY INVITATIONS

Let $s$ = the number of invitations she sent.
The number of invitations sent is the sum of the number delivered and the number lost.

$s = 59 + 3$

$s = 62$ invitations

## 79. FAST FOOD

Let $b$ = the amount of money the entrepreneur will need to borrow.
The amount of money she will need to borrow is the franchise fee and startup costs minus the amount she has to invest.

$b = 287,000 - 68,500$

$b = \$218,500$

## 81. CELEBRITY EARNINGS

Let $O$ = the amount that Oprah Winfrey earned in 1998
The amount Oprah earned is the amount the sum of the amount Celine Dion earned and \$69 million.

$O = 56,000,000 + 69,000,000$

$O = \$125,000,000$

## 83. POWER OUTAGE

Let $a$ = the amount the meter must increase to cause the system to shut down.
The amount the meter must increase for the system to shut down is 85 minus the current reading on the meter.

$a = 85 - 60$

$a = 25$ units

## 85. AUTO REPAIR

Let p = the amount she paid to have her car fixed.
The amount she paid to have her car fixed is the amount that the gas station would have charges minus \$29.

$p = 219 - 29$

$p = \$190$

## Writing

87. Answers will vary.
89. Answers will vary.
91. Answers will vary.

## Review

93. Rounded to the nearest ten 325,784 is 325,780.
95. $2 \cdot 3^2 \cdot 5 = 2 \cdot 9 \cdot 5 = 90$
97. $8 - 2(3) + 1^3 = 8 - 6 + 1 = 3$

**Section 1.7 Solving Equations by Division and Multiplication**

**Vocabulary**

1.  According to the **division** property of equality, "If equal quantities are divided by the same nonzero quantity, the results will be equal quantities."

**Concepts**

3.  $\dfrac{6x}{6} = x$

5.  If $x = y$, then $\dfrac{x}{z} = \dfrac{y}{z}$ $(z \neq 0)$.

7.  The variable is being multiplied by 4 and can be undo by dividing by 4.

9.  Name the first step in solving
    a.  Subtract 5 from both sides
    b.  Add 5 to both sides
    c.  Divide both sides by 5
    d.  Multiply both sides by 5

**Notation**

11. Complete each solution.
$$3x = 12$$
$$\frac{3x}{3} = \frac{12}{3}$$
$$x = 4$$
Check: $3x = 12$
$$3 \cdot 4 \overset{?}{=} 12$$
$$12 = 12$$
So 4 is a solution.

**Practice**

13. $3x = 3$
$$\frac{3x}{3} = \frac{3}{3}$$
$$x = 1$$
Check: $3x = 3$
$$3 \cdot 1 \overset{?}{=} 3$$
$$3 = 3$$
So 1 is a solution.

15. $2x = 192$

$$\frac{2x}{2} = \frac{192}{2}$$

$$x = 96$$

Check: $2x = 192$

$$2 \cdot 96 \overset{?}{=} 192$$

$$192 = 192$$

So 96 is a solution.

17. $17y = 51$

$$\frac{17y}{17} = \frac{51}{17}$$

$$y = 3$$

Check: $17y = 51$

$$17 \cdot 3 \overset{?}{=} 51$$

$$51 = 51$$

So 3 is a solution.

19. $34y = 204$

$$\frac{34y}{34} = \frac{204}{34}$$

$$y = 6$$

Check: $34y = 204$

$$34 \cdot 6 \overset{?}{=} 204$$

$$204 = 204$$

So 6 is a solution.

21. $100 = 100x$

$$\frac{100}{100} = \frac{100x}{100}$$

$$1 = x$$

Check: $100 = 100x$

$$100 \overset{?}{=} 100 \cdot 1$$

$$100 = 100$$

So 1 is a solution.

23. $16 = 8r$

$$\frac{16}{16} = \frac{8r}{16}$$

$$2 = r$$

Check: $16 = 8r$

$$16 \overset{?}{=} 8 \cdot 2$$

$$16 = 16$$

So 2 is a solution.

25. Solve

$$\frac{x}{7} = 2$$

$$7 \cdot \frac{x}{7} = 2 \cdot 7$$

$$x = 14$$

Check: $\dfrac{x}{7} = 2$

$$\frac{14}{7} \overset{?}{=} 2$$

$$2 = 2$$

So 14 is a solution.

27. Solve

$$\frac{y}{14} = 3$$

$$14 \cdot \frac{y}{14} = 3 \cdot 14$$

$$y = 42$$

Check: $\dfrac{y}{14} = 3$

$$\frac{42}{14} \overset{?}{=} 3$$

$$3 = 3$$

So 42 is a solution.

29. Solve

$$\frac{a}{15} = 5$$

$$15 \cdot \frac{a}{15} = 5 \cdot 15$$

$$a = 75$$

Check: $\dfrac{a}{15} = 5$

$$\frac{75}{15} \overset{?}{=} 5$$

$$5 = 5$$

So 75 is a solution.

31. Solve

$$\frac{c}{13} = 3$$

$$13 \cdot \frac{c}{13} = 3 \cdot 13$$

$$c = 39$$

Check: $\dfrac{c}{13} = 3$

$$\frac{39}{13} \overset{?}{=} 3$$

$$3 = 3$$

So 39 is a solution.

33. Solve

$$1 = \frac{x}{50}$$

$$50 \cdot 1 = \frac{x}{50} \cdot 50$$

$$50 = x$$

Check: $1 = \dfrac{x}{50}$

$$1 \overset{?}{=} \frac{50}{50}$$

$$1 = 1$$

So 50 is a solution.

35. Solve

$$7 = \frac{t}{7}$$

$$7 \cdot 7 = \frac{x}{7} \cdot 7$$

$$49 = x$$

Check: $7 = \dfrac{t}{7}$

$$1 \stackrel{?}{=} \frac{7}{7}$$

$$1 = 1$$

So 49 is a solution.

37. $9z = 90$

$$\frac{9z}{9} = \frac{90}{9}$$

$$z = 10$$

Check: $9z = 90$

$$9 \cdot 10 \stackrel{?}{=} 90$$

$$90 = 90$$

So 10 is a solution.

39. $7x = 21$

$$\frac{7x}{7} = \frac{21}{7}$$

$$x = 3$$

Check: $7x = 21$

$$7 \cdot 3 \stackrel{?}{=} 21$$

$$21 = 21$$

So 3 is a solution.

41. $86 = 43t$

$$\frac{86}{43} = \frac{43t}{43}$$

$$t = 2$$

Check: $86 = 43t$

$$86 \stackrel{?}{=} 43 \cdot 2$$

$$86 = 86$$

So 2 is a solution.

43. $21s = 21$

$$\frac{21s}{21} = \frac{21}{21}$$

$$s = 1$$

Check: $21s = 21$

$$21 \cdot 1 \overset{?}{=} 21$$

$$21 = 21$$

So 1 is a solution.

45. Solve

$$\frac{d}{20} = 2$$

$$20 \cdot \frac{d}{20} = 2 \cdot 20$$

$$d = 40$$

Check: $\dfrac{d}{20} = 2$

$$\frac{40}{20} \overset{?}{=} 2$$

$$2 = 2$$

So 40 is a solution.

47. Solve

$$400 = \frac{t}{3}$$

$$3 \cdot 400 = \frac{t}{3} \cdot 3$$

$$1200 = t$$

Check: $400 = \dfrac{t}{3}$

$$400 \overset{?}{=} \frac{1200}{3}$$

$$400 = 400$$

So 39 is a solution.

**Applications**

49. NOBEL PRIZE

**Analyze the problem**
- 3 people shared the cash award.
- Each person received $318,500.
- We are asked to find the Nobel Prize cash award.

**Form an equation** Since we want to find what the Nobel Prize cash award was, we let $c =$ the Nobel Prize cash award. To form an equation, we look for a key word or phrase in the problem.

    **Key phrase:** shared the prize money

    **Translation:** divide

We can now form the equation.

The Nobel Prize cash award divided by the number of recipients was $318,500.

$$x \div 3 = \$318,500$$

**Solve the equation**

$$\frac{x}{3} = 318,500$$

$$3 \cdot \frac{x}{3} = 3 \cdot 318,500$$

$$x = 955,500$$

**State the conclusion** The Nobel Prize cash award was $955,500.

**Check the result** If we divide the Nobel Prize award by 3, we have $\dfrac{\$955,500}{3} = \$318,500$. This was the amount each person received. The answer checks.

51. SPEED READING

Let $s =$ the speed Alicia can expect to read after taking the classes.

$$s = 3 \cdot 130$$

$$s = 390 \text{ words a minute}$$

53. STAMPS

Let $r =$ the number of rows on each sheet.

$$r = \frac{112}{8}$$

$$r = 14 \text{ rows of stamps per sheet}$$

55. PHYSICAL EDUCATION

Let $s =$ the number of students in the PE class.

$$s = 3 \cdot 32$$

$$s = 96 \text{ students}$$

57. ANIMAL SHELTER

Let $c =$ the number of calls received each day.

$$c = 4 \cdot 8$$

$$c = 32 \text{ calls per day}$$

59. GRAVITY

Let $w =$ the weight the scale would register.

$$w = \frac{330}{6}$$

$$w = 55 \text{ pounds}$$

**Writing**

61. Answers will vary.
63. Answers will vary.

**Review**

65. $P = 2l + 2w$

$P = 2(8) + 2(16)$

$P = 48 \text{ cm}$

67. The prime factorization of 120 is $2^3 \cdot 3 \cdot 5$.

69. $3^2 \cdot 2^3 = 9 \cdot 8 = 72$

71. FUEL ECONOMY

The average city mileage is $\dfrac{24 + 22 + 28 + 29 + 27}{5} = 26\,\text{mph}$.

**Chapter 1 Key Concept**

1. Let $x =$ the monthly cost to lease the van.

3. Let $x =$ the width of the field.

5. Let $x =$ the distance traveled by the motorist.

7. $a + b = b + a$

9. $\dfrac{b}{1} = b$

11. $n - 1 < n$

13. $(r + s) + t = r + (s + t)$

**Chapter 1 Review**

**Section 1.1**

1. Graph each set
    a. Natural numbers less than 5

    b. Whole numbers between 0 and 3

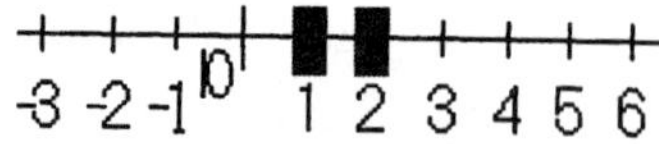

3. Consider 2,365,720
    a. 6 is in the ten thousands column
    b. 7 is in the hundreds column

5. Write in standard notation.
    a. 3,207
    b. 23,253,412
    c. 16,000,000,000

7. Round 2,507,348
    a. 2,507,300
    b. 2,510,000
    c. 2,507,350
    d. 2,500,000

**Section 1.2**

9. Do each addition.
    a.     236
        +282
         518
    b.    5,345
        + 655
         6,000
    c. $135 + 213 + 615 + 47 = 135 + 213 + 662 = 135 + 875 = 1,010$
    d. $4,447 + 7,478 + 13,061 = 4,447 + 20,539 = 24,986$

11. $P = 4s$

    $P = 4(24) = 96\,in$

13. Translate each figure.
    a. $4 + 2 = 6$
    b. $5 - 2 = 3$

15. SAVINGS ACCOUNTS

$931 + 271 - 37 - 380 = 931 + 271 - 417 = 1202 - 417 = 785$

The final balance is \$785.

**Section 1.3**

17. Do each multiplication.
    a. $8 \cdot 7 = 56$
    b. $7(8) = 56$
    c. $8 \cdot 0 = 0$
    d. $7 \cdot 1 = 7$
    e. $10 \cdot 8 \cdot 7 = 10 \cdot 56 = 560$
    f. $5 \cdot (7 \cdot 6) = 5 \cdot (42) = 210$

19. State the property.
    a. Associative property of multiplication
    b. Commutative property of multiplication

21. HORSESHOES

$Perimeter = 2l + 2w = 2(48) + 2(6) = 96 + 12 = 108 \text{ ft}$

$Area = lw = 48(6) = 288 \text{ ft}^2$

23. Do each division.
    a. $\dfrac{6}{3} = 2$
    b. $\dfrac{15}{1} = 15$
    c. $73 \div 0$ is undefined
    d. $\dfrac{0}{8} = 0$
    e. $357 \div 17 = 21$
    f. $1,443 \div 39 = 37$

$$
\begin{array}{r}
19 \\
21\overline{)405} \quad \text{Answer: 19 R6} \\
\underline{21}\phantom{00} \\
195 \\
\underline{189} \\
6
\end{array}
$$

g.

$$
\begin{array}{r}
23 \\
54\overline{)1269} \quad \text{Answer: 23 R27} \\
\underline{108}\phantom{0} \\
189 \\
\underline{162} \\
27
\end{array}
$$

h.

25. COPIES

$$\frac{84}{3} = 28$$

There were 28 copies of the test made.

**Section 1.4**

27. Identify
    a.    31 is prime
    b.    100 is composite
    c.    1 is neither
    d.    0 is neither
    e.    125 is composite
    f.    47 is prime

29. Find the prime factorization
    a.    The prime factorization of 42 is $2 \cdot 3 \cdot 7$.
    b.    The prime factorization of 375 is $3 \cdot 5^3$.

31. Evaluate
    a.    $5^3 = 125$
    b.    $11^2 = 121$
    c.    $2^3 \cdot 5^2 = 8 \cdot 25 = 200$
    d.    $2^2 \cdot 3^3 \cdot 5^2 = 4 \cdot 27 \cdot 25 = 2,700$

**Section 1.5**

33. DICE GAME

$$3(6) + 2(5) = 18 + 10 = 28$$

**Section 1.6**

35. Tell whether the given number is a solution.
    a.    5 is not a solution.

$$x + 2 = 13$$

$$5 + 2 \overset{?}{=} 13$$

$$7 \neq 13$$

    b.    4 is a solution.

$$x - 3 = 1$$

$$4 - 3 \overset{?}{=} 1$$

$$1 = 1$$

37. Solve

a.
$$x - 7 = 2$$
$$x - 7 + 7 = 2 + 7$$
$$x = 9$$
Check:
$$9 - 7 = 2$$

b.
$$x - 11 = 20$$
$$x - 11 + 11 = 20 + 11$$
$$x = 31$$
Check:
$$31 - 11 = 20$$

c.
$$225 = y - 115$$
$$225 + 115 = y - 115 + 115$$
$$340 = y$$
Check:
$$225 = 340 - 115$$

d.
$$101 = p - 32$$
$$101 + 32 = p - 32 + 32$$
$$133 = p$$
Check:
$$101 = 133 - 32$$

e.
$$x + 9 = 18$$
$$x + 9 - 9 = 18 - 9$$
$$x = 9$$
Check:
$$9 + 9 = 18$$

f.
$$b + 12 = 26$$
$$b + 12 - 12 = 26 - 12$$
$$b = 14$$
Check:
$$14 + 12 = 26$$

g.
$$175 = p + 55$$
$$175 - 55 = p + 55 - 55$$
$$120 = p$$
Check:
$$175 = 120 + 55$$

h.
$$212 = m + 207$$
$$212 - 207 = m + 207 - 207$$
$$5 = m$$
Check:
$$212 = 5 + 207$$

i.
$$x - 7 = 0$$
$$x - 7 + 7 = 0 + 7$$
$$x = 7$$
Check:
$$7 - 7 = 0$$

j.
$$x + 15 = 1,000$$
$$x + 15 - 15 = 1,000 - 15$$
$$x = 985$$
Check:
$$985 + 15 = 1,000$$

39. DOCTOR'S PATIENTS

Let $p =$ the number of patients originally.
$$p - 13 = 172$$
$$p - 13 + 13 = 172 + 13$$
$$p = 185$$
The doctor originally had 185 patients.

**Section 1.7**

41. CARPENTRY

Let $l =$ the length of each piece.
$$3l = 72$$
$$\frac{3l}{3} = \frac{72}{3}$$
$$l = 24$$
Each piece will be 24 inches long.

1.  Graph the whole numbers less than 5.

3.  7,507

5.  Bar Graph

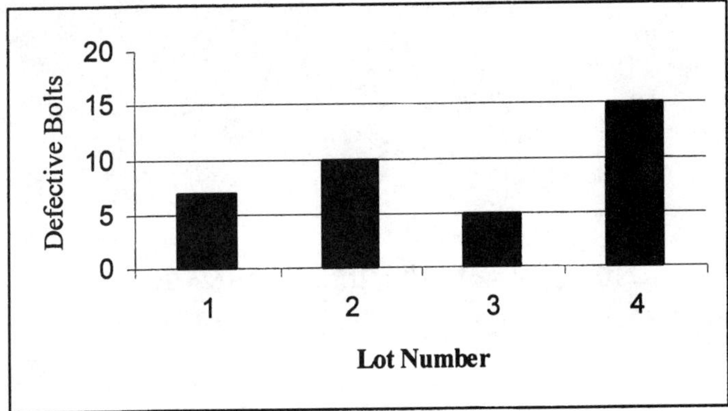



7.  $15 > 10$

9.  $327 + 435 + 123 + 606 = 327 + 435 + 729 = 327 + 1,164 = 1,491$

11.  $\begin{aligned} &44{,}526 \\ +\ &13{,}579 \\ \hline &58{,}105 \end{aligned}$

13. STOCKS
    Let $p =$ the price on Thursday.
    $p = 73 + 12 - 9 = 73 + 3 = \$76$

15.  $\begin{aligned} &53 \\ \times\ &8 \\ \hline &424 \end{aligned}$

17.  $63 \overline{)4{,}536}$   quotient 72

$$
\begin{array}{r}
72 \\
63\overline{)4{,}536} \\
\underline{441}\phantom{00} \\
126 \\
\underline{126} \\
0
\end{array}
$$

19. FURNITURE SALE

$Perimeter = 2l + 2w$

$$= 2(105) + 2(75)$$
$$= 210 + 150$$
$$= 360\,\text{ft}$$

The advertisement asks for square feet, which means we need to find the area.

$Area = lw$

$$= 105(75)$$
$$= 7,875 \text{ square feet}$$

21. COLLECTIBLES

There are $12 \cdot 24 \cdot 12 = 12 \cdot 288 = 3,456$ cards in a case.

23. $3 \cdot 4^2 - 2^2 = 3 \cdot 16 - 4 = 48 - 4 = 44$

25. $10 + 2\left[12 - 2(6 - 4)\right] = 10 + 2\left[12 - 2(2)\right] = 10 + 2\left[12 - 4\right] = 10 + 2\left[8\right] = 10 + 16 = 26$

27. $x + 13 = 16$

$$3 + 13 \overset{?}{=} 16$$
$$16 = 16$$

3 is a solution to the equation.

29. $\qquad y - 12 = 18$

$$y - 12 + 12 = 18 + 12$$
$$y = 30$$

$\qquad$ Check:

$$30 - 12 = 18$$

31. $\qquad \dfrac{q}{3} = 27$

$$3 \cdot \dfrac{q}{3} = 27 \cdot 3$$
$$q = 81$$

Check:

$$\dfrac{81}{3} = 27$$

33. LIBRARY

Let $b =$ the age of the building.

$$b + 6 = 200$$
$$b + 6 - 6 = 200 - 6$$
$$b = 194 \text{ years old}$$

## Section 2.1 An Introduction to the Integers

### Vocabulary

1. **Negative** numbers are less than 0.
3. Numbers can be represented by points equally spaced on a **number** line.
5. The symbols > and < are called **inequality** symbols.
7. Two numbers on a number line that are the same distance from zero, but on opposite sides of the origin, are called **opposites**.

### Concepts

9. The values of numbers get smaller as we move to the left on a number line.

11. Every integer has an opposite.

13. Fifteen subtract eight is represented with a subtraction sign.

15. $15 > 12$

17. a. $-225$
     b. $-10$
     c. $-3$
     d. $-12,000$

19. Negative four is three units to the right of negative seven.
21. Negative eight and two are five units from three on a number line.
23. Negative seven is closer to negative three; 2 is 5 units away and $-7$ is 4 units away.
25. Three examples could include using $-$ as a subtraction sign, as a negative or to show the opposite. For example: $3-1, -9, -(-3)$.

### Notation

27. a. $-(-8)$
     b. $|-8|$
     c. $8-8$
     d. $-|-8|$

### Practice

29. $|9| = 9$

31. $|-8| = 8$

33. $|-14| = 14$

35. $-|20| = -20$

37. $-|-6| = -6$

39. $|203| = 203$

41. $-0 = 0$

43. $-(-11) = 11$

45. $-(-4) = 4$

47. $-(-1,201) = 1,201$

49. $\{-3,-1,0,3,4\}$

-3   -2   -1   0   1   2   3   4

51. $\left\{-(-3),-(5),\left|-2\right|\right\}$

-5   -4   -3   -2   -1   0   1   2   3

53. $-5<5$

55. $-12<-6$

57. $-10>-11$

59. $\left|-2\right|>0$

61. $-1,255<-(-1,254)$

63. $-\left|-3\right|<4$

**Applications**

65. FLIGHT OF A BALL

| Time (sec) | Position |
|:---:|:---:|
| 1 | 2 |
| 2 | 3 |
| 3 | 2 |
| 4 | 0 |
| 5 | -3 |
| 6 | -7 |

67. TECHNOLOGY

The peaks occur at 2, 4, and 0. The valleys occur at –3, -5 and –2.

69. GOLF

    a.    –1, or one below par was most often shot on this hole.

    b.    –3, 3 below par was the best score on this hole.

    c.    This appears too easy as most scores are below par.

71. WEATHER MAP

    a.    The temperature range is negative ten to negative twenty.

    b.    Chicago will be approximately ten degrees colder than Denver.

    c.    The coldest it should get in Seattle is ten degrees.

73. HISTORICAL TIME LINE

    a.    The basic unit was 200 years.

    b.    A.D. could be thought of as positive numbers

    c.    B.C. could be thought of as negative numbers.

    d.    The birth of Christ distinguishes positive from negative.

75. LINE GRAPH

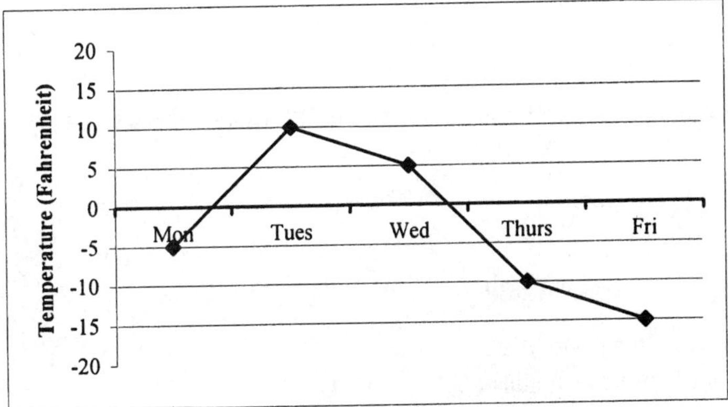


**Writing**

77. Answers may vary
79. Answers may vary
81. Answers may vary

**Review**

83. 23,500

85. $2x = 34$

$x = 17$

87. This illustrates the associative property of multiplication.

**Section 2.2 Addition of Integers**

**Vocabulary**

1.  When 0 is added to a number, the number remains the same. We call 0 the additive **identity**.

**Concepts**

3.  $-3+6=3$
5.  $-5+3=-2$

7.  a. The sum of two positive integers is always positive.
    b. The sum of two negative integers is always negative.

9.  a. $\left|-7\right|=7$

    b. $\left|10\right|=10$

11. To add two integers with unlike signs, **subtract** their absolute values, the smaller from the larger. Then attach to that result the sign of the number with the **larger** absolute value.

**Notation**

13. $-16+\left(-2\right)+\left(-1\right)=-18+\left(-1\right)$
$$=-19$$

15. $\left(-3+8\right)+\left(-3\right)=5+\left(-3\right)$
$$=2$$

17. $-6+-5$ could be written as $-6+\left(-5\right)$

**Practice**

19. $-11$ additive inverse $11$
21. $-23$ additive inverse $23$
23. $0$ additive inverse $0$
25. $99$ additive inverse $-99$
27. $-6+\left(-3\right)=-9$
29. $-5+\left(-5\right)=-10$
31. $-6+7=1$
33. $-15+8=-7$
35. $20+\left(-40\right)=-20$
37. $30+\left(-15\right)=15$
39. $-1+9=8$
41. $-7+9=2$
43. $5+\left(-15\right)=-10$
45. $24+\left(-15\right)=9$
47. $35+\left(-27\right)=8$
49. $24+\left(-45\right)=-21$
51. $-2+6+\left(-1\right)=4+\left(-1\right)=3$

53. $-9+1+(-2)=-8+(-2)=-10$

55. $6+(-4)+(-13)+7=2+(-13)+7=-11+7=-4$

57. $9+(-3)+5+(-4)=6+5+(-4)=11+(-4)=7$

59. $-6+(-7)+(-8)=-13+(-8)=-21$

61. $-7+0=-7$

63. $9+0=9$

65. $-4+4=0$

67. $2+(-2)=0$

69. $5+(-5)=0$

71. $2+(-10+8)=2+(-2)=0$

73. $(-4+8)+(-11+4)=4+(-7)=-3$

75. $[-3+(-4)]+(-5+2)=-7+(-3)=-10$

77. $[6+(-4)]+[8+(-11)]=2+(-3)=-1$

79. $-2+[-8+(-7)]=-2+(-15)=-17$

81. $789+(-9,135)=-8,346$

83. $-675+(-456)+99=-1,131+99=-1,032$

## Applications

85. G FORCES

$1G+2G=3G$

$1G+(-4G)=-3G$

87. CASH FLOW

$900-(450+380)=\$70$ shortfall each month.

89. MEDICAL QUESTIONNAIRE

This patient's risk is $-1+(-3)+(-2)+3+2+2=-6+7=1$ which places him in the 2% risk category.

91. ATOMS

The net charge of atom (a) is $(-1)+(-1)+(-1)+(-1)+1+1+1=-1$

The net charge of atom (b) is $(-1)+(-1)+(-1)+(-1)+1+1+1+1=0$

93. FLOODING

Consider four feet under flood stage as negative four so the river is now $-4+11=7$ feet above flood stage.

95. FILM PROFITS

The company earned a profit of 10 million $+(-5$ million$)+15$ million $+(-10$ million$)=\$10$ million

## Writing

97. Answers may vary

99. Answers may vary

101. The area of this rectangle is $5(3) = 15 \text{ ft}^2$.

103. $x - 7 = 20$

$\quad\quad x = 27$

105. The prime factorization of 125 is $125 = 25 \cdot 5 = 5^2 \cdot 5 = 5^3$

**Section 2.3 Subtraction of Integers**

**Vocabulary**

1. The answer to a subtraction problem is called the **difference**.

**Concepts**

3. **Subtraction** is the same as adding the opposite of the number to be subtracted.
5. Subtracting $-6$ is the same as adding 6.
7. For any numbers $x$ and $y$, $x - y = x + (-y)$.
9. After using parentheses as grouping symbols, if another set of grouping symbols is needed, we use **brackets**.
11. Negative eight minus negative four is $-8 - (-4)$
13. The distance can be found using $3 - (-4) = 3 + 4 = 7$ units
15. $8 - 3 = 5$ and $3 - 8 = 3 + (-8) = -5$ therefore they are not the same.

**Notation**

17. $1 - 3 - (-2) = 1 + (-3) + 2$
$$= -2 + 2$$
$$= 0$$

19. $(-8 - 2) - (-6) = [-8 + (-2)] - (-6)$
$$= -10 - (-6)$$
$$= -10 + 6$$
$$= -4$$

**Practice**

21. $8 - (-1) = 8 + 1 = 9$
23. $-4 - 9 = -4 + (-9) = -13$
25. $-5 - 5 = -5 + (-5) = -10$
27. $-5 - (-4) = -5 + 4 = -1$
29. $-1 - (-1) = -1 + 1 = 0$
31. $-2 - (-10) = -2 + 10 = 8$
33. $0 - (-5) = 0 + 5 = 5$
35. $0 - 4 = 0 + (-4) = -4$

37. $-2 - 2 = -2 + (-2) = -4$
39. $-10 - 10 = -10 + (-10) = -20$
41. $9 - 9 = 9 + (-9) = 0$
43. $-3 - (-3) = -3 + 3 = 0$

45. $-4-(-4)-15 = -4+4-15$
$$= 0-15$$
$$= -15$$

47. $-3-3-3 = -3+(-3)+(-3) = -9$

49. $5-9-(-7) = 5+(-9)+7$
$$= -4+7$$
$$= 3$$

51. $10-9-(-8) = 1+8$
$$= 9$$

53. $-1-(-3)-4 = -1+3+(-4)$
$$= 2+(-4)$$
$$= -2$$

55. $-5-8-(-3) = -5+(-8)+3$
$$= -13+3$$
$$= -10$$

57. $(-6-5)-3 = [-6+(-5)]-3$
$$= -11+(-3)$$
$$= -14$$

59. $(6-4)-(1-2) = 2-[1+(-2)]$
$$= 2-(-1)$$
$$= 2+1$$
$$= 3$$

61. $-9-(6-7) = -9-[6+(-7)]$
$$= -9-(-1)$$
$$= -9+1$$
$$= -8$$

63. $-8-[4-(-6)] = -8-[4+6]$
$$= -8-10$$
$$= -8+(-10)$$
$$= -18$$

65. $[-4+(-8)]-(-6) = -12+6 = -6$

67. $7-(-3) = 7+3 = 10$

69. $-10-(-6) = -10+6 = -4$

71. $-1,557-890 = -1,557+(-890) = -2,447$

73. $20,007 - (-496) = 20,007 + 496 = 20,503$

75. $-162 - (-789) - 2,303 = -162 + 789 - 2,303$
$$= 627 + (-2,303)$$
$$= -1,676$$

## Applications

77. SCUBA DIVING
$-50 - 70 = -50 + (-70) = -120$ feet

79. READING PROGRAM
The reading scores improved by
$-23 + x = -7$
$$x = 16 \text{ points}$$

81. AMPERAGE
$5 - 7 - 6 = 5 + (-7) + (-6) = -2 + (-6) = -8$

83. GEOGRAPHY
$-283 - (-1,290) = -283 + 1,290 = 1,007$ feet

85. FOOTBALL
After the third play $-1 + (-6) + (-5) + 8 = -12 + 8 = -4$ yards

87. DIVING
a.

Water
Bottom    Line                    Platform
-10   -5   0    5    10   15   20   25

b. The total dive length $25 - (-12) = 25 + 12 = 37$ feet

89. CHECKING ACCOUNT
$1,303 - 676 - 121 - 750 = 627 - 121 - 750$
$$= 506 - 750$$
$$= 506 + (-750)$$
$$= -244$$
Michael will be overdrawn by \$244.

## Writing

91. Answers may vary
93. Answers may vary

95. $5x = 15 \rightarrow x = 3$

97. Factors of twenty are $\{1, 2, 4, 5, 10, 20\}$

99. To make twelve cans of orange juice you need $12(13) = 156$ oranges

101. $4,502 = 4(1,000) + 5(100) + 0(10) + 2(1)$

**Section 2.4 Multiplication of Integers**

**Vocabulary**

1. In the multiplication $-5(-4)$, the integers $-5$ and $-4$ which are being multiplied, are called **factors**. The answer, 20, is called the **product**.
3. In the expression $-3^5$, **3** is the base and 5 is the **exponent**.

**Concepts**

5. The product of two integers with **unlike** signs is negative.
7. The **commutative** property of multiplication implies that $-2(-3) = -3(-2)$.
9. $-1(9) = -9$ ; In general the result of multiplication by negative one is the opposite of the number being multiplied.
11. The four possible combinations, not including multiplication by zero, are: positive(positive); positive(negative); negative(positive); negative(negative).

13. a. negative result
    b. positive result

15. a. $\left|-3\right| = 3$
    b. $\left|12\right| = 12$
    c. $\left|-5\right| = 5$
    d. $\left|9\right| = 9$
    e. $\left|10\right| = 10$
    f. $\left|-25\right| = 25$

17.

| Problem | Negative Factors | Answer |
| --- | --- | --- |
| $-2(-2)$ | 2 | 4 |
| $-2(-2)(-2)(-2)$ | 4 | 16 |
| $-2(-2)(-2)(-2)(-2)(-2)$ | 6 | 64 |

The product of an **even** number of negative integers is positive.

**Notation**

19. $-3(-2)(-4) = 6(-4)$
$$= -24$$

21. Accepted practice would have parenthesis about the negative five; $-6(-5)$.

**Practice**

23. $-9(-6) = 54$
25. $-3 \cdot 5 = -15$
27. $12(-3) = -36$
29. $(-8)(-7) = 56$

31. $(-2)10 = -20$

33. $-40 \cdot 3 = -120$

35. $-8(0) = 0$

37. $-1(-6) = 6$

39. $-7(-1) = 7$

41. $1(-23) = -23$

43. $-6(-4)(-2) = 24(-2) = -48$

45. $5(-2)(-4) = -10(-4) = 40$

47. $2(3)(-5) = 6(-5) = -30$

49. $6(-5)(2) = -30(2) = -60$

51. $(-1)(-1)(-1) = 1(-1) = -1$

53. $-2(-3)(3)(-1) = 6(3)(-1) = 18(-1) = -18$

55. $3(-4)(0) = 0$

57. $-2(0)(-10) = 0$

59. $-6(-10) = 60$

61. $(-4)^2 = 16$

63. $(-5)^3 = -125$

65. $(-2)^3 = -8$

67. $(-9)^2 = 81$

69. $(-1)^5 = -1$

71. $(-1)^8 = 1$

73. $(-7)^2 = 49$ and $-7^2 = -49$

75. $(-12)^2 = 144$ and $-12^2 = -144$

77. $-76(787) = -59,812$

79. $(-81)^4 = 43,046,721$

81. $(-32)(-12)(-67) = 384(-67) = -25,728$

83. $(-25)^4 = 390,625$

**Applications**

85. DIETING
    a.  Plan #1 has an expected weight loss of $10(3) = 30$ pounds, plan #2 has an expected weight loss of $14(2) = 28$ pounds.
    b.  Plan #1 is the plan for the most weight loss but may not be chosen since the exercise time is double that of plan #2.

87. MAGNIFICATION
    a.  The high value is two and the low value is negative three.
    b.  The new high would be four and the new low would be negative six.

89. TEMPERATURE CHANGES
    The total temperature change would be $-4(5) = -20°$

91. EROSION
    During the next decade the levy will decrease $-2(10) = -20$ feet

93. WOMEN'S NATIONAL BASKETBALL ASSOCIATION
The financial loss would be $11,906(-3) = -\$35,718$

**Writing**

95. Answers may vary
97. Answers may vary

**Review**

99. $3^2 \cdot 5 = 9(5) = 45$
101. The enrollment increase was $12,300 - 10,200 = 2,100$ students
103. $<$ means less than

**Section 2.5 Division of Integers**

**Vocabulary**

1. In $\dfrac{-27}{3} = -9$, the number –9 is called the **quotient**, and the number 3 is the **divisor**.

3. The **absolute value** of a number is the distance between it and 0 on the number line.

5. The quotient of two negative integers is **positive**.

**Concepts**

7. $5(-5) = -25$

9. $0(?) = -6$

11. There are two related division statements, $\dfrac{-20}{5} = -4$ or $\dfrac{-20}{-4} = 5$

13. a. This is always true.
    b. This is sometimes true.
    c. This is always true.

**Practice**

15. $\dfrac{-14}{2} = -7$

17. $\dfrac{-8}{-4} = 2$

19. $\dfrac{-25}{-5} = 5$

21. $\dfrac{-45}{-15} = 3$

23. $\dfrac{40}{-2} = -20$

25. $\dfrac{50}{-25} = -2$

27. $\dfrac{0}{-16} = 0$

29. $\dfrac{-6}{0} = $ undefined

31. $\dfrac{-5}{1} = -5$

33. $-5 \div (-5) = 1$

35. $\dfrac{-9}{9} = -1$

37. $\dfrac{-10}{-1} = 10$

39. $\dfrac{-100}{25} = -4$

41. $\dfrac{75}{-25} = -3$

43. $\dfrac{-500}{-100} = 5$

45. $\dfrac{-200}{50} = -4$

47. $\dfrac{-45}{9} = -5$

49. $\dfrac{8}{-2} = -4$

51. $\dfrac{-13,550}{25} = -542$

53. $\dfrac{272}{-17} = -16$

**Applications**

55. TEMPERATURE DROP

The average change in this time span was $\dfrac{-20}{5} = -4°$.

57. SUBMARINE DIVE

$\dfrac{-3,000}{3} = -1,000$ feet describes how deep each of the three dives will be.

59. BASEBALL TRADE

To make up being twelve games behind or $-12$ the team hopes to finish the season six games back or $\dfrac{-12}{2} = -6.$

.61. PRICE MARKDOWN

She can mark down each pair of jeans by $\dfrac{-300}{20} = -\$15$.

63. PAY CUT

$\dfrac{-9,135,000}{5,250} = -1,740$, each employee will experience a $-\$1,740$ pay cut.

**Writing**

65. Answers may vary

67. Answers may vary

**Review**

69. $3\left(\dfrac{18}{3}\right)^2 - 2(2) = 3(36) - 4 = 104$

71. The prime factorization of $210 = 21 \cdot 10 = 3 \cdot 7 \cdot 2 \cdot 5 = 2 \cdot 3 \cdot 5 \cdot 7$

73. $99 = r + 43$

$\quad\; 56 = r$

75. $3^4 = 81$

## Section 2.6 Order of Operations and Estimation

### Vocabulary

1. When asked to evaluate expressions containing more than one operation, we should apply the rules for the **order** of operations.
3. Absolute value symbols, parentheses, and brackets are types of **grouping** symbols.

### Concepts

5. Three operations need to be performed to evaluate this expression. First, the power, second multiplication and finally the subtraction.
7. In the numerator the multiplication should be performed first. In the denominator the subtraction should be performed first.
9. The difference of these two expressions is the base $-3^2$ has base 3 whereas $(-3)^2$ has base $(-3)$.

### Notation

11. $-8 - 5(-2)^2 = -8 - 5(4)$
$$= -8 - 20$$
$$= -8 + (-20)$$
$$= -28$$

13. $[-4(2+7)] - 6 = [-4(9)] - 6$
$$= -36 - 6$$
$$= -42$$

### Practice

15. $(-3)^2 - 4^2 = 9 - 16 = -7$

17. $3^2 - 4(-2)(-1) = 9 + (-4)(-2)(-1)$
$$= 9 + 8(-1)$$
$$= 1$$

19. $(2-5)(5+2) = -3(7) = -21$

21. $-10 - 2^2 = -10 - 4 = -14$

23. $\dfrac{-6-8}{2} = \dfrac{-14}{2} = -7$

25. $\dfrac{-5-5}{2} = \dfrac{-10}{2} = -5$

27. $-12 \div (-2)2 = 6 \cdot 2 = 12$

29. $-16 - 4 \div (-2) = -16 - (-2) = -16 + 2 = -14$

31. $|-5(-6)| = |30| = 30$

33. $|-4 - (-6)| = |2| = 2$

35. $5|3| = 5(3) = 15$

37. $-6|-7| = -6(7) = -42$

39. $(7-5)^2 - (1-4)^2 = 2^2 - (-3)^2 = 4 - 9 = -5$

41. $-1(2^2 - 2 + 1^2) = -1(4 - 2 + 1) = -1(3) = -3$

43. $-50 - 2(-3)^3 = -50 - 2(-27) = -50 + 54 = 4$

45. $-6^2 + 6^2 = -36 + 36 = 0$

47. $3\left(\dfrac{-18}{3}\right) - 2(-2) = 3(-6) + 4 = -18 + 4 = -14$

49. $6 + \dfrac{25}{-5} + 6 \cdot 3 = 6 + (-5) + 18 = 19$

51. $\dfrac{1-3^2}{-2} = \dfrac{1-9}{-2} = \dfrac{-8}{-2} = 4$

53. $\dfrac{-4(-5) - 2}{-6} = \dfrac{20-2}{-6} = \dfrac{18}{-6} = -3$

55. $-3\left(\dfrac{32}{-4}\right) - 1(-1)^5 = -3(-8) + 1 = 24 + 1 = 25$

57. $6(2^3)(-1) = 6(8)(-1) = 48(-1) = -48$

59. $2 + 3[5 - (1-10)] = 2 + 3[5 + 9] = 2 + 3(14) = 44$

61. $-7(2 - 3 \cdot 5) = -7(2 - 15) = -7(-13) = 91$

63. $-[6 - (1-4)^2] = -[6 - (-3)^2]$
$$= -[6 - 9]$$
$$= -(-3)$$
$$= 3$$

65. $15 + (-3 \cdot 4 - 8) = 15 + [-12 + (-8)]$
$$= 15 + (-20)$$
$$= -5$$

67. $|-3 \cdot 4 + (-5)| = |-12 + (-5)| = |-17| = 17$

69. $|(-5)^2 - 2 \cdot 7| = |25 - 14| = |11| = 11$

71. $-2 + |6 - 4^2| = -2 + |6 - 16|$
$$= -2 + |-10|$$
$$= -2 + 10$$
$$= 8$$

73. $2|1 - 8| \cdot |-8| = 2|-7| \cdot 8 = 2(7)(8) = 14(8) = 112$

75. $-2(-34)^2 - (-605) = -2(1,156) + 605 = -2,312 + 605 = -1,707$

77. $-60 - \dfrac{1,620}{-36} = -60 - (-45) = -60 + 45 = -15$

79. $-379 + (-103) + 287 \approx -200$

81. $-39 \cdot 8 \approx -320$

83. $-3,887 + (-5,106) \approx -9,000$

85. $\dfrac{6,267}{-5} \approx -1,200$

## Applications

87. TESTING

Twelve correct, three wrong and five unanswered results in a score of
$12(3) + 3(-4) + 5(-1) = 36 + (-12) + (-5) = 19$

89. SCOUTING REPORT

The average gain was $\dfrac{16 + 10 + (-2) + 0 + 4 + (-4) + 66 + (-2)}{8} = \dfrac{88}{8} = 11$ yards

91. OIL PRICES

There appears to be about a 102-cent loss for the week.
$(91 - 68) + (-47 - 91) + [-22 - (-47)]$
$= 23 + (-138) + 25 + (-12)$
$= -102$

## Writing

93. Answers may vary
95. Answers may vary

## Review

97. $8 = 2x$

$4 = x$

99. The perimeter of a rectangle is found be summing the lengths of all sides.
101. The elevator is not overloaded; the weight of the passengers is $7(140) = 980$ pounds.

**Section 2.7 Solving Equations Involving Integers**

**Vocabulary**

1.  To **solve** an equation, we isolate the variable on one side of the $=$ sign.

**Concepts**

3.  $\dfrac{x(-3)}{-3} = x$

5.  a. $x + 3 - 3 = 10 - 3$
    b. $x + 3 + (-3) = 10 + (-3)$

7.  a. $-2x = -100$, multiplication by $-2$
    b. $-6 + x = -9$ addition of $-6$
    c. $-4x - 8 = 12$ multiplication by $-4$ and subtraction of 8
    d. $-1 = -6 + (-5x)$ multiplication by $-5$ and addition of $-6$

9.  When solving the equation $-4 + t = -8 - 2$, it is best to **simplify** the right-hand side of the equation first before undoing any operations performed on the variable.

11. When solving an equation, we isolate the variable by undoing the operations performed on it in the **opposite** order.

13. a. $-2x - 3 = -19$, undo the subtraction of 3 first.
    b. $-6 + \dfrac{h}{-3} = -14$, undo the addition of $-6$ first.

**Notation**

15. $\begin{aligned}
y + (-7) &= -16 + 3 \\
y + (-7) &= -13 \\
y + (-7) + 7 &= -13 + 7 \\
y &= -6
\end{aligned}$

17. $\begin{aligned}
-13 &= -4y - 1 \\
-13 + 1 &= -4y - 1 + 1 \\
-12 &= -4y \\
\dfrac{-12}{-4} &= \dfrac{-4y}{-4} \\
3 &= y \\
y &= 3
\end{aligned}$

19. $-10x$ means $-10 \cdot x$

**Practice**

21. $-3x - 4 = 2; -2$

$$-3(-2) - 4 \overset{?}{=} 2$$

$$2 = 2$$

$-2$ is a solution

23. $-x + 8 = -4; 4$

$$-4 + 8 \overset{?}{=} -4$$

$$4 \neq -4$$

$4$ is not a solution

25. $\quad x + 6 = -12$

$$x = -12 - 6$$

$$x = -18$$

$$\text{check}$$

$$-18 + 6 = -12$$

27. $\quad -6 + m = -20$

$$m = -20 + 6$$

$$m = -14$$

$$\text{check}$$

$$-6 + (-14) = -20$$

29. $-5 + 3 = -7 + f$

$$-2 = -7 + f$$

$$-2 + 7 = f$$

$$5 = f$$

$$\text{check}$$

$$-2 = -7 + 5$$

31. $\quad h - 8 = -9$

$$h = -9 + 8$$

$$h = -1$$

$$\text{check}$$

$$-1 - 8 = -9$$

33. $\quad 0 = y + 9$

$$-9 = y$$

$$\text{check}$$

$$0 = -9 + 9$$

35.    $r - (-7) = -1 - 6$

$$r + 7 = -7$$
$$r = -7 - 7$$
$$r = -14$$

check
$$-14 - (-7) = -7$$

37.    $t - 4 = -8 - (-2)$

$$t - 4 = -6$$
$$t = -6 + 4$$
$$t = -2$$

check
$$-2 - 4 = -6$$

39.  $x - 5 = -5$

$$x = -5 + 5$$
$$x = 0$$

check
$$0 - 5 = -5$$

41.    $-2s = 16$

$$s = \frac{16}{-2}$$
$$s = -8$$

check
$$-2(-8) = 16$$

43.    $-5t = -25$

$$s = \frac{-25}{-5}$$
$$s = 5$$

check
$$-5(5) = -25$$

45.  $-2 + (-4) = -3n$

$$-6 = -3n$$
$$\frac{-6}{-3} = n$$
$$2 = n$$

check
$$-6 = -3(2)$$

47. $\quad -9h = -3(-3)$

$\quad\quad -9h = 9$

$$h = \frac{9}{-9}$$

$\quad\quad\quad h = -1$

$\quad\quad\quad$ check

$\quad -9(-1) = 9$

49. $\dfrac{t}{-3} = -2$

$\quad\quad t = -3(-2)$

$\quad\quad t = 6$

$\quad\quad$ check

$$\frac{6}{-3} = -2$$

51. $\quad 0 = \dfrac{y}{8}$

$\quad 0(8) = y$

$\quad\quad 0 = y$

$\quad\quad$ check

$$0 = \frac{0}{8}$$

53. $\dfrac{x}{-2} = -6 + 3$

$\quad \dfrac{x}{-2} = -3$

$\quad\quad x = -2(-3)$

$\quad\quad x = 6$

$\quad\quad$ check

$$\frac{6}{-2} = -3$$

55. $\quad \dfrac{x}{4} = -5 - 8$

$\quad\quad \dfrac{x}{4} = -13$

$\quad\quad\quad x = 4(-13)$

$\quad\quad\quad x = -52$

$\quad\quad\quad$ check

$$\frac{-52}{4} = -13$$

57. $$2y + 8 = -6$$
$$2y + 8 - 8 = -6 - 8$$
$$2y = -14$$
$$y = \frac{-14}{2}$$
$$y = -7$$
check
$$2(-7) + 8 = -6$$

59. $$-21 = 4h - 5$$
$$-21 + 5 = 4h - 5 + 5$$
$$-16 = 4h$$
$$\frac{-16}{4} = h$$
$$-4 = h$$
check
$$-21 = 4(-4) - 5$$

61. $$-3y + 1 = 16$$
$$-3y = 15$$
$$y = \frac{15}{-3}$$
$$y = -5$$
check
$$-3(-5) + 1 = 16$$

63. $$8 = -3x + 2$$
$$6 = -3x$$
$$\frac{6}{-3} = \frac{-3x}{-3}$$
$$-2 = x$$

65. $$-35 = 5 - 4x$$
$$-40 = -4x$$
$$\frac{-40}{-4} = \frac{-4x}{-4}$$
$$10 = x$$

67. $$4 - 5x = 34$$
$$-5x = 30$$
$$\frac{-5x}{-5} = \frac{30}{-5}$$
$$x = -6$$

69. $-5-6-5x=4$

$$-11-5x=4$$
$$-5x=15$$
$$\frac{-5x}{-5}=\frac{15}{-5}$$
$$x=-3$$

71. $4-6x=-5-9$

$$4-6x=-14$$
$$-6x=-18$$
$$\frac{-6x}{-6}=\frac{-18}{-6}$$
$$x=3$$

73. $\dfrac{h}{-6}+4=5$

$$\frac{h}{-6}=1$$
$$h=-6$$

75. $-2(4)=\dfrac{t}{-6}+1$

$$-9=\frac{t}{-6}$$
$$54=t$$

77. $0=6+\dfrac{c}{-5}$

$$-6=\frac{c}{-5}$$
$$30=c$$

79. $-1=-8+\dfrac{h}{-2}$

$$7=\frac{h}{-2}$$
$$-14=h$$

81. $2x+3(0)=-6$

$$2x=-6$$
$$x=-3$$

83. $2(0)-2y=4$

$$-2y=4$$
$$y=-2$$

85. $-x = 8$

$x = -8$

Check

$-(-8) = 8$

87. $-15 = -k$

$15 = k$

Check

$-15 = -(15)$

**Applications**

89. SHARKS

**Analyze the Problem**
- The first observations were at $-120$ feet.
- The next observations were at $-75$ feet.
- We must find <u>how many feet the cage was raised</u>.

**Form an Equation**
- Let $x = $ <u>the number of feet the cage was raised</u>.

**Key Word**: <u>Raised</u>    **Translation**: <u>Add</u>

$-120 + x = -75$

**Solve the Equation**

$$-120 + x = -75$$

$$-120 + x + 120 = -75 + 120$$

$$x = 45$$

**State the Conclusion**

<u>The shark cage was raised 45 feet.</u>

**Check the Result**

If we add the number of feet the cage was raised to the first position, we get $-120 + 45 = -75$. The answer checks.

91. DREDGING A HARBOR

Let $x$ be the amount the harbor is dredged.

$$-47 - x = -65$$

$$-x = -18$$

$$x = 18$$

The harbor was dredged another 18 feet.

93. FOOTBALL

Let $r$ be the rushing total in the second half.

$$43 + r = -8$$

$$43 + r - 43 = -8 - 43$$

$$r = -51$$

The rushing total in the second half was $-51$ yards.

95. MARKET SHARE

Let $m$ represent market share points.
$$-43 + m = -9$$
$$-43 + m + 43 = -9 + 43$$
$$m = 34$$
The company picked up 34 market share points.


97. PRICE REDUCTION

Let $d$ represent the price drop.
$$\frac{60}{12} = d$$
$$\$5 = d$$
The monthly price drop was $5.

99. ELECTION POLLS

Let $s$ represent the support gained.
$$-31 + s = -2$$
$$-31 + s + 31 = -2 + 31$$
$$s = 29$$
This candidate gained 29 points.

101. INTERNATIONAL TIME ZONES

Let $S$ represent Seattle's time zone.
$$S = 9 - 17$$
$$S = -8$$
Seattle is in zone $-8$.

**Writing**

103. Answers may vary

**Review**

105. $5^6 = 5 \cdot 5 \cdot 5 \cdot 5 \cdot 5 \cdot 5$

107. $7 + 3y = 43$
$$3y = 36$$
$$y = 12$$

109. $16 \div 8 = \dfrac{16}{8}$

**Chapter 2 Key Concepts**

1.  Stocks fell five points, –5.

3.  Thirty seconds before going on the air, –30.

5.  Ten degrees above normal, +10.

7.  Two hundred five dollars overdrawn, –$205.

9.

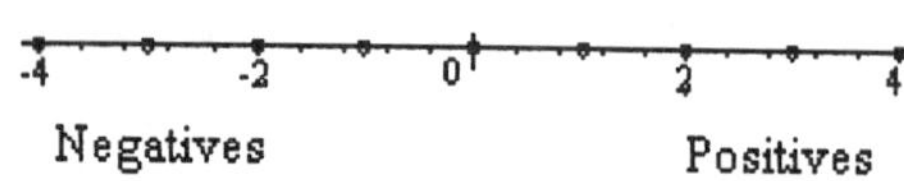



11. $x < y$

13. Addition
    Like Signs – Add the absolute values and maintain the common sign.
    Unlike Signs – Subtract the absolute values and maintain the sign of the number with the largest absolute value.

15. Multiplication
    Like Signs – The product is positive.
    Unlike Signs – The product is negative.

**Chapter 2 Review**

**Section 2.1 An Introduction to the Integers**

1. a. $\{-3, -1, 0, 4\}$

   b. Integers greater than $-3$ but less than $4$

3. WATER PRESSURE
   A depth of thirty-three feet may be represented as $-33$.

5. a. $|-4| = 4$

   b. $|0| = 0$

   c. $|-43| = 43$

   d. $-|12| = -12$

**Section 2.2 Addition of Integers**

7. a. $-(-12) = 12$

   b. $-(8) = -8$

   c. $-(-8) = 8$

   d. $-0 = 0$

9. a. $-6 + (-4) = -10$

   b. $-23 + (-60) = -83$

   c. $-1 + (-4) + (-3) = -8$

   d. $-4 + 3 = -1$

   e. $-28 + 140 = 112$

   f. $9 + (-20) = -11$

   g. $3 + (-2) + (-4) = -3$

   h. $(-2 + 1) + [(-5) + 4] = -1 + (-1) = -2$

11. a. The additive inverse of $-11$ is $11$.

    b. The additive inverse of $4$ is $-4$.

**Section 2.3. Subtraction of Integers**

13. a. $5 - 8 = -3$

    b. $-9 - 12 = -21$

    c. $-4 - (-8) = -4 + 8 = 4$

    d. $-6 - 106 = -112$

    e. $-8 - (-2) = -8 + 2 = -6$

    f. $7 - 1 = 6$

    g. $0 - 37 = -37$

    h. $0 - (-30) = 0 + 30 = 30$

15. a. $-9-7+12=-16+12=-4$
    b. $7-[(-6)-2]=7-(-8)=7+8=15$
    c. $1-(2-7)=1-(-5)=1+5=6$
    d. $-12-(6-10)=-12-(-4)=-12+4=-8$

17. GOLD MINING
    The depth of the second discovery is $150+75=225$ feet down or $-225$ feet.

19. RECORD TEMPERATURES
    Alaska $100-(-80)=100+80=180$ degrees
    Virginia $110-(-30)=110+30=140$ degrees

**Section 2.4 Multiplication of Integers**

21. a. $(-6)(-2)(-3)=12(-3)=-36$
    b. $4(-3)3=-12(3)=-36$
    c. $0(-7)=0$
    d. $(-1)(-1)(-1)(-1)=1(-1)(-1)=-1(-1)=1$

23. a. $(-5)^2=25$
    b. $(-2)^5=-32$
    c. $(-8)^2=64$
    d. $(-4)^3=-64$

25. The difference between $-2^2$ and $(-2)^2$ is the base, the first expression has two as a base and the second
    expression has negative two as a base.
    $-2^2=-4$ and $(-2)^2=4$

**Section 2.5 Division of Integers**

27. a. $\dfrac{-14}{7}=-2$
    b. $\dfrac{25}{-5}=-5$
    c. $-64\div 8=-8$
    d. $\dfrac{-202}{-2}=101$

29. PRODUCTION TIME
    $\dfrac{12\text{ minutes}}{6\text{ months}}=2$ minutes per month or $-2$

**Section 2.6 Order of Operations and Estimation**

31. a. $-4\left(\dfrac{15}{-3}\right)-2^3 = -4(-5)-8 = 20-8 = 12$

  b. $-20+2(12-5\cdot2) = -20+2(2) = -20+4 = -16$

  c. $-20+2[12-(-7+5)^2] = -20+2[12-(-2)^2]$
  $$= -20+2[12-4]$$
  $$= -20+2[8]$$
  $$= -20+16$$
  $$= -4$$

  d. $8-|-3\cdot4+5| = 8-|-12+5|$
  $$= 8-|-7|$$
  $$= 8-7$$
  $$= 1$$

33. a. $-89+57+(-42) \approx -70$

  b. $\dfrac{-507}{-24} \approx 20$

  c. $(-681)(9) \approx -7{,}000$

  d. $317-(-775) \approx 1{,}100$

**Section 2.7 Solving Equations Involving Integers**

35. a. $t+(-8) = -18$
  $$t = -18+8$$
  $$t = -10$$

  b. $\dfrac{x}{-3} = -4$
  $$x = -3(-4)$$
  $$x = 12$$

  c. $y+8 = 0$
  $$y = -8$$

  d. $-7m = -28$
  $$m = \dfrac{-28}{-7}$$
  $$m = 4$$

37. a. $-5t + 1 = -14$

$-5t = -15$

$t = 3$

Check

$-5(3) + 1 = -15 + 1 = -14$

b. $3(2) = 2 - 2x$

$6 - 2 = -2x$

$4 = -2x$

$-2 = x$

Check

$6 = 2 - 2(-2)$

c. $\dfrac{x}{-4} - 5 = -1 - 1$

$\dfrac{x}{-4} = 3$

$x = -4(3)$

$x = -12$

Check

$\dfrac{-12}{-4} - 5 = -2$

d. $c - (-5) = 5$

$c + 5 = 5$

$c = 0$

Check

$0 - (-5) = 5$

39. CREDIT CARD PROMOTION

Let $p$ = the number of customers

$8p = 968$

$p = 121$

121 customers applied for credit.

1.  a. $-8 > -9$
    b. $-8 < |-8|$ since $|-8| = 8$
    c. The opposite of $5 < 0$ since the opposite of 5 is $-5$.

3.  SCHOOL ENROLLMENT
    Monroe will face the greatest shortage, -2,488.

5.  a. $-65 + 31 = -34$
    b. $-17 + (-17) = -34$
    c. $[6 + (-4)] + [-6 + (-4)] = 2 + (-10) = -8$

7.  a. $-10 \cdot 7 = -70$
    b. $-4(-2)(-6) = 8(-6) = -48$
    c. $(-2)(-2)(-2)(-2) = 4(4) = 16$
    d. $-55(0) = 0$

9.  a. $\dfrac{-32}{4} = -8$

    b. $\dfrac{8}{6-6} = \dfrac{8}{0}$ is undefined

    c. $\dfrac{-5}{1} = -5$

    d. $\dfrac{0}{-6} = 0$

11. GEOGRAPHY
    $436 - 282 = 154$ feet difference in elevation.

13. a. $(-4)^2 = 16$
    b. $-4^2 = -16$
    c. $(-4 - 3)^2 = (-7)^2 = 49$

15. $4 - (-3)^2 + 6 = 4 - 9 + 6 = 1$

17. $-10 + 2[6 - (-2)^2(-5)] = -10 + 2[6 - 4(-5)]$
$$= -10 + 2[6 + 20]$$
$$= -10 + 2(26)$$
$$= 42$$

19. $c - (-7) = -8$
$$c + 7 = -8$$
$$c = -15$$

21. $\dfrac{x}{-4} = 10$

$x = -40$

23. $-5 = -6a + 7$

$-12 = -6a$

$2 = a$

Check

$-5 = -6(2) + 7$

25. CHECKING ACCOUNT

$b + 225 = -19$

$b = -\$244$

Before the deposit the balance was -\$244.

27. $5(-4) = (-4) + (-4) + (-4) + (-4) + (-4) = -20$

**Chapters 1 – 2 Cumulative Review**

1. The natural numbers are 1, 2, 5, 9.

3. The negative numbers are –2, –1.

5. 6 is in the thousands column.

7. To the nearest hundred 7,326,500.

9. BIDS
   If the award goes to the lowest bid then the contract should be given to CRF Cable.

11. $237 + 549 = 786$

13. $5,369 - 685 = 4,684$

15. The perimeter is $2(17) + 2(35) = 34 + 70 = 104$ feet .
    The area is $35(17) = 595 \text{ ft}^2$ .

17. $435 \cdot 27 = 11,745$

19. $4,587 \times 67 = 307,329$

21. SHIPPING
    In 12 gross there would be $12 \cdot 12 \cdot 12 = 1,728$ tennis balls.

23. 17 is prime and odd.

25. 0 is even.

27. $504 = 3(168) = 3(3)(56) = 3^2(2)(28) = 3^2(2)(2)(14) = 3^2 \cdot 2^2 \cdot (2)(7) = 2^3 \cdot 3^2 \cdot 7$

29. $5^2 \cdot 7 = 25(7) = 175$

31. $25 + 5 \cdot 5 = 25 + 25 = 50$

33. SPEED CHECK
    The average speed was $\dfrac{38 + 42 + 36 + 38 + 48 + 44}{6} = \dfrac{246}{6} = 41$ mph .
    The drivers were not obeying the speed limit.

35. $50 = x + 37$
    $13 = x$

37. $5p = 135$
    $p = 27$

39. $\{-2,-1,0,2\}$

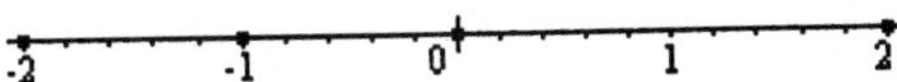

41. $-17 < -16$ is a true statement.

43. $-2 + (-3) = -5$

45. $-3 - 5 = -8$

47. $(-8)(-3) = 24$

49. $\dfrac{-14}{-7} = 2$

51. $5 + (-3)(-7) = 5 + 21 = 26$

53. $\dfrac{10 - (-5)}{1 - 2 \cdot 3} = \dfrac{10 + 5}{1 - 6} = \dfrac{15}{-5} = -3$

55. $-5t + 1 = -14$

$-5t = -15$

$t = 3$

Check $-5(3) + 1 = -14$

57. BUYING A BUSINESS

Each person's share of the debt was $\dfrac{1,512,444}{12} = \$126,037$.

**Section 3.1 The Fundamental Property of Fractions**

**Vocabulary**

1. For the fraction $\dfrac{7}{8}$, 7 is the **numerator** and 8 is the **denominator**.

3. A **proper** fraction is less than 1. An **improper** fraction is greater than or equal to 1.

5. Two fractions are **equivalent** if they have the same value.

7. Multiplying the numerator and denominator of a fraction by a number to obtain an equivalent fraction that involves larger numbers is called expressing the fraction in **higher** terms or **building** up the fraction.

**Concepts**

9. a. 2 is the common factor
   b. 3 is the common factor
   c. 5 is the common factor
   d. 7 is the common factor

11. This demonstrates equivalent fractions $\dfrac{2}{6} = \dfrac{1}{3}$.

13. a. The difference is the first example is factored while the second approach uses prime factorization.
    b. The results are the same.

15. This does not demonstrate factored numerator and denominator.

17. a. $8 = \dfrac{8}{1}$

    b. $-25 = -\dfrac{25}{1}$

**Notation**

19. $\dfrac{18}{24} = \dfrac{3 \cdot 3 \cdot 2}{3 \cdot 2 \cdot 2 \cdot 2}$

    $= \dfrac{\cancel{3} \cdot 3 \cdot \cancel{2}}{\cancel{3} \cdot 2 \cdot 2 \cdot \cancel{2}}$

    $= \dfrac{3}{4}$

**Practice**

21. $\dfrac{3}{9} = \dfrac{1}{3}$

23. $\dfrac{7}{21} = \dfrac{1}{3}$

25. $\dfrac{20}{30} = \dfrac{2}{3}$

27. $\dfrac{15}{6} = \dfrac{5}{2}$

29. $-\dfrac{28}{56} = -\dfrac{1}{2}$

31. $-\dfrac{90}{105} = -\dfrac{6}{7}$

33. $\dfrac{60}{108} = \dfrac{5}{9}$

35. $\dfrac{180}{210} = \dfrac{6}{7}$

37. $\dfrac{55}{67}$ is in lowest terms

39. $\dfrac{36}{96} = \dfrac{3}{8}$

41. $\dfrac{25}{35} = \dfrac{5}{7}$

43. $\dfrac{12}{15} = \dfrac{4}{5}$

45. $\dfrac{6}{7} = \dfrac{6}{7}$

47. $\dfrac{7}{8} = \dfrac{7}{8}$

49. $-\dfrac{10}{30} = -\dfrac{1}{3}$

51. $\dfrac{15}{25} = \dfrac{3}{5}$

53. $\dfrac{35}{28} = \dfrac{5}{4}$

55. $\dfrac{56}{28} = 2$

57. $\dfrac{7}{8}\left(\dfrac{5}{5}\right) = \dfrac{35}{40}$

59. $\dfrac{4}{5}\left(\dfrac{7}{7}\right) = \dfrac{28}{35}$

61. $\dfrac{5}{6}\left(\dfrac{9}{9}\right) = \dfrac{45}{54}$

63. $\dfrac{1}{2}\left(\dfrac{15}{15}\right) = \dfrac{15}{30}$

65. $\dfrac{2}{7}\left(\dfrac{2}{2}\right) = \dfrac{4}{14}$

67. $\dfrac{9}{10}\left(\dfrac{6}{6}\right) = \dfrac{54}{60}$

69. $\dfrac{5}{4}\left(\dfrac{5}{5}\right) = \dfrac{25}{20}$

71. $\dfrac{2}{15}\left(\dfrac{3}{3}\right) = \dfrac{6}{45}$

73. $3\left(\dfrac{5}{5}\right) = \dfrac{15}{5}$

75. $6\left(\dfrac{8}{8}\right) = \dfrac{48}{8}$

77. $4\left(\dfrac{9}{9}\right) = \dfrac{36}{9}$

79. $-2\left(\dfrac{2}{2}\right) = -\dfrac{4}{2}$

**Applications**

81. COMMUTING

The motorist has covered $\dfrac{3}{5}$ of the commute.

83. SINK HOLE

$-\dfrac{15}{16}$ inch

85. PERSONNEL RECORDS

| Name | Total Time | Time Alone | Amount Completed |
|---|---|---|---|
| Bob | 10 hours | 7 hours | $\dfrac{7}{10}$ |
| Ali | 8 hours | 1 hour | $\dfrac{1}{8}$ |

87. MUSIC

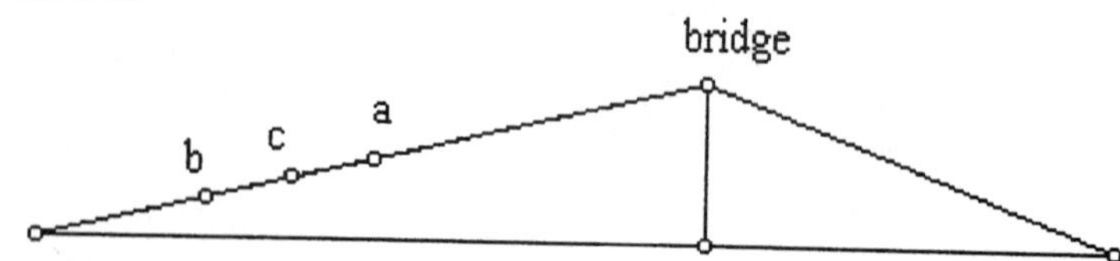


89. MACHINERY

There are two directions to turn the dial, one quarter turn to the left or three-quarter turn to the right.

91. SUPERMARKET DISPLAY

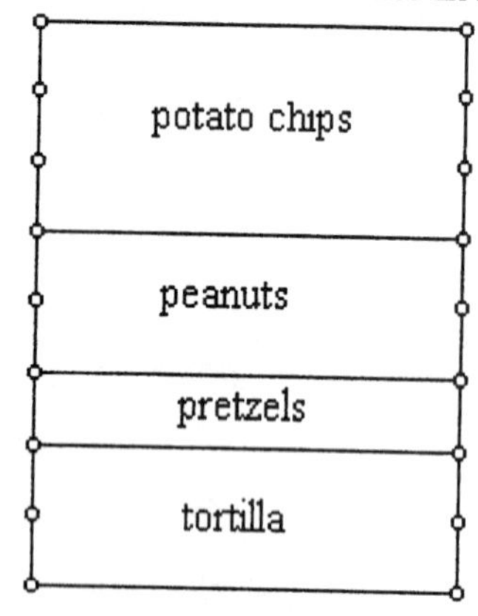


93. CAMERA

She should select a shutter speed of $\dfrac{1}{250}$.

**Writing**

95. Answers may vary
97. Answers may vary

**Review**

99. $-5x + 1 = 16$

$-5x = 15$

$x = -3$

101. 564,112 to the nearest thousand is 564,000

**Section 3.2 Multiplying Fractions**

**Vocabulary**

1. The word *of* in mathematics usually means **multiply**.
3. The result of a multiplication problem is called the **product**.
5. In a triangle, $b$ stands for the length of the **base** and $h$ stands for the **height**.

**Concepts**

7. $\dfrac{a}{b} \cdot \dfrac{c}{d} = \dfrac{ac}{bd}$

9. a. $\dfrac{1}{4}$ will be shaded

   b. The rectangle is divided into 12 equal parts. One part has been shaded twice. $\dfrac{1}{3} \cdot \dfrac{1}{4} = \dfrac{1}{12}$

11. a. The product of numbers with unlike signs is negative.
    b. The product of two numbers with like signs is positive.

13. a. This is a true statement.
    b. This is a true statement.
    c. This is a false statement since $-\dfrac{3}{8}a = -\dfrac{3a}{8}$ .
    d. This is a true statement.

**Notation**

15. $\dfrac{5}{8} \cdot \dfrac{7}{15} = \dfrac{5 \cdot 7}{8 \cdot 15}$

$$= \dfrac{5 \cdot 7}{8 \cdot 5 \cdot 3}$$

$$= \dfrac{\cancel{5} \cdot 7}{8 \cdot \cancel{5} \cdot 3}$$

$$= \dfrac{7}{24}$$

**Practice**

17. $\dfrac{1}{4} \cdot \dfrac{1}{2} = \dfrac{1}{8}$

19. $\dfrac{3}{8} \cdot \dfrac{7}{16} = \dfrac{21}{128}$

21. $\dfrac{2}{3} \cdot \dfrac{6}{7} = \dfrac{12}{21} = \dfrac{4}{7}$

23. $\dfrac{2(7)}{15} \cdot \dfrac{11}{2(4)} = \dfrac{77}{60}$

25. $-\dfrac{5(3)}{3(8)} \cdot \dfrac{8}{5(5)} = -\dfrac{1}{5}$

27. $\left(-\dfrac{11}{7(3)}\right)\left(-\dfrac{7(2)}{11(3)}\right) = \dfrac{2}{9}$

29. $\dfrac{7}{10}\left(\dfrac{10(2)}{7(3)}\right) = \dfrac{2}{3}$

31. $\dfrac{3}{4} \cdot \dfrac{4}{3} = 1$

33. $\dfrac{1}{3} \cdot \dfrac{3(5)}{4(4)} \cdot \dfrac{4}{5(5)} = \dfrac{1}{20}$

35. $\left(\dfrac{\cancel{2}}{3}\right)\left(-\dfrac{1}{\cancel{2}(2)(\cancel{4})}\right)\left(-\dfrac{\cancel{4}}{5}\right) = \dfrac{1}{30}$

37. $\dfrac{5}{6} \cdot 18 = 5(3) = 15$

39. $15\left(-\dfrac{4}{5}\right) = 3(-4) = -12$

41. $\dfrac{5x}{12} \cdot \dfrac{1}{6} = \dfrac{5x}{72}$

43. $\dfrac{b}{12} \cdot \dfrac{3}{10} = \dfrac{b}{40}$

45. $\dfrac{1}{3} \cdot 3d = d$

47. $\dfrac{2}{3} \cdot \dfrac{3s}{2} = s$

49. $\dfrac{5}{6} \cdot x = \dfrac{5x}{6}$ or $\dfrac{5}{6}x$

51. $-\dfrac{8}{9} \cdot v = -\dfrac{8v}{9}$ or $-\dfrac{8}{9}v$

53. $\left(\dfrac{2}{3}\right)^2 = \dfrac{4}{9}$

55. $\left(-\dfrac{5}{9}\right)^2 = \dfrac{25}{81}$

57. $\left(\dfrac{4}{3}\right)^2 = \dfrac{16}{9}$

59. $\left(-\dfrac{3}{4}\right)^3 = -\dfrac{27}{64}$

61. Illustration 1

|  | $\dfrac{1}{2}$ | $\dfrac{1}{3}$ | $\dfrac{1}{4}$ | $\dfrac{1}{5}$ | $\dfrac{1}{6}$ |
| --- | --- | --- | --- | --- | --- |
| $\dfrac{1}{2}$ | $\dfrac{1}{4}$ | $\dfrac{1}{6}$ | $\dfrac{1}{8}$ | $\dfrac{1}{10}$ | $\dfrac{1}{12}$ |
| $\dfrac{1}{3}$ | $\dfrac{1}{6}$ | $\dfrac{1}{9}$ | $\dfrac{1}{12}$ | $\dfrac{1}{15}$ | $\dfrac{1}{18}$ |
| $\dfrac{1}{4}$ | $\dfrac{1}{8}$ | $\dfrac{1}{12}$ | $\dfrac{1}{16}$ | $\dfrac{1}{20}$ | $\dfrac{1}{24}$ |
| $\dfrac{1}{5}$ | $\dfrac{1}{10}$ | $\dfrac{1}{15}$ | $\dfrac{1}{20}$ | $\dfrac{1}{25}$ | $\dfrac{1}{30}$ |
| $\dfrac{1}{6}$ | $\dfrac{1}{12}$ | $\dfrac{1}{18}$ | $\dfrac{1}{24}$ | $\dfrac{1}{30}$ | $\dfrac{1}{36}$ |

63. Using $A = \dfrac{1}{2}bh$, $A = \dfrac{1}{2}(10)(3) = 15$ ft$^2$

65. Using $A = \dfrac{1}{2}bh$, $A = \dfrac{1}{2}(3)(5) = \dfrac{15}{2}$ yd$^2$

**Applications**

67. THE CONSTITUTION

$\dfrac{2}{3}(435) = 2(145) = 290$

69. TENNIS BALL

$\dfrac{1}{3}(54) = 18$ inches, bounce 1

$\dfrac{1}{3}(18) = 6$ inches, bounce 2

$\dfrac{1}{3}(6) = 2$ inches, bounce 3

71. COOKING

$\dfrac{1}{2}\left(\dfrac{3}{4}\right) = \dfrac{3}{8}$ cup of sugar, $\dfrac{1}{2}\left(\dfrac{1}{3}\right) = \dfrac{1}{6}$ cup of molasses

Note: the recipe is for 2 dozen so use ½ of each ingredient to make 1 dozen.

73. BOTANY

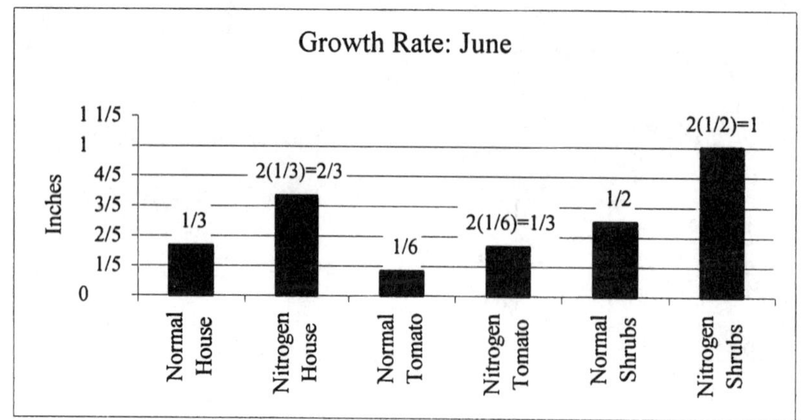


75. THE STARS AND STRIPES

Using $A = \frac{1}{2}bh$, $A = \frac{1}{2}(22)(11) = 121$ in$^2$

77. TILE DESIGN

Using $A = \frac{1}{2}bh$, $A = \frac{1}{2}(3)(3) = \frac{9}{2}$ in$^2$.

Now multiply this by four, $4\left(\frac{9}{4}\right) = 18$ in$^2$ of the tile is blue.

**Writing**

79. Answers may vary
81. Answers may vary

**Review**

83. 987,000

85. $2(-6) + 6 \neq 6$, so $x = -6$ is not a solution

87. $125 = 5^3$

**Vocabulary**

1.  Two numbers are called **reciprocals** if their product is 1.

**Concepts**

3.  $\dfrac{1}{2} \div \dfrac{2}{3} = \dfrac{1}{2} \cdot \dfrac{3}{2}$

5.  Dividing each rectangle into thirds represents $4 \div \dfrac{1}{3}$ and the quotient is 12.

7.  The product of $\dfrac{4}{5}$ and it's reciprocal is one; $\dfrac{4}{5}\left(\dfrac{5}{4}\right) = 1$ .

9.  a. $15 \div 3 = 5$

    b. $15\left(\dfrac{1}{3}\right) = 5$

    c. Division by 3 is the same as multiplication by $\dfrac{1}{3}$ .

**Notation**

11.
$$\dfrac{25}{36} \div \dfrac{10}{9} = \dfrac{25}{36} \cdot \dfrac{9}{10}$$
$$= \dfrac{25 \cdot 9}{36 \cdot 10}$$
$$= \dfrac{5 \cdot 5 \cdot 9}{4 \cdot 9 \cdot 2 \cdot 5}$$
$$= \dfrac{\cancel{5} \cdot 5 \cdot \cancel{9}}{4 \cdot \cancel{9} \cdot 2 \cdot \cancel{5}}$$
$$= \dfrac{5}{8}$$

**Practice**

13.  $\dfrac{1}{2} \div \dfrac{3}{5} = \dfrac{1}{2} \cdot \dfrac{5}{3} = \dfrac{5}{6}$

15.  $\dfrac{3}{16} \div \dfrac{1}{9} = \dfrac{3}{16} \cdot \dfrac{9}{1} = \dfrac{27}{16}$

17.  $\dfrac{4}{5} \div \dfrac{4}{5} = \dfrac{4}{5} \cdot \dfrac{5}{4} = 1$

19.  $\left(-\dfrac{7}{4}\right) \div \left(-\dfrac{21}{8}\right) = -\dfrac{7}{4} \cdot \left(-\dfrac{4(2)}{3(7)}\right) = \dfrac{2}{3}$

21. $3 \div \dfrac{1}{12} = 3(12) = 36$

23. $120 \div \dfrac{12}{5} = 120\left(\dfrac{5}{12}\right) = 50$

25. $-\dfrac{4}{5} \div (-6) = -\dfrac{4}{5} \cdot \left(\dfrac{1}{-6}\right) = \dfrac{2}{15}$

27. $\dfrac{15}{16} \div 180 = \dfrac{15}{16} \cdot \left(\dfrac{1}{12(15)}\right) = \dfrac{1}{192}$

29. $-\dfrac{9}{10} \div \dfrac{4}{15} = -\dfrac{9}{2(5)} \cdot \dfrac{3(5)}{4} = -\dfrac{27}{8}$

31. $\dfrac{9}{10} \div \left(-\dfrac{3}{25}\right) = \dfrac{3(3)}{2(5)} \cdot \left(-\dfrac{5(5)}{3}\right) = -\dfrac{15}{2}$

33. $-\dfrac{1}{8} \div 8 = -\dfrac{1}{8} \cdot \dfrac{1}{8} = -\dfrac{1}{64}$

35. $\dfrac{15}{32} \div \dfrac{15}{32} = \dfrac{15}{32} \cdot \dfrac{32}{15} = 1$

37. $\dfrac{4}{5} \div \dfrac{3}{2} = \dfrac{4}{5} \cdot \dfrac{2}{3} = \dfrac{8}{15}$

39. $\dfrac{1}{8} \div \dfrac{3}{4} = \dfrac{1}{4(2)} \cdot \dfrac{4}{3} = \dfrac{1}{6}$

41. $\dfrac{13}{16} \div \dfrac{1}{2} = \dfrac{13}{2(8)} \cdot \dfrac{2}{1} = \dfrac{13}{8}$

43. $-\dfrac{15}{32} \div \dfrac{3}{4} = -\dfrac{3(5)}{(8)4} \cdot \dfrac{4}{3} = -\dfrac{5}{8}$

**Applications**

45. MARATHON

$26 \div \dfrac{1}{4} = 26(4) = 104$ laps

47. LASER TECHNOLOGY

$\dfrac{7}{8} \div \dfrac{1}{64} = \dfrac{7}{8} \cdot \dfrac{8^2}{1} = 56$ slices

49. UNDERGROUND CABLE

Route One: $15 \div \dfrac{3}{5} = 15\left(\dfrac{5}{3}\right) = 25$ days

Route Two: $12 \div \dfrac{2}{5} = 12\left(\dfrac{5}{2}\right) = 30$ days

The fewest installation days is Route one.

51. $3 \times 5$ CARDS

a. One inch on this ruler is divided into 16 parts.

b. The stack of cards is $\dfrac{3}{4}$ inch

c. One card is $\dfrac{3}{4} \div 90 = \dfrac{3}{4}\left(\dfrac{1}{90}\right) = \dfrac{1}{120}$ inch thick

53. FORESTRY

This map contains $6,284 \div \dfrac{4}{5} = 6,284\left(\dfrac{5}{4}\right) = 7,855$ sections .

**Writing**

55. Answers may vary
57. Answers may vary

**Review**

59. $4x - 2 = -18$

$\quad\quad 4x = -16$

$\quad\quad\; x = -4$

61. $\dfrac{5}{6} \div \dfrac{7}{12} = \dfrac{5}{6} \cdot \dfrac{12}{7} = \dfrac{10}{7}$

63. This is a false statement.

65. 637,500

## Section 3.4 Adding and Subtracting Fractions

### Vocabulary

1. The **least** common denominator for a set of fractions is the smallest number each denominator will divide exactly.
3. To express a fraction in **higher** terms, we multiply the numerator and denominator by the same number.

### Concepts

5. This rule tells us how to add fractions having like **denominators**. To find the sum, we add the **numerators** and then write that result over the **common** denominator.

7. Since the denominators are not the same some preliminary work must be done.

9. The numerator and denominator are being multiplied by 4.

11. a. a 5 appears once
    b. a 3 appears twice
    c. a 2 appears three times

13. The LCD for these two fractions is $2 \cdot 2 \cdot 3 \cdot 5 = 60$.

15. a. $\dfrac{1}{3} > \dfrac{1}{4}$

    b. $\dfrac{1}{3} = \dfrac{4}{12} > \dfrac{1}{4} = \dfrac{3}{12}$

### Notation

17. $\dfrac{2}{5} + \dfrac{1}{3} = \dfrac{2 \cdot 3}{5 \cdot 3} + \dfrac{1 \cdot 5}{3 \cdot 5}$

$$= \dfrac{6}{15} + \dfrac{5}{15}$$

$$= \dfrac{6 + 5}{15}$$

$$= \dfrac{11}{15}$$

### Practice

19. LCD $(18, 6) = 2(3)(3) = 18$
21. LCD $(8, 6) = 2(2)(2)(3) = 24$
23. LCD $(8, 20) = 2(2)(2)(5) = 40$
25. LCD $(15, 12) = 2(2)(3)(5) = 60$
27. $\dfrac{3}{7} + \dfrac{1}{7} = \dfrac{4}{7}$

29. $\dfrac{37}{103} - \dfrac{17}{103} = \dfrac{20}{103}$

31. $\dfrac{11}{25} - \dfrac{1}{25} = \dfrac{10}{25} = \dfrac{2}{5}$

33. $\dfrac{5}{7} + \dfrac{3}{7} = \dfrac{8}{7}$

35. $\dfrac{1}{4} + \dfrac{3}{8} = \dfrac{2}{8} + \dfrac{3}{8} = \dfrac{5}{8}$

37. $\dfrac{13}{20} - \dfrac{1}{5} = \dfrac{13}{20} - \dfrac{4}{20} = \dfrac{9}{20}$

39. $\dfrac{4}{5} + \dfrac{2}{3} = \dfrac{12}{15} + \dfrac{10}{15} = \dfrac{22}{15}$

41. $\dfrac{1}{8} + \dfrac{2}{7} = \dfrac{7}{56} + \dfrac{16}{56} = \dfrac{23}{56}$

43. $\dfrac{3}{4} - \dfrac{2}{3} = \dfrac{9}{12} - \dfrac{8}{12} = \dfrac{1}{12}$

45. $\dfrac{5}{6} - \dfrac{3}{4} = \dfrac{10}{12} - \dfrac{9}{12} = \dfrac{1}{12}$

47. $\dfrac{16}{25} - \left(-\dfrac{3}{10}\right) = \dfrac{32}{50} + \dfrac{15}{50} = \dfrac{47}{50}$

49. $-\dfrac{7}{16} + \dfrac{1}{4} = -\dfrac{7}{16} + \dfrac{4}{16} = -\dfrac{3}{16}$

51. $\dfrac{1}{12} - \dfrac{3}{4} = \dfrac{1}{12} - \dfrac{9}{12} = -\dfrac{8}{12} = -\dfrac{2}{3}$

53. $-\dfrac{5}{8} - \dfrac{1}{3} = -\dfrac{15}{24} - \dfrac{8}{24} = -\dfrac{23}{24}$

55. $-3 + \dfrac{2}{5} = -\dfrac{15}{5} + \dfrac{2}{5} = -\dfrac{13}{5}$

57. $-\dfrac{3}{4} - 5 = -\dfrac{3}{4} - \dfrac{20}{4} = -\dfrac{23}{4}$

59. $\dfrac{1}{3} + \dfrac{1}{4} + \dfrac{1}{5} = \dfrac{20}{60} + \dfrac{15}{60} + \dfrac{12}{60} = \dfrac{47}{60}$

61. $-\dfrac{2}{3} + \dfrac{5}{4} + \dfrac{1}{6} = -\dfrac{8}{12} + \dfrac{15}{12} + \dfrac{2}{12} = \dfrac{9}{12} = \dfrac{3}{4}$

63. $\dfrac{5}{24} + \dfrac{3}{16} = \dfrac{10}{48} + \dfrac{9}{48} = \dfrac{19}{48}$

65. $-\dfrac{11}{15} - \dfrac{2}{9} = -\dfrac{33}{45} - \dfrac{10}{45} = -\dfrac{43}{45}$

67. $\dfrac{7}{25} + \dfrac{1}{15} = \dfrac{21}{75} + \dfrac{5}{75} = \dfrac{26}{75}$

69. $\dfrac{4}{27} + \dfrac{1}{6} = \dfrac{8}{54} + \dfrac{9}{54} = \dfrac{17}{54}$

71. $\dfrac{11}{60} - \dfrac{2}{45} = \dfrac{33}{180} - \dfrac{8}{180} = \dfrac{25}{180} = \dfrac{5}{36}$

73. $\dfrac{2}{15} - \dfrac{5}{12} = \dfrac{8}{60} - \dfrac{25}{60} = -\dfrac{17}{60}$

**Applications**

75. BOTANY

    a.   The growth was $\dfrac{5}{32} + \dfrac{1}{16} = \dfrac{5}{32} + \dfrac{2}{32} = \dfrac{7}{32}$ inches

    b.   The difference was $\dfrac{5}{32} - \dfrac{1}{16} = \dfrac{5}{32} - \dfrac{2}{32} = \dfrac{3}{32}$ inches

77. FAMILY DINNER

$\dfrac{3}{8} + \dfrac{2}{6} = \dfrac{9}{24} + \dfrac{8}{24} = \dfrac{17}{24}$ of the pizza was left.

They ate $\dfrac{5}{8} + \dfrac{4}{6} = \dfrac{15}{24} + \dfrac{16}{24} = \dfrac{31}{24}$ pizza, which is more than a whole pizza, so the family could not have been fed one pizza.

79. WEIGHTS AND MEASURES

The scale is off $\dfrac{1}{16}$ of a pound. This results in undercharging customers.

81. HIKING

From longest to shortest $\dfrac{4}{5}, \dfrac{3}{4}, \dfrac{5}{8}$, since $\dfrac{4}{5} = \dfrac{32}{40}, \dfrac{3}{4} = \dfrac{30}{40}$, and $\dfrac{5}{8} = \dfrac{25}{40}$.

83. STUDY HABITS

The fraction of students that study 2 hours or more are $\dfrac{2}{5} + \dfrac{3}{10} = \dfrac{4}{10} + \dfrac{3}{10} = \dfrac{7}{10}$.

85. GARAGE DOOR OPENER

The difference in strength is $\dfrac{1}{2} - \dfrac{1}{3} = \dfrac{3}{6} - \dfrac{2}{6} = \dfrac{1}{6}$ hp.

**Writing**

87. Answers may vary
89. Answers may vary

**Review**

91. $20 = 2^2 \cdot 5$

93. $A = lw$

1.  LCM (3, 5) = 15
3.  LCM (8, 14) = 56
5.  LCM (14, 21) = 42
7.  LCM (6, 18) = 18
9.  LCM (44, 60) = 660
11. LCM (100, 120) = 660
13. LCM (6, 24, 36) = 72
15. LCM (18, 54, 63) = 378
17. GCF (6, 9) = 3
19. GCF (22, 33) = 11
21. GCF (16, 20) = 4
23. GCF (25, 100) = 25
25. GCF (100, 120) = 20
27. GCF (48, 108) = 12
29. GCF (18, 24, 36) = 6
31. GCF (18, 54, 63) = 9

33. NURSING
    2 hours is 120 minutes; LCM (45, 120) is 360 minutes or 6 hours

**Section 3.5 Multiplying and Dividing Mixed Numbers**

**Vocabulary**

1. A **mixed** number is the sum of a whole number and a proper fraction.
3. To **graph** a number means to locate its position on a number line and highlight it using a heavy dot.

**Concepts**

5. a. $-5\dfrac{1}{2}^{\,\circ}$

   b. $-1\dfrac{7}{8}$ in

7. a. The arrow is registering $-2\dfrac{2}{3}$ .

   b. It will register $-3\dfrac{1}{3}$ .

9. $-\dfrac{4}{5}, -\dfrac{2}{5}, \dfrac{1}{5}$

11. DIVING

   Forward $2\dfrac{1}{2}$ somersaults from the pike position.

13.

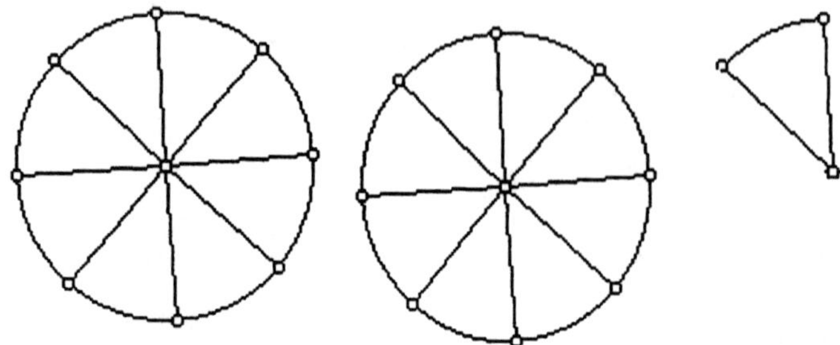

**Notation**

15. $-5\dfrac{1}{4}\cdot 1\dfrac{1}{7} = -\dfrac{21}{4}\cdot\dfrac{8}{7}$

$\phantom{15.}\quad = -\dfrac{21\cdot 8}{4\cdot 7}$

$\phantom{15.}\quad = -\dfrac{\cancel{7}\cdot 3\cdot \cancel{4}\cdot 2}{\cancel{4}\cdot \cancel{7}}$

$\phantom{15.}\quad = -\dfrac{6}{1}$

$\phantom{15.}\quad = -6$

17. $\dfrac{15}{4} = 3\dfrac{3}{4}$

19. $\dfrac{29}{5} = 5\dfrac{4}{5}$

21. $-\dfrac{20}{6} = -3\dfrac{2}{6} = -3\dfrac{1}{3}$

23. $\dfrac{127}{12} = 10\dfrac{7}{12}$

25. $6\dfrac{1}{2} = \dfrac{13}{2}$

27. $20\dfrac{4}{5} = \dfrac{104}{5}$

29. $-6\dfrac{2}{9} = -\dfrac{56}{9}$

31. $200\dfrac{2}{3} = \dfrac{602}{3}$

33. $\left\{-2\dfrac{8}{9}, 1\dfrac{2}{3}, \dfrac{16}{5}\right\}$

$$-5\ -4\ -3\ -2\ -1\ \ 0\ \ 1\ \ 2\ \ 3\ \ 4\ \ 5$$

35. $\left\{-\dfrac{10}{3}, -\dfrac{98}{99}, 3\dfrac{1}{7}\right\}$

$$-5\ -4\ -3\ -2\ -1\ \ 0\ \ 1\ \ 2\ \ 3\ \ 4\ \ 5$$

37. $1\dfrac{2}{3} \cdot 2\dfrac{1}{7} = \dfrac{5}{3} \cdot \dfrac{15}{7} = \dfrac{5}{3} \cdot \dfrac{3 \cdot 5}{7} = \dfrac{25}{7} = 3\dfrac{4}{7}$

39. $-7\dfrac{1}{2}\left(-1\dfrac{2}{5}\right) = \dfrac{15}{2} \cdot \dfrac{7}{5} = \dfrac{3 \cdot 5}{2} \cdot \dfrac{7}{5} = \dfrac{21}{2} = 10\dfrac{1}{2}$

41. $3\dfrac{1}{16} \cdot 4\dfrac{4}{7} = \dfrac{49}{16} \cdot \dfrac{32}{7} = \dfrac{7 \cdot 7}{16} \cdot \dfrac{16 \cdot 2}{7} = 14$

43. $-6 \cdot 2\dfrac{7}{24} = -\dfrac{6}{1} \cdot \dfrac{55}{24} = -\dfrac{6}{1} \cdot \dfrac{55}{6 \cdot 4} = -\dfrac{55}{4} = -13\dfrac{3}{4}$

45. $2\dfrac{1}{2}\left(-3\dfrac{1}{3}\right) = -\dfrac{5}{2} \cdot \dfrac{10}{3} = -\dfrac{5}{2} \cdot \dfrac{5 \cdot 2}{3} = -\dfrac{25}{3} = -8\dfrac{1}{3}$

47. $2\dfrac{5}{8}\cdot\dfrac{5}{27}=\dfrac{21}{8}\cdot\dfrac{5}{27}=\dfrac{3\cdot7}{8}\cdot\dfrac{5}{3\cdot9}=\dfrac{35}{72}$

49. $\left(1\dfrac{2}{3}\right)(6)\left(-\dfrac{1}{8}\right)=\left(\dfrac{5}{3}\right)\left(\dfrac{6}{1}\right)\left(-\dfrac{1}{8}\right)=-\dfrac{30}{24}=-\dfrac{5}{4}=-1\dfrac{1}{4}$

51. $\left(1\dfrac{2}{3}\right)^2=\left(\dfrac{5}{3}\right)^2=\dfrac{25}{9}=2\dfrac{7}{9}$

53. $\left(-1\dfrac{1}{3}\right)^3=\left(-\dfrac{4}{3}\right)^3=-\dfrac{64}{27}=-2\dfrac{10}{27}$

55. $3\dfrac{1}{3}\div1\dfrac{5}{6}=\dfrac{10}{3}\div\dfrac{11}{6}=\dfrac{10}{3}\cdot\dfrac{6}{11}=\dfrac{20}{11}=1\dfrac{9}{11}$

57. $-6\dfrac{3}{5}\div7\dfrac{1}{3}=-\dfrac{33}{5}\div\dfrac{22}{3}=-\dfrac{33}{5}\cdot\dfrac{3}{22}=-\dfrac{9}{10}$

59. $-20\dfrac{1}{4}\div\left(-1\dfrac{11}{16}\right)$

$=-\dfrac{81}{4}\div\left(-\dfrac{27}{16}\right)$

$=\dfrac{81}{4}\cdot\dfrac{16}{27}$

$=\dfrac{3^4}{4}\cdot\dfrac{4^2}{3^3}$

$=12$

61. $6\dfrac{1}{4}\div20=\dfrac{25}{4}\cdot\dfrac{1}{20}$

$=\dfrac{5^2}{4}\cdot\dfrac{1}{4(5)}$

$=\dfrac{5}{16}$

63. $1\dfrac{2}{3}\div\left(-2\dfrac{1}{2}\right)=\dfrac{5}{3}\cdot-\dfrac{2}{5}$

$=-\dfrac{2}{3}$

65. $8 \div \left(3\dfrac{1}{5}\right) = \dfrac{8}{1} \cdot \dfrac{5}{16}$

$$= \dfrac{8}{1} \cdot \dfrac{5}{2(8)}$$

$$= \dfrac{5}{2}$$

$$= 2\dfrac{1}{2}$$

67. $-4\dfrac{1}{2} \div 2\dfrac{1}{4} = -\dfrac{9}{2} \cdot \dfrac{4}{9} = -2$

**Applications**

69. CALORIES

$$20\left(3\dfrac{1}{5}\right) = \dfrac{20}{1} \cdot \dfrac{16}{5} = 64 \text{ calories}$$

71. SHOPPING

$$4\dfrac{1}{4} \cdot \dfrac{64}{100} = \dfrac{17}{4} \cdot \dfrac{4^3}{4(25)} = \dfrac{68}{25} = \$2.72 \quad \left(\text{Note: } \$0.64 = \dfrac{64}{100}\right)$$

73. SUBDIVISION

The remaining acreage is $1,000 - 100 = 900$ acres. $900 \div 1\dfrac{1}{3} = \dfrac{900}{1} \cdot \dfrac{3}{4} = \dfrac{225(4)}{1} \cdot \dfrac{3}{4} = 675$ lots

75. GRAPH PAPER

The dimensions of this paper are $11\left(\dfrac{1}{4}\right) = \dfrac{11}{4} = 2\dfrac{3}{4}$ inches by $5\left(\dfrac{1}{4}\right) = \dfrac{5}{4} = 1\dfrac{1}{4}$ inches.

77. EMERGENCY EXIT

$$A = \dfrac{1}{2}bh$$

$$A = \dfrac{1}{2}\left(8\dfrac{1}{4}\right)\left(10\dfrac{1}{3}\right)$$

$$A = \dfrac{1}{2}\left(\dfrac{3(11)}{4}\right)\left(\dfrac{31}{3}\right)$$

$$A = \dfrac{341}{8}$$

$$A = 42\dfrac{5}{8} \text{ in}^2$$

79. FIRE ESCAPE

The total height is $43(105) = 4,515$ inches, so the total number of steps is

$$4,515 \div 7\dfrac{1}{2} = 4,515 \cdot \dfrac{2}{15} = 301(2) = 602 \text{ steps}.$$

81. SHOPPING ON THE INTERNET
  She should choose size 14, slim cut.

**Writing**

83. Answers may vary
85. Answers may vary

**Review**

87. $3^2 \cdot 2^3 = 72$
89. 4(8)
91. Division by 2 must be undone.

81. SHOPPING ON THE INTERNET

**Section 3.6 Adding and Subtracting Mixed Numbers**

**Vocabulary**

1.  By the **commutative** property of addition, we can add numbers in any order.

3.  To do the subtraction, we **borrow** 1 in the form of $\dfrac{3}{3}$.

**Concepts**

5.  a. The whole number part is 76 the fractional part is $\dfrac{3}{4}$.

  b. $76 + \dfrac{3}{4}$

7.  The fundamental property of fractions is being highlighted.

9.  a. $9\dfrac{17}{16} = 9 + \dfrac{16}{16} + \dfrac{1}{16} = 10\dfrac{1}{16}$

  b. $1,288\dfrac{7}{3} = 1,288 + \dfrac{6}{3} + \dfrac{1}{3} = 1,290\dfrac{1}{3}$

  c. $16\dfrac{12}{8} = 16 + \dfrac{8}{8} + \dfrac{4}{8} = 17\dfrac{1}{2}$

  d. $45\dfrac{24}{20} = 45 + \dfrac{20}{20} + \dfrac{4}{20} = 46\dfrac{1}{5}$

**Notation**

11.

$$70\dfrac{3}{5} + 39\dfrac{2}{7} = 70 + \dfrac{3}{5} + 39 + \dfrac{2}{7}$$

$$= 70 + 39 + \dfrac{3}{5} + \dfrac{2}{7}$$

$$= 109 + \dfrac{3}{5} + \dfrac{2}{7}$$

$$= 109 + \dfrac{3 \cdot 7}{5 \cdot 7} + \dfrac{2 \cdot 5}{7 \cdot 5}$$

$$= 109 + \dfrac{21}{35} + \dfrac{10}{35}$$

$$= 109 + \dfrac{31}{35}$$

$$= 109\dfrac{31}{35}$$

**Practice**

13. $2\dfrac{1}{5}+2\dfrac{1}{5}=4\dfrac{2}{5}$

15. $8\dfrac{2}{7}-3\dfrac{1}{7}=5\dfrac{1}{7}$

17. $3\dfrac{1}{4}+4\dfrac{1}{4}=7\dfrac{2}{4}=7\dfrac{1}{2}$

19. $4\dfrac{1}{6}+1\dfrac{1}{5}=5+\dfrac{5}{30}+\dfrac{6}{30}=5\dfrac{11}{30}$

21. $2\dfrac{1}{2}-1\dfrac{1}{4}=2\dfrac{2}{4}-1\dfrac{1}{4}=1\dfrac{1}{4}$

23. $2\dfrac{5}{6}-1\dfrac{3}{8}=2\dfrac{20}{24}-1\dfrac{9}{24}=1\dfrac{11}{24}$

25. $5\dfrac{1}{2}+3\dfrac{4}{5}=5\dfrac{5}{10}+3\dfrac{8}{10}=8\dfrac{13}{10}=8+\dfrac{10}{10}+\dfrac{3}{10}=9\dfrac{3}{10}$

27. $7\dfrac{1}{2}-4\dfrac{1}{7}=7\dfrac{7}{14}-4\dfrac{2}{14}=3\dfrac{5}{14}$

29. $56\dfrac{2}{5}+73\dfrac{1}{3}=129+\dfrac{6}{15}+\dfrac{5}{15}=129\dfrac{11}{15}$

31. $380\dfrac{1}{6}+17\dfrac{1}{4}=397+\dfrac{2}{12}+\dfrac{3}{12}=397\dfrac{5}{12}$

33. $228\dfrac{5}{9}+44\dfrac{2}{3}=272+\dfrac{15}{27}+\dfrac{18}{27}=272\dfrac{33}{27}=273\dfrac{6}{27}=273\dfrac{2}{9}$

35. $778\dfrac{5}{7}-155\dfrac{1}{3}=778\dfrac{15}{21}-155\dfrac{7}{21}=623\dfrac{8}{21}$

37. $140\dfrac{5}{6}-129\dfrac{4}{5}=140\dfrac{25}{30}-129\dfrac{24}{30}=11\dfrac{1}{30}$

39. $422\dfrac{13}{16}-321\dfrac{3}{8}=422\dfrac{13}{16}-321\dfrac{6}{16}=101\dfrac{7}{16}$

41. $16\dfrac{1}{4}-13\dfrac{3}{4}=15\dfrac{5}{4}-13\dfrac{3}{4}=2\dfrac{2}{4}=2\dfrac{1}{2}$

43. $76\dfrac{1}{6}-49\dfrac{7}{8}=76\dfrac{4}{24}-49\dfrac{21}{24}=75\dfrac{28}{24}-49\dfrac{21}{24}=26\dfrac{7}{24}$

45. $140\dfrac{3}{16}-129\dfrac{3}{4}=140\dfrac{3}{16}-129\dfrac{12}{16}=139\dfrac{19}{16}-129\dfrac{12}{16}=10\dfrac{7}{16}$

47. $334\dfrac{1}{9}-13\dfrac{5}{6}=334\dfrac{2}{18}-13\dfrac{15}{18}=333\dfrac{20}{18}-13\dfrac{15}{18}=320\dfrac{5}{18}$

49. $7 - \dfrac{2}{3} = 6\dfrac{3}{3} - \dfrac{2}{3} = 6\dfrac{1}{3}$

51. $9 - 8\dfrac{3}{4} = 8\dfrac{4}{4} - 8\dfrac{3}{4} = \dfrac{1}{4}$

53. $4\dfrac{1}{7} - \dfrac{4}{5} = 4\dfrac{5}{35} - \dfrac{28}{35} = 3\dfrac{40}{35} - \dfrac{28}{35} = 3\dfrac{12}{35}$

55. $6\dfrac{5}{8} - 3 = 3\dfrac{5}{8}$

57. $\dfrac{7}{3} + 2 = 2\dfrac{1}{3} + 2 = 4\dfrac{1}{3}$

59. $2 + 1\dfrac{7}{8} = 3\dfrac{7}{8}$

61. $12\dfrac{1}{2} + 5\dfrac{3}{4} + 35\dfrac{1}{6} = 52 + \dfrac{6}{12} + \dfrac{9}{12} + \dfrac{2}{12} = 52 + \dfrac{17}{12} = 53\dfrac{5}{12}$

63. $58\dfrac{7}{8} + 340 + 61\dfrac{1}{4} = 459 + \dfrac{7}{8} + \dfrac{2}{8} = 459 + \dfrac{9}{8} = 460\dfrac{1}{8}$

65. $-3\dfrac{3}{4} + \left(-1\dfrac{1}{2}\right) = -3\dfrac{3}{4} + \left(-1\dfrac{2}{4}\right) = -4\dfrac{5}{4} = -5\dfrac{1}{4}$

67. $-4\dfrac{5}{8} - 1\dfrac{1}{4} = -4\dfrac{5}{8} - 1\dfrac{2}{8} = -5\dfrac{7}{8}$

**Applications**

69. FREEWAY TRAVEL

Grand Avenue and Citrus Avenue are $3\dfrac{1}{2} - \dfrac{3}{4} = 3\dfrac{2}{4} - \dfrac{3}{4} = 2\dfrac{6}{4} - \dfrac{3}{4} = 2\dfrac{3}{4}$ miles apart.

71. TRAIL MIX

The recipe will yield $2\dfrac{3}{4} + 2\left(\dfrac{1}{2}\right) + \dfrac{2}{3} + \dfrac{1}{3} + 2\dfrac{2}{3} + \dfrac{1}{4} = 5 + \dfrac{9}{12} + \dfrac{8}{12} + \dfrac{4}{12} + \dfrac{8}{12} + \dfrac{3}{12}$

$$= 5 + \dfrac{32}{12}$$

$$= 7\dfrac{8}{12}$$

$$= 7\dfrac{2}{3} \text{ cups}$$

73. HOSE REPAIR

The hose is $50 - 1\dfrac{1}{2} = 49\dfrac{2}{2} - 1\dfrac{1}{2} = 48\dfrac{1}{2}$ feet long.

75. SHIPPING

a.

| | Rate | Time | Distance = Rate (Time) |
|---|---|---|---|
| Passenger | $16\dfrac{1}{2}$ | 1 | $16\dfrac{1}{2}$ |
| Cargo | $5\dfrac{1}{5}$ | 1 | $5\dfrac{1}{5}$ |

b.  At 1:00 a.m. the ships are $16\dfrac{1}{2}+5\dfrac{1}{5}=21+\dfrac{5}{10}+\dfrac{2}{10}=21\dfrac{7}{10}$ miles apart.

77. SERVICE STATION

a.  The difference in price is $179\dfrac{9}{10}-159\dfrac{9}{10}=20$ cents.

b.  The cost per gallon for the full serve is 30 cents more, $199\dfrac{9}{10}-169\dfrac{9}{10}=30$.

79. WATER SLIDE

The slide was
$$311\dfrac{5}{12}-119\dfrac{3}{4}=311\dfrac{5}{12}-119\dfrac{9}{12}$$
$$=310\dfrac{17}{12}-119\dfrac{9}{12}$$
$$=191\dfrac{8}{12}$$
$$=191\dfrac{2}{3} \text{ feet long.}$$

**Writing**

81. Answers may vary
83. Answers may vary

**Review**

85. $2x-1=3x-8$

   $7=x$

87. $-2-(-8)=-2+8=6$

89. Area measures the amount of surface a figure encloses.

## Section 3.7 Order of Operations and Complex Fractions

**Vocabulary**

1. $\dfrac{\frac{1}{2}}{\frac{3}{4}}$ is a **complex** fraction.

**Concepts**

3. $\dfrac{\frac{2}{3}}{\frac{1}{5}} = \dfrac{2}{3} \div \dfrac{1}{5}$

5. The common denominator of all the fractions is 15.
7. When this complex fraction is simplified the result will be negative.
9. Subtraction should be performed first.

**Notation**

11.

$$\dfrac{\frac{1}{8}}{\frac{3}{4}} = \dfrac{1}{8} \div \dfrac{3}{4}$$

$$= \dfrac{1}{8} \cdot \dfrac{4}{3}$$

$$= \dfrac{1 \cdot 4}{8 \cdot 3}$$

$$= \dfrac{1 \cdot \cancel{4}}{2 \cdot \cancel{4} \cdot 3}$$

$$= \dfrac{1}{6}$$

**Practice**

13. $\dfrac{2}{3}\left(-\dfrac{1}{4}\right) + \dfrac{1}{2} = -\dfrac{1}{6} + \dfrac{3}{6} = \dfrac{2}{6} = \dfrac{1}{3}$

15. $\dfrac{4}{5} - \left(-\dfrac{1}{3}\right)^2 = \dfrac{4}{5} - \dfrac{1}{9}$

$$= \dfrac{36}{45} - \dfrac{5}{45}$$

$$= \dfrac{31}{45}$$

17. $-4\left(-\dfrac{1}{5}\right)-\left(\dfrac{1}{4}\right)\left(-\dfrac{1}{2}\right)=\dfrac{4}{5}+\dfrac{1}{8}$

$$=\dfrac{32}{40}+\dfrac{5}{40}$$

$$=\dfrac{37}{40}$$

19. $1\dfrac{3}{5}\left(\dfrac{1}{2}\right)^{2}\left(\dfrac{3}{4}\right)=\dfrac{8}{5}\left(\dfrac{1}{4}\right)\left(\dfrac{3}{4}\right)$

$$=\dfrac{3}{10}$$

21.

$\dfrac{7}{8}-\left(\dfrac{4}{5}+1\dfrac{3}{4}\right)=\dfrac{7}{8}-\left(\dfrac{4}{5}+\dfrac{7}{4}\right)$

$$=\dfrac{7}{8}-\left(\dfrac{16}{20}+\dfrac{35}{20}\right)$$

$$=\dfrac{7}{8}-\dfrac{51}{20}$$

$$=\dfrac{35}{40}-\dfrac{102}{40}$$

$$=-\dfrac{67}{40}$$

$$=-1\dfrac{27}{40}$$

23.

$\left(\dfrac{9}{20}\div 2\dfrac{2}{5}\right)+\left(\dfrac{3}{4}\right)^{2}=\left(\dfrac{9}{20}\div\dfrac{12}{5}\right)+\left(\dfrac{9}{16}\right)$

$$=\left(\dfrac{9}{20}\bullet\dfrac{5}{12}\right)+\left(\dfrac{9}{16}\right)$$

$$=\dfrac{3}{16}+\dfrac{9}{16}$$

$$=\dfrac{12}{16}$$

$$=\dfrac{3}{4}$$

25. $\left(-\dfrac{3}{4}\cdot\dfrac{9}{16}\right)+\left(\dfrac{1}{2}-\dfrac{1}{8}\right)=-\dfrac{27}{64}+\dfrac{3}{8}$

$$=-\dfrac{27}{64}+\dfrac{24}{64}$$

$$=-\dfrac{3}{64}$$

27.

$$\left|\dfrac{2}{3}-\dfrac{9}{10}\right|\div\left(-\dfrac{1}{5}\right)=\left|\dfrac{20}{30}-\dfrac{27}{30}\right|\cdot(-5)$$

$$=\left|-\dfrac{7}{30}\right|\cdot(-5)$$

$$=\dfrac{7}{30}\cdot(-5)$$

$$=-\dfrac{7}{6}$$

$$=-1\dfrac{1}{6}$$

29.

$$\left(2-\dfrac{1}{2}\right)^2+\left(2+\dfrac{1}{2}\right)^2=\left(\dfrac{3}{2}\right)^2+\left(\dfrac{5}{2}\right)^2$$

$$=\dfrac{9}{4}+\dfrac{25}{4}$$

$$=\dfrac{34}{4}$$

$$=\dfrac{17}{2}$$

$$=8\dfrac{1}{2}$$

31. $\dfrac{1}{2}(-7)=-\dfrac{7}{2}$

$$\left(-\dfrac{7}{2}\right)^2=\dfrac{49}{4}$$

33. $\dfrac{1}{2}\left(\dfrac{11}{2}\right)=\dfrac{11}{4}$

$$\left(\dfrac{11}{4}\right)^2=\dfrac{121}{16}$$

35.

$$P = 2l + 2w$$

$$P = 2\left(2\frac{7}{8}\right) + 2\left(1\frac{1}{4}\right)$$

$$P = 2\left(\frac{23}{8}\right) + 2\left(\frac{5}{4}\right)$$

$$P = \frac{23}{4} + \frac{5}{2}$$

$$P = \frac{23}{4} + \frac{10}{4}$$

$$P = \frac{33}{4}$$

$$P = 8\frac{1}{4} \text{ inches}$$

37. $\dfrac{\dfrac{2}{3}}{\dfrac{4}{5}} = \dfrac{2}{3} \cdot \dfrac{5}{4} = \dfrac{5}{6}$

39. $\dfrac{-\dfrac{14}{15}}{\dfrac{7}{10}} = -\dfrac{14}{15} \cdot \dfrac{10}{7} = -\dfrac{4}{3} = -1\dfrac{1}{3}$

41. $\dfrac{5}{\dfrac{10}{21}} = 5 \cdot \dfrac{21}{10} = \dfrac{21}{2} = 10\dfrac{1}{2}$

43. $\dfrac{-\dfrac{5}{6}}{-1\dfrac{7}{8}} = \dfrac{5}{6} \cdot \left(\dfrac{8}{15}\right) = \dfrac{\cancel{5}}{\cancel{2}\cdot 3} \cdot \left(\dfrac{4\cdot\cancel{2}}{3\cdot\cancel{5}}\right) = \dfrac{4}{9}$

45. $\dfrac{\dfrac{1}{2} + \dfrac{1}{4}}{\dfrac{1}{2} - \dfrac{1}{4}} = \dfrac{\dfrac{3}{4}}{\dfrac{1}{4}} = \dfrac{3}{4} \cdot \dfrac{4}{1} = 3$

47. $\dfrac{\dfrac{3}{8} + \dfrac{1}{4}}{\dfrac{3}{8} - \dfrac{1}{4}} = \dfrac{\dfrac{3}{8} + \dfrac{2}{8}}{\dfrac{3}{8} - \dfrac{2}{8}} = \dfrac{\dfrac{5}{8}}{\dfrac{1}{8}} = \dfrac{5}{8} \cdot \dfrac{8}{1} = 5$

49. $\dfrac{\dfrac{1}{5}+3}{-\dfrac{4}{25}} = \dfrac{3\dfrac{1}{5}}{-\dfrac{4}{25}} = \dfrac{\dfrac{16}{5}}{-\dfrac{4}{25}} = \dfrac{16}{5}\cdot\left(-\dfrac{25}{4}\right) = -20$

51. $\dfrac{5\dfrac{1}{2}}{-\dfrac{1}{4}+\dfrac{3}{4}} = \dfrac{\dfrac{11}{2}}{\dfrac{1}{2}} = \dfrac{11}{2}\cdot\dfrac{2}{1} = 11$

53. $\dfrac{\dfrac{1}{5}-\left(-\dfrac{1}{4}\right)}{\dfrac{1}{4}+\dfrac{4}{5}} = \dfrac{\dfrac{1}{5}+\dfrac{1}{4}}{\dfrac{5}{20}+\dfrac{16}{20}} = \dfrac{\dfrac{4}{20}+\dfrac{5}{20}}{\dfrac{21}{20}} = \dfrac{\dfrac{9}{20}}{\dfrac{21}{20}} = \dfrac{9}{20}\cdot\dfrac{20}{21} = \dfrac{3}{7}$

55. $\dfrac{\dfrac{1}{3}+\left(-\dfrac{5}{6}\right)}{1\dfrac{1}{3}} = \dfrac{\dfrac{2}{6}+\left(-\dfrac{5}{6}\right)}{\dfrac{4}{3}} = \dfrac{-\dfrac{1}{2}}{\dfrac{4}{3}} = -\dfrac{1}{2}\cdot\dfrac{3}{4} = -\dfrac{3}{8}$

**Applications**

57. SANDWICH SHOP

$\dfrac{1\dfrac{3}{4}+2\dfrac{1}{2}}{\dfrac{1}{2}} = \dfrac{\dfrac{7}{4}+\dfrac{5}{2}}{\dfrac{1}{2}} = \dfrac{\dfrac{7}{4}+\dfrac{10}{4}}{\dfrac{1}{2}} = \dfrac{\dfrac{17}{4}}{\dfrac{1}{2}} = \dfrac{17}{4}\cdot 2 = \dfrac{17}{2} = 8\dfrac{1}{2}$ sandwiches

59. PHYSICAL FITNESS

|  | Rate | Time | Distance = Rate(Time) |
|---|---|---|---|
| Jogger | $2\dfrac{1}{2}$ | $1\dfrac{1}{2}$ | $2\dfrac{1}{2}\left(1\dfrac{1}{2}\right) = \dfrac{5}{2}\left(\dfrac{3}{2}\right) = \dfrac{15}{4} = 3\dfrac{3}{4}$ |
| Cyclist | $7\dfrac{1}{5}$ | $1\dfrac{1}{2}$ | $7\dfrac{1}{5}\left(1\dfrac{1}{2}\right) = \dfrac{36}{5}\left(\dfrac{3}{2}\right) = \dfrac{54}{5} = 10\dfrac{4}{5}$ |

The total distance apart is $3\dfrac{3}{4}+10\dfrac{4}{5} = 13+\dfrac{3}{4}+\dfrac{4}{5} = 13+\dfrac{15}{20}+\dfrac{16}{20} = 13+\dfrac{31}{20} = 14\dfrac{11}{20}$ miles .

61. POSTAGE RATES

$\dfrac{1}{16}+\dfrac{5}{8}+3\left(\dfrac{1}{16}\right) = \dfrac{1}{16}+\dfrac{10}{16}+\dfrac{3}{16} = \dfrac{14}{16} = \dfrac{7}{8}$ oz which is less than one ounce.

It can be mailed for the 1 oz rate.

63. PHYSICAL THERAPY

The total distance was $7\left(\dfrac{1}{4}\right)+7\left(\dfrac{1}{2}\right)+7\left(\dfrac{3}{4}\right) = \dfrac{7}{4}+\dfrac{7}{2}+\dfrac{21}{4} = \dfrac{7}{4}+\dfrac{14}{4}+\dfrac{21}{4} = \dfrac{42}{4} = \dfrac{21}{2} = 10\dfrac{1}{2}$ miles .

## 65. AMUSEMENT PARK

$$\frac{1}{\dfrac{1}{10}+\dfrac{1}{15}}=\frac{1}{\dfrac{3}{30}+\dfrac{2}{30}}=\frac{1}{\dfrac{5}{30}}=\frac{30}{5}=6 \text{ seconds}$$

**Writing**

67. Answers may vary
69. Answers may vary

**Review**

71. $\dfrac{2(15)+6}{2\cdot 3^2}=\dfrac{30+6}{2\cdot 9}=\dfrac{36}{18}=2$

73. $-5x+3=28$

$\qquad -5x=25$

$\qquad\quad x=-5$

75. $\begin{aligned}2+3[-3-(-4-1)]&=2+3[-3-(-5)]\\&=2+3[-3+5]\\&=2+3[2]\\&=8\end{aligned}$

**Section 3.8 Solving Equations Containing Fractions**

**Vocabulary**

1.  To find the **reciprocal** of a fraction, invert the numerator and the denominator.
3.  The **least common denominator** of a set of fractions is the smallest number each denominator will divide exactly.

**Concepts**

5.  Let $x = 40$, then $\dfrac{5}{8}x = \dfrac{5}{8}(40) = 5^2 = 25$, so this is a solution.

7.  The product of reciprocals is one.

9.  a. $\dfrac{4}{5}p$

    b. $\dfrac{1}{4}t$

**Notation**

11.    $\dfrac{7}{8}x = 21$

$$\dfrac{8}{7}\left(\dfrac{7}{8}x\right) = \dfrac{8}{7}(21)$$

$$x = 24$$

13. a. This is a **true** statement.
    b. The is a **false** statement.
    c. This is a **true** statement.
    d. This is a **true** statement.

**Practice**

15. $\dfrac{4}{7}x = 16$

$$x = \dfrac{7}{4}(16)$$

$$x = 28$$

17. $\dfrac{7}{8}t = -28$

$$x = \dfrac{8}{7}(-28)$$

$$x = -32$$

19. $-\dfrac{3}{5}h = 4$

$$h = -\dfrac{5}{3}(4)$$

$$h = -\dfrac{20}{3}$$

21. $\dfrac{2}{3}x = \dfrac{4}{5}$

$$x = \dfrac{3}{2}\left(\dfrac{4}{5}\right)$$

$$x = \dfrac{6}{5}$$

23. $\dfrac{2}{5}y = 0$

$$y = \dfrac{5}{2}(0)$$

$$y = 0$$

25. $-\dfrac{5c}{6} = -25$

$$c = -\dfrac{6}{5}(-25)$$

$$c = 30$$

27. $\dfrac{-5f}{7} = -2$

$$c = -\dfrac{7}{5}(-2)$$

$$c = \dfrac{14}{5}$$

29. $\dfrac{5}{8}y = \dfrac{1}{10}$

$$y = \dfrac{8}{5}\left(\dfrac{1}{10}\right)$$

$$y = \dfrac{4}{25}$$

31. $x - \dfrac{1}{9} = \dfrac{7}{9}$

$$x = \dfrac{8}{9}$$

33. $x + \dfrac{1}{9} = \dfrac{4}{9}$

$x = \dfrac{3}{9}$

$x = \dfrac{1}{3}$

35. $x - \dfrac{1}{6} = \dfrac{2}{9}$

$x = \dfrac{2}{9} + \dfrac{1}{6}$

$x = \dfrac{4}{18} + \dfrac{3}{18}$

$x = \dfrac{7}{18}$

37. $y + \dfrac{7}{8} = \dfrac{1}{4}$

$y = \dfrac{1}{4} - \dfrac{7}{8}$

$y = \dfrac{2}{8} - \dfrac{7}{8}$

$y = -\dfrac{5}{8}$

39. $\dfrac{5}{4} + t = \dfrac{1}{4}$

$t = \dfrac{1}{4} - \dfrac{5}{4}$

$t = -\dfrac{4}{4} = -1$

41. $x + \dfrac{3}{4} = -\dfrac{1}{2}$

$x = -\dfrac{3}{4} - \dfrac{1}{2}$

$x = -\dfrac{3}{4} - \dfrac{2}{4}$

$x = -\dfrac{5}{4}$

43. $\dfrac{-x}{4}+1=10$

$$-\dfrac{x}{4}=9$$
$$x=-36$$

45.

$$2x-\dfrac{1}{2}=\dfrac{1}{3}$$
$$2x=\dfrac{1}{2}+\dfrac{1}{3}$$
$$2x=\dfrac{3}{6}+\dfrac{2}{6}$$
$$2x=\dfrac{5}{6}$$
$$x=\dfrac{5}{12}$$

47. $\dfrac{1}{2}x-\dfrac{1}{9}=\dfrac{1}{3}$

$$\dfrac{1}{2}x=\dfrac{3}{9}+\dfrac{1}{9}$$
$$\dfrac{1}{2}x=\dfrac{4}{9}$$
$$x=\dfrac{8}{9}$$

49. $5+\dfrac{x}{3}=\dfrac{1}{2}$

$$\dfrac{1}{3}x=-4\dfrac{1}{2}$$
$$\dfrac{1}{3}x=-\dfrac{9}{2}$$
$$x=-\dfrac{27}{2}$$

51. $\dfrac{2}{5}x+1=\dfrac{1}{3}$

$$\dfrac{2}{5}x=-\dfrac{2}{3}$$
$$x=-\dfrac{10}{6}$$
$$x=-\dfrac{5}{3}$$

53. $\dfrac{x}{3}+\dfrac{1}{4}=-2$

$\qquad \dfrac{x}{3}=-\dfrac{9}{4}$

$\qquad x=-\dfrac{27}{4}$

55. $4+\dfrac{s}{3}=8$

$\qquad \dfrac{s}{3}=4$

$\qquad s=12$

57. $\dfrac{5h}{6}-8=12$

$\qquad \dfrac{5h}{6}=20$

$\qquad h=\dfrac{6}{5}(20)$

$\qquad h=24$

59. $-4+9+\dfrac{5t}{12}=0$

$\qquad \dfrac{5t}{12}=-5$

$\qquad t=\dfrac{12}{5}(-5)$

$\qquad t=-12$

61. $-3-2+\dfrac{4x}{15}=0$

$\qquad \dfrac{4x}{15}=5$

$\qquad y=\dfrac{15}{4}(5)$

$\qquad y=\dfrac{75}{4}$

**Applications**

63. TRANSMISSION REPAIR
*Analyze the problem*

- Only $\dfrac{1}{3}$ of the customers needed new transmissions.
- The shop installed 32 new transmissions last year.
- Find the number of <u>customers</u> the shop had last year.

*Form an equation*
Let $x$ = <u>the number of customers last year</u>
*Key phrase:* one-third of    *Translation:* <u>multiply</u>

$\dfrac{1}{3}$ of the number of customers last year was 32.

$$\dfrac{1}{3}x = 32$$

*Solve the equation*

$$\dfrac{1}{3}x = 32$$

$$3\left(\dfrac{1}{3}x\right) = 3(32)$$

$$x = 96$$

*State the conclusion* <u>The shop had 96 customers last year.</u>
*Check the result* If we find 1/3 of 96, we get 32. The answer checks.

65. TOOTH DEVELOPMENT
Let $t$ represent teeth

$$\dfrac{4}{5}t = 16$$

$$\dfrac{5}{4}\left(\dfrac{4}{5}t\right) = \dfrac{5}{4}(16)$$

$$t = 20$$

The child will eventually have 20 teeth.

67. HOME SALES
Let $h$ represent homes, 9 homes represent the remaining one-fourth of the homes.

$$\dfrac{1}{4}h = 9$$

$$h = 4(9)$$

$$h = 36$$

There are 36 homes in the subdivision.

69. TELEPHONE BOOK
Let $p$ represent total pages, 150 is one-third of the phone book.

$$\dfrac{1}{3}p = 150$$

$$p = 450$$

There are 450 pages in the telephone book.

71. SAFETY REQUIREMENT

Let $w$ represent width

$$A = lw$$

$$30 = 3\frac{3}{4}w$$

$$30 = \frac{15}{4}w$$

$$\frac{4}{15}(30) = w$$

$$8 = w$$

The taillight must be 8 inches wide.

73. CPR CLASS

Let $c$ represent the time for the complete class

$$\frac{1}{4}c + \frac{2}{3}c + 30 = c$$

$$\frac{3}{12}c + \frac{8}{12}c + 30 = c$$

$$30 = \frac{1}{12}c$$

$$360 = c$$

The course is 360 minutes.

**Writing**

75. Answers may vary
77. Answers may vary

**Review**

79. $(-2)^5 = -32$

81. $3x - 2 = 7$

$$3x = 9$$

$$x = 3$$

83. 12,590,767 to the nearest million is 13,000,000.

1. $\dfrac{15}{25} = \dfrac{15 \div 5}{25 \div 5} = \dfrac{3}{5}$

3. $\dfrac{1}{5} = \dfrac{1 \cdot 7}{5 \cdot 7} = \dfrac{7}{35}$

**Section 3.1 The Fundamental Property of Fractions**

1. This woman spends $\dfrac{7}{24}$ of her day sleeping.

3. $\dfrac{2}{-3} = -\dfrac{2}{3} = \dfrac{-2}{3}$

5. The numerator and denominator are divided by two.

7. a. $\dfrac{15}{45} = \dfrac{1}{3}$

   b. $\dfrac{20}{48} = \dfrac{5}{12}$

   c. $-\dfrac{63}{84} = -\dfrac{3}{4}$

   d. $\dfrac{66}{108} = \dfrac{11}{18}$

9. a. $\dfrac{2}{3} \cdot \dfrac{6}{6} = \dfrac{12}{18}$

   b. $-\dfrac{3}{8} \cdot \dfrac{2}{2} = -\dfrac{6}{16}$

   c. $\dfrac{7}{15} \cdot \dfrac{3}{3} = \dfrac{21}{45}$

   d. $\dfrac{4}{1} \cdot \dfrac{9}{9} = \dfrac{36}{9}$

**Section 3.2 Multiplying Fractions**

11. a. $\dfrac{3}{4}x = \dfrac{3x}{4}$ is a true statement.

    b. $-\dfrac{5}{9}e \neq -\dfrac{5}{9e}$, so it is a false statement.

13. a. $\left(\dfrac{3}{4}\right)^2 = \dfrac{9}{16}$

    b. $\left(-\dfrac{5}{2}\right)^3 = -\dfrac{125}{8}$

    c. $\left(\dfrac{2}{3}\right)^2 = \dfrac{4}{9}$

    d. $\left(-\dfrac{2}{5}\right)^3 = -\dfrac{8}{125}$

15. $A = \dfrac{1}{2}bh$

    $A = \dfrac{1}{2}(15)(8)$

    $A = 60 \text{ in}^2$

## Section 3.3 Dividing Fractions

17. a. $\dfrac{1}{6} \div \dfrac{11}{25} = \dfrac{1}{6} \cdot \dfrac{25}{11} = \dfrac{25}{66}$

    b. $-\dfrac{7}{8} \div \dfrac{1}{4} = -\dfrac{7}{8} \cdot \dfrac{4}{1} = -\dfrac{7}{2}$

    c. $-\dfrac{15}{16} \div (-10) = -\dfrac{15}{16} \left(-\dfrac{1}{10}\right) = \dfrac{3}{32}$

    d. $8 \div \dfrac{16}{5} = 8 \cdot \dfrac{5}{16} = \dfrac{5}{2}$

    e. $\dfrac{1}{8} \div \dfrac{1}{4} = \dfrac{1}{8} \cdot \dfrac{4}{1} = \dfrac{1}{2}$

    f. $\dfrac{4}{5} \div \dfrac{1}{2} = \dfrac{4}{5} \cdot \dfrac{2}{1} = \dfrac{8}{5}$

## Section 3.4 Adding and Subtracting Fractions

19. a. $\dfrac{2}{7} + \dfrac{3}{7} = \dfrac{5}{7}$

    b. $-\dfrac{3}{5} - \dfrac{3}{5} = -\dfrac{6}{5}$

    c. $\dfrac{3}{8} - \dfrac{1}{8} = \dfrac{2}{8}$

    d. $\dfrac{7}{8} + \dfrac{7}{8} = \dfrac{14}{8} = \dfrac{7}{4} = 1\dfrac{3}{4}$

21. $30 = 2 \cdot 3 \cdot 5$

$45 = 3 \cdot 3 \cdot 5$

$\text{LCD}(30, 45) = 2 \cdot 3 \cdot 3 \cdot 5 = 90$

23. MACHINE SHOP

$\dfrac{3}{4} - \dfrac{17}{32} = \dfrac{24}{32} - \dfrac{17}{32} = \dfrac{7}{32}$ inch must be milled away .

## Section 3.5 Multiplying and Dividing Fractions

25. a. $2\dfrac{1}{6}$

b. $\dfrac{13}{6}$

27. a. $9\dfrac{3}{8} = \dfrac{75}{8}$

b. $-2\dfrac{1}{5} = -\dfrac{11}{5}$

c. $100\dfrac{1}{2} = \dfrac{201}{2}$

d. $1\dfrac{99}{100} = \dfrac{199}{100}$

29. a. $-5\dfrac{1}{4} \cdot \dfrac{2}{35} = -\dfrac{21}{4} \cdot \dfrac{2}{35} = -\dfrac{3}{10}$

b. $\left(-3\dfrac{1}{2}\right) \div \left(-3\dfrac{2}{3}\right) = -\dfrac{7}{2} \div \left(-\dfrac{11}{3}\right) = -\dfrac{7}{2} \cdot \left(-\dfrac{3}{11}\right) = \dfrac{21}{22}$

c. $\left(-6\dfrac{2}{3}\right)(-6) = \dfrac{20}{3}\left(\dfrac{6}{1}\right) = 40$

d. $-8 \div 3\dfrac{1}{5} = -8 \div \left(\dfrac{16}{5}\right) = -8 \cdot \left(\dfrac{5}{16}\right) = -\dfrac{5}{2} = -2\dfrac{1}{2}$

**Section 3.6 Adding and Subtracting Mixed Numbers**

31. a. $1\dfrac{3}{8} + 2\dfrac{1}{5} = 3 + \dfrac{3}{8} + \dfrac{1}{5} = 3 + \dfrac{15}{40} + \dfrac{8}{40} = 3\dfrac{23}{40}$

 b. $3\dfrac{1}{2} + 2\dfrac{2}{3} = 5 + \dfrac{1}{2} + \dfrac{2}{3} = 5 + \dfrac{3}{6} + \dfrac{4}{6} = 5\dfrac{7}{6} = 6\dfrac{1}{6}$

 c. $2\dfrac{5}{6} - 1\dfrac{3}{4} = 2\dfrac{10}{12} - 1\dfrac{9}{12} = 1\dfrac{1}{12}$

 d. $3\dfrac{7}{16} - 2\dfrac{1}{8} = 3\dfrac{7}{16} - 2\dfrac{2}{16} = 1\dfrac{5}{16}$

33. a. $133\dfrac{1}{9} + 49\dfrac{1}{6} = 182 + \dfrac{1}{9} + \dfrac{1}{6} = 182 + \dfrac{2}{18} + \dfrac{3}{18} = 182\dfrac{5}{18}$

 b. $98\dfrac{11}{20} + 14\dfrac{3}{5} = 112 + \dfrac{11}{20} + \dfrac{3}{5} = 112 + \dfrac{11}{20} + \dfrac{12}{20} = 112\dfrac{23}{20} = 113\dfrac{3}{20}$

 c. $50\dfrac{5}{8} - 19\dfrac{1}{6} = 50\dfrac{15}{24} - 19\dfrac{4}{24} = 31\dfrac{11}{24}$

 d. $375\dfrac{3}{4} - 59 = 316\dfrac{3}{4}$

**Section 3.7 Order of Operations and Complex Fractions**

35. a. $\dfrac{3}{4} + \left(-\dfrac{1}{3}\right)^2\left(\dfrac{5}{4}\right) = \dfrac{3}{4} + \left(\dfrac{1}{9}\right)\left(\dfrac{5}{4}\right) = \dfrac{3}{4} + \dfrac{5}{36} = \dfrac{27}{36} + \dfrac{5}{36} = \dfrac{32}{36} = \dfrac{8}{9}$

 b. $\left(\dfrac{2}{3} \div \dfrac{16}{9}\right) - \left(1\dfrac{2}{3} \cdot \dfrac{1}{15}\right) = \left(\dfrac{2}{3} \cdot \dfrac{9}{16}\right) - \left(\dfrac{5}{3} \cdot \dfrac{1}{15}\right)$

$= \left(\dfrac{3}{8}\right) - \left(\dfrac{1}{9}\right)$

$= \dfrac{27}{72} - \dfrac{8}{72}$

$= \dfrac{19}{72}$

37. a. $\dfrac{2}{3}x = 16$

$x = \dfrac{3}{2}(16)$

$x = 24$

b. $-\dfrac{7s}{4} = -49$

$s = \dfrac{4}{7}(49)$

$s = 28$

c. $\dfrac{y}{5} = -\dfrac{1}{15}$

$y = -\dfrac{5}{15}$

$y = -\dfrac{1}{3}$

d. $2x - 3 = 8$

$2x = 11$

$x = \dfrac{11}{2}$

39. HISTORY TEXTBOOK

Let $p$ represent the total number of pages

$\dfrac{2}{3}p = 220$

$p = \dfrac{3}{2}(220)$

$p = 330$ pages in this textbook

**Chapter 3 Test**

1.  a.  $\dfrac{4}{5}$ of the plant is above ground.

    b.  $\dfrac{1}{5}$ of the plant is below ground.

3.  $-\dfrac{3}{4}\left(\dfrac{1}{5}\right)=-\dfrac{3}{20}$

5.  $\dfrac{4}{3}\div\dfrac{1}{9}=\dfrac{4}{3}\cdot\dfrac{9}{1}=12$

7.  $\dfrac{7}{8}\cdot\dfrac{3}{3}=\dfrac{21}{24}$

9.  SPORTS CONTRACT

$$13\dfrac{1}{2}\div 9=\dfrac{27}{2}\cdot\dfrac{1}{9}=\dfrac{3}{2}=\$1\dfrac{1}{2}\text{ million/year}$$

11. $67\dfrac{1}{4}-29\dfrac{5}{6}=67\dfrac{3}{12}-29\dfrac{10}{12}$

$$=66\dfrac{15}{12}-29\dfrac{10}{12}$$

$$=37\dfrac{5}{12}$$

13. $-\dfrac{3}{7}+2=-\dfrac{3}{7}+\dfrac{14}{7}=\dfrac{11}{7}=1\dfrac{4}{7}$

15. The perimeter is $20+22\dfrac{2}{3}+10\dfrac{2}{3}=52\dfrac{4}{3}=53\dfrac{1}{3}$ inches .

    The area is

$$A=\dfrac{1}{2}bh$$

$$A=\dfrac{1}{2}(20)\left(10\dfrac{2}{3}\right)$$

$$A=10\left(\dfrac{32}{3}\right)$$

$$A=\dfrac{320}{3}$$

$$A=106\dfrac{2}{3}\text{ inches}^2$$

17. $\dfrac{-\dfrac{5}{6}}{\dfrac{7}{8}} = -\dfrac{5}{6}\left(\dfrac{8}{7}\right) = -\dfrac{20}{21}$

19. a. $\dfrac{x}{3} = 14$

$x = 3(14)$

$x = 52$

 b. $-\dfrac{5}{2}t + 2 = 20$

$-\dfrac{5}{2}t = 18$

$t = -\dfrac{2}{5}(18)$

$t = -\dfrac{36}{5}$

21. The parts of a fraction are the numerator, denominator and fraction bar. A fraction represents equal parts of a whole, or a division.

23. a. The numerator and denominator are factored and the common factor is divided out.

 b. This illustration demonstrates equivalent fractions, $\dfrac{1}{2} = \dfrac{2}{4}$.

 c. The fraction is rewritten with a denominator of twenty.

**Chapters 1 – 3 Cumulative Review**

1. $5,434,700$

3. THE STOCK MARKET
   The highest mark was approximately $11,555$ at 10:30 a.m.

5. $4,679 + 3,457 = 8,136$

7. $5,345 \times 56 = 299,320$

9. The pool perimeter is $2(150) + 2(75) = 250$ feet .

11. $84 = 2^2 \cdot 3 \cdot 7$

13. $360 = 2^3 \cdot 3^2 \cdot 5$

15. $6 + (-2)(-5) = 6 + 10 = 16$

17. $\dfrac{2(-7) + 3(2)}{2(-2)} = \dfrac{-14 + 6}{-4} = \dfrac{-8}{-4} = 2$

19. $3x + 2 = -13$
    $3x = -15$
    $x = -5$

21. $\dfrac{y}{4} - 1 = -5$

    $\dfrac{y}{4} = -4$

    $y = -16$

23. OBSERVATION HOURS

   The student must observe $100 - 37 = 63$ hours, or $\dfrac{63}{3} = 21$ three-hour sessions.

25. $\dfrac{21}{28} = \dfrac{3}{4}$

27. $\dfrac{6}{5}\left(-\dfrac{2}{3}\right) = -\dfrac{4}{5}$

29. $\dfrac{2}{3} + \dfrac{3}{4} = \dfrac{8}{12} + \dfrac{9}{12} = \dfrac{17}{12} = 1\dfrac{5}{12}$

31. $3\dfrac{5}{6} = \dfrac{23}{6}$

33. $4\dfrac{2}{3} + 5\dfrac{1}{4} = 9 + \dfrac{2}{3} + \dfrac{1}{4}$

$\qquad\qquad = 9 + \dfrac{8}{12} + \dfrac{3}{12}$

$\qquad\qquad = 9\dfrac{11}{12}$

35. FIRE HAZARD

The distance between the ground terminal and the hot terminal increased by $\dfrac{3}{4} - \dfrac{1}{16} = \dfrac{12}{16} - \dfrac{1}{16} = \dfrac{11}{16}$ inch.

37. $\left(\dfrac{1}{4} - \dfrac{7}{8}\right) \div \left(-2\dfrac{3}{16}\right) = \left(\dfrac{2}{8} - \dfrac{7}{8}\right) \div \left(-\dfrac{35}{16}\right)$

$\qquad\qquad\qquad\qquad = \left(-\dfrac{5}{8}\right) \cdot \left(-\dfrac{16}{35}\right)$

$\qquad\qquad\qquad\qquad = \dfrac{2}{7}$

39. $x + \dfrac{1}{5} = -\dfrac{14}{15}$

$\quad x = -\dfrac{17}{15}$

Check

$-\dfrac{17}{15} + \dfrac{3}{15} = -\dfrac{14}{15}$

41. $\dfrac{2}{3}x = -10$

$\quad x = -15$

Check

$\dfrac{2}{3}(-15) = -10$

43. The difference is the equal sign in an equation.

**Section 4.1 An Introduction to Decimals**

**Vocabulary**

1.  From left to right, **thousands, hundreds, tens, ones, decimal point, tenths, hundredths, thousandths, and ten thousandths**.
3.  We can approximate a decimal number using the process called **rounding**.

**Concepts**

5.  a. 32.415 is thirty-two and four hundred fifteen thousandths.
    b. The whole number part is 32.

    c. The fractional part is $\dfrac{415}{1,000}$.

    d. In expanded notation $30 + 2 + \dfrac{4}{10} + \dfrac{1}{100} + \dfrac{5}{1,000}$.

7.  From left to right $\left\{-3\dfrac{1}{100}, -0.7, \dfrac{7}{10}, 3.01\right\}$

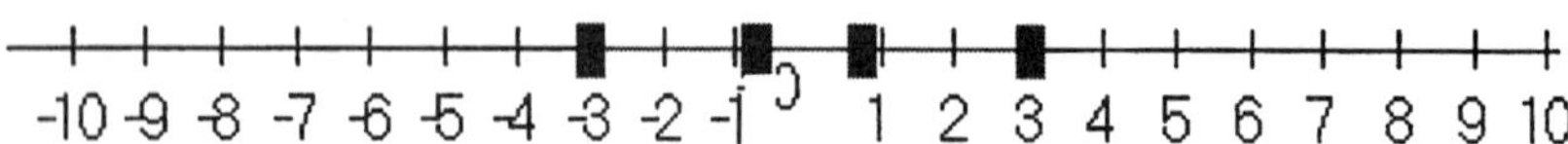

9.  a. True
    b. False
    c. True
    d. True

11. The shaded part is $\dfrac{47}{100}$ or 0.47.

13.

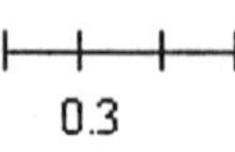

0.3

**Notation**

15. 9,816.0245

**Practice**

17. fifty and one tenth; $50\dfrac{1}{10}$

19. negative one hundred thirty-seven ten thousandths; $-\dfrac{137}{10,000}$

21. three hundred four and three ten thousandths; $304\dfrac{3}{10,000}$

23. negative seventy-two and four hundred ninety-three thousandths; $-72\dfrac{493}{1,000}$

25. $-0.39$
27. 6.187
29. 506.1

31. 2.7

33. −0.14

35. 33.00

37. 3.142

39. 1.414

41. 39

43. 2,988

45. a. Nearest dollar $3,090

b. Nearest ten cents $3,090.30

47. $-23.45 < -23.1$

49. $-0.065 > -0.066$

51. From least to greatest $132.64, 132.6401, 132.6499$.

**Applications**

53. WRITING A CHECK

The amount is $1,025.78.

55. INJECTIONS

The arrow should be placed two increments to the left of 0.4.

57. METRIC SYSTEM

a. 0.30

b. 1,609.34

c. 453.59

d. 3.79

59. GEOLOGY

| Sample | Location | Size | Classification |
|---|---|---|---|
| A | Riverbank | 0.009 | Sand |
| B | Pond | 0.0007 | Silt |
| C | NE corner | 0.095 | Granule |
| D | Dry lake | 0.00003 | Clay |

61. AIR QUALITY

From the highest reading to the smallest: Texas City; Houston; Westport; Galveston; White Plains; Crestline.

63. OLYMPICS

The gold went to Retton, the silver to Szabo, and the bronze to Pauca.

65. E-COMMERCE

For the third quarter of 1997 the loss was approximately - $0.07 and for the last quarter of 1998 the loss was approximately - $0.30.

**Writing**

67. Answers may vary.

69. Answers may vary.

71. Answers may vary.

**Review**

73. $75\dfrac{3}{4} + 88\dfrac{4}{5} = 75\dfrac{15}{20} + 88\dfrac{16}{20} = 163\dfrac{31}{20} = 164\dfrac{11}{20}$

75. $A = \dfrac{1}{2}(16)(9)$

$A = 72 \text{ in}^2$

77. $-2 + (-3) + 4 = -5 + 4 = -1$

**Section 4.2 Addition and Subtraction with Decimals**

**Vocabulary**

1.  The answer to an addition problem is called the **sum**.
3.  Every whole number has an unwritten decimal **point** to its right.

**Concepts**

5.  a. $0.3 + 0.17 = 0.47$

   b. $0.3 = \dfrac{3}{10}; 0.17 = \dfrac{17}{100}$

   $$\dfrac{3}{10} + \dfrac{17}{100} = \dfrac{30}{100} + \dfrac{17}{100} = \dfrac{47}{100}$$

   c. $\dfrac{47}{100} = 0.47$

   d. The results are the same.

**Practice**

7.  $32.5 + 7.4 = 39.9$

9.  $21.6 + 33.12 = 54.72$

11. $12 + 3.9 = 15.9$

13. $0.03034 + 0.2003 = 0.23064$

15. $247.9 + 40 + 0.56 = 288.46$

17. $45 + 9.9 + 0.12 + 3.02 = 58.04$

19. $12.98 - 3.45 = 9.53$

21. $78.1 - 7.81 = 70.29$

23. $5 - 0.023 = 4.977$

25. $24 - 23.81 = 0.19$

27. $-45.6 + 34.7 = -10.9$

29. $46.09 + (-7.8) = 38.29$

31. $-7.8 + (-6.5) = -14.3$

33. $-0.0045 + (-0.031) = -0.0355$

35. $-9.5 - 7.1 = -16.6$

37. $30.03 - (-17.88) = 47.91$

39. $-2.002 - (-4.6) = 2.598$

41. $-7 - (-18.01) = 11.01$

43. $3.4 - 6.6 + 7.3 = 4.1$

45. $(-9.1 - 6.05) - (-51) = 35.85$

47. $16 - (67.2 + 6.27) = -57.47$

49. $(-7.2 + 6.3) - (-3.1 - 4) = 6.2$

51. $\left| -14.1 + 6.9 \right| + 8 = \left| -7.2 \right| + 8 = 15.2$

53. $2.43 + 5.6 = 8.03$

## Applications

55. SPORTS PAGE
    a. The Italian team's time was $53.03 + 0.014 = 53.044$ seconds.
    b. The second place score was $102.71 - 0.33 = 102.38$ points.

57. VEHICLE SPECIFICATIONS
    The wheelbase of this car is $187.8 - (43.5 + 40.9) = 187.8 - 84.4 = 103.4$ inches.

59. BAROMETRIC PRESSURE
    The difference between the areas of high and low pressure is $30.7 - 28.9 = 1.8$.
    We could expect fair weather in Texas.

61. OFFSHORE DRILLING

|  | Underwater | Underground | Total |
|---|---|---|---|
| Design 1 | 1.74 miles | 2.32 miles | $1.74 + 2.32 = 4.06$ miles |
| Design 2 | 2.90 miles | 0 miles | $2.90 + 0 = 2.90$ miles |

63. AMERICAN RECORDHOLDERS
    The difference in the two times is $54.48 - 10.49 = 43.99$ seconds.

65. DEPOSIT SLIP
    The subtotal is $242.50 + 116.10 + 47.93 + 359.16 = \$765.69$.
    The total deposit is $\$765.69 - 25 = \$740.69$.

67. THE HOME SHOPPING NETWORK
    a. The difference of the MSRP and the sale price is $149.79 - 47.85 = \$101.94$.
    b. The set will cost $47.85 + 7.95 = \$55.80$.

69. $2,367.909 + 5,789.0253 = 8,156.9343$

71. $9,000.09 - 7,067.445 = 1,932.645$

73. $3,434.768 - (908 - 2.3 + 0.0098) = 3,434.768 - 905.7098 = 2,529.0582$

## Writing

75. Answers may vary.
77. Answers may vary.

## Review

79. $44\dfrac{3}{8} + 66\dfrac{1}{5} = 44\dfrac{15}{40} + 66\dfrac{8}{40} = 110\dfrac{23}{40}$

*Section 4.2 Addition & Subtraction with Decimals   128*

81. $-\dfrac{15}{26} \cdot 1\dfrac{4}{9} = -\dfrac{15}{26} \cdot \dfrac{13}{9} = -\dfrac{5}{6}$

**Section 4.3 Multiplication with Decimals**

**Vocabulary**

1. In the multiplication problem $2.89 \cdot 15.7$, the numbers 2.89 and 15.7 are called **factors**. The answer 45.373, is called the **product**.

**Concepts**

3. To multiply decimals, multiply them as if they were **whole** numbers. The number of decimal places in the product is the same as the **sum** of the decimal places of the factors.
5. When we move the decimal point to the right the decimal gets **larger**.

7. a. $\dfrac{3}{10} \cdot \dfrac{7}{100} = \dfrac{21}{1,000}$

   b. $\dfrac{3}{10} = 0.3; \dfrac{7}{100} = 0.07$

   $0.3(0.07) = 0.021 = \dfrac{21}{1,000}$

**Practice**

9. $0.4(0.2) = 0.08$

11. $-0.5(0.3) = -0.15$

13. $1.4(0.7) = 0.98$

15. $0.08(0.9) = 0.072$

17. $-5.6(-2.2) = 12.32$

19. $-4.9(0.001) = -0.0049$

21. $-0.35(0.24) = -0.084$

23. $-2.13(4.05) = -8.6265$

25. $16(0.6) = 9.6$

27. $-7(8.1) = -56.7$

29. $0.04(306) = 12.24$

31. $60.61(-0.3) = -18.183$

33. $-0.2(0.3)(-0.4) = 0.024$

35. $5.5(10)(-0.3) = -16.5$

37. $4.2(10) = 42$

39. $67.164(100) = 6,716.4$

41. $-0.056(10) = -0.56$

43. $1,000(8.05) = 8,050$

45. $0.098(10,000) = 980$

47. $-0.2(1,000) = -200$

49.

| Decimal | Its Square |
|---------|------------|
| 0.1 | 0.01 |
| 0.2 | 0.04 |
| 0.3 | 0.09 |
| 0.4 | 0.16 |
| 0.5 | 0.25 |
| 0.6 | 0.36 |
| 0.7 | 0.49 |
| 0.8 | 0.64 |
| 0.9 | 0.81 |

51. $(1.2)^2 = 1.44$

53. $(-1.3)^2 = 1.69$

55. $-4.6(23.4 - 19.6) = -4.6(3.8) = -17.48$

57. $(-0.2)^2 + 2(7.1) = 0.04 + 14.2 = 14.24$

59. $(-0.7 - 0.5)(2.4 - 3.1) = -1.2(-0.7) = 0.84$

61. $(0.5 + 0.6)^2(-3.2) = (1.1)^2(-3.2) = 1.21(-3.2) = -3.872$

63. $|-2.6| \cdot |-7.2| = 2.6(7.2) = 18.72$

65. $(|-2.6 - 6.7|)^2 = (|-9.3|)^2 = (9.3)^2 = 86.49$

**Applications**

67. CONCERT SEATING

a.

| Ticket | Price | Number | Receipts |
|--------|-------|--------|----------|
| Floor | $12.50 | 1,000 | $12.50(1,000) = \$12,500$ |
| Balcony | $15.75 | 100 | $15.75(100) = \$1,575$ |

b. The total receipts were $12,500 + 1,575 = \$14,075$.

69. STORM DAMAGE

During the three-week period the house fell $0.57 + 2(0.09) = 0.75$ inches.

71. WEIGHTLIFTING

There are $2(45.5 + 20.5 + 2.2) = 136.4$ pounds on this bar.

73. BAKERY SUPPLIES

| Type | Price | Pounds | Cost |
|---|---|---|---|
| Almonds | $3.25 | 16 | $16(3.25) = \$52$ |
| Walnuts | $2.10 | 25 | $25(2.10) = \$52.50$ |
| Peanuts | $1.85 | 17 | $17(1.85) = \$31.45$ |

75. SWIMMING POOL CONSTRUCTION

The perimeter of the pool is $2(50) + 2(30.3) = 160.6$ meters, which is the amount of coping needed.

77. BIOLOGY

The three dimensions from top to bottom are
$34(0.000000004) = 0.000000136$ inch;

$3.4(0.000000004) = 0.0000000136$ inch;

$10(0.000000004) = 0.00000004$ inch.

79. $(-9.0089 + 10.0087)(15.3) = 15.29694$

81. $(18.18 + 6.61)^2 + (5 - 9.09)^2 = 631.2722$

83. ELECTRIC BILL

To the nearest cent the cost would be $\$0.14277(719) = \$102.65$.

**Writing**

85. Answers may vary.
87. Answers may vary.

**Review**

89. $7x + 5 = 54$

$\qquad 7x = 49$

$\qquad x = 7$

91. The absolute value of negative three.

93. $-\dfrac{8}{8} = -1$

## Section 4.4 Division with Decimals

**Vocabulary**

1.  In the division $2.5\overline{)4.075} = 1.63$, the decimal 4.075 is called the **dividend**, the decimal 2.5 is the **divisor**, and 1.63 is the **quotient**.

**Concepts**

3.  To divide by a decimal, move the decimal point of the divisor so that it becomes a **whole** number. The decimal point of the dividend is then moved the same number of places to the **right**. The decimal point in the quotient is written directly **above** the decimal point of the dividend.

5.  $45 = 45.0 = 45.000$ is a true statement.

7. Moving the decimal point one place to the right is the same as multiplying by ten.

9. Check the result by $0.9(2.13) = 1.917$

11. This is the correct result since $2.07(4.6) = 9.522$.

**Notation**

13. The arrows indicated decimal shifts two places to the right or multiplying by 100.

**Practice**

15. $8\overline{)36}$ quotient $4.5$

17. $-39 \div 4 = -9.75$

19. $49.6 \div 8 = 6.2$

21. $9\overline{)288.9}$ quotient $32.1$

23. $(-14.76) \div (-6) = 2.46$

25. $\dfrac{-55.02}{7} = -7.86$

27. $45\overline{)119.7}$ quotient $2.66$

29. $250.95 \div 35 = 7.17$

31. $41.6 \div 0.32 = 130$

33. $(-199.5) \div (-0.19) = 1,050$

35. $\dfrac{0.0102}{0.017} = 0.6$

37. $\dfrac{0.0186}{0.031} = 0.6$

39. $3\overline{)16} \approx 5.3$

41. $-5.714 \div 2.4 \approx -2.4$

43. $12.243 \div 0.9 \approx 13.60$

45. $0.04\overline{)0.03164} \approx 0.79$

47. $7.895 \div 100 = 0.07895$

49. $0.064 \div (-100) = -0.00064$

51. $1{,}000\overline{)34.8} = 0.0348$

53. $\dfrac{45.04}{10} = 4.504$

55. $\dfrac{-1.2 - 3.4}{3(1.6)} = \dfrac{-4.6}{4.8} \approx -0.96$

57. $\dfrac{40.7(-5.3)}{0.4 - 0.61} = \dfrac{-215.71}{-0.21} \approx 1{,}027.19$

59. $\dfrac{5(48.38 - 32)}{9} = \dfrac{81.9}{9} = 9.1$

61. $\dfrac{6.7 - (0.3)^2 + 1.6}{0.3^3} = \dfrac{8.21}{0.027} \approx 304.07$

**Applications**

63. BUTCHER SHOP

There will be $\dfrac{14}{0.05} = 280$ slices.

65. HIKING

The hiker will hike for $\dfrac{27.5}{2.5} = 11$ hours, arriving at 6:00 p.m.

67. SPRAY BOTTLE

In an 8.5 ounce bottle there would be $\dfrac{8.5}{0.015} \approx 566.67$ or 567 sprays.

69. HOURLY PAY

In 1988 the average pay was $\dfrac{322.02}{34.7} = \$9.28$ per hour and in 1998 the average pay was

$\dfrac{441.84}{34.6} = \$12.77$ per hour.

71. OIL WELL

The total they must drill is $0.68 + 0.36 + 0.44 = 1.48$ miles.

To do this in four weeks they need to drill $1.48 \div 4 = 0.37$ miles per week.

**Writing**

73. Answers may vary.

75. Answers may vary.

**Review**

77. $\dfrac{\dfrac{7}{8}}{\dfrac{3}{4}} = \dfrac{7}{8} \cdot \dfrac{4}{3} = \dfrac{7}{6}$

79. The integers are $\{...,-3,-2,-1,0,1,2,3,...\}$.

81. $-\dfrac{3}{4}A = -9$

$\qquad A = -\dfrac{4}{3}(-9)$

$\qquad A = 12$

83. $-2.35 + 0.4 + 27.375 = -2.35 + 27.775 = 25.425$

**Estimation Study Set**

1. The deluxe model is approximately $240 more expensive.
3. There is approximately 2 cubic feet less.
5. For this amount she can purchase approximately 30 standard models.
7. To split the cost three ways each will pay approximately $330.
9. There is about $520 left to finance.
11. This is not reasonable as the sum is about 26 more than 345.
13. This seems reasonable.
15. This seems reasonable.
17. This is not reasonable as 53 multiplied by 5 is 265.

**Section 4.5 Fractions and Decimals**

**Vocabulary**

1. The decimal form of the fraction $\frac{1}{3}$ is a **repeating** decimal, which is written $0.\overline{3}$ or $0.333....$

3. The **decimal** equivalent of $\frac{1}{16}$ is 0.0625.

**Concepts**

5. $7 \div 8$ is indicated.

7. When rounding 0.272727... to the nearest hundredth the result is smaller.

9. From left to right $\{-3.8\overline{3}, -0.75, 0.\overline{6}, 1\frac{3}{4}\}$

11. a. $\frac{1}{3} = 0.3$ is false

    b. $\frac{3}{4} = 0.75$ is true

    c. $20\frac{1}{2} = 20.5$ is true

    d. $\frac{1}{16} = 0.1\overline{6}$ is false

**Notation**

13. a. The remainder will never be zero.

    b. The decimal equivalent of $\frac{5}{6}$ is a repeating decimal.

**Practice**

15. $\frac{1}{2} = 0.5$

17. $-\frac{5}{8} = -0.625$

19. $\frac{9}{16} = 0.5625$

21. $-\frac{17}{32} = -0.53125$

23. $\frac{11}{20} = 0.55$

25. $\dfrac{31}{40} = 0.775$

27. $-\dfrac{3}{200} = -0.015$

29. $\dfrac{1}{500} = 0.002$

31. $\dfrac{2}{3} = 0.\overline{6}$

33. $\dfrac{5}{11} = 0.\overline{45}$

35. $-\dfrac{7}{12} = -0.58\overline{3}$

37. $\dfrac{1}{30} = 0.0\overline{3}$

39. $\dfrac{7}{30} \approx 0.23$

41. $\dfrac{17}{45} \approx 0.38$

43. $\dfrac{5}{33} \approx 0.152$

45. $\dfrac{10}{27} \approx 0.370$

47. $\dfrac{4}{3} \approx 1.33$

49. $-\dfrac{34}{11} \approx -3.09$

51. $3\dfrac{3}{4} = 3.75$

53. $-8\dfrac{2}{3} \approx -8.67$

55. $12\dfrac{11}{16} = 12.6875$

57. $203\dfrac{11}{15} \approx 203.73$

59. $\dfrac{7}{8} = 0.875 < 0.895$

61. $-\dfrac{11}{20} = -0.55 < -0.\overline{4}$

63. $\dfrac{1}{9} + \dfrac{3}{10} = \dfrac{10}{90} + \dfrac{27}{90} = \dfrac{37}{90}$

65. $\dfrac{9}{10} - \dfrac{7}{12} = \dfrac{54}{60} - \dfrac{35}{60} = \dfrac{19}{60}$

67. $\dfrac{5}{11}\left(\dfrac{3}{10}\right) = \dfrac{3}{22}$

69. $\dfrac{1}{3}\left(-\dfrac{1}{15}\right)\left(\dfrac{1}{2}\right) = -\dfrac{1}{45}\left(\dfrac{1}{2}\right) = -\dfrac{1}{90}$

71. $0.24 + \dfrac{1}{3} \approx 0.24 + 0.\overline{3} \approx 0.57$

73. $5.69 - \dfrac{5}{12} \approx 5.69 - 0.41\overline{6} \approx 5.27$

75. $(3.5 + 6.7)\left(-\dfrac{1}{4}\right) = 10.2(-0.25) = -2.55$

77. $\left(\dfrac{1}{5}\right)^2 (1.7) = 0.04(1.7) = 0.068$

79. $7.5 - (0.78)\left(\dfrac{1}{2}\right) = 7.5 - 0.39 = 7.11$

81. $\dfrac{3}{8}(-3.2) + (4.5)\left(-\dfrac{1}{9}\right) = -1.2 - 0.5 = -1.7$

83. $\dfrac{3}{4}(3.14)(3)^3 = \dfrac{3}{4}(3.14)(27) = 63.585$

85. $\dfrac{23}{101} = 0.\overline{2277}$

87. $\dfrac{1,736}{50} = 34.72$

**Applications**

89. DRAFTING

$\dfrac{1}{16} = 0.0625;\ \dfrac{6}{16} = 0.375;\ \dfrac{9}{16} = 0.5625;\ \dfrac{15}{16} = 0.9375$

91. GARDENING

$0.065 = \dfrac{65}{1,000} = \dfrac{13}{200};\ \dfrac{3}{40} = \dfrac{15}{200}$ thus the $\dfrac{3}{40}$ inch line is thicker.

Or $\dfrac{3}{40} = 0.075 > 0.65$ so the $\dfrac{3}{40}$ inch line is thicker.

93. HORSE RACING

$$23^2 = 23\frac{2}{5} = 23.4 \text{ sec}; \ 23^4 = 23\frac{4}{5} = 23.8 \text{ sec}; \ 24^1 = 24\frac{1}{5} = 24.2 \text{ sec}; \ 32^3 = 32\frac{3}{5} = 32.6 \text{ sec}$$

95. WINDOW REPLACEMENT

The area of the window is found by multiplying the area of one triangular panel,

$$A = \frac{1}{2}(6)(5.2) = 15.6 \text{ in}^2, \text{ by six since there are six panels, } 6(15.6) = 93.6 \text{ in}^2$$

**Writing**

97. Answers may vary.
99. Answers may vary.

**Review**

101. $-2 + (-3) + 10 + (-6) = -5 + 4 = -1$

103. $\dfrac{5}{2} + \dfrac{2}{3} = \dfrac{15}{6} + \dfrac{4}{6} = \dfrac{19}{6}$

105. $\dfrac{3}{4} \cdot \dfrac{8}{9} = \dfrac{2}{3}$

**Section 4.6 Solving Equations Containing Decimals**

**Vocabulary**

1.  To **solve** an equation, we isolate the variable on one side of the equals sign.
3.  The property $(a \cdot b) \cdot c = a \cdot (b \cdot c)$ is called the **associative** property of multiplication.

**Concepts**

5.  $2.1(1.7) - 6.3 \stackrel{?}{=} -2.73$

$3.57 - 6.3 \stackrel{?}{=} -2.73$

$-2.73 = -2.73$

**Notation**

7.  $0.6s - 2.3 = -1.82$

$0.6s - 2.3 + 2.3 = -1.82 + 2.3$

$0.6s = 0.48$

$\dfrac{0.6s}{0.6} = \dfrac{0.48}{0.6}$

$s = 0.8$

**Practice**

9.  $x + 8.1 = 9.8$

$x + 8.1 - 8.1 = 9.8 - 8.1$

$x = 1.7$

11. $7.08 = t - 0.03$

$7.08 + 0.03 = t - 0.03 + 0.03$

$7.11 = t$

13. $-5.6 + h = -17.1$

$-5.6 + 5.6 + h = -17.1 + 5.6$

$h = -11.5$

15. $7.75 = t - (-7.85)$

$7.75 = t + 7.85$

$7.75 - 7.85 = t + 7.85 - 7.85$

$-0.1 = t$

17. $2x = -8.72$

$x = \dfrac{-8.72}{2}$

$x = -4.36$

19. $-3.51 = -2.7x$

$$\frac{-3.51}{-2.7} = x$$

$$1.3 = x$$

21. $\dfrac{x}{2.04} = -4$

$$x = -4(2.04)$$

$$x = -8.16$$

23. $\dfrac{-x}{5.1} = -4.4$

$$x = -5.1(-4.4)$$

$$x = 22.44$$

25. $\dfrac{1}{3}x = -7.06$

$$x = 3(-7.06)$$

$$x = -21.18$$

27. $\dfrac{x}{100} = 0.004$

$$x = 100(0.004)$$

$$x = 0.4$$

29. $2x + 7.8 = 3.4$

$$2x = -4.4$$

$$x = -2.2$$

31. $-0.8 = 5y + 9.2$

$$-10 = 5y$$

$$-2 = y$$

33. $0.3x - 2.1 = 7.2$

$$0.3x = 9.3$$

$$x = 31$$

35. $-1.5b + 2.7 = 1.2$

$$-1.5b = -1.5$$

$$b = 1$$

37. $0.9a - 6 = -5.73$

$$0.9a = 0.27$$

$$a = 0.3$$

**Applications**

39. PETITION DRIVE

**Analyze the problem**
- Her base pay is 15 dollars a day.
- She makes 30 cents for each signature.
- She wants to make 60 dollars a day.
- Find the number of signatures she needs to get.

**Form an equation**

Let $x =$ the number of signatures she needs to collect.

We need to work in terms of the same units, so we write 30 cents as $0.30.

If we multiply the pay per signature by the number of signatures, we get the money she makes just from collecting signatures. Therefore, $0.30x =$ total amount (in dollars) made from collecting signatures.

Base pay plus 0.30 times the number of signatures is 60.

$$15 + 0.30x = 60$$

**Solve the equation**

$$15 + 0.30x = 60$$
$$0.30x = 45$$
$$x = 150$$

**State the conclusion**

She needs to collect 150 signatures to make $60.

**Check the result**

If she collects 150 signatures, she will make $0.30(150) = \$45$ from signatures. If we add the $15, we get $60. The answer checks.


41. DISASTER RELIEF

Let $d$ represent the amount of relief to request at the federal level, then $6.8 + 12.5 + d = 27.9$
$$19.3 + d = 27.9$$
$$d = \$8.6$$

The county should request $8.6 million from the federal government.


43. GPA

Let $g$ represent her grade point before the decline.

$$g - 0.18 = 3.09$$
$$g = 3.27$$

Before the drop in grade point her average was 3.27.


45. POINTS PER GAME

Let $j$ represent her scoring average as a junior.

$$2j = 21.4$$
$$j = \frac{21.4}{2}$$
$$j = 10.7$$

As a junior her average was 10.7 points per game.

47. FUEL EFFICIENCY

Let $s$ represent the fuel average in 1960.

$$s - 0.4 + 1.3 + 3.1 + 0.3 = 16.7$$

$$s + 4.3 = 16.7$$

$$s = 12.4$$

In 1960 the average miles per gallon was 12.4.

49. CALLIGRAPHY

Let $w$ represent words.

$$20 + 0.15w = 50$$

$$0.15w = 30$$

$$w = 200$$

The maximum number of words is 200.

**Writing**

51. Answers may vary.

**Review**

53. $-\dfrac{2}{3} + \dfrac{3}{4} = -\dfrac{8}{12} + \dfrac{9}{12} = \dfrac{1}{12}$

55. $\dfrac{7}{8} \div \dfrac{13}{16} = \dfrac{7}{8} \cdot \dfrac{16}{13} = \dfrac{14}{13}$

57. $\dfrac{-3-3}{-3+4} = \dfrac{-6}{1} = -6$

59. $\dfrac{6}{5}x = 10$

$$x = \dfrac{50}{6}$$

$$x = \dfrac{25}{3}$$

**Vocabulary**

1. When we find what number is squared to obtain a given number, we are finding the square **root** of the given number.

3. The symbol $\sqrt{\phantom{x}}$ is called a **radical** sign. It indicates that we are to find a **positive** square root.

5. In $\sqrt{26}$, 26 is called the **radicand**.

**Concepts**

7. The square of 5 is 25, because $5^2 = 25$.

9. The two square roots of 49 are 7 and $-7$ because $7^2 = 49$ and $(-7)^2 = 49$.

11. Since $\left(\dfrac{3}{4}\right)^2 = \dfrac{9}{16}$, we know that $\sqrt{\dfrac{9}{16}} = \dfrac{3}{4}$.

13. From smallest to largest, $\sqrt{6}, \sqrt{11}, \sqrt{23}, \sqrt{27}$.

15. a. $\sqrt{1} = 1$
   b. $\sqrt{0} = 0$

17. a. $\sqrt{6} \approx 2.4$.
   b. $2.4^2 = 5.76$
   c. $6 - 5.76 = 0.24$

19. From left to right, $-\sqrt{5}, \sqrt{9}$.

$$\begin{array}{cccccccccccc} + & + & + & \blacksquare & + & + & 0 & + & + & + & \blacksquare & + & + \\ -5 & -4 & -3 & -2 & -1 & & & 1 & 2 & 3 & 4 & 5 \end{array}$$

21. a. $\sqrt{19}$ is between 4 and 5.
   b. $\sqrt{87}$ is between 9 and 10.

**Notation**

23. $-\sqrt{49} + \sqrt{64} = -7 + 8$
$$= 1$$

**Practice**

25. $\sqrt{16} = 4$

27. $-\sqrt{121} = -11$

29. $-\sqrt{0.49} = -0.7$

31. $\sqrt{0.25} = 0.5$

33. $\sqrt{0.09} = 0.3$

35. $-\sqrt{\dfrac{1}{81}} = -\dfrac{1}{9}$

37. $-\sqrt{\dfrac{16}{9}} = -\dfrac{4}{3}$

39. $\sqrt{\dfrac{4}{25}} = \dfrac{2}{5}$

41. $5\sqrt{36} + 1 = 5(6) + 1 = 31$

43. $-4\sqrt{36} + 2\sqrt{4} = -4(6) + 2(2) = -20$

45. $\sqrt{\dfrac{1}{16}} - \sqrt{\dfrac{9}{25}} = \dfrac{1}{4} - \dfrac{3}{5} = \dfrac{5}{20} - \dfrac{12}{20} = -\dfrac{7}{20}$

47. $5(\sqrt{49})(-2) = 5(7)(-2) = -70$

49. $\sqrt{0.04} + 2.36 = 0.2 + 2.36 = 2.56$

51. $-3\sqrt{1.44} = -3(1.2) = -3.6$

53.

| Number | Square Root |
| --- | --- |
| 1 | 1.000 |
| 2 | 1.414 |
| 3 | 1.732 |
| 4 | 2.000 |
| 5 | 2.236 |
| 6 | 2.449 |
| 7 | 2.646 |
| 8 | 2.828 |
| 9 | 3.000 |
| 10 | 3.162 |

55. $\sqrt{1,369} = 37$

57. $\sqrt{3,721} = 61$

59. $\sqrt{15} \approx 3.87$

61. $\sqrt{66} \approx 8.12$

63. $\sqrt{24.05} \approx 4.904$

65. $-\sqrt{11.1} \approx -3.332$

67. $\sqrt{24,000,201} = 4,899$

69. $-\sqrt{0.00111} \approx -0.0333$

**Applications**

71. CARPENTRY
    a.  The slanted part can be considered the hypotenuse of a right triangle so its length is
    $\sqrt{3^2 + 4^2} = \sqrt{25} = 5$ feet.
    b.  The slanted part can be considered the hypotenuse of a right triangle so its length is
    $\sqrt{6^2 + 8^2} = \sqrt{100} = 10$ feet.

73. BASEBALL DIAMOND
    From home plate to second base is $\sqrt{16,200} \approx 127.3$ feet.

75. BIG-SCREEN TV
    The screen size is $\sqrt{1,681} = 41$ inches.

**Writing**

77. Answers may vary.
79. Answers may vary.
81. Answers may vary.

**Review**

83. To isolate the variable we must first undo the subtraction then undo the multiplication.

85. $5(-2)^2 - \dfrac{16}{4} = 5(4) - 4 = 16$

87. $\dfrac{5}{8} \div \dfrac{3}{4} = \dfrac{5}{8} \cdot \dfrac{4}{3} = \dfrac{5}{6}$

89. $8 + \dfrac{a}{5} = 14$

$\quad\quad \dfrac{a}{5} = 6$

$\quad\quad a = 5(6)$

$\quad\quad a = 30$

1. From left to right $\{-4, -2\frac{1}{2}, -0.1, \frac{99}{100}, 1.\overline{3}, \frac{13}{4}, \sqrt{17}\}$.

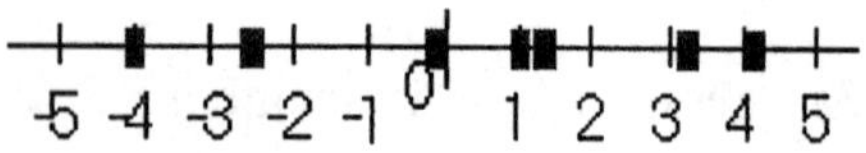

$$-5 \quad -4 \quad -3 \quad -2 \quad -1 \quad 0 \quad 1 \quad 2 \quad 3 \quad 4 \quad 5$$

3. Natural Numbers $\{1, 2, 3, 4, 5, ...\}$.

5. Integers $\{..., -3, -2, -1, 0, 1, 2, 3, 4, 5, ...\}$.

7. Irrational numbers are non-repeating, non-terminating decimals that cannot be expressed in the form $\frac{a}{b}$.

9. Every fraction can be written as a terminating decimal is **false**.

11. Some irrational numbers are integers is **false**.

13. No numbers are both rational and irrational numbers is **true**.

15. The set of whole numbers is a subset of the irrational numbers is **false**.

17. Every natural number is an integer is **true**.

19. There is no real number such that when it is squared the result is a negative number.

**Section 4.1 An Introduction to Decimals**

1.  As a decimal the shaded region is 0.67, as a fraction the shaded region is $\dfrac{67}{100}$ .

3.  $16.4523 = 10 + 6 + \dfrac{4}{10} + \dfrac{5}{100} + \dfrac{2}{1,000} + \dfrac{3}{10,000}$ .

5.  From left to right $\{-2.7, -0.8, 1.55\}$ .

7.  This is a true statement.

9.  a. $4.578 \approx 4.58$
    b. $3,706.0895 \approx 3,706.090$
    c. $-0.0614 \approx -0.1$
    d. $88.12 \approx 88.1$

**Section 4.2 Addition and Subtraction with Decimals**

11. a. $-16.1 + 8.4 = -7.7$
    b. $-4.8 - (-7.9) = 3.1$
    c. $-3.55 + (-1.25) = -4.8$
    d. $-15.1 - 13.99 = -29.09$
    e. $-8.8 + (-7.3 - 9.5) = -8.8 + (-16.8) = -25.6$
    f. $(5 - 0.096) - (-0.035) = 4.904 + 0.035 = 4.939$

13. MICROWAVE OVEN
    The window is $13.4 - (2.5 + 2.75) = 13.4 - 5.25 = 8.15$ inches tall.

**Section 4.3 Multiplication with Decimals**

15. a. $1,000(90.1452) = 90,145.2$
    b. $(-10)(-2.897)(100) = 28.97(100) = 2,897$

17. a. $(0.6 + 0.7)^2 - 12.3 = 1.3^2 - 12.3 = -10.61$
    b. $3(7.8) + 2(1.1)^2 = 23.4 + 2(1.21) = 23.4 + 2.42 = 25.82$

19. WORD PROCESSOR
    The height of the printed portion will be $11 - (1.0 + 0.6) = 9.4$ inches.
    The width of the printed portion will be $8.5 - (0.5 + 0.7) = 7.3$ inches.
    The area of the printed portion will be $A = 9.4(7.3) = 68.62$ square inches.

**Section 4.4 Division with Decimals**

21. a. $12\overline{)15}$ gives $1.25$

   b. $-41.8 \div 4 = -10.45$

   c. $\dfrac{-29.67}{-23} = 1.29$

   d. $24.618 \div 6 = 4.103$

23. a. $78.98 \div 6.1 \approx 12.9$

   b. $\dfrac{-5.338}{0.008} \approx -667.3$

25. THANKSGIVING DINNER

   The cost per person for dinner was $\dfrac{\$41.70}{5} = \$8.34$ .

27. $\dfrac{(1.4)^2 + 2(4.6)}{0.5 + 0.3} = \dfrac{1.96 + 9.2}{0.8} = 13.95$

29. TELESCOPE

   This adjustment requires $\dfrac{0.2375}{0.025} = 9.5$ revolutions.

**Section 4.5 Fractions and Decimals**

31. a. $\dfrac{6}{11} = 0.\overline{54}$

   b. $-\dfrac{2}{3} = -0.\overline{6}$

33. a. $\dfrac{13}{25} = 0.52 > 0.499$

   b. $-0.\overline{26} > -\dfrac{4}{15} = -0.2\overline{7}$

35. a. $\dfrac{1}{3} + 0.4 = \dfrac{1}{3} + \dfrac{4}{10} = \dfrac{1}{3} + \dfrac{2}{5} = \dfrac{5}{15} + \dfrac{6}{15} = \dfrac{11}{15}$

   b. $\dfrac{4}{5}(-7.8) = 0.8(-7.8) = -6.24$

   c. $\dfrac{1}{2}(9.7 + 8.9)(10) = 5(18.6) = 93$

   d. $\dfrac{1}{3}(3.14)(3)^2(4.2) = (3.14)(3)(4.2) = 39.564$

37. ROADSIDE EMERGENCY

The area of this reflector is $A = \dfrac{1}{2}(10.9)(6.4) = 34.88$ square inches.

## Section 4.6 Solving Equations Containing Decimals

39. $-1.3 \overset{?}{=} 1.2(-1.1) + 0.02$
$-1.3 = -1.3$
So $r = -1.1$ is a solution.

## Section 4.7 Square Roots

41. Two square roots of 64 are 8 and –8 because $8^2 = 64$ and $(-8)^2 = 64$.

43. When graphed on a number line $\sqrt{83}$ would lie between 9 and 10.

45. From left to right $\{-\sqrt{2}, \sqrt{0}, \sqrt{3}\}$

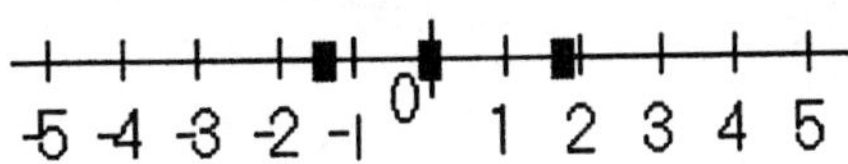

47. a. $\sqrt{19} \approx 4.36$
b. $\sqrt{59} \approx 7.68$

## Chapter 4 Test

1.  The amount of shaded squares is $\dfrac{79}{100} = 0.79$.

3.  $0.271 = \dfrac{271}{1,000}$

5.  SKATING RECEIPTS
    For the two days the total income was $30.25 + 62.25 + 40.50 + 75.75 = \$208.75$.

7.  EARTHQUAKE FAULT LINE
    The total drop was $0.83 + 0.19 = 1.02$ inches.

9.  NEW YORK CITY
    The area of central park is $2.5(0.5) = 1.25 \text{ mi}^2$.

11. $4.1 - (3.2)(0.4)^2 = 4.1 - (3.2)(0.16) = 3.588$

13. $\dfrac{12.146}{-5.3} \approx -2.29$

15. From left to right $\left\{ -\dfrac{4}{5}, \dfrac{3}{8} \right\}$, the decimal equivalents are $\{-0.8, 0.375\}$.

17. a.  $-2.4t = 16.8$

    $t = -7$

    b.  $-0.008 + x = 6$

    $x = 6.008$

19. CHEMISTRY
    The weight of compound C is $1.86 + 2.09 + C = 4.37$

    $C = 0.42$ grams

21. From left to right $\{-\sqrt{5}, \sqrt{2}\}$.

23. a.  $-6.78 > -6.79$

    b.  $\dfrac{3}{8} = 0.375 > 0.3$

    c.  $\sqrt{\dfrac{16}{81}} = \dfrac{4}{9} = \dfrac{36}{81} > \dfrac{16}{81}$

    d.  $0.\overline{45} > 0.45$

1. THE EXECUTIVE BRANCH
   During a four year term the president earns $400,000(4) = \$1,600,000$ and the vice-president earns $181,000(4) = \$724,000$. The difference in earnings is $1,600,000 - 724,000 = \$876,000$.

3. $43\overline{)1,203}$ is 27 with 42 remaining.

5. The prime factorization of 220 is $2^2 \bullet 5 \bullet 11$.

7. The whole numbers are $\{0,1,2,3,4,5,....\}$.

9. Subtraction is the same as **adding** the opposite.

11. The equivalent multiplication statement is $-15 = -5 \bullet 3$.

13. $8 - 2d = -10$
$$-2d = -18$$
$$d = 9$$

15. $\left|-7(5)\right| = \left|-35\right| = 35$

17. $3.61$

19. $7x - 38 = -3$
$$7x = 35$$
$$x = 5$$

21. This represents equivalent fractions.

23. $\dfrac{3}{8} \bullet \dfrac{7}{16} = \dfrac{21}{128}$

25. $\dfrac{4}{3} + \dfrac{2}{7} = \dfrac{28}{21} + \dfrac{6}{21} = \dfrac{34}{21}$

27. $76\dfrac{1}{6} - 49\dfrac{7}{8} = 76\dfrac{4}{24} - 49\dfrac{21}{24} = 75\dfrac{28}{24} - 49\dfrac{21}{24} = 26\dfrac{7}{24}$

29. $\dfrac{2}{3}y = -30$
$$y = -\dfrac{90}{2}$$
$$y = -45$$

31. KITE
    The area of the kite is found by considering the kite as two triangles,
$$A = \dfrac{1}{2}\left(7\dfrac{1}{2}\right)(21)$$
    $A = 78.75$ doubling this yields $157.5$ in$^2$ which is the area of the kite.

33. GLASS
    To the nearest thousandths 0.001 inch.

35. $-1.8(4.52) = -8.136$

37. $56.012(100) = 5,601.2$

39. $-9.1 - (-6.05 - 51) = -9.1 - (-57.05) = 47.95$

41. LITERATURE

$$C = \frac{5(451 - 32)}{9} = 232.\overline{7} \approx 232.8$$

43. $\quad -3 = 0.5t + 1.5$

$\quad -4.5 = 0.5t$

$\quad\; -9 = t$

45. $-4\sqrt{36} + 2\sqrt{81} = -4(6) + 2(9) = -24 + 18 = -6$

**Section 5.1 Percents, Decimals, and Fractions**

**Vocabulary**

1.  **Percent** means part per one hundred.

**Concepts**

3.  To write a percent as a fraction, drop the percent symbol and write the given number over **100**.
5.  To change a decimal to a percent, multiply the decimal by 100 by moving the decimal point two places to the **right**, and then insert a % symbol.

7.  a. $0.84$, $84\%$, $\dfrac{84}{100} = \dfrac{21}{25}$

   b. 16% of the figure is not shaded.

**Practice**

9.  $17\% = \dfrac{17}{100}$

11. $5\% = \dfrac{5}{100} = \dfrac{1}{20}$

13. $60\% = \dfrac{60}{100} = \dfrac{3}{5}$

15. $125\% = \dfrac{125}{100} = \dfrac{5}{4}$

17. $\dfrac{2}{3}\% = \dfrac{\frac{2}{3}}{100} = \dfrac{2}{300} = \dfrac{1}{150}$

19. $5\dfrac{1}{4}\% = \dfrac{\frac{21}{4}}{100} = \dfrac{21}{400}$

21. $0.6\% = \dfrac{\frac{6}{10}}{100} = \dfrac{3}{500}$

23. $1.9\% = \dfrac{1\frac{9}{10}}{100} = \dfrac{\frac{19}{10}}{100} = \dfrac{19}{1,000}$

25. $19\% = 0.19$

27. $6\% = 0.06$

29. $40.8\% = 0.408$

31. $250\% = 2.50$

33. $0.79\% = 0.0079$

35. $\dfrac{1}{4}\% = 0.25\% = 0.0025$

37. $0.93 = 93\%$

39. $0.612 = 61.2\%$

41. $0.0314 = 3.14\%$

43. $8.43 = 843\%$

45. $50 = 5,000\%$

47. $9.1 = 910\%$

49. $\dfrac{17}{100} = 17\%$

51. $\dfrac{4}{25} = \dfrac{16}{100} = 16\%$

53. $\dfrac{2}{5} = \dfrac{40}{100} = 40\%$

55. $\dfrac{21}{20} = \dfrac{105}{100} = 105\%$

57. $\dfrac{5}{8} = 62.5\%$

59. $\dfrac{3}{16} = 18.75\%$

61. $\dfrac{2}{3} = 66\dfrac{2}{3}\%$

63. $\dfrac{5}{6} = 83\dfrac{1}{3}\%$

65. $\dfrac{1}{9} \approx 11.11\%$

67. $\dfrac{5}{9} \approx 55.56\%$

**Applications**

69. U.N. SECURITY COUNCIL

    a. $\dfrac{15}{188}$ belong to the Security Council

    b. $\dfrac{15}{188} \approx 8\%$

71. PIANO KEYS

    a. $\dfrac{36}{88} = \dfrac{9}{22}$

    b. $\dfrac{9}{22} \approx 41\%$

73. a. There are 29 bones in the spinal column, $\dfrac{5}{29}$ are lumbar.

   b. $\dfrac{5}{29} \approx 17\%$

   c. $\dfrac{7}{29} \approx 24\%$

## 75. STEEP GRADE

Over 100 feet, a 5% grade rises 5 feet since $\dfrac{5}{100} = 5\%$..

## 77. IVORY SOAP

$99\dfrac{44}{100}\% = 99.44\% = 0.9944$

## 79. BASKETBALL STANDINGS

The winning percentage is presented as a decimal.  It is equal to 89.6%.

## 81. SKIN

The torso is 27.5 % since the total percents of the entire body should sum to 100%.

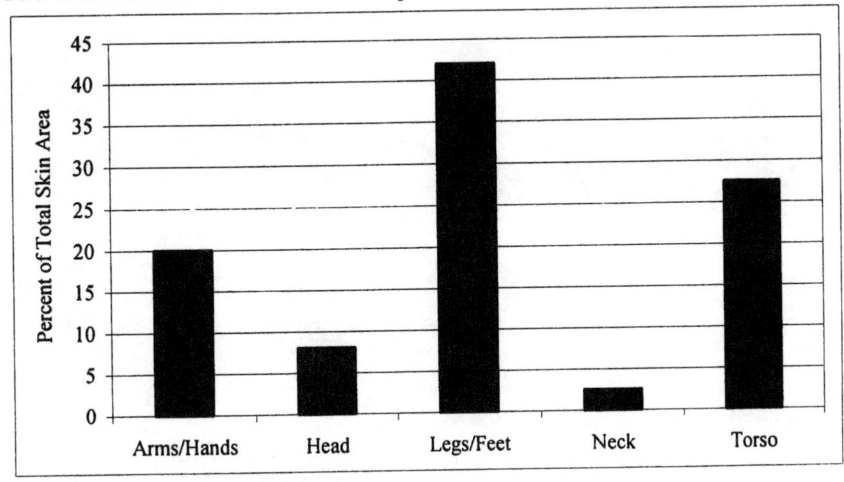


## 83. CHARITY

$\dfrac{92}{100} = 92\%$ of the money spent by the Red Cross went to programs and services.

## 85. BIRTHDAY

$\dfrac{1}{365} \approx 0.27\%$

**Writing**

87. Answers may vary
89. Answers may vary
91. Answers may vary

**Review**

93. $-\dfrac{2}{3}x = -6$

$\qquad x = \dfrac{18}{2}$

$\qquad x = 9$

95. $\dfrac{7}{11} - \dfrac{2}{9} = \dfrac{63}{99} - \dfrac{22}{99} = \dfrac{41}{99}$

97. $3.875 + 23.2 = 27.075$

**Section 5.2 Solving Percent Problems**

**Vocabulary**

1. $n = 0.10(50)$
3. $48 = n \cdot 47$
5. In a circle **graph**, pie-shaped wedges are used to show the division of a whole quantity into its component parts.

**Concepts**

7. Change to a decimal.
   a. $12\% = 0.12$
   b. $5.6\% = 0.056$
   c. $125\% = 1.25$
   d. $\dfrac{1}{4}\% = 0.25\% = 0.0025$

9. 120% of 55 is more than 55

11. a. 100% of 25 is 25
    b. 132 of 132 is 100%
    c. 87 is 87% of 100

13. VIDEO GAMES
    Nintendo 64 has $100\% - 4\% - 63\% = 33\%$ of the market.

**Notation**

15. Translate each.
    a. "of" translates to multiply
    b. "is" translates to equal
    c. "what number" translates to a variable, often denoted with $x$

**Practice**

17. $x = 0.36(250)$
    $x = 90$

19. $16 = x(20)$
    $0.08 = x$
    $x = 80\%$

21. $7.8 = 0.12x$
    $65 = x$

23. $x = 0.008(12)$
    $x = 0.096$

25. $0.5 = x(40,000)$
    $0.0000125 = x$
    $x = 0.00125\%$

27. $3.3 = 0.075x$

$\quad 44 = x$

29. $x = 0.0725(600)$

$\quad x = 43.5$

31. $1.02(105) = x$

$\quad 107.1 = x$

33. $0.\overline{3}x = 33$

$\quad x = 99$

35. $0.095x = 5.7$

$\quad x = 60$

37. $x(8,000) = 2,500$

$\quad x = 0.3125$

$\quad x = 31.25\%$

39.

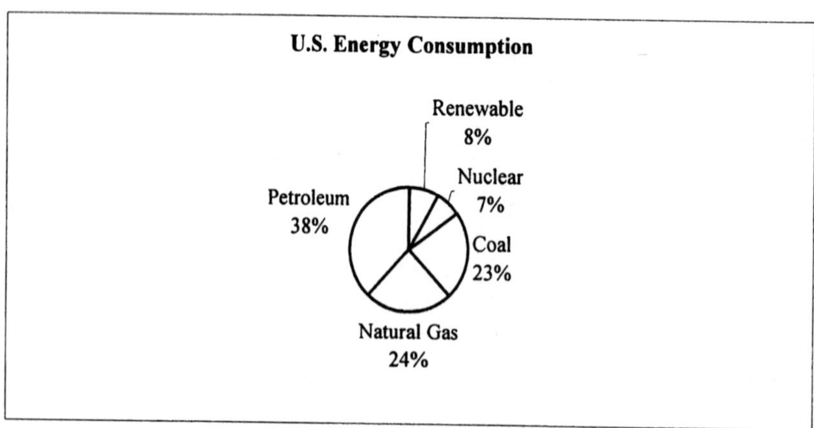


## Applications

41. CHILD CARE

$0.70c = 84$

$c = 120$

The daycare center could enroll 120 youngsters.

43. GOVERNMENT SPENDING

$0.37(1,650 \text{ billion}) = 610.5 \text{ billion}$

$610.5 billion was spent on Social Security, Medicare and other retirement programs

45. THE INTERNET

$0.24(50,000) = 12,000 = 12\text{K}$

$50\text{K} - 12\text{K} = 38\text{K}$

There are 38K bytes left to download.

47. PRODUCTION PROMOTION

$0.25s = 6$

$s = 24$

There are 24 ounces in the large bottle.

49. DRIVER'S LICENSE

$0.7(40) = 28$

To pass the test 28 questions must be answered correctly, so he passed.

51. MIXTURES

| Gallons Solution | % Acid | Gallons Acid |
|---|---|---|
| 60 | 50% | $0.5(60) = 30$ |
| 40 | 30% | $0.3(40) = 12$ |

53. MAKING COPIES

$1.80(1.5) = 2.7$

The height of the print on the copy will be 2.7 inches.

55. INSURANCE

$200 is what percentage of $4,000

$200 = d(4,000)$

$0.05 = d$ or 5%

57. A MAJORITY

The total votes cast were $8,501 + 3,614 + 2,630 + 2,432 = 17,177$.

50% of 17,177 is 8,589, so there must be a runoff election.

**Writing**

59. Answers may vary
61. Answers may vary

**Review**

63. $2.78 + 6 + 9.09 + 0.3 = 18.17$

65. 5.001 is closer to 5 on a number line since $5.001 - 5 = 0.001$ and $5 - 4.9 = 0.10$.

67. $(0.2)^3 = 0.008$

## Section 5.3 Applications of Percents

### Vocabulary

1.  Some salespeople are paid on **commission**. It is based on a percent of the total dollar amount of the goods or services they sell.
3.  The difference between the original price and the sale price of an item is called the **discount**.

### Concepts

5.  If there has been a 100% increase then the membership has doubled.

### Applications

7.  STATE SALES TAX
    The sales tax is $0.0475(900) = \$42.75$.

9.  ROOM TAX
    $10.32 is what percent of $129
    $10.32 = t(129)$

    $0.08 = t$ an 8% room tax

11. SALES RECEIPT
    Subtotal $8.97 + 9.87 + 28.50 = \$47.34$

    Sales Tax $0.06(47.34) = \$2.84$

    Total $\$47.34 + \$2.84 = \$50.18$

13. SALES TAX HIKE
    1% of $15,000 is $0.01(15,000) = \$150$

15. PAYCHECK
    Federal Tax, $28.80 is what percent of $360
    $28.80 = t(360)$

    $0.08 = t$ or 8% tax

    Medicare Taxes, $5.04 is what percent of $360
    $5.04 = m(360)$

    $0.014 = m$ or 1.4% tax

    Worker's Compensation, $4.32 is what percent of $360
    $4.32 = w(360)$

    $0.012 = w$ or 1.2% tax

17. OVERTIME
    480 less 25% of 480 is $480 - 0.25(480) = 360$.
    The company target is 360 hours of overtime next month.

19. REDUCED CALORIES
    150 calories less 36% of 150 calories is $150 - 0.36(150) = 96$.
    There are 96 calories in the new product.

21. ENDANGERED SPECIES

There was a decline during 1995-1996.

$1,599 - 1,523 = 76$

$76 = x(1,599)$, $x$ represents the percent decrease

$0.04753 \approx x$

During this time frame the decrease was approximately 5%.

23. CAR INSURANCE

The difference in the premium was $40.

$40 is what percent of $400

$40 = c(400)$

$0.10 = c$, a 10% discount

25. LAKE SHORELINE

The shoreline increased by 1.8 miles.

1.8 miles is what percent of 5.8 miles

$1.8 = s(5.8)$

$0.31034 \approx s$

There was a 31% increase.

27. EARTH MOVING

a. The soil increases from 1 yd$^3$ to 1.25 yd$^3$, a difference of 0.25 yd$^3$.

0.25 yd$^3$ is what percent of 1 yd$^3$

$0.25 = x(1)$

$0.25 = d$ or a 25% increase

b. The soil decreases from 1.25 yd$^3$ to 0.80 yd$^3$, a difference of 0.45 yd$^3$.

0.45 yd$^3$ is what percent of 1.25 yd$^3$

$0.45 = d(1.25)$

$0.36 = d$ or a 36% decrease

29. REAL ESTATE

6% of $98,500 is $0.06(98,500) = \$5,910$.

Each agent received half of this amount or $2,955.

31. SPORTS AGENT

$37,500 is what percent of $2,500,000

$37,500 = f(2,500,000)$

$0.015 = f$ or a rate of 1.5%

33. CONCERT PARKING

The total parking intake is $\$6(6000) = \$36,000$.

$0.\overline{3}(36,000) = \$12,000$

The vendor can expect $12,000.

35. **WATCH SALE**

The regular price is $\$29.95 + \$10 = \$39.95$.
The discount is found considering $10 is what percent of $39.95

$$10 = w(39.95)$$

$0.25 \approx w$ or a 25% discount

37. **RING SALE**

80% of what is $149.99
$$0.8r = 149.99$$

$$r \approx 187.49$$
The original price was $187.49.

39. **VCR SALE**

The sale price is $\$399.97 - \$50 = \$349.97$.
The percent decrease is found considering $50 is what percent of $399.97.

$$50 = v(399.97)$$

$0.125 \approx v$ or a 12.5% discount

41. **REBATE**

The discount is $3.60.
The new price is $\$15.48 - \$3.60 = \$11.88$.
The percent decrease is found considering $3.60 is what percent of $ 15.48.

$$3.60 = x(15.48)$$

$0.2326 \approx x$ or a 23% decrease

43. **TV SHOPPING**

The ring sells for 45% of $170 or $0.45(170) = \$76.50$.

## Writing

45. Answers may vary
47. Answers may vary

## Review

49. $-5(-5)(-2) = -50$

51. $-4 - (-7) = -4 + 7 = 3$

53. In a 40-hour week she earns $40(12.50) = \$500$.

55. $\dfrac{4}{9} - \dfrac{3}{5} = \dfrac{20}{45} - \dfrac{27}{45} = -\dfrac{7}{45}$

57. $\dfrac{6}{7} \div \dfrac{3}{5} = \dfrac{6}{7} \cdot \dfrac{5}{3} = \dfrac{10}{7} = 1\dfrac{3}{7}$

**Estimation**
An estimated solution is given as well as the exact solution for comparison.

**Study Set**
1. COLLEGE COURSES
   $0.2(800) = 160$; 20% of 815 students is 163 students enrolled in science

3. DISCOUNT
   $0.3(200) = 60$; 30% of $196.88 is $59.06

5. FIRE DAMAGE
   $0.5(108,000) = 54,000$; 50% of $107,809 is $53,904.50

7. WEIGHTLIFTING
   $2(160) = 320$; 200% of 158 is 316

9. TRAFFIC STUDY
   20% of 650 is 130 since 10% is 65, double that for 20%

11. NO-SHOWS
   $0.30(68) = 20.4$ or 21 people; 31% of 68 is 21.08 or 22 people

13. INTERNET SURVEY
   $0.60(30,000) = 18,000$; 58% of 28,657 is 16,621.06 or 16,622 people

15. VOTING
   50% of 6,200 is 3,100; 48% of 6,200 is 2,976

## Vocabulary

1. In banking the original amount of money borrowed or deposited is known as the **principal**.
3. The percent that is used to calculate the amount of interest to be paid is called the **interest** rate.
5. Interest computed only on the original principal is called **simple** interest.

## Concepts

7. a. $7\% = 0.07$
   b. $9.8\% = 0.098$
   c. $6\frac{1}{4}\% = 0.0625$

9. $I = Prt$
   $I = (10,000)(0.06)(3)$
   $I = \$1,800$

11. a. Compound interest is illustrated.
    b. The original principal was $1,000.
    c. Interest was calculated four times.
    d. The first compounding earned $50.
    e. The money was invested for one year.

## Notation

13. Multiplication is implied by the *Prt*.

## Applications

15. RETIREMENT INCOME
    At the end of year one there will be the original principal plus the interest,
    $5,000 + 0.06(5,000) = \$5,300$

17. REMODELING
    In two years the homeowner will pay $8,000(0.092)(2) = \$1,472$.

19. MEETING A PAYROLL
    The business had to repay
    $$P + I = 4,200 + 4,200(0.18)\left(\frac{30}{365}\right)$$
    $$= \$4,200 + \$62.14$$
    $$= \$4,262.14$$

21. SAVINGS ACCOUNT

| $P$ | $r$ | $t$ | $I = Prt$ |
|---|---|---|---|
| $10,000 | 0.0725 | 2 | $10,000(0.0725)(2) = \$1,450$ |

### 23. LOAN APPLICATION

Amount    $1,200
Length    2 years
Rate      8%
Interest    $1,200(0.08)(2) = \$192$

Repayment    $1,200 + 192 = \$1,392$

Payments    monthly

Borrower agrees to pay 24 equal payments of $\dfrac{1,392}{24} = \$58$ to repay the loan.

### 25. LOW-INTEREST LOAN

The total repayment is
$$P + I = 18,000,000 + 18,000,000(0.023)(2)$$
$$= 18,000,000 + 828,000$$
$$= \$18,828,000$$

### 27. COMPOUNDING ANNUALLY

Using $A = P\left(1 + \dfrac{r}{n}\right)^{nt}$, the account balance will be $A = 600\left(1 + \dfrac{0.08}{1}\right)^{1(3)}$

$$A = \$755.83$$

### 29. COLLEGE FUND

Using $A = P\left(1 + \dfrac{r}{n}\right)^{nt}$, the account balance will be $A = 1,000\left(1 + \dfrac{0.06}{365}\right)^{365(4)}$

$$A = \$1,271.22$$

### 31. TAX REFUND

Using $A = P\left(1 + \dfrac{r}{n}\right)^{nt}$, the account balance will be $A = 545\left(1 + \dfrac{0.046}{365}\right)^{365(1)}$

$$A = \$570.65$$

### 33. LOTTERY

Using $A = P\left(1 + \dfrac{r}{n}\right)^{nt}$, the account balance will $A = 500,000\left(1 + \dfrac{0.06}{365}\right)^{365(1)}$

$$A = \$530,915.66$$

The amount earned will be $\$530,915.66 - \$500,000 = \$30,915.66$.

**Writing**

35. Answers may vary
37. Answers may vary

**Review**

39. $\sqrt{\dfrac{1}{4}} = \dfrac{1}{2}$

41. 23.0

43. $\dfrac{2}{3}x = -2$

$x = \dfrac{-6}{2}$

$x = -3$

45. $8\dfrac{1}{3} \bullet 6 = \dfrac{25}{3} \bullet 6 = 50$

## Chapter 5 Key Concepts

1. $0.32(620) = 198.4$

3. $25 = 0.40x$

$$x = \frac{25}{0.40}$$

$$x = 62.5$$

5. $106.25 = 625x$

$$x = \frac{106.25}{625}$$

$$x = 0.17 \text{ or } 17\%$$

7. $I = Prt$

$$I = 10,000(0.06)5$$

$$I = \$3,000$$

**Section 5.1 Percents, Decimals, and Fractions**

1.  a. $39\%, 0.39, \dfrac{39}{100}$

    b. $111\%, 1.11, \dfrac{111}{100}$

3.  a. $15\% = \dfrac{15}{100} = \dfrac{3}{20}$

    b. $120\% = \dfrac{120}{100} = \dfrac{6}{5}$

    c. $9\dfrac{1}{4}\% = \dfrac{37}{4}\% = \dfrac{\frac{37}{4}}{100} = \dfrac{37}{400}$

    d. $0.1\% = \dfrac{\frac{1}{10}}{100} = \dfrac{1}{1,000}$

5.  a. $0.83 = 83\%$
    b. $0.625 = 6.25\%$
    c. $0.051 = 5.1\%$
    d. $6 = 600\%$

7.  a. $\dfrac{1}{3} = 33\dfrac{1}{3}\%$

    b. $\dfrac{5}{6} = 0.8\overline{3} = 83\dfrac{1}{3}\%$

9.  BILL OF RIGHTS
    17 of 27 were adopted after the Bill of Rights
    $\dfrac{17}{27} = 0.6\overline{29} \approx 63\%$

**Section 5.2 Solving Percent Problems**

11. The amount is 15, the base is 45 and the percent is $33\dfrac{1}{3}$.

13. a. 200 is 40% of 500; $n = 0.40(500) = 200$
    b. 16% of 125 is 20; $(0.16)x = 20$
    c. 1.4 is 1.75% of 80; $1.4 = x(80)$
    d. $66\dfrac{2}{3}\%$ of 3,150 is 2,100; $x = 0.6\overline{6}(3,150) = 2,100$
    e. 220% of 55 is 121; $x = 2.20(55) = 121$
    f. 0.05% of 60,000 is 30; $x = 0.0005(60,000) = 30$

15. HOME SALES

$0.75h = 51$

$h = 68$

There were 68 homes in this subdivision.

17. TIPPING

15% of \$36.20 is $0.15(36.30) = \$5.43$.

19. EARTH'S SURFACE

70.9% of 196,800,000 is $0.709(196,800,000) = 139,531,200 \text{ mi}^2$

**Section 5.3 Applications of Percent**

21. SALES TAX RATE

$12,300t = 492$

$t = 0.04$

The tax rate is 4%.

23. TROOP SIZE

The increase was $12,500 - 10,000 = 2,500$ troops.

$2,500 = x(10,000)$

$x = 0.25$ or a 25% increase

25. TOOL CHEST

The discount was \$50 so the original price was $\$139.99 + \$50 = \$189.99$.
The discount rate was found answering \$50 is what percent of \$189.99.

$50 = 189.99t$

$0.263172 \approx t$ or approximately a 26% discount

**Section 5.4 Interest**

27. CODE VIOLATIONS

The interest was $I = 10,000(0.125)\left(\dfrac{90}{365}\right) \approx \$308.22$.

The total repayment was $\$10,000 + \$308.22 = \$10,308.22$.

29. Using $A = P(1 + \dfrac{r}{n})^{nt}$, $A = 2,000\left(1 + \dfrac{0.07}{2}\right)^{2(1)}$

$$A = \$2,142.45$$

31. CASH GRANT

Using $A = P(1 + \dfrac{r}{n})^{nt}$, $A = 500,000\left(1 + \dfrac{0.083}{365}\right)^{365(1)}$.

$$A \approx \$543,265.78$$

The interest earned is $\$543,265.78 - \$500,000 = \$43,265.78$.

1. 61% of the figure is shaded, 0.61, $\dfrac{61}{100}$

3. a. 67% = 0.67
   b. 12.3% = 0.123
   c. $9\dfrac{3}{4}\% = 0.0975$

5. a. 0.19 = 19%
   b. 3.47 = 347 %
   c. 0.005 = 0.5%

7. $\dfrac{7}{30} = 0.2\overline{3} = 23.33\%$

9. $\dfrac{2}{3} = 0.\overline{6} = 66\dfrac{2}{3}\%$

11. SHRINKAGE
    a. 3% of 34 is 0.03(34) = 1.02 inches
    b. $34 - 1.02 = 32.98$ inches

13. TIPPING
    0.15(25.40) = $3.81 tip

15. SWIMMING WORKOUT
    20% of the total number of laps is 18
    $0.2L = 18$
    $L = 90$ laps

17. $x = 0.24(600)$
    $x = 144$

19. HOMEOWNER'S INSURANCE
    4% of \$898 is 0.04(898) = \$35.92

21. CAR WAX SALE
    The sale price is $\$14.95 - \$3 = \$11.95$.
    The discount is \$3.
    The discount rate is $3 = 14.95w$

    $0.2007 \approx w$, approximately 20%

23. $I = 3{,}000(0.05)(1) = \$150$

25. POLITICAL AD
    37% of what is not clear in this advertisement.

**Chapters 1 – 5 Cumulative Review**

1. SHAQUILLE

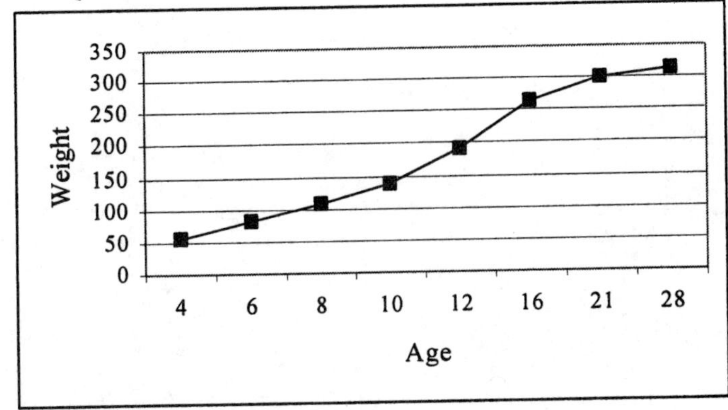

3. a. The factors of 40 are {1, 2, 4, 5, 8, 10, 20, 40}
   b. The prime factorization of 40 is $40 = 2^3 \cdot 5$

5. PAINTING
   An 8-foot square tarp will cover an area of $8 \cdot 8 = 64 \text{ ft}^2$.

7. $12 - 2[-8 - 2^4(-1)]$
   $= 12 - 2[-8 - 16(-1)]$
   $= 12 - 2[-8 + 16]$
   $= 12 - 2[8]$
   $= 12 - 16$
   $= -4$

9. $3x - 4 = 2$
   $3x = 6$
   $x = 2$

11. $-x = -1$
    $x = 1$

13. SPELLING
    $\dfrac{4}{11}$ of the letters in Mississippi are vowels

15. $-\dfrac{16}{35} \cdot \dfrac{25}{48}$

    $= -\dfrac{16}{7 \cdot 5} \cdot \dfrac{5 \cdot 5}{16 \cdot 3}$

    $= -\dfrac{1}{7} \cdot \dfrac{5}{3}$

    $= -\dfrac{5}{21}$

17. $\dfrac{4}{3}+\dfrac{2}{7}=\dfrac{28}{21}+\dfrac{6}{21}=\dfrac{34}{21}=1\dfrac{13}{21}$

19. $\dfrac{5}{6}y=-25$

$\qquad y=-25\left(\dfrac{6}{5}\right)$

$\qquad y=-30$

21. $78.1-7.81=70.29$

23. $0.752(1,000)=752$

25. $\dfrac{3.6-(-1.5)}{0.5(-1.5)-0.4(3.6)}=\dfrac{5.1}{-2.19}\approx-2.33$

27. $\dfrac{11}{15}=0.7\overline{3}$

29. $3\sqrt{81}-8\sqrt{49}$

$\quad =3(9)-8(7)$

$\quad =27-56$

$\quad =-29$

31.

| Percent | Decimal | Fraction |
|---------|---------|----------|
| 29% | 0.29 | $\dfrac{29}{100}$ |
| 47.3% | 0.473 | $\dfrac{473}{1,000}$ |
| 87.5% | 0.875 | $\dfrac{875}{1,000}=\dfrac{7}{8}$ |

33. $0.16(400)=64$

35. TIPPING

15% of $75.18 is $0.15(75.18)=\$11.277$ so the total gratuity is \$11.28, and rounded up to the nearest dollar is \$12. Therefore the total to be paid is $\$75.18+\$12=\$87.18$.

37. SAVING ACCOUNT

$\quad I=10,000(0.0725)(2)=\$1,450$

### Section 6.1 Reading Graphs and Tables

**Vocabulary**

1. In illustration 1, figure **(a)** is a bar graph.
3. In illustration 1, figure **(c)** is a pictograph.
5. In illustration 1, figure **(d)** is a histogram.

**Concepts**

7. The bars of a **histogram** touch.

**Applications**

9. PRIORITY MAIL
   The cost of mailing a $7\frac{1}{4}$ pound package is $6.95.

11. COMPARING PACKAGES
   Juan would save $\$6.65 - \$1.79 = \$4.86$

13. FILING A JOINT RETURN
   Raul's tax amount would be $\$5,850 + 0.28(\$18,100) = \$10,918$. Remember the 28% is multiplied by
   the amount over $39,000 or $\$57,100 - \$39,000 = \$18,100$.

15. TAX-SAVINGS STRATEGIES
   Angelina's tax amount would be, filing jointly, $\$5,850 + 0.28(\$12,000) = \$9,210$. Remember the 28%
   is the amount over $39,000 or $\$51,000 - \$39,000 = \$12,000$. If she were to file under single status her
   tax amount would be $\$3,502.50 + 0.28(\$29,650) = \$11,804.50$. Remember the 28% is the amount over
   $23,350 or $\$53,000 - \$23,350 = \$29,650$. Her tax savings would be $\$11,804.50 - \$9,210 = \$2,594.50$.

17. Nuclear energy supplied the least amount of energy in 1975.
19. Oil produced 49% of the electrical energy in 1975.
21. There was an increase of approximately 60%.
23. The production of lead in 1970 was approximately equal to zinc in 1980.
25. In 1970 the production of zinc was less than one-half that of lead.
27. The production of zinc increased by 320 metric tons.
29. Reckless driving and failure to yield have seen decreases in arrests.
31. Reckless driving occurred the least often.
33. Seniors spent the most on ice cream at Barney's.
35. Seniors spent $750, or $50 more than parents.
37. Groups that are about the same size speak French and German languages.
39. More people speak English.
41. $100 - 2.4 - 2.4 = 51.4\%$ of the world's population speaks other languages.
43. About $5 \div (19 + 18 + 5 + 3 + 20) = 5 \div 65 \approx 0.08$ or 8% of energy sources are nuclear.
45. About $20 \div 18 \approx 0.11$ or 11% more energy is derived from coal than from crude.
47. Solar energy accounts for less than $3 \div 65 \approx 0.046$ or 4.6%.
49. In 1975 the average weekly earnings were $190.
51. Miner's salaries were increasing the most rapidly.
53. Miners received the greatest increase in wages.
55. Runner 1 ran faster at the beginning.
57. Runner 1 dropped the baton.

59. Runner 2 was stopped while runner 1 was running.

61. COMMUTING MILES
    27 employees commute between 14.5 and 19.5 miles to work.

63. NIGHT SHIFT STAFFING
    90 nights had about 30 room calls.

65. MAKING A BAR GRAPH

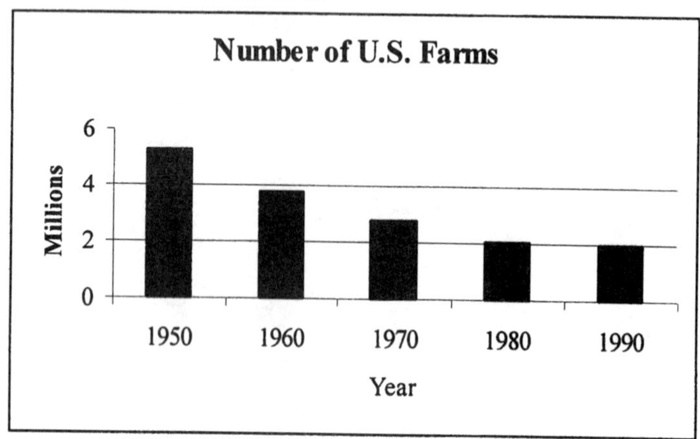



67. MAKING A LINE GRAPH

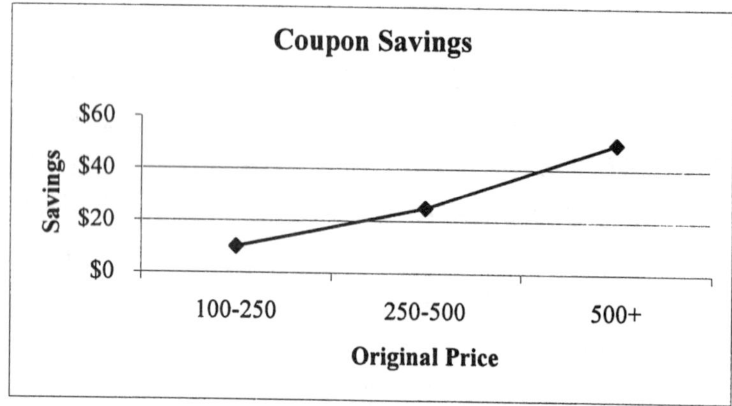



**Writing**

69. Answers may vary.

**Review**

71. $5 - 3 \bullet 4 = 5 - 12 = -7$

73. $\left(\dfrac{1}{2} + \dfrac{1}{3}\right)^2 = \left(\dfrac{3}{6} + \dfrac{2}{6}\right)^2 = \left(\dfrac{5}{6}\right)^2 = \dfrac{25}{36}$

75. The prime numbers between 10 and 30 are $\{11, 13, 17, 19, 23, 29\}$.

77. The even number less than 6 that is not prime is 4.

**Vocabulary**

1.  The sum of the values in a distribution divided by the number of values in the distribution is called the **mean** of the distribution.
3.  The middle value in a distribution is called the **median** of the distribution.

**Concepts**

5.  The mean of several values is given by $\text{Mean} = \dfrac{\text{the sum of the values}}{\text{the number of values}}$.

7.  $\dfrac{3+4+7+7+8+11+16}{7} = \dfrac{56}{7} = 8$

9.  $\dfrac{5+9+12+35+37+45+60+77}{8} = \dfrac{280}{8} = 35$

11. $\dfrac{15+7+12+19+27+17+19+35+20}{9} = \dfrac{171}{9} = 19$

13. The median is 9.

15. The median is $\dfrac{5+7}{2} = 6$.

17. The median is $\dfrac{17+18}{2} = 17.5$.

19. The mode is 3.
21. The mode does not exist.
23. The mode is 22.7.

**Applications**

25. SOFT DRINK PRICES
    The mean price of a soft drink is
    $$\dfrac{50+60+50+50+70+75+50+45+50+50+65+75+60+75+100+50+80+75}{18} = \dfrac{1{,}130}{18} = 62.\overline{7}$$
    or approximately 63 cents.

27. SOFT DRINK PRICES
    The median price for a soda is 60 cents.

29. SOFT DRINK PRICES
    The modal price for a soda is 50 cents.

31. CHANGING TEMPERATURES
    The mean temperature is approximately from midnight until 11:00 a.m. is
    $\dfrac{729}{12} = 60.75$ degrees.

33. AVERAGE TEMPERATURE
    The mean temperature is approximately
    $\dfrac{1538}{24} \approx 64.1$ degrees.

35. **FLEET MILEAGE**

On average each car traveled $\dfrac{98,790}{37} = 2,670$ miles.

37. **DAILY MILEAGE**

The average miles per day per vehicle in Exercise 35 was $\dfrac{2,670}{30} = 89$.

39. **EXAM AVERAGES**
The median and mode are 85.

41. **COMPARING GRADES**

The first average is $\dfrac{37+53+78}{3} = \dfrac{168}{3} = 56$ and the sister's average is $\dfrac{53+57+58}{3} = \dfrac{168}{3} = 56$. Therefore, they had the same average although the sister's grades were more consistent.

43. **OCTUPLETS**

The mean birth weight was $\dfrac{24+27+28+26+11.2+17.5+28.5+18}{8} = \dfrac{180.2}{8} = 22.525$ ounces and the

median weight was $\dfrac{24+26}{2} = \dfrac{50}{2} = 25$ ounces.

45. **COMPARISON SHOPPING**

The mean price was $\dfrac{49.78}{12} \approx \$4.15$.

The mode price was \$4.29 and the median was $\dfrac{4.19+4.19}{2} = \$4.19$.

47. **HOURLY PAY**

In 1988 the average production worker earned $\dfrac{322.02}{34.7} \approx \$9.28$.

In 1998 the average was $\dfrac{441.84}{34.6} \approx \$12.77$.

49. **OIL WELL**

In order for this to be a four-week project the group must drill $\dfrac{0.68+0.36+0.44}{4} = 0.37$ miles per week.

**Writing**

51. Answers may vary.

**Review**

53. The prime factorization of 81 is $3^4$.

55. $\dfrac{3}{4} \cdot \dfrac{2}{9} = \dfrac{\cancel{3}}{\cancel{2}(2)} \cdot \dfrac{\cancel{2}}{\cancel{3}(3)} = \dfrac{1}{6}$

57. $\dfrac{18}{5} + \dfrac{12}{5} = \dfrac{30}{5} = 6$

59. $\dfrac{8}{5} + \dfrac{3}{10} = \dfrac{16}{10} + \dfrac{3}{10} = \dfrac{19}{10} = 1\dfrac{9}{10}$

1. The mean of the distribution is $\dfrac{137}{10} = 13.7$.
3. The mode of the distribution is 15.
5. The mean of the distribution is $\dfrac{36}{6} = 6$.
7. The mode of the distribution is 6.
9. Answers will vary; one such distribution could be 1, 1, 2, 3.
11. Answers will vary; one such distribution could be 1, 1, 2, 3.

**Section 6.1 Reading Graphs and Tables**

1. WIND-CHILL TEMPERATURE
   The wind-chill is $-18^0$.

3. a. Almost 5 billion coupons were redeemed in 1987.
   b. Between 1989 and 1991 the number of redeemed coupons remained unchanged.
   c. In 1991 – 1993 the number of redeemed coupons increased the most.
   d. The percent increase was approximately $6 \div 10 = 0.60$ or 60%.

5. About $100 + 80 = 180$ households watch between 6 and 15 hours of TV weekly.

**Section 6.2 Mean, Median, and Mode**

7. EARNING AN A
   Jose's average was $\dfrac{762}{8} = 95.25$ so he earned an A in class.

9. PRETZEL PACKAGING
   The modal weight is 1.2 ounces.

11. BLOOD SAMPLES
    The mean is $\dfrac{65.7}{9} = 7.3$ microns, the median is 7.2 microns, and the mode is 6.9 microns.

**Chapter 6 Test**

1.  The staff amount per prisoner is approximately $0.732(2,266) \approx \$1,658.71$.

3.  Prisoner's food cost approximately $\dfrac{93}{2,266} \approx 0.04104$ or $4.1\%$.

5.  The percentage of employees in food and clothing is approximately $\dfrac{3.4}{18} = 0.1\overline{8}$ or $19\%$.

7.  In 1995 there were approximately 270,000 delays.

9.  1985 was the worst for air traffic delays.

11. At time A both bicyclists are moving, but cyclist one is going faster (answer A).

13. At time C both cyclists are stopped (answer E).

15. The median is $\dfrac{7+8}{2} = 7.5$.

17. The mode is 5.

19. RATINGS

    The median is 3.6, the mode is 3.1 and the mean is $\dfrac{25.5}{7} \approx 3.6429$ or approximately 3.6.

1.  GASOLINE
    358,600,000

3.  $38,908 + 15,696 = 54,604$

5.  $345 \times 67 = 23,115$

7.  $683 + 459 = 1,142$

9.  $4 \cdot 5 = 5 + 5 + 5 + 5 = 20$

11. a. $F_{18} = \{1, 2, 3, 6, 9, 18\}$
    b. $18 = 2 \cdot 3^2$

13. 27 is not prime since the factors are 1, 3, 9, 27.

15. $\dfrac{-315}{-1} = 315$

17.

19. This is a false statement.

21. $-25 + 5 = -20$

23. $-25(5)(-1) = 125$

25. $\dfrac{(-6)^2 - 1^5}{-4 - 3} = \dfrac{36 - 1}{-7} = \dfrac{35}{-7} = -5$

27. $-3^2 = -9$
    $(-3)^2 = 9$

29. $360 = 0.45x$
    $\dfrac{360}{0.45} = x$
    $800 = x$

31. Division by 0 is undefined. $\dfrac{7}{0} ; \dfrac{0}{7}$

33. $\dfrac{1}{2} - \dfrac{2}{3} = \dfrac{3}{6} - \dfrac{4}{6} = -\dfrac{1}{6}$

35. $\dfrac{4}{5} \cdot \dfrac{2}{7} = \dfrac{8}{35}$

37. $\dfrac{7}{13} = 0.538$ or $54\%$

39. $0.31(2.4) = 0.744$

41. $\dfrac{8}{11} = 0.\overline{72}$

43. WEEKLY SCHEDULE

Time watching TV is $168 - (48.3 + 34.5 + 21 + 3.1 + 27.5) = 33.6$ hours.

**Section 7.1 Some Basic Definitions**

**Vocabulary**

1.  A line **segment** has two endpoints.
3.  A **midpoint** divides a line segment into two parts of equal length.
5.  A **protractor** is used to measure angles.
7.  A **right** angle measures $90^0$.
9.  The measure of a straight angle is $\mathbf{180^0}$.
11. The sum of two **supplementary** angles is $180^0$.

**Concepts**

13. $\overrightarrow{GF}$ has point G as its endpoint is a **true** statement.
15. Line CD has three endpoints is a **false** statement.  It has 2.
17. $m(\angle AGC) = m(\angle BGD)$ is a **true** statement.
19. $\angle FGB \cong \angle EGA$ is a **true** statement.
21. $\angle AGC$ is an **acute** angle
23. $\angle FGD$ is an **obtuse** angle.
25. $\angle BGE$ is a **right** angle.
27. $\angle DGC$ is a **straight** angle.
29. $\angle AGF$ and $\angle DGC$ are vertical angles is a **true** statement.
31. $m(\angle AGB) = m(\angle BGC)$ is a **false** statement.
33. $\angle 1$ and $\angle 2$ **are** congruent angles.
35. $\angle AGB$ and $\angle DGE$ **are** congruent angles.
37. $\angle AGF$ and $\angle FGE$ are **not** congruent angles.
39. $\angle 1$ and $\angle CGD$ are adjacent angles is a **true** statement.
41. $\angle FGA$ and $\angle AGC$ are supplementary angles is a **true** statement.
43. $\angle AGF$ and $\angle 2$ are complementary angles is a **true** statement.
45. $\angle EGD$ and $\angle DGB$ are supplementary angles is a **true** statement.

**Notation**

47. The symbol $\angle$ means **angle**.
49. The symbol $\overrightarrow{AB}$ is read as "**ray AB.**"

**Practice**

51. $\overline{AC}$ has length **3**.
53. $\overline{CE}$ has length **3**.
55. $\overline{CD}$ has length **1**.
57. The midpoint of $\overline{AD}$ is B.
59. The measure of this angle is $40^0$.
61. The measure of this angle is $135^0$.
63. The measure of angle $x$ is $55^0 - 45^0 = 10^0$.
65. The measure of angle $x$ is $50^0 - 22.5^0 = 27.5^0$.
67. Since these are vertical angles, $2x = x + 30 \rightarrow x = 30^0$.

69. Since these are vertical angles,

$$4x + 15 = 7x - 60.$$
$$75 = 3x$$
$$25^0 = x$$

71. The complement of a $30^0$ angle is $60^0$ since $30 + x = 90$.

73. The supplement of a $105^0$ angle is $75^0$ since $105 + x = 180$.

75. $\angle 4$ is the supplement of $\angle 1$, therefore $\angle 4 = 180 - 50 = 130^0$.

77. $m(\angle 1) + m(\angle 2) + m(\angle 3) = 50 + 130 + 50 = 230^0$, since $\angle 2 = \angle 4$ by vertical angles.

79. $\angle 1 = 100^0$ by vertical angles.

81. Since we are told that $\angle 3 \cong \angle 4$, $\angle 3 + \angle 4 = 80^0$ since $\angle 1 = 100^0$, therefore $\angle 3 = 40^0$.

**Applications**

83. BASEBALL

85. SYNTHESIZER

$x$ is the supplement of $115^0$, so $x = 180 - 115 = 65^0$

$y$ is a vertical angle to $115^0$, so $y = 115^0$.

87. GARDENING

The angle of the handle to the ground is the supplement of $150^0$, or $180 - 150 = 30^0$.

**Writing**

89. Answers will vary.

91. Answers will vary.

93. Answers will vary.

**Review**

95. $2^4 = 2 \bullet 2 \bullet 2 \bullet 2 = 16$

97. $\dfrac{3}{4} - \dfrac{1}{8} - \dfrac{1}{3} = \dfrac{18}{24} - \dfrac{3}{24} - \dfrac{8}{24} = \dfrac{7}{24}$

99. 7% of 7 is $(0.07)(7) = 0.49$

### Vocabulary

1. Two lines in the same plane are **coplanar**.
3. If two lines intersect and form right angles, they are **perpendicular**.
5. In illustration 1, $\angle 4$ and $\angle 6$ are **alternate** interior angles.

### Concepts

7. $\angle 4$ and $\angle 6$, $\angle 3$ and $\angle 5$ are pairs of alternating interior angles.
9. $\angle 3, \angle 4, \angle 5, \angle 6$ are each interior angles.
11. $l_1$ and $l_3$ are parallel lines.

### Notation

13. The symbol $\lrcorner$ represents **a right angle**.

15. The symbol $\perp$ is read as **"is perpendicular to."**

### Practice

17. $m(\angle 1) = 130^0; m(\angle 2) = 180 - 30 = 50^0; m(\angle 3) = 50^0; m(\angle 5) = 130^0; m(\angle 6) = 50^0;$

$\quad m(\angle 7) = 50^0; m(\angle 8) = 130^0$

19. $m(\angle A) = 50^0; m(\angle 1) = 85^0; m(\angle 2) = 45^0; m(\angle 3) = 180 - 45 = 135^0$

21. $6x - 10 = 50$

$\qquad 6x = 60$

$\qquad x = 10$

23. $70 + 4x - 10 = 180$

$\qquad 4x = 120$

$\qquad x = 30^0$

25. $40 + 3x + 20 = 180$

$\qquad 3x = 120$

$\qquad x = 40^0$

27. $9x - 38 = 70$

$\qquad 9x = 108$

$\qquad x = 12^0$

### Applications

29. CONSTRUCTING PYRAMIDS
    If the stones are level the plum bob string should pass through the midpoint of the crossbar of the A frame.

31. LOGO

The perpendicular lines form a cross-hair in the center of the logo.

33. HANGING WALLPAPER

Answers will vary.

**Writing**

35. Answers will vary.
37. Answers will vary.
39. Answers will vary.

**Review**

41. 60% of 120 is $(0.60)(120) = 72$

43. $\dfrac{225}{500} = 0.45$ , 225 is 45% of 500.

45. Every whole number is an integer.

**Section 7.3 Polygons**

**Vocabulary**

1.   A **regular** polygon has sides that are all the same length and angles that all have the same measure.
3.   A **hexagon** is a polygon with six sides.
5.   An eight-sided polygon is an **octagon**.
7.   A triangle with three sides of equal length is called an **equilateral** triangle.
9.   The longest side of a right triangle is the **hypotenuse**.
11. A **parallelogram** with a right angle is a rectangle.
13. A **rhombus** is a parallelogram with four sides of equal length.
15. The legs of an **isosceles** trapezoid have the same length.

**Concepts**

17. 4, quadrilateral, 4
19. 3, triangle, 3
21. 5, pentagon, 5
23. 6, hexagon, 6
25. This is a scalene triangle.
27. This is a right triangle.
29. This is an equilateral triangle.
31. This is an isosceles triangle.
33. This is a square.
35. This is a rhombus.
37. This is a rectangle.
39. This is a trapezoid.

**Notation**

41. The symbol $\triangle$ means **triangle**.

**Practice**

43. $m(\angle C) = 180 - 60 - 30 = 90^0$

45. $m(\angle C) = 180 - 35 - 100 = 45^0$

47. $m(\angle C) = 180 - 25.5 - 63.8 = 90.7^0$

49. $m(\angle 1) = 90 - 60 = 30^0$

51. $m(\angle 2) = 90 - 30 = 60^0$

53. The sum of the angle measures of a hexagon is $180(6-2) = 720^0$.

55. The sum of the angle measures of a decagon is $180(10-2) = 1440^0$.

**Applications**

57. Answers will vary.
59. Answers will vary.

61. POLYGONS IN NATURE
    The lemon appears as an octagon.
    The chili pepper appears as a triangle.
    The apple appears as a pentagon.

63. CHEMISTRY

Pentagons and hexagons are used to represent methylprednisolone.

65. EASEL

The two outside legs form the beginning of an isosceles triangle.

**Writing**

67. Answers will vary.

**Review**

69. 20% of 110 is $(0.20)(110) = 22$

71. $\dfrac{80}{200} = 0.40$, 80 is 40% of 200.

73. $0.85 \div 2(0.25) = 0.425(0.25) = 0.10625$

## Section 7.4 Properties of Triangles

### Vocabulary

1.  **Congruent** triangles are the same size and the same shape.
3.  If a triangle has an angle that measures $90^0$, it is a **right** triangle.

### Concepts

5.  This is a true statement.
7.  This is a false statement.
9.  These are congruent triangles.
11. For a right triangle, $a$ and $b$ represent the legs of the triangle and $c$ represents the hypotenuse.

### Notation

13. The symbol $\cong$ is read as "**is congruent to.**"

### Practice

15. $\overline{AC}$ corresponds to $\overline{DF}$

$\overline{DE}$ corresponds to $\overline{AB}$

$\overline{BC}$ corresponds to $\overline{EF}$

$\angle A$ corresponds to $\angle D$

$\angle E$ corresponds to $\angle B$

$\angle F$ corresponds to $\angle C$

17. These are congruent triangles by SSS.
19. These are not necessarily congruent.
21. These are congruent triangles by SSS.
23. These are congruent triangles by SAS.
25. $x = 6$ cm
27. $x = 50^0$
29. Using $c^2 = a^2 + b^2$, $c = \sqrt{3^2 + 4^2} : c = 5$
31. Using $c^2 = a^2 + b^2$, $b = \sqrt{c^2 - a^2} : b = \sqrt{17^2 - 15^2} : b = 8$
33. Using $c^2 = a^2 + b^2$, $b = \sqrt{c^2 - a^2} : b = \sqrt{9^2 - 5^2} : a = \sqrt{56}$
35. This is a right triangle since $17^2 = 15^2 + 8^2$
37. This is not a right triangle since $26^2 \neq 7^2 + 24^2$.

### Applications

39. ADJUSTING A LADDER
    Use Pythagorean's theorem with the ladder representing the hypotenuse:
    $$20^2 = 16^2 + b^2 : b = \sqrt{20^2 - 16^2} : b = 12$$
    The ladder is 12 feet from the wall.

41. PICTURE FRAME

$$c^2 = 15^2 + 20^2 : w = \sqrt{625} : c = 25$$

If the sides of the frame form a right angle then the frame maker should read 25 inches on the yardstick.

43. BASEBALL

The distance from home plate to second base is the hypotenuse:

$$d^2 = 90^2 + 90^2 : d = \sqrt{16200} : d \approx 127.3$$

The distance is approximately 127.3 feet.

**Writing**

45. Answers will vary.

**Review**

47. Estimate would be $\dfrac{1(4)}{3} = 1\dfrac{1}{3}$; Actual answer is $\dfrac{0.95(3.89)}{2.997} \approx 1.233$

49. Estimate would be $0.3(60) = 18$; Actual answer is $0.32(60) = 19.2$

51. Estimate would be $0.5(18) = 9$; Actual answer is $0.495(18.1) = 8.9595$

## Section 7.5 Perimeters and Areas of Polygons

### Vocabulary

1. The distance around a polygon is called the **perimeter**.
3. The measure of the surface enclosed by a polygon is called its **area**.
5. The area of a polygon is measured in **square** units.

### Concepts

7. Two possible rectangles: length of 15 inches and width of 5 inches or length of 16 inches and width of 4 inches.
9. A square with area of 25 m$^2$ has side length 5 m.
11. A parallelogram with an area of 15 yd$^2$ may have a base of 5 yd and a height of 3 yd.
13. One of the rectangles of this figure may have a length of 5 ft and a width of 4 ft and the other rectangle may have a length of 20 ft and a width of 3 ft.

### Notation

15. The formula for the perimeter of a square is $P = 4s$.
17. The symbol 1 in$^2$ means one square inch.
19. The formula for the area of a square is $A = s^2$.
21. The formula $A = \dfrac{1}{2}bh$ gives the area of a triangle.

### Practice

23. $P = 4(8) = 32$ inches

25. $P = 6 + 4 + 2 + 2 + 2 + 4 + 6 + 10 = 36$ inches

27. $P = 7 + 6 + 6 + 8 + 10 = 37$ cm

29. $P = 21 + 32 + 32 = 85$ cm

31. $\dfrac{85}{3} = 28\dfrac{1}{3}$

33. $A = 4^2 = 16$ cm$^2$

35. $A = 4(15) = 60$ cm$^2$

37. $A = \dfrac{1}{2}(5)(10) = 25$ in$^2$

39. $A = \dfrac{1}{2}(13)(9 + 17) = 169$ mm$^2$

41. $A = \dfrac{1}{2}(4)(8) + 8^2 = 80$ m$^2$

43. $A = 10^2 - \dfrac{1}{2}(5)(10) = 75$ yd$^2$

45. $A = 6(14) - 3^2 = 75$ m$^2$

47. There are $12^2 = 144$ square inches in 1 square foot.

**Applications**

49. FENCING A YARD

The perimeter of the yard is $P = 2(110) + 2(85) = 390$ ft.

The cost of the fence is $C = 390(12.50) = \$4,875$.

51. PLANTING A SCREEN

The perimeter of her yard to be planted is $P = 2(70) + 100 = 240$ feet.

Planting a tree every 3 feet results in $\dfrac{240}{3} = 80$ trees plus the first tree planted for a total of 81 trees.

53. BUYING A FLOOR

Since there are 9 square feet in a square yard the cost of the ceramic tile is $C = 3.75(9) = \$33.75$ per square yard, therefore the linoleum is more expensive.

55. CARPETING A ROOM

The area of this room is $A = 24(15) = 360$ square feet or $\dfrac{360}{9} = 40$ square yards. The cost to carpet this room is $C = 40(30) = \$1,200$.

57. TILING A FLOOR

The area of this room is $A = 14(20) = 280$ square feet. At a cost of $\$1.29$ per square foot, it will cost $C = 1.29(280) = \$361.20$ to tile this floor.

59. MAKING A SAIL

Since the nylon is priced by the square yard, convert the dimensions of the sail to yards. The dimensions of the sail are 4 yards by 8 yards. The area of this triangular sail is $A = \dfrac{1}{2}(4)(8) = 16$ square yards. The cost to make this sail is $C = 16(12) = \$192$.

61. GEOGRAPHY

The approximate area of Nevada is $A = \dfrac{1}{2}(315)(505 + 205) = 111,825$ square miles.

63. CARPENTRY

Assuming the drywall is hung horizontally, each long side will need $\dfrac{(48)(12)}{(4)(8)} = 18$ sheets, for a total of 36 sheets. The two ends of the barn will have the drywall hung vertically and each end will need $\dfrac{20(12)}{(4)(8)} = 7.5$ sheets, for a total of 15 sheets of drywall. The total is 51 sheets.

65. DRIVING SAFETY

Spot one has length of 20 feet and width of 10 feet, for total area of 200 square feet. Spot two has base dimensions of 20 feet and 16 feet with height of 10 feet, thus the area is 180 square feet. Spot three is triangular with a height of 28 feet and a base of 28 feet for an area of 392 square feet.

**Writing**

67. Answers will vary

**Review**

69. $\dfrac{3}{4} + \dfrac{2}{3} = \dfrac{9}{12} + \dfrac{8}{12} = 1\dfrac{5}{12}$

71. $3\dfrac{3}{4} + 2\dfrac{1}{3} = 3\dfrac{9}{12} + 2\dfrac{4}{12} = 5 + 1\dfrac{1}{12} = 6\dfrac{1}{12}$

73. $7\dfrac{1}{2} \div 5\dfrac{2}{5} = \dfrac{15}{2}\left(\dfrac{5}{27}\right) = \dfrac{25}{18} = 1\dfrac{7}{18}$

## Section 7.6 Circles

### Vocabulary

1. A segment drawn from the center of a circle to a point on the circle is called a **radius**.
3. A **diameter** is a chord that passes through the center of a circle.
5. An arc that is shorter than a semicircle is called a **minor** arc.
7. The distance around a circle is called its **circumference**.

### Concepts

9. Radii of this circle are $\overline{OA}, \overline{OC}, \overline{OB}$.

11. The three chords of this circle are $\overline{DA}, \overline{DC}, \overline{AC}$.

13. The two semicircles are $\overparen{ABC}, \overparen{ADC}$.

15. To find the diameter of a circle whose radius you know, double the radius.

17. a. The radius of this circle will be 1 inch.
    b. The diameter of this circle will be 2 inches.
    c. The circumference will be $2\pi r = 2\pi(1) = 2\pi$ inches.
    d. The area of the circle will be $\pi r^2 = \pi(1)^2 = \pi$ square inches

19. Squaring 6 should be the first operation.

### Notation

21. The symbol $\overparen{AB}$ is read as **arc AB**.

23. The formula for the circumference of a circle is $C = \pi D$ or $C = 2\pi r$.

25. If C is the circumference of a circle and D is its diameter, then $\dfrac{C}{D} = \pi$.

27. $8\pi$

### Practice

29. $C = \pi D$
$$C = \pi(12)$$
$$C \approx 37.70 \text{ inches}$$

31. $C = \pi D$
$$D = \frac{C}{\pi}$$
$$D = \frac{36\pi}{\pi}$$
$$D = 36 \text{ meters}$$

33. The perimeter contributed by the two edges of the rectangular region is 16 feet. Consider the two semicircle regions together as one circle, $C = \pi D : C = \pi(3) : C \approx 9.42$. The total circumference is the sum of these or 25.42 feet.

35. The perimeter contributed by the rectangular region is $P = 2(8) + 6 = 22$ inches. The perimeter of the semicircle is $C = \frac{1}{2}\pi D : C = \frac{1}{2}\pi(6) : C \approx 9.42$. The total perimeter is the sum of these or 31.42 inches.

37. $A = \pi r^2 = \pi(3)^2 \approx 28.3$ square inches.

39. The area of the rectangular region is $A = 6(10) = 60$ square inches. Consider the two semicircles as a circle, the area of this circle is then $A = \pi r^2 = \pi(3)^2 \approx 28.3$ square inches. The total area is the sum of these or 88.3 square inches.

41. The area of the triangular region is $A = \frac{1}{2}bh = \frac{1}{2}(12)(12) = 72$ cm$^2$. The area of the semicircle is

$$A = \frac{1}{2}\pi r^2 = \frac{1}{2}\pi \left(\frac{12}{2}\right)^2 \approx 56.5 \text{ cm}^2.$$ The total area is the sum of these or $128.5$ cm$^2$.

43. The area of the rectangular region is $A = 10(4) = 40$ in$^2$. The area of the circular region is $A = \pi r^2 = \pi(2)^2 \approx 12.6$ in$^2$. The area of the shaded region is the difference of these or 27.4 in$^2$.

45. The area of the parallelogram is $A = 13(9) = 117$ in$^2$. The area of the circle is $A = \pi r^2 = \pi(4)^2 \approx 50.3$ in$^2$. The area of the shaded region is the difference of these or $66.7$ in$^2$.

**Applications**

47. AREA OF ROUND LAKE

The area of the lake is approximately $A = \pi r^2$ where $r = \frac{1}{2}(2)$, which is $A = \pi(1)^2 \approx 3.14$ mi$^2$.

49. GIANT SEQUOIA

The diameter of this tree is $C = \pi D : D = \frac{C}{\pi} : D = \frac{102.6}{\pi} : D \approx 32.66$ feet.

51. JOGGING

One lap around the track is the circumference of the circular running surface or $C = \pi D : C = \pi(.25) : C \approx 0.7854$ miles. To run 10 miles on this track Joan must run

$$\frac{10}{0.7854} \approx 12.73 \text{ laps.}$$

53. BANDING THE EARTH

The circumference of the earth is $C = 2\pi r$, the new circumference will be $C + 10 = 2\pi r_1$, where $r_1$ is the new radius associated with the larger circumference. The difference between the two radii will yield the height of the band above the Earth's surface. $\dfrac{C+10}{2\pi} = r_1$ and $\dfrac{C}{2\pi} = r$

$$r_1 - r = \frac{C+10}{2\pi} - \frac{C}{2\pi}$$

$$r_1 - r = \frac{10}{2\pi} \approx 1.59 \text{ feet}$$

55. ARCHERY

The area of the entire target is $A = \pi r^2 = \pi(2)^2 \approx 12.57 \text{ ft}^2$. The area of the bull's eye is $A = \pi r^2 = \pi(0.5)^2 \approx 0.79 \text{ ft}^2$. The bull's eye constitutes approximately $\dfrac{0.79}{12.57} \approx 0.0628$ or 6.28%.

**Writing**

57. Answers will vary.
59. Answers will vary.
61. Answers will vary.

**Review**

63. $\dfrac{9}{10} = 0.90$ or 90%

65. A pentagon has five sides.

**Section 7.7 Surface Area and Volume**

**Vocabulary**

1.  The space contained within a geometric solid is called its **volume**.
3.  A **cube** is a rectangular solid with all sides of equal length.
5.  The **surface** area of a rectangular solid is the sum of the areas of its faces.
7.  A **cylinder** is a hollow figure like a drinking straw.
9.  A **cone** looks like a witch's pointed hat.

**Concepts**

11. The volume of a rectangular solid is found using $V = lwh$.

13. The volume of a sphere is found using $V = \frac{4}{3}\pi r^3$.

15. The volume of a cone is found using $V = \frac{1}{3}\pi r^2 h$.

17. The surface area of a rectangular solid is found using $SA = 2lw + 2lh + 2hw$.

19. There are 27 ft$^3$ in a cubic yard. One cubic yard is $3 \times 3 \times 3 = 27$ ft$^3$.

21. There are 1,000 dm$^3$ in a cubic meter.

23. a. Volume should be applied for the size of a room to be air-conditioned.
   b. Area should be applied for the amount of land in a national park.
   c. Volume should be applied for the amount of space in a freezer.
   d. Surface area should be applied for the amount of cardboard in a shoebox.
   e. Perimeter should be applied for the distance around a checkerboard.
   f. Surface area should be applied for the amount of material to make a basketball.

25. a. The volume of this shape is $V = 3 \times 4 \times 6 = 72$ in$^3$.
   b. The area of the front of this shape is $A = 3 \times 6 = 18$ in$^2$.
   c. The area of the base of this shape is $A = 4 \times 6 = 24$ in$^2$.

**Notation**

27. The notation 1 in$^3$ is read as **1 cubic inch**.

**Practice**

29. $V = 3(4)(5) = 60$ cm$^3$.

31. $V = \frac{1}{2}(3)(4)(8)$

   $V = 48$ m$^3$

33. $V = \frac{4}{3}\pi(9^3)$

   $V \approx 3,053.63$ in$^3$

35. $V = \pi(6^2)(12)$

$V \approx 1{,}357.17 \text{ m}^3$

37. $V = \dfrac{1}{3}\pi(5^2)(12)$

$V \approx 314.16 \text{ cm}^3$

39. $V = \dfrac{1}{3}(10^2)(12)$

$V = 400 \text{ m}^3$

41. $SA = 2(4)(5) + 2(3)(4) + 2(3)(5)$

$SA = 40 + 24 + 30$

$SA = 94 \text{ cm}^2$

43. $SA = 4\pi(10^2)$

$SA \approx 1{,}256.64 \text{ in}^2$

45. Consider this as a cube with a square pyramid. The volume of the cube is $V = 8^3 = 512 \text{ cm}^3$. The volume of the square pyramid is $V = \dfrac{1}{3}(8^2)(3) = 64 \text{ cm}^3$. The total volume is $512 + 64 = 576 \text{ cm}^3$.

47. Consider this as two identical cones. The volume of one of the cones is $V = \dfrac{1}{3}\pi(4^2)(10) \approx 167.55 \text{ in}^3$. Doubling this amount yields $335.103 \text{ in}^3$.

**Applications**

49. VOLUME OF A SUGAR CUBE

$V = \left(\dfrac{1}{2}\right)^3 = \dfrac{1}{8} \text{ in}^3$

51. WATER HEATER

Over 200 gallons of hot water from $V = 27(17)(8) = 3{,}672 \text{ in}^3$. Converting to feet, $\dfrac{3{,}672}{12^3} = 2.125$ cubic feet of space.

53. VOLUME OF AN OIL TANK

$V = \pi(3^2)(7) \approx 197.92 \text{ ft}^3$

55. HOT-AIR BALLOON

$V = \dfrac{4}{3}\pi(20^3) \approx 33{,}510.32 \text{ ft}^3$

57. ENGINE

The compression ratio is $\dfrac{30.4}{3.8} = 8$, so the ratio is 8 : 1.

**Writing**

59. Answers may vary.
61. Answers may vary.

**Review**

63. $-5(5-2)^2 + 3 = -5(3)^2 + 3 = -45 + 3 = -42$

65. $38 = 40x$

$0.95 = x$

38 is 95% of 40.

1. $d = rt$

3. $P = 2l + 2w$

5. $A = \dfrac{1}{2}(700)(600) = 210,000 \text{ ft}^2$

7. $retail = 45.50 + 35 = 80.50$

9. $A = \pi\left(\dfrac{14}{2}\right)^2 = 153.94 \text{ feet}^2$

11.

| Type | Principal | Rate | Time | Interest |
|---|---|---|---|---|
| Savings | $5,000 | 5% | 3 yr | $I = 5000(0.05)(3) = \$750$ |
| Passbook | 2,250 | 2% | 1 | $I = 2250(0.02)(1) = \$45$ |
| Trust Fund | 10,000 | 6.25% | 10 | $I = 10000(0.0625)(10) = \$6,250$ |

**Chapter 7 Review**

**Section 7.1 Some Basic Definitions**

1.  The points are C and D, the line is CD and the plane is GHI.

3.  $\angle ABC, \angle CBA, \angle B, \angle 1$

5.  Acute angles: $\angle 1, \angle 2$; Right angles $\angle ABD, \angle CBD$; Obtuse angle $\angle CBE$; Straight angle $\angle ABC$.

7.  $x + 35 = 50$

    $x = 15^0$

9.  a. $m\angle 1 = 65^0$
    b. $m\angle 2 = 180 - 65 = 115^0$

11. The supplement of a $140^0$ angle is $180 - 40 = 40^0$.

**Section 7.2 Parallel and Perpendicular Lines**

13. Part $a$ represents parallel lines.

15. The corresponding angles are $\angle 1$ and $\angle 5$, $\angle 4$ and $\angle 8$, $\angle 2$ and $\angle 6$, and $\angle 3$ and $\angle 7$.

17. $m\angle 1 = 70^0; m\angle 2 = 110^0; m\angle 3 = 70^0; m\angle 4 = 110^0; m\angle 5 = 70^0;$

    $m\angle 6 = 110^0; m\angle 7 = 70^0$

19. $50 = 2x - 30$

    $80 = 2x$

    $40 = x$

**Section 7.3 Polygons**

21. a. This is an octagon.
    b. This is a pentagon.
    c. This is a triangle.
    d. This is a hexagon.
    e. This is a quadrilateral.

23. a. This is an isosceles triangle.
    b. This is a scalene triangle.
    c. This is an equilateral triangle.
    d. This is a right triangle.

25. a. $x + 70 + 20 = 180$

        $x = 90$
    b. $70 + 60 + x = 180$

        $x = 50$

27. If one of the base angle is $60^0$ then this is an equilateral triangle.

29. a. $m\overline{BD} = 15$ cm
    b. $m\angle 1 = 40^0$
    c. $m\angle 2 = 180 - 40 - 40 = 100^0$

31. a. $m\angle B = 65^0$

   b. $m\angle C = 180 - 65 = 115^0$

**Section 7.4 Properties of Triangles**

33. $\angle A$ corresponds to $\angle D$
   $\angle B$ corresponds to $\angle E$
   $\angle C$ corresponds to $\angle F$
   $\overline{AC}$ corresponds to $\overline{DF}$
   $\overline{AB}$ corresponds to $\overline{DE}$
   $\overline{BC}$ corresponds to $\overline{EF}$

35. a. $a^2 + b^2 = c^2$

$$5^2 + 12^2 = c^2$$
$$c^2 = 169$$
$$c = 13$$

   b. $a^2 + b^2 = c^2$

$$8^2 + b^2 = 17^2$$
$$b^2 = 17^2 - 8^2$$
$$b^2 = 225$$
$$b = 15$$

**Section 7.5 Perimeters and Areas of Polygons**

37. $P = 4(18) = 72$ inches

39. a. $P = 8 + 4 + 4 + 8 + 6 = 30$ m
   b. $P = 10 + 4 + 4 + 4 + 6 + 8 = 36$ m

41. There are nine square feet in one square yard since 3 ft = 1 yd.

**Section 7.6 Circles**

43. a. The chords are $\overline{CD}$ and $\overline{AB}$.
   b. The diameter is $\overline{AB}$.
   c. The radii are $\overline{OA}$, $\overline{OC}$, $\overline{OD}$, and $\overline{OB}$.
   d. The center is O.

45. $P = 2(10) + 8\pi$

   $P \approx 45.1$ cm

47. $A = 10(8) + \pi(4^2)$

   $A \approx 130.3$ cm$^2$

49. There are $12^3 = 1,728$ in$^3$ in one cubic foot.

51. a. $SA = 2(4.4)(3.1) + 2(2.3)(4.4) + 2(3.1)(2.3)$

    $SA = 61.78$ ft$^2$

    b. $SA = 4\pi(5^2)$

    $SA \approx 314.2$ in$^2$

1. $m(\overline{AB}) = 4$ units

3. This is true.

5. This is false.

7. $x + 17 = 67$

   $x = 50^0$

9. $40 = 5y - 20$

   $60 = 5y$

   $12 = y$

11. The complement of $67^0$ is $90 - 67 = 23^0$.

13. $m\angle 1 = 70^0$

15. $m\angle 3 = 70^0$

17.

| Polygon | Sides |
|---|---|
| Triangle | 3 |
| Quadrilateral | 4 |
| Hexagon | 6 |
| Pentagon | 5 |
| Octagon | 8 |

19. Since this is an isosceles triangle $m\angle A = 57^0$.

21. $x + 65 + 85 = 180$

   $x + 150 = 180$

   $x = 30^0$

23. $m(\overline{AB}) = m(\overline{DC})$

   $m(\overline{AD}) = m(\overline{BC})$

   $m(\overline{AC}) = m(\overline{BD})$

25. $m(\overline{DE}) = m(\overline{AB}) = 8$ in

27. The straight line distance from third to first is

   $90^2 + 90^2 = d^2$

   $16,200 = d^2$

   $127.28 \text{ ft} = d$

29. $A = \dfrac{1}{2}(6)(12.2 + 15.7)$

   $A = 83.7 \text{ ft}^2$

31. $A = \pi(3^2)$

   $A \approx 28.27 \text{ ft}^2$

33. $V = \dfrac{4}{3}\pi(4^3)$

$V \approx 268.08 \text{ m}^3$

35. Answers may vary.

1.  AMUSEMENT PARKS

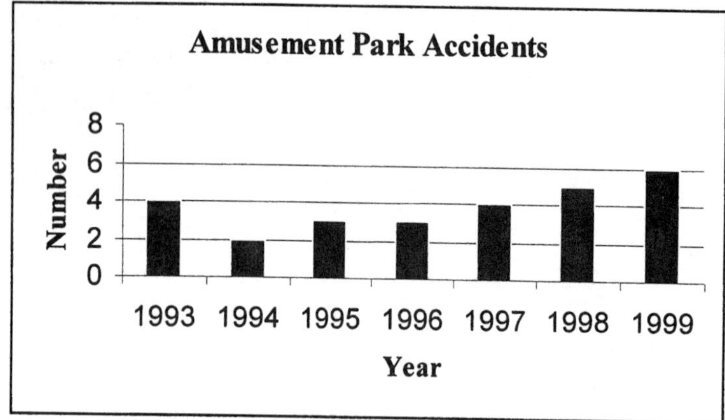


3.  $35,021 - 23,999 = 11,022$

5.  $2,110,000$

7.  $F_{24} = \{1, 2, 3, 4, 6, 8, 12, 24\}$

9.  $\begin{aligned} -10(-2) - 2^3 + 1 &= 20 - 2^3 + 1 \\ &= 20 - 8 + 1 \\ &= 13 \end{aligned}$

11.  $\begin{aligned} \left| -6 - (-3) \right| &= \left| -6 + 3 \right| \\ &= \left| -3 \right| \\ &= 3 \end{aligned}$

13.  $\begin{aligned} -x + 2 &= 13 \\ -x &= 11 \\ x &= -11 \end{aligned}$

15.  $\dfrac{35}{28} = \dfrac{5}{4}$

17.  $\dfrac{3}{4} - \dfrac{3}{5} = \dfrac{15}{20} - \dfrac{12}{20} = \dfrac{3}{20}$

19.  $\begin{aligned} -\dfrac{6}{25}\left( 2\dfrac{7}{24} \right) &= -\dfrac{6}{25}\left( \dfrac{55}{24} \right) \\ &= -\dfrac{\cancel{6}}{\cancel{5}(5)}\left( \dfrac{\cancel{5}(11)}{\cancel{6}(4)} \right) \\ &= -\dfrac{11}{20} \end{aligned}$

21. **VETERINARY MEDECINE**

A single dose is $\dfrac{\frac{3}{4}}{8} = \dfrac{3}{4} \bullet \dfrac{1}{8} = \dfrac{3}{32}$ ounce.

23. $\dfrac{3}{4} + \left(-\dfrac{1}{3}\right)^2 \dfrac{5}{4} = \dfrac{3}{4} + \left(\dfrac{1}{9}\right)\dfrac{5}{4} = \dfrac{3}{4} + \dfrac{5}{36} = \dfrac{27}{36} + \dfrac{5}{36} = \dfrac{32}{36} = \dfrac{8}{9}$

25. **GLOBAL WARMING**
   a.  The greatest rise in temperature was recorded in 1991. The rise was about $0.3^0$.
   b.  The greatest decline in temperature was recorded in 1993. The decline was about $-0.5^0$.

27. $\pi \approx 3.1416$

29. $3.4 + 106.78 + 35 + 0.008 = 145.188$

31. $(89.9708)(1,000) = 89,970.8$

33. $-8.8 + (-7.3 - 9.5) = -8.8 + (-16.8) = -25.6$

35. $\dfrac{2}{15} = 0.1\overline{3}$

37. **DECORATIONS**
   Let $b$ represent the number of balloons.
   $$20 = 15.15 + 0.05b$$
   $$4.85 = 0.05b$$
   $$97 = b$$
   She can buy 97 balloons.

39. $\sqrt{\dfrac{49}{81}} = \dfrac{7}{9}$

41. $0.15(450) = 67.5$

43.

| Percent | Decimal | Fraction |
|---|---|---|
| 57% | 0.57 | $\dfrac{57}{100}$ |
| 0.1% | 0.001 | $\dfrac{1}{1,000}$ |
| $33\dfrac{1}{3}\%$ | $0.\overline{3}$ | $\dfrac{1}{3}$ |

45. $m\angle 1 = 180 - 130 = 50^0$

47. $m\angle 3 = 180 - 130 = 50^0$

49. $m\angle 1 = 75^0$

51. $m\angle 2 = 180 - 75 = 105^0$

53. JAVELIN THROW

$x = 90 - 44 = 46^0$

$y = 180 - 46 = 134^0$

55. The hypotenuse is $c^2 = 12^2 + 5^2$

$$c^2 = 169$$

$$c = 13 \text{ m}$$

57. $A = \dfrac{1}{2}(14)(18)$

$A = 7(18)$

$A = 126 \text{ ft}^2$

59. $C = \pi(14)$

$C \approx 43.98 \text{ cm}$

$A = \pi(7^2)$

$A \approx 153.94 \text{ cm}^2$

61. $V = 5(6)(7) = 210 \text{ m}^3$

63. $V = \dfrac{1}{3}(9)\pi(4)^2$

$V \approx 150.80 \text{ m}^3$

65. $SA = 2(15)(24) + 2(15)(18) + 2(24)(18)$

$SA = 2{,}124 \text{ in}^2$

# Section 8.1 Algebraic Expressions

## Vocabulary

1. To **evaluate** an algebraic expression, we substitute the values for the variables and then apply the rules for the order of operations.

3. $2x + 5$ is an example of an algebraic **expression** whereas $2x = 5 = 7$ is an example of an **equation**.

## Concepts

5. Answers may vary. Examples: $6 + 20x$; $\frac{6-x}{20}$

7. $3x - 6$ evaluated for $x = 4$, could be misunderstood as $34 - 6$ without parentheses, $3(4) - 6$.

9. a. Let $x$ be the weight of the car; $2x - 500 =$ the weight of the van.
   b. If the car weighs 2,000 pounds then the van weighs $2(2,000) - 500 = 4,000 - 500 = 3500$ pounds.

11.

| Coin | Number | Value | Total (cents) |
|---|---|---|---|
| Penny | 6 | 5 | 30 |
| Dime | $d$ | 10 | $10d$ |
| Half dollar | $x + 5$ | 50 | $50(x + 5)$ |

## Notation

13. $9a - a^2$ evaluated for $a = 5$ is
$$= 9(5) - 5^2$$
$$= 9(5) - 25$$
$$= 45 - 25$$
$$= 20$$

## Practice

15. The sum of the length $l$ and 15 is $l + 15$.

17. The product of a number and 50 is $50x$.

19. The ratio of the amount won $w$ and lost $l$ is $\frac{w}{l}$.

21. P increased by p is $P + p$.

23. The square of k minus 2,005 is $k^2 - 2,005$.

25. J reduced by 500 is $J - 500$.

27. 1,000 split $n$ equal ways is $\frac{1000}{n}$.

29. 90 more than the current price $p$, is $p + 90$.

31. The total of 35, $h$, and 300 is $35 + h + 300$.

33. 680 fewer than the entire population $p$ is $p - 680$.

35. The product of $d$ and 4 decreased by 15 is $4d - 15$.

37. Twice the sum of 200 and $t$ is $2(200 + t)$.

39. The absolute value of the difference of $a$ and 2 is $|a - 2|$.

41. 7 less than a number

43. The product of 7 and a number increased by 4.

45. a. In 5 hours there are $5(60)$ or 300 minutes
    b. In $h$ hours there are $60h$ minutes

47. a. There are $3y$ feet in a $y$ yards
    b. There are $\frac{f}{3}$ yards in $f$ feet

49. At 29¢ per mile for driving $x$ miles the rental fee is $29x$¢

51. For a production of $c$ cans, $\frac{c}{6}$ six packs were produced

53. With 5 people in each car the caravan carried $5b$ people

55. Tickets for a family and two neighbors will cost $\$5(x + 2)$ where $x + 2$ represents the  total number of people

57.

| $x$ | $x^3 - 1$ |
|---|---|
| 0 | $0^3 - 1 = 1$ |
| $-1$ | $(-1)^3 - 1 = -2$ |
| $-3$ | $(-3)^3 - 1 = -28$ |

59.

| $s$ | $\frac{5s+36}{s}$ |
|---|---|
| 1 | $\frac{5(1)+36}{1} = 41$ |
| 6 | $\frac{5(6)+36}{6} = 11$ |
| $-12$ | $\frac{5(-12)+36}{-12} = 2$ |

61.

| $x$ | $2x - \frac{x}{2}$ |
|---|---|
| 100 | $2(100) - \frac{100}{2} = 150$ |
| $-300$ | $2(-300) - \frac{-300}{2} = -450$ |

63.

| $x$ | $(x + 1)(x + 5)$ |
|---|---|
| $-1$ | $(-1 + 1)(-1 + 5) = 0$ |
| $-5$ | $(-5 + 1)(-5 + 5) = 0$ |
| $-6$ | $(-6 + 1)(-6 + 5) = 5$ |

65. $x = 3$, $y = -2$, $z = -4$

$3(-2)^2 - 6(-2) - 4$

$= 3(4) + 12 - 4$

$= 24 - 4$

$= 20$

67. $x = 3$, $y = -2$, $z = -4$

$(3 + 3)(-2)$

$= 6(-2)$

$= -12$

69. $x = 3$, $y = -2$, $z = -4$

$(3 + (-2))^2 - |(-4) + (-2)|$

$= 1^2 - |-6|$

$= 1 - 6$

$= -5$

71. $x = 3$, $y = -2$, $z = -4$

$(4(3))^2 + 3(-2)^2$

$= 12^2 + 3(4)$

$= 144 + 12$

$= 156$

73. $x = 3$, $y = -2$, $z = -4$

$-\dfrac{2(3)+(-2)^3}{(-2)+(2)(-4)}$

$= -\dfrac{6+(-8)}{-2+(-8)}$

$= -\dfrac{-2}{-10}$

$= -\dfrac{2}{10}$

$= -\dfrac{1}{5}$

75. $(5)^2 - 4(-1)(-2)$

$= 25 + (-4)(-1)(-2)$

$= 25 + 4(-2)$

$= 25 + (-8)$

$= 17$

77. $(-5)^2 + 2(-5)(-1) + (-1)^2$

$= 25 + 2(5) + 1$

$= 25 + 10 + 1$

$= 36$

79. $\dfrac{10}{2}[2(-4) + (10 - 1)(6)]$

$= 5[-8 + (9)(6)]$

$= 5[-8 + 54]$

$= 5(46)$

$= 230$

81. $\dfrac{1.8^2+(-7.6)^2}{2}$

$= \dfrac{3.24+57.76}{2}$

$= \dfrac{61}{2}$

$= 30.5$

**Applications**

83. ROCKETRY

| $t$ | $h = 64t - 16t^2$ |
|---|---|
| 0 | $h = 64(0) - 16(0)^2 = 0$ |
| 0.5 | $h = 64(0.5) - 16(0.5)^2 = 28$ |
| 1 | $h = 64(1) - 16(1)^2 = 48$ |
| 1.5 | $h = 64(1.5) - 16(1.5)^2 = 60$ |
| 2 | $h = 64(2) - 16(2)^2 = 64$ |
| 2.5 | $h = 64(2.5) - 16(2.5)^2 = 60$ |
| 3 | $h = 64(3) - 16(3)^2 = 48$ |
| 3.5 | $h = 64(3.5) - 16(3.5)^2 = 28$ |
| 4 | $h = 64(4) - 16(4)^2 = 0$ |

85. TRANSLATION

$C = \dfrac{5(F-32)}{9}$

For $F = -34$,

$C = \dfrac{5(-34-32)}{9}$

$C = \dfrac{5(-34+(-32))}{9}$

$C = \dfrac{5(-66)}{9}$

$C = \dfrac{-330}{9}$

$C \approx -36.6666..$

To the nearest degree, $C = -37°$

$C = \dfrac{5(F-32)}{9}$

For $F = -84$,

$C = \dfrac{5(-84-32)}{9}$

$C = \dfrac{5(-84+(-32))}{9}$

$C = \dfrac{5(-116)}{9}$

$C = \dfrac{-580}{9}$

$C \approx -64.444...$

To the nearest degree, $C = -64°$

87. $A = \frac{1}{2}h(b_1 + b_2)$

$A = \frac{1}{2}(\frac{3}{4})(1\frac{1}{4} + 2\frac{3}{8})$

$A = \frac{3}{8}(3\frac{5}{8})$

$A = 1\frac{23}{64}$ in$^2$

89. LANDSCAPING

Using $\pi(R^2 - r^2)$ where R is 9.75 feet and r is 4.5 feet, then the number of square feet of sod is
$\pi[(9.75)^2 - (4.5)^2] = 235$ ft$^2$

**Writing**

91. Answers will vary
93. Answers will vary

**Review**

95. Simplify $-0 = 0$

97. $\left| -\frac{2}{3} \right| = \frac{2}{3}$

99. $5 \cdot 5 \cdot 5 \cdot 5 = 5^4$

101. $\frac{84+93+72}{3} = 83$

# Section 8.2 Simplifying Algebraic Expressions

## Vocabulary

1. To **simplify** an algebraic expression, we use properties of algebra to write the expression in a less complicated form.

3. In the term $3x^2$, the number factor 3 is called the numerical **coefficient** and $x^2$ is called the **variable** part.

5. We can use the distributive property to **remove** the parentheses when an expression is multiplied by a quantity.

## Concepts

7. $a(b + c) = ab + ac$ illustrates the distributive property.

9. $2(3 + 4) = 2 \cdot 3 + 2 \cdot 4$
   $2(3 + 4)$ represents the product of 2 rows and 3 columns plus 4 columns.

11. a. $2(x + 4) = 2x + 8$
    b. $2(x - 4) = 2x - 8$
    c. $-2(x + 4) = -2x - 8$
    d. $-2(x - 4) = -2x + 8$
    e. $-2(-x + 4) = 2x - 8$
    f. $-2(-x - 4) = 2x + 8$

13. $x + (20 - x) = 20$, the original board length was 20 feet.

15. a. $a^\circ + 90^\circ - a^\circ = 90^\circ$ These are complementary angles.
    b. $a^\circ + 180^0 - a^\circ = 180^\circ$ These are supplementary angles.

## Notation

17. $7(a + 2) = 7a + 7(2)$
    $= 7a + 14$

19. a. $2K$ and $3k$ are not like terms since K is represented in upper and lower case.
    b. $-d$ and $d$ are like terms.

21. a. $5x - (-1) = 5x + 1$
    b. $16t + (-6) = 16t - 6$

## Practice

23. $9(7m) = 63m$

25. $5(-7q) = -35q$

27. $12\left(\frac{5}{12}x\right) = 5x$

29. $8\left(\frac{3}{4}y\right) = 2(3y) = 6y$

31. $(-5p)(-4b) = 20pb$

33. $-5(4r)(-2r)$
$= (-5)(4)(-2)(r)(r)$
$= 40r^2$

35. $5(x + 3) = 5x + 5(3) = 5x + 15$

37. $-2(b - 1) = -2b + (-2)(-1) = -2b + 2$

39. $(3t - 2)8 = 3t(8) + (-2)(8) = 24t - 16$

41. $(2y - 1)6 = 2y(6) + (-1)(6) = 12y - 6$

43. $0.4(x - 4) = 0.4x + 0.4(-4) = 0.4x - 1.6$

45. $-\frac{2}{3}(3w - 6)$
$= -\frac{2}{3}(3w) + (-\frac{2}{3})(-6)$
$= -\frac{6}{3}w + \frac{12}{3}$
$= -2w + 4$

47. $r(r - 10) = r(r) + r(-10) = r^2 - 10$

49. $-(x - 7)$
$= (-1)(x) + (-1)(-7)$
$= -x + 7$

51. $17(2x - y + 2)$
$= 17(2x) + 17(-y) + 17(2)$
$= 34x - 17y + 34$

53. $-(-14 + 3p - t)$
$= (-1)(-14) + (-1)(3p) + (-1)(-t)$
$= 14 - 3p + t$

55. a. $-b$ has coefficient $-1$
b. $-9.9x^3$ has coefficient $-9.9$
c. $\frac{1}{4}x$ has coefficient $\frac{1}{4}$
d. $-\frac{2x}{3}$ has coefficient $-\frac{2}{3}$

57. $-5r + 4s$ has two terms, $-5r$ and $4s$. The numerical coefficients are $-5$ and $4$.

59. $-15r^2s$ has one term, itself, $-15r^2s$. The numerical coefficient is $-15$.

61. $50a + 2$ has two terms, $50a$ and $2$. The numerical coefficients are $50$ and $2$.

63. $x^3 - 125$ has two terms, $x^3$ and $-125$. The numerical coefficients are $1$ and $-125$.

65. $3x + 17x = 20x$

67. $8x^2 - 5x^2 = 3x^2$

69. $-4x + 4x = 0$

71. $-7b^2 + 7b^2 = 0$

73. $a + a + a = 3a$

75. $0 - 3x = -3x$

77. $0 - (-t) = t$

79. $3x + 5x - 7x = 8x - 7x = x$

81. $-13x^2 + 2x^2 - 5x^2 = -11x^2 - 5x^2 = -16x^2$

83. $1.8h - 0.7h = 1.1h$

85. $\frac{3}{5}t + \frac{1}{5}t = \frac{4}{5}t$

87. $-0.2r - (-0.6r) = -0.2r + 0.6r = 0.4r$

89. $2z + 5(z - 3)$
$\quad = 2z + 5z + 5(-3)$
$\quad = 7z + (-15)$
$\quad = 7z - 15$

91. $-(c + 7) - 2(c - 3)$
$\quad = -c - 7 + (-2)c + (-2)(-3)$
$\quad = -c - 7 + (-2)c + 6$
$\quad = -3c - 1$

93. $2x + 4(X - x) + 3X$
$\quad = 2x + 4X + 4(-x) + 3X$
$\quad = 2x + 4X + + (-4x) + 3X$
$\quad = -2x + 7X$
$\quad = 7X - 2x$

95. $(a + 2) - (a - b)$
$\quad = a + 2 + (-1)a + (-1)(-b)$
$\quad = a + 2 + (-a) + b$
$\quad = b + 2$

97. $x(x + 3) - 3x^2$
$\quad = x(x) + x(3) - 3x^2$
$\quad = x^2 + 3x + (-3x^2)$
$\quad = -2x^2 + 3x$

**Applications**

99. THE AMERICAN RED CROSS

The perimeter of the red cross is $12x$, where each of the 12 edges has a length of $x$.

101. PING-PONG

This table has two sides of length $(x + 4)$ and two sides of width $x$, so the perimeter is

$$2(x + 4) + 2(x)$$
$$= 2x + 2(4) + 2x$$
$$= (4x + 8) \text{ feet}$$

**Writing**

103. Answers will vary.

105. Answers will vary.

**Review**

107. $x = -3$, $y = -5$, $z = 0$

$$(-3)^2(0)((-5)^3 - 0)$$
$$= (0)(-125)$$
$$= 0$$

109. $x = -3$, $y = -5$, $z = 0$

$$\frac{(-3) - (-5)^2}{2(-5) - 1 + (-3)}$$
$$= \frac{(-3) - 25}{-10 - 1 + (-3)}$$
$$= \frac{(-3) + (-25)}{-10 + (-1) + (-3)}$$
$$= \frac{-28}{-14}$$
$$= 2$$

**Section 8.3 Solving Equations**

**Vocabulary**

1. An **equation** is a statement that two quantities are equal.
3. If a number is a solution of an equation, the number is said to **satisfy** the equation.
5. The product of a number and its **reciprocal** is 1.

**Concepts**

7. To solve the equation $2x - 7 = 21$, we must first undo the **subtraction** of 7 by adding 7 to both sides. We then undo the **multiplication** by 2 by dividing both sides by 2.

9. To solve the equation $\frac{x}{-2} + 3 = 5$, we first undo the **addition** of 3 by subtracting 3 from both sides. We then undo the **division** by $-2$ by multiplying both sides by $-2$.

11. The reciprocal of $-\frac{4}{5}$ is $-\frac{5}{4}$.

13. The $\text{LCD}(2, 3, 5)$ is $2 \cdot 3 \cdot 5 = 30$.

15. a. $3x + 5 - x = 3x - x + 5 = 2x + 5$
    b. $3x + 5 - x = 9$
    $2x + 5 = 9$
    $2x + 5 - 5 = 9 - 5$
    $2x = 4$
    $\frac{2x}{2} = \frac{4}{2}$
    $x = 2$
    c. For $x = 9$,
    $3x + 5 - x$
    $= 3(9) + 5 - 9$
    $= 27 + 5 + (-9)$
    $= 32 + (-9)$
    $= 23$
    d. $3(-1) + 5 - (-1) = -3 + 6 = 3 \neq 9$
    $-1$ is not a solution.

**Notation**

17. $2x - 7 = 21$
    $2x - 7 + 7 = 21 + 7$
    $2x = 28$
    $\frac{2x}{2} = \frac{28}{2}$
    $x = 14$

19. a. $-x = (-1)x$
    b. $\frac{3x}{5} = \frac{3}{5}x$
    c. $-31 = x$
    $x = -31$

**Practice**

21. $2x + 5 = 17$
    $2x + 5 - 5 = 17 - 5$
    $2x = 12$
    $\frac{2x}{2} = \frac{12}{2}$
    $x = 6$

23. $-5q - 2 = 1$
    $-5q + (-2) + 2 = 1 + 2$
    $-5q = 3$
    $\frac{-5q}{-5} = \frac{3}{-5}$
    $q = -\frac{3}{5}$

25. $0.6 = 4.1 - x$
    $0.6 + (-4.1) = 4.1 + (-4.1) + (-x)$
    $-3.5 = -x$
    $\frac{-3.5}{-1} = \frac{-x}{-1}$
    $3.5 = x$

27. $-g = -4$
    $(-1)(g) = -4$
    $\frac{-g}{-1} = \frac{-4}{-1}$
    $g = 4$

29. $-8 - 3c = 0$
    $-8 + 8 - 3c = 0 + 8$
    $-3c = 8$
    $\frac{-3c}{-3} = \frac{8}{-3}$
    $c = -\frac{8}{3}$

31. $-\frac{5}{6}k = 10$
    $(-\frac{6}{5})(-\frac{5}{6}k) = 10(-\frac{6}{5})$
    $k = -\frac{60}{5}$
    $k = -12$

33. $-\frac{t}{3} + 2 = 6$
    $-\frac{t}{3} + 2 - 2 = 6 - 2$
    $-\frac{t}{3} = 4$
    $(-\frac{3}{1})(-\frac{t}{3}) = 4(-\frac{3}{1})$
    $t = -12$

35. $\frac{2x}{3} - 2 = 4$
    $\frac{2}{3}x + (-2) + 2 = 4 + 2$
    $\frac{2}{3}x = 6$
    $\frac{3}{2}(\frac{2}{3}x) = \frac{3}{2}(6)$
    $x = \frac{18}{2}$
    $x = 9$

37. $\frac{x+5}{3} = 11$

$\quad 3\left(\frac{x+5}{3}\right) = 3(11)$

$\quad x + 5 = 33$

$\quad x + 5 - 5 = 33 - 5$

$\quad x = 28$

39. $\frac{y-2}{7} = -3$

$\quad 7\left(\frac{y-2}{7}\right) = -3(7)$

$\quad y - 2 = -21$

$\quad y + (-2) + 2 = -21 + 2$

$\quad y = -19$

41. $2(-3) + 4y = 14$

$\quad -6 + 4y = 14$

$\quad -6 + 6 + 4y = 14 + 6$

$\quad 4y = 20$

$\quad \frac{4y}{4} = \frac{20}{4}$

$\quad y = 5$

43. $-2x - 4(1) = -6$

$\quad -2x - 4 = -6$

$\quad -2x - 4 + 4 = -6 + 5$

$\quad -2x = -2$

$\quad \frac{-2x}{-2} = \frac{-2}{-2}$

$\quad x = 1$

45. $3(x + 2) - x = 12$

$\quad 3x + 3(2) - x = 12$

$\quad 2x + 6 = 12$

$\quad 2x + 6 - 6 = 12 - 6$

$\quad 2x = 6$

$\quad \frac{2x}{2} = \frac{6}{2}$

$\quad x = 3$

47. $-3(2y - 2) - y = 5$

$\quad -3(2y) + (-3)(-2) - y = 5$

$\quad -6y + 6 - y = 5$

$\quad -7y + 6 + (-6) = 5 + (-6)$

$\quad -7y = -1$

$\quad \frac{-7y}{-7} = \frac{-1}{-7}$

$\quad y = \frac{1}{7}$

49. $0 - 2y = 8$

$\quad -2y = 8$

$\quad \frac{-2y}{-2y} = \frac{8}{-2}$

$\quad y = -4$

51. $5x + 7.2 = 4x$

$5x + 7.2 - 7.2 = 4x - 7.2$

$5x = 4x - 7.2$

$5x - 4x = 4x - 4x - 7.2$

$x = -7.2$

53. $8y + 4 = 4y$

$8y + 4 - 4 = 4y - 4$

$8y = 4y -$

$8y - 4y = 4y - 4y - 4$

$4y = -4$

$\frac{4y}{4} = \frac{-4}{4}$

$y = -1$

55. $15x = x$

$15x - x = x - x$

$15x - x = 0$

$14x = 0$

$x = 0$

57. $4 + \frac{y}{2} = \frac{3}{5}$

$10\left(4 + \frac{y}{2}\right) = 10\left(\frac{3}{5}\right)$

$40 + 5y = 6$

$5y = -34$

$y = -\frac{34}{5}$

59. $\frac{1}{3} + \frac{c}{5} = -\frac{3}{2}$

$30\left(\frac{1}{3} + \frac{c}{5}\right) = 30\left(-\frac{3}{2}\right)$

$\frac{30}{3} + \frac{30c}{5} = \frac{-90}{2}$

$10 + 6c = -45$

$10 - 10 + 6c = -45 - 10$

$6c = -55$

$\frac{6c}{6} = -\frac{55}{6}$

$c = -\frac{55}{6}$

61. $\frac{y}{6} + \frac{y}{4} = -1$

$12\left(\frac{y}{6} + \frac{y}{4}\right) = 12(-1)$

$2y + 3y = -12$

$5y = -12$

$y = -\frac{12}{5}$

63. $-\frac{2}{9} = \frac{5x}{6} - \frac{1}{3}$

$18\left(-\frac{2}{9}\right) = 18\left(\frac{5x}{6} - \frac{1}{3}\right)$

$-\frac{36}{9} = \frac{90x}{6} - \frac{18}{3}$

$-4 = 15x - 6$

$-4 + 6 = 15x + (-6) + 6$

$2 = 15x$

$\frac{2}{15} = \frac{15x}{15}$

$\frac{2}{15} = x$

65. $\frac{1}{2}x - \frac{1}{9} = \frac{1}{3}$

$\quad 18\left(\frac{1}{2}x - \frac{1}{9}\right) = 18\left(\frac{1}{3}\right)$

$\quad 9x - 2 = 6$

$\quad 9x = 8$

$\quad x = \frac{8}{9}$

67. $\frac{2}{5}x + 1 = \frac{1}{3} + x$

$\quad 1 - \frac{1}{3} = x - \frac{2}{5}x$

$\quad \frac{2}{3} = \frac{3}{5}x$

$\quad \frac{5}{3} \cdot \frac{2}{3} = x$

$\quad \frac{10}{9} = x$

69. $3(a + 2) = 2(a - 7)$

$\quad 3a + 3(2) = 2a + 2(-7)$

$\quad 3a + 6 = 2a + (-14)$

$\quad 3a - 2a + 6 = 2a - 2a + (-14)$

$\quad a + 6 = -14$

$\quad a + 6 + (-6) = -14 + (-6)$

$\quad a = -20$

71. $9(x + 11) + 5(13 - x) = 0$

$\quad 9x + 9(11) + 5(13) + 5(-x) = 0$

$\quad 9x - 5x + 99 + 65 = 0$

$\quad 4x + 164 = 0$

$\quad 4x + 164 - 164 = 0 - 164$

$\quad 4x = -164$

$\quad \frac{4x}{4} = \frac{-164}{4}$

$\quad x = -41$

73. $\frac{3t - 21}{2} = t - 6$

$\quad 2\left(\frac{3t - 21}{2}\right) = 2(t - 6)$

$\quad 3t - 21 = 2t + 2(-6)$

$\quad 3t - 21 = 2t + (-12)$

$\quad 3t + (-21) + 21 = 2t + (-12) + 21$

$\quad 3t = 2t + 9$

$\quad 3t - 2t = 2t - 2t + 9$

$\quad t = 9$

75. $\frac{10 - 5s}{3} = s + 6$

$\quad 3\left(\frac{10 - 5s}{3}\right) = 3(s + 6)$

$\quad 10 - 5s = 3s + 3(6)$

$\quad 10 - 5s = 3s + 18$

$\quad 10 + (-5s) + 5s = 3s + 5s + 18$

$\quad 10 = 8s + 18$

$\quad 10 + (-18) = 8s + 18 + (-18)$

$\quad -8 = 8s$

$\quad \frac{-8}{8} = \frac{8s}{8}$

$\quad -1 = s$

77. $2 - 3(x - 5) = 4(x - 1)$

$2 + (-3)x + (-3)(-5) = 4x + 4(-1)$

$2 + (-3x) + 15 = 4x + (-4)$

$-3x + 17 = 4x + (-4)$

$-3x + 3x + 17 = 4x + 3x + (-4)$

$17 = 7x + (-4)$

$17 + 4 = 7x + (-4) + 4$

$21 = 7x$

$\frac{21}{7} = \frac{7x}{7}$

$3 = x$

79. $8x + 3(2 - x) = 5(x + 2) - 4$

$8x + 6 - 3x = 5x + 10 - 4$

$5x + 6 = 5x + 6$

Identity

81. $-3(s + 2) = -2(s + 4) - s$

$-3s - 6 = -2s - 8 - s$

$-3s - 6 = -3s - 8$

$-6 \neq -8$

Impossible

83. $2(3z + 4) = 2(3z - 2) + 13$

$6z + 8 = 6z - 4 + 13$

$6z + 8 \neq 6z + 9$

Impossible

85. $4(y - 3) - y = 3(y - 4)$

$4y - 12 - y = 3y - 12$

$3y - 12 = 3y - 12$

Identity

87. $1.73x = -4.952 - 2.27x$

$4x = -4.952$

$x = \frac{-4.952}{4}$

$x = -1.238$

89. $20(x - 3.7) = 32,832$

$x - 3.7 = 1,641.6$

$x = 1,645.3$

**Writing**

91. Answers will vary.

93. Answers will vary.

**Review**

95. $-(-8) = 8$

97. $-8(-8) = 64$

99. $\frac{1}{8} \cdot \frac{1}{8} = \frac{1}{64}$

101. $8x + 8 + 8x - 8 = 16x$

**Section 8.4 Formulas**

**Vocabulary**

1.  A **formula** is an equation that is used to state a known relationship between two or more variables.
3.  The distance around a geometric figure is called its **perimeter**.
5.  A segment drawn from the center of a circle to a point on the circle is called a **radius**.
7.  The perimeter of a circle is called its **circumference**.

**Concepts**

9.  a. $d = rt$, where $d$ is distance, $r$ is rate, and $t$ is time.
    b. $r = c + m$, where $r$ is the retail cost, $c$ is the cost, and $m$ is the markup.
    c. $p = r - c$, where $p$ is profit, $r$ is the revenue, and $c$ is the costs.
    d. $I = Prt$, where $I$ is the interest, $P$ is the principal, $r$ is the interest rate, and $t$ is the amount of time the money is invested.
    e. $C = 2\pi r$, where $C$ is the circumference and $r$ is the radius

11.

|  | rate (mi/sec) | time (sec) | distance $d = rt$ |
|---|---|---|---|
| light | 186,282 | 60 | $d = 11,176,920$ |
| sound | 1,088 | 60 | $d = 65,280$ |

13. a. Volume is used to find the amount of storage in a freezer.
    b. Circumference is used to find how far a bicycle tire rolls in one revolution.
    c. Area is used to find the amount of land making up the Sahara Desert.
    d. Perimeter is used to find the distance around a Monopoly board.

15. The area of a rectangle is the product of length and width.
    $(x + 3)2 = (2x + 6)$ cm$^2$

**Notation**

17. $V = \frac{1}{3}Bh$
    $3V = 3(\frac{1}{3}Bh)$
    $3V = Bh$
    $\frac{3V}{h} = \frac{Bh}{h}$
    $\frac{3V}{h} = B$ or $B = \frac{3V}{h}$

19. a. $\pi \approx 3.14$
    b. $98 \cdot \pi$
    c. $r$ represents radius of the cylinder. $h$ represents the height of the cylinder.

**Practice**

21. SWIMMING

Using $d = rt$, let $d = 1826$ and $t = 742$.
Find $r$.
$d = rt$, solving for $r$, $\frac{d}{t} = \frac{rt}{t}$, $\frac{d}{t} = r$
$r = \frac{1826}{742}$
$r \approx 2.4609$
This swimmers average rate was approximately 2.5 miles per hour.

23. HOLLYWOOD

Let $p =$ the production cost on this movie.
$190,000,000 - p = 125,000,000$
$190,000,000 - p + p = 125,000,000 + p$
$190,000,000 = 125,000,000 + p$
$190,000,000 - 125,000,000$
$\quad = 125,000,000 - 125,000,000 + p$
$65,000,000 = p$
The production costs of this movie were \$65 million.

25. ENTREPRENEURS

Using $I = Prt$, where $P = 2500, t = 2, I = 175$, find $r$.
First solve $I = Prt$ for $r$.
$\frac{I}{Pt} = \frac{Prt}{Pt}$
$\frac{I}{pt} = r$
now substitute the above values,
$\frac{175}{(2500)(2)} = r$
$\frac{175}{5000} = r$
$0.035 = r$
The simple interest rate on this loan was 3.5%.

27. METALLURGY

Using the formula for temperature conversion $C = \frac{5(F-32)}{9}$ and substituting $C = 2212$.
$2212 = \frac{5(F-32)}{9}$
$9(2212) = 9\left(\frac{5(F-32)}{9}\right)$
$19,908 = 5(F - 32)$
$19,908 = 5F - 160$
$19,908 + 160 = 5F - 160 + 160$
$20,068 = 5F$
$\frac{20,068}{5} = \frac{5F}{5}$
$4013.6 = F$
To the nearest degree, the Fahrenheit temperature at which silver boils is 4,014°.

29. VALENTINES DAY

Let $m =$ roses markup.
$12.95 + m = 37.50$
$12.95 - 12.95 + m = 37.50 - 12.95$
$m = 24.55$
The markup on these roses is \$24.55.

31. YO-YO

In one "around the world" trick a yo-yo travels approximately $C = 2\pi r$, where $r = 21$ inches, $C = 2\pi(21) \approx 132$ inches.

33. $E = IR$

$$\frac{E}{I} = \frac{IR}{I}$$

$$\frac{E}{I} = R \text{ or } R = \frac{E}{I}$$

35. $V = lwh$

$$\frac{V}{lh} = \frac{lwh}{lh}$$

$$\frac{V}{lh} = w \text{ or } w = \frac{V}{lh}$$

37. $C = 2\pi r$

$$\frac{C}{2\pi} = \frac{2\pi r}{2\pi}$$

$$\frac{C}{2\pi} = r$$

39. $a + b + c = 180$

$$a + b + c - b - c = 180 - b - c$$

$$a = 180 - b - c$$

41. $y = mx + b$

$$y - b = mx + b - b$$

$$y - b = mx$$

$$\frac{y-b}{m} = \frac{mx}{m}$$

$$\frac{y-b}{m} = x \text{ or } x = \frac{y-b}{m}$$

43. $A = P + Prt$

$$A - P = P - P + Prt$$

$$A - P = Prt$$

$$\frac{A-P}{Pr} = \frac{Prt}{Pr}$$

$$\frac{A-P}{Pr} = t \text{ or } t = \frac{A-P}{Pr}$$

45. $V = \frac{1}{3}\pi r^2 h$

$$3V = 3\left(\frac{1}{3}\pi r^2 h\right)$$

$$3V = \pi r^2 h$$

$$\frac{3V}{\pi r^2} = \frac{\pi r^2 h}{\pi r^2}$$

$$\frac{3V}{\pi r^2} = h \text{ or } h = \frac{3V}{\pi r^2}$$

47. $x = \frac{a+b}{2}$

$$2x = a + b$$

$$2x - a = a + b - a$$

$$2x - a = b$$

49. $D = \frac{C-s}{n}$

$$nD = C - s$$

$$nD - nD = C - s - nD$$

$$0 + s = C - s - nD + s$$

$$s = C - nD$$

51. $E = mc^2$

$\frac{E}{m} = \frac{mc^2}{m}$

$\frac{E}{m} = c^2$

53. $c^2 = a^2 + b^2$

$c^2 - b^2 = a^2 + b^2 - b^2$

$c^2 - b^2 = a^2$

55. $A = \frac{1}{2}h(b + d)$

$2A = h(b + d)$

$\frac{2A}{h} = \frac{h(b+d)}{h}$

$\frac{2A}{h} = b + d$

$\frac{2A}{h} - d = b + d - d$

$\frac{2A}{h} - d = b$

57. $3y - 9 = x$

$3y - 9 + 9 = x + 9$

$3y = x + 9$

$\frac{3y}{3} = \frac{x+9}{3}$

$y = \frac{x+9}{3}$

$y = \frac{1}{3}x + 3$

59. $4y + 16 = -3x$

$4y + 16 - 16 = -3x - 16$

$4y = -3x - 16$

$\frac{4y}{4} = \frac{-3x-16}{4}$

$y = \frac{-3x-16}{4}$

$y = -\frac{3}{4}x - 4$

**Applications**

61. PROPERTIES OF WATER

$C = 100^0$

$F = \frac{9}{5}(100) + 32$

$F = 212^0$

$F = 32^0$

$C = \frac{5(32-32)}{9}$

$C = \frac{5(0)}{9}$

$C = 0^0$

63. AVON PRODUCTS

| Quarterly Financials | Mar 00 | Dec 99 |
| --- | --- | --- |
| Revenue | $1,324.9$ | $1,566.6$ |
| Cost | $497.3$ | $606.6$ |
| Profit = Revenue-Cost | $827.6$ | $960.0$ |

65. CARPENTRY

Let $P =$ the perimeter
$P = 10 + 10 + 16$
$P = 36$ ft.
Let $A =$ the area
$A = \frac{1}{2}bh$
$A = \frac{1}{2}(16)(6)$
$A = 8(6)$
$A = 48$ ft$^2$

67. ARCHERY

Circumference is $c = \pi d$ where $d = 2(8) = 16$ therefore $c = \pi(16)$
$c = 16\pi$
$c \approx 50.26548$
The circumference of this target to the nearest tenth is 50.3 inches.
Area is $a = \pi r^2$, where $r = 8$,
$a = \pi(8)^2$
$a = 64\pi$
$a \approx 201.0619298$
The area of this target to the nearest tenth is 201.1 in$^2$.

69. LANDSCAPING

The perimeter of the trellis is the sum of the lengths of the four sides,
$P = 12 + 24 + 10 + 10$
$P = 56$
The perimeter is 56 inches.
The area of the trellis is found using $A = \frac{1}{2}h(b + d)$, where $b = 24, h = 8$ and $d = 12$
$A = \frac{1}{2}(8)(24 + 12)$
$A = \frac{1}{2}(8)(36)$
$A = 4(36)$
$A = 144$
The area of this trellis is 144 in$^2$.

71. THE WALL

The height of each of the triangles is 245 feet and the base is 10 feet. Since there are two identical triangles, doubling the area for one of the triangles will give the total area.
$A = \frac{1}{2}bh$
$A = \frac{1}{2}(10)(245)$
$A = (5)245$
$A = 1225$
The surface area of one side of "the wall" is 1,225 ft$^2$, the total surface area is double that or 2,450 ft$^2$.

73. RUBBER MEETS THE ROAD

The tire's footprint approximates a rectangle of length 6.375 inches and width of 7.5 inches.

$P = 2l + 2w$

$P = 2(6.375) + 2(7.5)$

$P = 12.75 + 15$

$P = 27.75$

The perimeter of the tire's footprint is 27.75 inches.

$A = lw$

$A = (6.375)(7.5)$

$A = 47.8125$

The area of this tire's footprint is 47.8125 in$^2$.

75. FIREWOOD

The base of the firewood stack is a rectangle that is 8 feet in length and 4 feet wide.

$A = lw$

$A = (8)(4)$

$A = 32$

This stack of wood covers a base of 32 ft$^2$.

The volume of this stack of wood is found using $V = lwh$

$V = (8)(4)(4)$

$V = 32(4)$

$V = 128$

The volume of this stack of wood is 128 ft$^3$.

77. IGLOO

This igloo can be represented by a hemisphere. The volume of a sphere is $V = \frac{4}{3}\pi r^3$

The volume of this igloo is one-half the volume of a sphere.

$\frac{1}{2}V = \frac{1}{2}[\frac{4}{3}\pi r^3]$

$\frac{1}{2}V = \frac{4}{6}\pi r^3$

$\frac{1}{2}V = \frac{4}{6}\pi(5.5)^3$

$\frac{1}{2}V \approx 110.91666\pi$

$\frac{1}{2}V \approx 348.4549852$

The volume of this igloo to the nearest cubic foot is 348 ft$^3$.

79. BARBECUING

Use the fish length as an approximation to the diameter of the circular grill.

$A = \pi r^2$, where $r = \frac{1}{2}d = \frac{1}{2}(18) = 9$

$A = \pi(9)^2$

$A = 81\pi$

$A = 254.4690049$

The area of this grill to the nearest square inch is 254 in$^2$.

81. GEOMETRY

$$a = 180\left(1 - \frac{2}{n}\right)$$

$$\frac{a}{180} = \frac{180\left(1 - \frac{2}{n}\right)}{180}$$

$$\frac{a}{180} = \left(1 - \frac{2}{n}\right)$$

$$\frac{a}{180} - 1 = 1 - 1 - \frac{2}{n}$$

$$\frac{a}{180} - 1 = -\frac{2}{n}$$

$$n\left(\frac{a}{180} - 1\right) = n\left(-\frac{2}{n}\right)$$

$$n\left(\frac{a}{180} - 1\right) = -2$$

$$\frac{n\left(\frac{a}{180} - 1\right)}{\left(\frac{a}{180} - 1\right)} = \frac{-2}{\left(\frac{a}{180} - 1\right)}$$

$$n = \frac{-2}{\left(\frac{a}{180} - 1\right)}$$

Eliminating the fraction in the denominator reduces this equation to,

$$n = \frac{-2}{\left(\frac{a}{180} - \frac{180}{180}\right)}$$

$$n = \frac{-2}{\left(\frac{a - 180}{180}\right)}$$

$$n = -2\left(\frac{180}{a - 180}\right)$$

$$n = \frac{-360}{a - 180}$$

$$n = \frac{360}{180 - a}$$

If $a = 108$ then

$$n = \frac{360}{180 - 108}$$

$$n = \frac{360}{72}$$

$$n = 5$$

If the interior angles are 108 degrees then there are 5 sides to the regular polygon.

**Writing**

83. Answers will vary.
85. Answers will vary.

**Review**

87. 82% of 168 is $p$.

$$0.82(168) = p$$

$$137.76 = p$$

89. $p$ percent of 200 is 30

$$200p = 30$$

$$\frac{200p}{200} = \frac{30}{200}$$

$$p = 0.15$$

15% of 200 is 30.

**Vocabulary**

1.   The **perimeter** of a triangle or a rectangle is the distance around it.
3.   The equal sides of an isoceles triangle meet to form the **vertex** angle.

**Concepts**

5.   PLUMBING
   a. $x = $ length of shortest section
   $x + 2 = $ length of middle sized section
   $3x = $ length of longest section
   b. If $x = 3$ feet, then $x + 2 = 3 + 2 = 5$ feet and $3x = 3(5) = 15$ feet.

7.

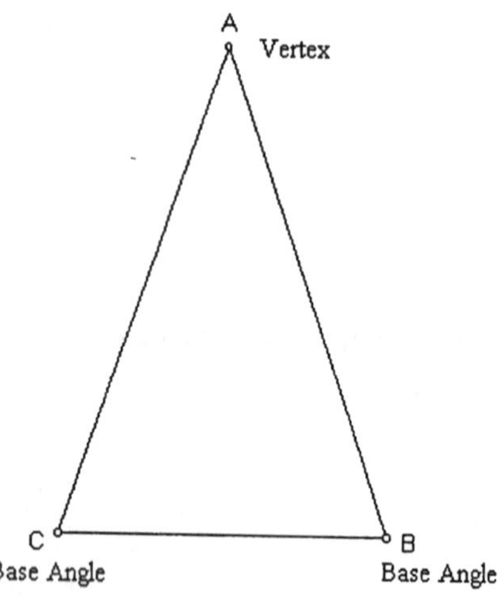


9.   Principal is \$30,000
   rate is 14%
   time is 1 year

11.

|  | $r$ | $t$ | $d = rt$ |
|---|---|---|---|
| Husband | 35 mph | $t$ hr | $35t$ |
| Wife | 45 mph | $t$ hr | $45t$ |

13. a. $0.40(42) = 16.8$ gallons
   b. The third barrel will contain $x + 42$ gallons, the sum of the two barrels.
   c. A reasonable estimate of the acid concentration from combining these two barrels would be 32% because it is between the two given concentrations.

**Practice**

15. $0.08x + 0.07(15,000 - x) = 1,110$
$100(0.08x + 0.07(15,000 - x))$
$\quad = 100(1110)$
$8x + 7(15,000 - x) = 111,000$
$8x + 7(15,000) + 7(-x) = 111,000$
$8x + 105,000 + (-7x) = 111,000$
$x + 105,000 = 111,000$
$x + 105,000 - 105,000$
$\quad = 111,000 - 105,000$
$x = 6,000$

17. $(6x + 2) + 2x = 90$
$6x + 2 + 2x = 90$
$8x + 2 = 90$
$8x + 2 - 2 = 90 - 2$
$8x = 88$
$\frac{8x}{8} = \frac{88}{8}$
$x = 11$
The angle measures are $6(11) + 2 = 68°$, and $2(11) = 22°$.

19. Since vertical angles have the same measure, $(2x + 5) = (3x - 10)$
$2x - 2x + 5 = 3x - 2x - 10$
$5 = x - 10$
$5 = x + (-10)$
$5 + 10 = x + (-10) + 10$
$15 = x$
The angle measures are $2(15) + 5 = 35$, and $3(15) - 10 = 35$, identical.

**Applications**

21. CARPENTRY
Let $x$ be the length of the shorter piece, then the longer piece has length $2x$.
$x + 2x = 12$
$3x = 12$
$x = 4$
The shorter board is 4 feet long and the longer board is 8 feet.

23. SOLAR HEATING
Let $w$ be the width of the first panel then $w + 3.4$ is the length of the second panel.
$w + (w + 3.4) = 18$
$2w + 3.4 = 18$
$2w = 14.6$
$w = 7.3$ feet
$w + 3.4 = 7.3 + 3.4 = 10.7$ feet

25. TOURING

Let $A$ be the length of stay in Australia, then the stay in Japan will be $A + 4$ weeks and Sweden will be $A - 2$ weeks.

$$A + (A + 4) + (A - 2) = 38$$
$$3A + 2 = 38$$
$$3A = 36$$
$$A = 12 \text{ weeks}$$
$$A + 4 = 12 + 4 = 16 \text{ weeks}$$
$$A - 2 = 12 - 2 = 10 \text{ weeks}$$

27. COUNTING CALORIES

Let $c$ be the number of calories in the ice cream, then the calories in the pie are $2c + 100$.

$$c + 2c + 100 = 850$$
$$3c + 100 = 850$$
$$3c = 750$$
$$c = 250 \text{ calories in the ice cream}$$
$$2c + 100 = 2(250) + 100 = 600 \text{ calories in the slice of pie.}$$

29. NET INCOME

Let $q$ be the income in quarter one, quarter two is $q + 182$, quarter three is $(q + 182 - 95)$, and the fourth quarter is $5q$.

$$q + (q + 182) + (q + 182 - 95) + 5q = 1453$$
$$8q + 269 = 1453$$
$$8q = 1184$$
$$q = \$148 \text{ million}$$
$$q + 182 = 148 + 182 = \$330 \text{ million}$$
$$q + 182 - 95 = 330 - 95 = \$235 \text{ million}$$
$$5q = 5(148) = \$740 \text{ million}$$

31. TRUSS

Let $t =$ the length of the third side, then the two equal sides are of length $t - 4$. Since we are told the perimeter is 25 feet,

$$(t - 4) + (t - 4) + t = 25$$
$$3t - 8 = 25$$
$$3t - 8 + 8 = 25 + 8$$
$$3t = 33$$
$$\frac{3t}{3} = \frac{33}{3}$$
$$t = 11, \text{ then } t - 4 = 11 - 4 = 7$$

The three sides of this truss are 11 feet, 7 feet and 7 feet.

33. **SWIMMING POOL**

Let $l$ = the length of this pool and $w$ = the width of this pool. We are told that $l = 6w + 30$ and that the perimeter is 1110.

Using $P = 2l + 2w$ and substituting we get,

$1110 = 2(6w + 30) + 2w$

$1110 = 2(6w) + 2(30) + 2w$

$1110 = 12w + 60 + 2w$

$1110 = 14w + 60$

$1110 - 60 = 14w + 60 - 60$

$1050 = 14w$

$\frac{1050}{14} = \frac{14w}{14}$

$75 = w$

This pool is 75 meters wide and $(6(75) + 30)$ or 480 meters long.

35. **GUY WIRES**

Since the sum of the three angles is $180°$ and we are told that the equal angles are 4 times the measure of the vertex angle $v$, we have,

$4v + 4v + v = 180$

$9v = 180$

$\frac{9v}{9} = \frac{180}{9}$

$v = 20$

The vertex angle is $20°$.

37. **WAREHOUSING COSTS**

Let $P$ = portables, $B$ = big screen and $C$ = consoles.

$P = B + 40, \ C = P - 25$

$276 = 1.5P + 4C + 7.5B$

$276 = 1.5(B + 40) + 4(P - 25) + 7.5B$

$276 = 1.5B + 60 + 4P - 100 + 7.5B$

$276 = 9B - 40 + 4(B + 40)$

$276 = 9B - 40 + 4B + 160$

$276 = 13B + 120$

$276 - 120 = 13B + 120 - 120$

$156 = 13B$

$\frac{156}{13} = \frac{13B}{13}$

$12 = B$

There are 12 big-screen sets in stock.

39. SOFTWARE SALES

Let $S$ = spreadsheet sales, $D$ = database sales and $W$ = word processing sales. Since we are told that $S = D$, and $W = (S + D) + 15$, the sales are:

$72,000 = 150S + 195D + 210[(S + D) + 15]$

Substitute $S = D$

$72,000 = 150S + 195S + 210[(S + S) + 15]$

$72,000 = 345S + 210(2S + 15)$

$72,000 = 345S + 420S + 3150$

$72,000 = 765S + 3150$

$72,000 - 3150 = 765S + 3150 - 3150$

$68,850 = 765S$

$\frac{68,850}{765} = \frac{765S}{765}$

$90 = S$

There were 90 spreadsheet programs sold.

41. INTEREST INCOME

Using $I = Prt$ and knowing that \$12,000 was initially deposited for one year we can set up two equations. Let $x$ = the amount deposited in the 6% account, therefore $(12000 - x)$ represents the amount deposited in the 4.5% account.

$637.50 = x(.06)(1) + (12000 - x)(.045)(1)$

$637.50 = 0.06x + 540 - 0.045x$

$637.50 = 540 + 0.015x$

$637.50 - 540 = 540 - 540 + 0.015x$

$97.5 = 0.015x$

$\frac{97.5}{0.015} = \frac{0.015x}{0.015}$

$6500 = x$

The amount deposited in account 721-94 was $12,000 - x$ or $12,000 - 6500 = \$5500$.

43. INVESTMENTS

Let $p$ = the amount invested in each of the three accounts. Using $I = Prt$ and substituting,

$1249.50 = .07x(1) + .08x(1) + .105x(1)$

$1249.50 = 0.255x$

$\frac{1249.50}{0.255} = \frac{0.255x}{0.255}$

$4900 = x$

There was \$4900 invested in each account.

45. **FINANCIAL PLANNING**

Looking at the 11% account first, using simple interest $I = Prt$ for one year, then

$I = P(0.11)(1)$

$I = 0.11P$

Now look at the 13% account, again using simple interest $I = Prt$, but this account produces $150 more in interest,

$I + 150 = P(0.13)(1)$

$I + 150 = 0.13P$

$I + 150 - 150 = 0.13P - 150$

$I = 0.13P - 150$

Notice in these two equations they are each solved for $I$, now set these equations equal,

$0.11P = 0.13P - 150$

$0.11P - 0.13P = 0.13P - 0.13P - 150$

$-0.02P = -150$

$\dfrac{-0.02P}{-0.02} = \dfrac{-150}{-0.02}$

$P = 7500$

This plumber had $7500 to invest.


47. **TORNADO**

|        | $r$ | $t$ | $d = rt$ |
|--------|-----|-----|----------|
| truck 1 | 25 | $t$ | $25t$ |
| truck 2 | 20 | $t$ | $20t$ |

$25t + 20t = 90$

$45t = 90$

$\dfrac{45t}{45} = \dfrac{90}{45}$

$t = 2$

It will take 2 hours for the trucks to lose contact.


49. **SPEED OF TRAINS**

|        | $r$ | $t$ | $d = rt$ |
|--------|-----|-----|----------|
| train 1 | $r - 20$ | 3 | $3(r - 20)$ |
| train 2 | $r$ | 3 | $3r$ |

Since we are told the trains are 330 miles apart,

$3(r - 20) + 3r = 330$

$3r - 60 + 3r = 330$

$6r - 60 = 330$

$6r - 60 + 60 = 330 + 60$

$6r = 390$

$\dfrac{6r}{6} = \dfrac{390}{6}$

$r = 65$

The speed of train 2 is 65 mph and the speed of train 1 is $r - 20$ or $65 - 20 = 45$ mph.

51. AIR TRAFFIC CONTROL

|            | $r$ | $t$ | $d = rt$ |
|------------|-----|-----|----------|
| to Montreal | 450 | $t$ | $450t$ |
| to Berlin   | 500 | $t$ | $500t$ |

Since there is 3,800 miles between the two cities,

$450t + 500t = 3800$

$950t = 3800$

$\dfrac{950t}{950} = \dfrac{3800}{950}$

$t = 4$

The traffic controllers will have to make the pilots aware of their passing one another 4 hours into the flight.

53. SALT SOLUTION

| % salt | amount |
|--------|--------|
| 3% | $a$ |
| 7% | 50 |
| 5% | $50 + a$ |

$0.03a + 0.07(50) = 0.05(50 + a)$

$0.03a + 3.5 = 2.5 + 0.05a$

$0.03a + 3.5 - 2.5 = 2.5 - 2.5 + 0.05a$

$0.03a + 1 = 0.05a$

$0.03a - 0.03a + 1 = 0.05a - 0.03a$

$1 = 0.02a$

$\dfrac{1}{0.02} = \dfrac{0.02a}{0.02}$

$50 = a$

50 gallons of the 3% solution should be mixed with the 7% solution to create the 5% solution.

55. ANTISEPTIC SOLUTION

| solution | amount |
|----------|--------|
| water (0%) | $w$ |
| 10% | 30 |
| 8% | $w + 30$ |

Note:  Water has 0% solution.

$(0)w + 0.1(30) = 0.08(w + 30)$

$(0)w + 3 = 0.08w + 2.4$

$3 - 2.4 = 0.08w + 2.4 - 2.4$

$0.6 = 0.08w$

$\dfrac{0.6}{0.8} = \dfrac{0.8w}{0.8}$

$w = 7.5$

She must add 7.5 ounces of water.

57. MIXING FUELS

| cost   | amount   |
|--------|----------|
| $1.15  | $g$      |
| $0.85  | 20       |
| $1     | $g + 20$ |

$1.15g + 0.85(20) = 1(g + 20)$

$1.15g + 17 = g + 20$

$1.15g + 17 - 17 = g + 20 - 17$

$1.15g = g + 3$

$1.15g - g = g - g + 3$

$0.15g = 3$

$\dfrac{0.15g}{0.15} = \dfrac{3}{0.15}$

$g = 20$

20 gallons of the $1.15 gasoline mixed with 20 gallons of the $0.85 gasoline yields an average price of $1.00 per gallon of gasoline.

59. MIXING CANDY

| candy       | cost   | amount |
|-------------|--------|--------|
| lemon-drops | $1.90  | $d$    |
| jelly beans | $1.20  | $j$    |
| mix         | $1.48  | 100    |

There are two equations to consider in this problem, the sum of the lemon drops and jelly beans is 100 pounds and then consider the cost equation, $\$1.90d + \$1.20j = \$1.48(100)$.

$(1)\, d + j = 100$

$(2)\, 1.90d + 1.20j = 1.48(100)$

Solving (1) for $d$ yields, $d = 100 - j$, now substitute this into (2).

$1.90(100 - j) + 1.20j = 1.48(100)$

$190 - 1.9j + 1.20j = 148$

$190 - 0.7j = 148$

$190 - 190 - 0.7j = 148 - 190$

$-0.7j = -42$

$\dfrac{-0.7j}{-0.7} = \dfrac{-42}{-0.7}$

$j = 60$

The mixed candy should contain 60 pounds of jelly beans and $d = 100 - 60 = 40$ pounds of lemon-drops.

61. BLENDING COFFEE

| cost | amount   |
|------|----------|
| $4   | $C$      |
| $7   | 40       |
| $5   | $C + 40$ |

$4C + 7(40) = 5(C + 40)$

$4C + 280 = 5C + 200$

$4C + 280 - 200 = 5C + 200 - 200$

$4C + 80 = 5C$

$4C - 4C + 80 = 5C - 4C$

$80 = C$

The shopkeeper should have mixed 80 pounds of the $4 coffee with the $7 coffee to have a $5 per pound blend.

**Writing**

63. Answers will vary.
65. Answers will vary.

**Review**

67. $-25(2x - 5)$
$= -25(2x) - 25(-5)$
$= -50x + 125$

69. $-(-3x - 3)$
$= -(-3x) - (-3)$
$= 3x + 3$

71. $8p - 9q + 11p + 20q$
$= 19p + 11q$

**Section 8.6 Inequalities**

**Vocabulary**

1.  An expression containing one of the symbols $>$, $<$, $\leq$, $\geq$, or $\neq$ is called an **inequality**.
3.  A **solution** of an inequality is an real number that makes the inequality true.

**Concepts**

5.  a. True
    b. True
    c. False
    d. False
    e. True
    f. False

7.  If a quantity is added to or subtracted from both sides of an inequality, the resulting inequality will have the **same** direction as the original one.

9.  If both sides of an inequality are multiplied or divided by a negative number, the resulting inequality will have the **opposite** direction as the original one.

11. a. A true statement will result if $3$ is substituted.
    b. A false statement will result if $-3$ is substituted.

13. a. $2x - 4 > 12$
    $$2x - 4 + 4 > 12 + 4$$
    $$2x > 16$$
    $$\frac{2x}{2} > \frac{16}{2}$$
    $$x > 8$$
    Any real number greater than eight will satisfy this inequality.

    b. 

    c. $(8, \infty)$

**Notation**

15. The symbol $<$ means " **is less than**" and the symbol $>$ means " **is greater than**."

17. The symbol $\neq$ means " **is not equal to**."

19. $x > -2$

21. $-2 \leq 17$

23. $4x - 5 \geq 7$
    $$4x - 5 + 5 \geq 7 + 5$$
    $$4x \geq 12$$
    $$\frac{4x}{4} \geq \frac{12}{4}$$
    $$x \geq 3$$

**Practice**

25. $x < 5$

$$\xleftarrow{\qquad}\overset{5}{)}\xrightarrow{\qquad}$$

$(-\infty, 5)$

27. $-3 < x \le 1$

$$\xleftarrow{\quad}\overset{-3}{(}\rule{2cm}{0.4pt}\overset{1}{]}\xrightarrow{\quad}$$

$(-3, 1]$

29. $x < -1$, $(-\infty, -1)$

31. $-7 < x \le 2$, $(-7, 2]$

33. $x + 2 > 5$

$x + 2 - 2 > 5 - 2$

$x > 3$

$(3, \infty)$

$$\xleftarrow{\quad}\overset{3}{(}\xrightarrow{\quad}$$

35. $-x - 3 \le 7$

$-x - 3 + 3 \le 7 + 3$

$-x \le 10$

$\frac{-x}{-1} \ge \frac{10}{-1}$

$x \ge -10$

$[-10, \infty)$

$$\xleftarrow{\quad}\overset{-10}{[}\xrightarrow{\quad}$$

37. $3 + x < 2$

$3 - 3 + x < 2 - 3$

$x < -1$

$(-\infty, -1)$

$$\xleftarrow{\quad}\overset{-1}{)}\xrightarrow{\quad}$$

39. $2x - 0.3 \le 0.5$

$2x - 0.3 + 0.3 \le 0.5 + 0.3$

$2x \le 0.8$

$\frac{2x}{2} \le \frac{0.8}{2}$

$x \le 0.4$

$(-\infty, 0.4]$

$$\xleftarrow{\quad}\overset{0.4}{]}\xrightarrow{\quad}$$

41. $-3x - 7 > -1$
$-3x - 7 + 7 > -1 + 7$
$-3x > 6$
$\frac{-3x}{-3} < \frac{6}{-3}$
$x < -2$
$(-\infty, -2)$

$$\xleftarrow{\qquad} \overset{-2}{)} \xrightarrow{\qquad}$$

43. $-4x + 6 > 17$
$-4x + 6 - 6 > 17 - 6$
$-4x > 11$
$\frac{-4x}{-4} < \frac{11}{-4}$
$x < -\frac{11}{4}$
$(-\infty, -\frac{11}{4})$

$$\xleftarrow{\qquad} \overset{-\frac{11}{4}}{)} \xrightarrow{\qquad}$$

45. $\frac{y}{4} + 1 \le -9$
$\frac{y}{4} \le -10$
$y \le -40$
$(\infty, -40]$

$$\xleftarrow{\qquad} \overset{-40}{)} \xrightarrow{\qquad}$$

47. $-\frac{1}{2}n \ge -1$
$n \le 2$
$(\infty, 2]$

$$\xleftarrow{\qquad} \overset{2}{]} \xrightarrow{\qquad}$$

49. $\frac{x}{-42} - 1 > -1$
$\frac{x}{-42} > 0$
$x > 0$
$(0, \infty)$

$$\xleftarrow{\qquad} \overset{0}{(} \xrightarrow{\qquad}$$

51. $\frac{2}{3}x \ge 2$
$\frac{3}{2}(\frac{2}{3}x) \ge \frac{3}{2}(2)$
$x \ge 3$
$[3, \infty)$

$$\xleftarrow{\qquad} \overset{3}{[} \xrightarrow{\qquad}$$

53. $-\frac{7}{8}x \le 21$
$(-\frac{8}{7})(-\frac{7}{8}x) \ge (-\frac{8}{7})21$
$x \ge -8(3)$
$x \ge -24$
$[-24, \infty)$

$$\xleftarrow{\qquad} \overset{-24}{[} \xrightarrow{\qquad}$$

55. $2x + 9 \leq x + 8$
$2x + 9 - 9 \leq x + 8 - 9$
$2x \leq x - 1$
$2x - x \leq x - x - 1$
$x \leq -1$
$(-\infty, -1]$

57. $9x + 13 \geq 8x$
$9x - 8x + 13 \geq 8x - 8x$
$x + 13 \geq 0$
$x + 13 - 13 \geq -13$
$x \geq -13$
$[-13, \infty)$

59. $8x + 4 > 3x + 4$
$8x - 3x + 4 > 3x - 3x + 4$
$5x + 4 > 4$
$5x + 4 - 4 > 4 - 4$
$5x > 0$
$\frac{5x}{5} > \frac{0}{5}$
$x > 0$
$(0, \infty)$

61. $5x + 7 < 2x + 1$
$5x - 2x + 7 < 2x - 2x + 1$
$3x + 7 < 1$
$3x + 7 - 7 < 1 - 7$
$3x < -6$
$\frac{3x}{3} < \frac{-6}{3}$
$x < -2$
$(-\infty, -2)$

63. $7 - x \leq 3x - 2$
$7 - x + x \leq 3x + x - 2$
$7 \leq 4x - 2$
$7 + 2 \leq 4x - 2 + 2$
$9 \leq 4x$
$\frac{9}{4} \leq \frac{4x}{4}$
$\frac{9}{4} \leq x$
$[\frac{9}{4}, \infty)$

65. $3(x - 8) < 5x + 6$

$3x + 3(-8) < 5x + 6$

$3x - 24 < 5x + 6$

$3x - 3x - 24 < 5x - 3x + 6$

$-24 < 2x + 6$

$-24 - 6 < 2x + 6 - 6$

$-30 < 2x$

$\frac{-30}{2} < \frac{2x}{2}$

$-15 < x$

$(-15, \infty)$

67. $8(5 - x) \leq 10(8 - x)$

$8(5) - 8x \leq 10(8) - 10x$

$40 - 8x \leq 80 - 10x$

$40 - 8x + 10x \leq 80 - 10x + 10x$

$40 + 2x \leq 80$

$40 - 40 + 2x \leq 80 - 40$

$2x \leq 40$

$\frac{2x}{2} \leq \frac{40}{2}$

$x \leq 20$

$(-\infty, 20]$

69. $\frac{1}{2} + \frac{x}{5} > \frac{3}{4}$

$\frac{x}{5} > \frac{1}{4}$

$x > \frac{5}{4}$

$\left(\frac{5}{4}, \infty\right)$

71. $-\frac{2}{3} \geq \frac{2y}{3} - \frac{3}{4}$

$\frac{1}{12} \geq \frac{2y}{3}$

$\frac{1}{8} \geq y$

73. $2 < x - 5 < 5$

$2 + 5 < x - 5 + 5 < 5 + 5$

$7 < x < 10$

$(7, 10)$

75. $-5 < x + 4 \leq 7$

$-5 - 4 < x + 4 - 4 \leq 7 - 4$

$-9 < x \leq 3$

$(-9, 3]$

77. $0 \le x + 10 \le 10$
$0 - 10 \le x + 10 - 10 \le 10 - 10$
$-10 \le x \le 0$
$[-10, 0]$

79. $4 < -2x < 10$
$\frac{4}{-2} > \frac{-2x}{-2} > \frac{10}{-2}$
$-2 > x > -5$
$-5 < x < -2$
$(-5, -2)$

81. $-3 \le \frac{x}{2} \le 5$
$2(-3) \le 2\left(\frac{x}{2}\right) \le 2(5)$
$-6 \le x \le 10$
$[-6, 10]$

83. $3 \le 2x - 1 < 5$
$3 + 1 \le 2x - 1 + 1 < 5 + 1$
$4 \le 2x < 6$
$\frac{4}{2} \le \frac{2x}{2} < \frac{6}{2}$
$2 \le x < 3$
$[2, 3)$

85. $0 < 10 - 5x \le 15$
$0 - 10 < 10 - 10 - 5x \le 15 - 10$
$-10 < -5x \le 5$
$\frac{-10}{-5} > \frac{-5x}{-5} \ge \frac{5}{-5}$
$2 > x \ge -1$
$-1 \le x < 2$
$[-1, 2)$

87. $0.6(0.5x - 2.94) < -1.353$
$0.5x - 2.94 < -2.255$
$.5x < 0.685$
$x < 1.37$
$(\infty, 1.37)$

89. $9(0.05 - 0.3x) + 0.162 \leq 0.081 + 15x$
$0.45 - 2.7x + 0.162 \leq 0.081 + 15x$
$0.531 \leq 17.7x$
$0.03 \leq x$
$[0.03, \infty)$

<br>

**Applications**

91. CALCULATING GRADES

Let $t$ represent the next test score;
$\frac{68+75+79+t}{4} \geq 80$
$4\left(\frac{68+75+79+t}{4}\right) \geq 4(80)$
$68 + 75 + 79 + t \geq 320$
$222 + t \geq 320$
$222 - 222 + t \geq 320 - 222$
$t \geq 98$
In order for this student to earn a B, her last exam score must be greater than or equal to 98%.

93. FLEET AVERAGES

Let $f$ represent the third model;
$\frac{17+19+f}{3} \geq 21$
$3\left(\frac{17+19+f}{3}\right) \geq 3(21)$
$17 + 19 + f \geq 63$
$36 + f \geq 63$
$36 - 36 + f \geq 63 - 36$
$f \geq 27$
For the fleet average to be at least 21 mpg the third model must have an economy rating of 27 mpg or more.

95. DOING HOMEWORK

Since there are 7 days in a week, 1 hours per day of Spanish homework is 60 minutes per day.
Therefore, the weeks Spanish work is
$S \geq 7(60)$
$S \geq 420$ minutes

97. SAFETY CODE

a. Ramps or inclines, let $R$ represent the angle, $0° < R \leq 18°$
b. Stairs, let $S$ represent the stair angle, $18° \leq S \leq 50°$
c. Preferred range for stairs, let $P$ represent this range; $30° \leq P \leq 37°$
d. Ladders with cleats, let $L$ represent this range; $75° \leq L < 90°$

99. LAND ELEVATIONS

Let $L$ represent the land elevations in Nevada.
a. $470 \leq L \leq 13,143$
b. Using 5,280 feet per mile,
$470$ feet $= \frac{470}{5280} \approx 0.08901515$ miles
$13,143$ feet $= \frac{13,143}{5280} \approx 2.4892045$ miles
To the nearest tenth
$0.1 \leq L \leq 2.5$

101. DRAFTING

Let $p$ represent the width of the manufactured plug;

$1.497 + 0.001 = 1.498$

$1.497 - 0.001 = 1.496$

$1.496 \le p \le 1.498$

Let $O$ represent the plug opening;

$1.5005 + 0.0005 = 1.5010$

$1.5005 - 0.0005 = 1.5000$

$1.5000 \le O \le 1.5010$

**Writing**

103. Answers will vary

**Review**

105. $-5^3 = -(5)(5)(5) = -125$

107.

| $x$ | $x^2 - 3$ |
| --- | --- |
| $-2$ | $(-2)^2 - 3 = 4 - 3 = 1$ |
| $0$ | $0^2 - 3 = -3$ |
| $3$ | $3^2 - 3 = 9 - 3 = 6$ |

1.  a. $-3x + 2 + 5x - 10$
    $-3x + 5x + 2 - 10$
    $2x - 8$
    b. $-3x + 2 + 5x - 10 = 4$
    $2x - 8 = 4$
    $2x - 8 + 8 = 4 + 8$
    $2x = 12$
    $\frac{2x}{2} = \frac{12}{2}$
    $x = 6$

3.  a. $\frac{1}{3}a + \frac{1}{3}a = \frac{2}{3}a$
    b. $\frac{1}{3}a + \frac{1}{3} = \frac{1}{2}$
    $\frac{1}{3}a + \frac{1}{3} - \frac{1}{3} = \frac{1}{2} - \frac{1}{3}$
    $\frac{1}{3}a = \frac{3}{6} - \frac{2}{6}$
    $\frac{1}{3}a = \frac{1}{6}$
    $3\left(\frac{1}{3}a\right) = 3\left(\frac{1}{6}\right)$
    $a = \frac{1}{2}$

5.  $2(x + 3) - x - 12$
    $2x + 2(3) - x - 12$
    $2x + 6 - x - 12$
    $2x - x + 6 - 12$
    $x - 6$
    This is an expression that the student began to simplify until the third line where the student set the expression equal to zero, thus creating an incorrect equation.

**Section 8.1**

1.   a. 25 more than the height $h$ is $h + 25$
    b. 15 less than the cutoff score $s$ is $s - 15$
    c. $\frac{1}{2}$ of the time $t$ is $\frac{1}{2}t$
    d. the product of 6 and $x$ is $6x$

3.   a. There are $10d$ years in $d$ decades.
    b. $x$ donuts is $\frac{x}{12}$ dozens of donuts
    c. If the house is $x$ years old then the patio is $x - 5$ years old.

5.

| $x$ | $20x - x^3$ |
|---|---|
| 0 | $20(0) - 0^3 = 0$ |
| 1 | $20(1) - 1^3 = 19$ |
| 4 | $20(4) - 4^3 = 80 - 64 = 16$ |

7.   The volume of the waffle cone using $\frac{\pi r^2 h}{3}$ is
$$V = \frac{\pi(1.5)^2(7.5)}{3}$$
$$V = \frac{\pi(2.25)(7.5)}{3}$$
$$V = \frac{53.01437603}{3}$$
$$V = 17.67145868 \text{ rounded to the nearest tenth is } V = 17.7 \text{ in}^3$$

**Section 8.2**

9.   a. $5(x + 3) = 5x + 5(3) = 5x + 15$
    b. $-2(2x + 3 - y)$
$$= -2(2x) - 2(3) - 2(-y)$$
$$= -4x - 6 + 2y$$
    c. $-(a - 4)$
$$= (-1)(a) - (-1)(4)$$
$$= -a + 4$$
    d. $\frac{3}{4}(4c - 8)$
$$= \frac{3}{4}(4c) - \frac{3}{4}(8)$$
$$= 3c - 6$$

11.  a. The coefficients of $2x - 5$ are 2 and $-5$.
    b. The coefficients of $16x^2 - 5x + 25$ are $16, -5$, and 25.
    c. The coefficients of $\frac{1}{2}x + y$ are $\frac{1}{2}$ and 1.
    d. The coefficients of $9.6t^2 - t$ are 9.6 and $-1$.

13. The perimeter of the triangle is $(x + 7) + (2x - 3) + x$
$$= x + x + 2x + 7 - 3$$
$$= (4x + 4) \text{ feet.}$$

15. Let $m$ represent the CD player markup, then $r = c + m$
$$395 = 219 + m$$
$$395 - 219 = 219 - 219 + m$$
$$176 = m$$
The markup on this CD player is $176.

17. Let $t$ represent the time to complete the race, then, $d = rt$ becomes, $t = \frac{d}{r}$
$$t = \frac{500}{147.956}$$
$$t = 3.37938306$$
To the nearest hundredth, it took the winner 3.38 hours to complete the race.

19. Using the formula for perimeter, $P = 2l + 2w$ where $l = 60$ in. and $w = 24$in.
$$P = 2(60) + 2(24)$$
$$P = 120 + 48$$
$$P = 168$$
The perimeter of this mattress is 168 inches.

21. Using the formula for the area of a triangle, $A = \frac{1}{2}bh$, where $b = 17$m and $h = 9$m,
$$A = \frac{1}{2}(17)(9)$$
$$A = \frac{1}{2}(153)$$
$$A = 76.5$$
The area for this triangle is 76.5 m$^2$.

23. Using the formula for the circumference of a circle, $C = 2\pi r$, where $r = 8$ is,
$$C = 2\pi(8)$$
$$C = 16\pi$$
$$C \approx 50.26548246$$
To the nearest hundredth the circumference of this circle is 50.27 cm.

25. Using the formula for volume of a rectangular solid, $V = bwh$, where $b = 24$, $w = 60$, and $h = 3$
$$B = 24(60)(3)$$
$$B = 4320$$
The volume of this air mattress is 4,320 in$^3$.

27. Using the formula for volume of a pyramid, $V = \frac{1}{3}Bh$, where $B = s^2$ and $s = 6$ and $h = 10$,
$$V = \frac{1}{3}(6^2)(10)$$
$$V = \frac{1}{3}(360)$$
$$V = 120$$
The volume of this pyramid is 120 ft$^3$.

29. a. $A = 2\pi rh$ for $h$
$$\frac{A}{2\pi r} = \frac{2\pi rh}{2\pi r}$$
$$\frac{A}{2\pi r} = h$$
b. $P = 2l + 2w$ for $l$
$$P - 2w = 2l + 2w - 2w$$
$$P - 2w = 2l$$
$$\frac{P-2w}{2} = \frac{2l}{2}$$
$$\frac{P-2w}{2} = l$$

31. UTILITY BILLS

    Let $k$ represent the kilowatt hour of energy used, then

    $43.96 = 17.50 + 0.18k$

    $43.96 - 17.50 = 17.50 - 17.50 + 0.18k$

    $26.46 = 0.18k$

    $\dfrac{26.46}{0.18} = \dfrac{0.18k}{0.18}$

    $147 = k$

    This resident used 147 kilowatts of energy.


33. A property of isoceles triangles is that measures of the base angles are equal. Use this fact with the sum of the internal angles of a triangle being $180°$. Let the measure of the base angles be represented by $b$,

    $180 = b + b + 27$

    $180 = 2b + 27$

    $180 - 27 = 2b + 27 - 27$

    $153 = 2b$

    $\dfrac{153}{2} = \dfrac{2b}{2}$

    $76.5 = b$

    The measure of the base angles is $76.5°$.


35. INVESTMENT INCOME

    Using the formula $I = Prt$, where $I = 2110$, $P_{mm}$ represents the amount put in the money market account and $27000 - P_{mm}$ is the amount placed in the cash management account, then

    $2110 = P_{mm}(0.07)(1) + (27000 - P_{mm})(0.09)(1)$

    $2110 = 0.07P_{mm} + (27000)(0.09) - P_{mm}(0.09)$

    $2110 = 0.07P_{mm} + 2430 - 0.09P_{mm}$

    $2110 = -0.02P_{mm} + 2430$

    $2110 - 2430 = -0.02P_{mm} + 2430 - 2430$

    $-320 = -0.02P_{mm}$

    $\dfrac{-320}{-0.02} = \dfrac{-0.02P_{mm}}{-0.02}$

    $16000 = P_{mm}$

    $\$16,000$ was invested in the 7% account and $27000 - 16000 = \$11,000$ invested in the 9% account.

37. MIXTURE

| Type | Cost | Amount |
|---|---|---|
| mix | $0.90 | $m$ |
| gumdrops | $1.50 | $g$ |
| new mixture | $1.20 | 20 |

(1) $m + g = 20$

(2) $0.90m + 1.50g = 1.20(20)$

Solve equation (1) for $g$ and substitute in (2),

$m + g - g = 20 - g$

$m = 20 - g$

Equation (2) becomes,

$0.90(20 - g) + 1.50g = 1.20(20)$

$0.90(20) - 0.90(g) + 1.50g = 24$

$18 - 0.9g + 1.50g = 24$

$18 + 0.6g = 24$

$18 - 18 + 0.6g = 24 - 18$

$0.6g = 6$

$\frac{0.6g}{0.6} = \frac{6}{0.6}$

$g = 10$

The new mixture is comprised of 10 pounds of gumdrops and 10 pounds of the ninety cents mix.

**Section 8.6**

39. a. $3x + 2 < 5$

$3x + 2 - 2 < 5 - 2$

$3x < 3$

$\frac{3x}{3} < \frac{3}{3}$

$x < 1$

$(-\infty, 1)$

b. $-5x - 8 > 7$

$-5x - 8 + 8 > 7 + 8$

$-5x > 15$

$\frac{-5x}{-5} > \frac{15}{-5}$

$x < -3$

$(-\infty, 3)$

c. $5x - 3 \geq 2x + 9$

$5x - 3 + 3 \geq 2x + 9 + 3$

$5x \geq 2x + 12$

$5x - 2x \geq 2x - 2x + 12$

$3x \geq 12$

$\frac{3x}{3} \geq \frac{12}{3}$

$x \geq 4$

$[4, \infty)$

d. $7x + 1 \leq 8x - 5$
$7x - 7x + 1 \leq 8x - 7x - 5$
$1 \leq x - 5$
$1 + 5 \leq x - 5 + 5$
$6 \leq x$
$[6, \infty)$

e. $5(3 - x) \leq 3(x - 3)$
$5(3) - 5(x) \leq 3(x) - 3(3)$
$15 - 5x \leq 3x - 9$
$15 - 5x + 5x \leq 3x + 5x - 9$
$15 \leq 8x - 9$
$15 + 9 \leq 8x - 9 + 9$
$24 \leq 8x$
$\frac{24}{8} \leq \frac{8x}{8}$
$3 \leq x$
$[3, \infty)$

f. $-\frac{3}{4}x \geq -9$
$-\frac{4}{3}\left(-\frac{3}{4}x\right) \geq -\frac{4}{3}(-9)$
$x \leq \frac{36}{3}$
$x \leq 12$
$(-\infty, 12]$

g. $8 < x + 2 < 13$
$8 - 2 < x + 2 - 2 < 13 - 2$
$6 < x < 11$
$(6, 11)$

h. $0 \leq 2 - 2x < 6$
$0 - 2 \leq 2 - 2 - 2x < 6 - 2$
$-2 \leq -2x < 4$
$\frac{-2}{-2} \geq \frac{-2x}{-2} > \frac{4}{-2}$
$1 \geq x > -2 \text{ or}$
$-2 < x \leq 1$
$(-2, 1]$

41. SPORTS EQUIPMENT

Let $w$ represent the weight of a ping-pong ball, 2.40 grams $< w <$ 2.53 grams

**Chapter 8 Test**

1. $x - 2 =$ is the number of songs on the CD.

3.

| $x$ | $2x - \dfrac{30}{x}$ |
|---|---|
| 5 | $2(5) - \dfrac{30}{5} = 10 - 6 = 4$ |
| 10 | $2(10) - \dfrac{30}{10} = 20 - 3 = 17$ |
| $-30$ | $2(-30) - \dfrac{30}{-30} = -60 + 1 = -59$ |

5. 6

7. $x$ is used as a factor

9. $5(-4x) = -20x$

11. $3(x + 2) + 3(4 - x)$
$= 3x + 3(2) + 3(4) + 3(-x)$
$= 3x + 6 + 12 - 3x$
$= 3x - 3x + 6 + 12$
$= 18$

13. $12x + 4 = -140$
$12x = -144$
$\dfrac{12x}{12} = \dfrac{-144}{12}$
$x = -12$

15. $6m + 2 = -14 - 2m$
$8m = -16$
$m = -2$

17. $\dfrac{m}{2} - \dfrac{1}{3} = \dfrac{1}{4}$
$12\left(\dfrac{m}{2} - \dfrac{1}{3}\right) = 12\left(\dfrac{1}{4}\right)$
$12\left(\dfrac{m}{2}\right) - 12\left(\dfrac{1}{3}\right) = 3$
$6m - 4 = 3$
$6m + (-4) + 4 = 3 + 4$
$6m = 7$
$\dfrac{6m}{6} = \dfrac{7}{6}$
$m = \dfrac{7}{6}$

19. $A = P + Prt$; for $r$
$A - P = P - P + Prt$
$A - P = Prt$
$\dfrac{A-P}{Pt} = \dfrac{Prt}{Pt}$
$\dfrac{A-P}{Pt} = r$

21. $F = \dfrac{9}{5}C + 32$
$\dfrac{5(F-32)}{9} = C$
$\dfrac{5(14-32)}{9} = C$
$-10^0 = C$

23. **TRAVEL TIMES**

Using $d = rt$

|       | $r$ | $t$ | $d$  |
|-------|-----|-----|------|
| Car   | 65  | $t$ | $65t$ |
| Truck | 55  | $t$ | $55t$ |

$65t + 55t = 72$

$120t = 72$

$t = \frac{72}{120}$

$t = \frac{3}{5}$

In $\frac{3}{5}$ of an hour the car from Rockford will meet the truck traveling from Madison.

25. **GEOMETRY**

The sum of the interior angles of a triangle is $180^o$. A property of isoceles triangles is that the base angles are equal, let $b$ represent the measure of the base angles.

$b + b + 44 = 180$

$2b + 44 - 44 = 180 - 44$

$2b = 136$

$\frac{2b}{2} = \frac{136}{2}$

$b = 68$

The measure of the base angles of this isoceles triangle is $68^o$.

27. $-8x - 20 \leq 4$

$-8x - 20 + 20 \leq 4 + 20$

$-8x \leq 24$

$\frac{-8x}{-8} \geq \frac{24}{-8}$

$x \geq -3 \text{ or } -3 \leq x$

$[-3, \infty)$

29. If a solved value is a true solution then substituting that value into the original problem will result in a true statement.

1.  a. This is an expression.
    b. This is an equation.

3.  $200 = 2^3 \cdot 5^2$

5.  $\frac{11}{21}\left(-\frac{14}{33}\right) = \frac{11}{3 \cdot 7}\left(-\frac{2 \cdot 7}{3 \cdot 11}\right) = -\frac{2}{9}$

7.  $\frac{4}{5} + \frac{2}{3} = \frac{12}{15} + \frac{10}{15} = \frac{22}{15} = 1\frac{7}{15}$

9.  $\frac{15}{16} = 0.9375$

11. a. $\left|-65\right| = 65$
    b. $-\left|-12\right| = -12$

13. 3 is real, natural, whole, rational, integer.

15. $\frac{17}{20}$ is a real, rational.

17. a. $4 \cdot 4 \cdot 4 = 4^3$
    b. $\pi r \cdot r \cdot h = \pi r^2 h$

19. a. The sum of the width $w$ and 12 is $w + 12$.
    b. Four less than a number $n$ is $n - 4$.

21.

| $x$ | $x^2 - 3$ |
|---|---|
| $-2$ | $(-2)^2 - 3 = 4 - 3 = 1$ |
| $0$ | $0^2 - 3 = -3$ |
| $3$ | $3^2 - 3 = 9 - 3 = 6$ |

23. LAND OF THE RISING SUN
    a. The area of the flag is $A = 3(2) = 6 \text{ ft}^2$.
    b. The area of the red disc is
    $A = \pi(0.625)^2$
    $A \approx 1.2 \text{ ft}^2$
    c. $\frac{1.2}{6} = 0.2$, Approximately 20% of the flag is occupied by the red disc.

25. $x = -5, y = 3, x = 0$
    $(3(-5) - 2(3))0 = 0$

27. $x = -5, y = 3, x = 0$
    $(-5)^2 - 3^2 + 0^2 = 25 - 9 = 16$

29. $-8(4d) = -32d$

31. $2x + 3x = 5x$

33. $q(q - 5) + 7q^2 = q^2 - 5q + 7q^2 = 8q^2 - 5q$

35. The length of the longest side is $(x+3)$ feet.

37. $3x - 4 = 23$
$\quad 3x = 27$
$\quad x = 9$

39. $-5p + 0.7 = 3.7$
$\quad -5p = 3$
$\quad p = -\frac{3}{5}$

41. $-\frac{4}{5}x = 16$
$\quad x = -\frac{16 \cdot 5}{4}$
$\quad x = -20$

43. $9y - 3 = 6y$
$\quad 3y - 3 = 0$
$\quad 3y = 3$
$\quad y = 1$

45. $A = 5(13)$
$\quad A = 65 \text{ ft}^2$

47. $A = P + Prt$
$\quad A - P = Prt$
$\quad \frac{A-P}{Pr} = \frac{Prt}{Pr}$
$\quad \frac{A-P}{Pr} = t$

49. WORK
$\quad W = Fd$
$\quad 28.35 = F(3)$
$\quad 9.45 \text{ lb} = F$
The paint can weights 9.45 pounds.

51. INVESTING

| Account | rate | amount | I |
|---|---|---|---|
| 9% | 0.09 | $x$ | $0.09x$ |
| 8% | 0.08 | $10000 - x$ | $0.08(10000 - x)$ |

$0.09x + 0.08(10000 - x) = 860$
$0.09x + 800 - 0.08x = 860$
$0.01x = 60$
$x = 6000$
They invested \$6000 in the 9% account and $10000 - 6000 = \$4000$ in the 8% account.

53. $x - 4 > -6$
$\quad x > -2$
$\quad (-2, \infty)$

55. $8x + 4 \geq 5x + 1$

$3x \geq -3$

$x \geq -1$

$[-1, \infty)$

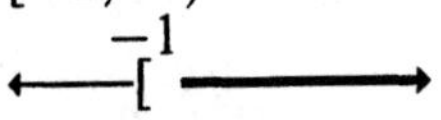

### Vocabulary

1. The pair of numbers $(-1, -5)$ is called an **ordered** pair.
3. The point with coordinates $(0, 0)$ is called the **origin**.
5. The point with coordinates $(4, 2)$ can be graphed on a **rectangular** coordinate system.

### Concepts

7. To plot the point with coordinates$(-5, 4.5)$, we start at the **origin** and move 5 units to the **left** and then move 4.5 units **up**.

9. $(3, 2)$ and $(2, 3)$ do not represent the same point.
11. Points with a negative $x$-coordinate and a positive $y$-coordinate lie in quadrant II.

13.

| $x$ | $y$ |
| --- | --- |
| 4 | 3 |
| 3 | $-4$ |
| 5 | 0 |
| $-3$ | $-4$ |
| $-5$ | 0 |
| 0 | 5 |

15. The point $(-10, 60)$ tells us that 10 minutes prior to the workout this woman's heart rate was 60 beats per minute.

17. One-half hour after starting this workout the woman's heart rate was 150 beats per minute.

19. There were two times that her heart rate was 100 beats per minute, about 5 minutes into the workout and 50 minutes into the workout.

21. The woman's heart rate before the workout was 60 beats per minute and after the workout it was 70 beats per minute, therefore the difference was 10 beats per minute.

### Notation

23. $(3, 5)$ is an ordered pair
    $3(5)$ denotes multiplication
    $5(3 + 5)$ is a mathematical expression

25. These three ordered pairs do name the same point.

**Practice**

27.

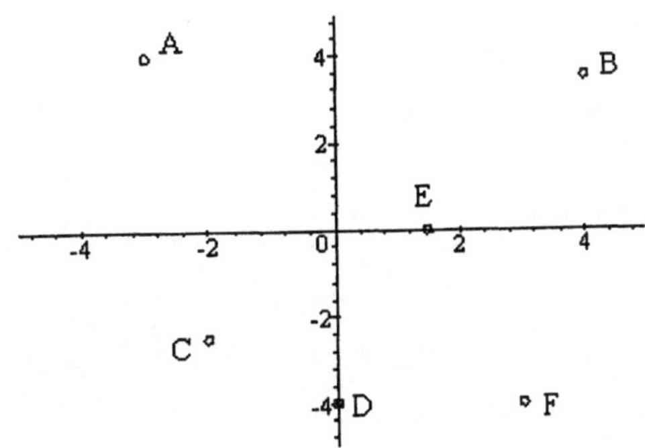

29. $\frac{5+7}{2} = 6$ and $\frac{3+9}{2} = 6$ therefore the midpoint is $(6, 6)$.

31. $\frac{2+(-3)}{2} = -\frac{1}{2}$ and $\frac{-7+12}{2} = \frac{5}{2}$ therefore the midpoint is $\left(-\frac{1}{2}, \frac{5}{2}\right)$.

33. $\frac{4+10}{2} = 7$ and $\frac{6+6}{2} = 6$ therefore the midpoint is $(7, 6)$.

**Applications**

35. CONSTRUCTION

| rivets | welds | anchors |
|--------|-------|---------|
| $(-6, 0)$ | $(-4, 3)$ | $(-6, -3)$ |
| $(-2, 0)$ | $(4, 3)$ | $(6, -3)$ |
| $(2, 0)$ | $(0, 3)$ | |
| $(6, 0)$ | | |

37. GOLF SWING
The points highlighted, starting from the left and moving right are
$(-3, 10), (-2, 7), (-1, 4.9), (0, 3), (1, 1.8), (2.5, 0.5), (4, 0)$

39. VIDEO RENTAL
a. The charge for a 1 day rental is $2.
b. The charge for a 2 day rental is $4.
c. The charge for a 5 day rental is $7.
d. The charge for a 7 day rental is $9

41. GAS MILEAGE

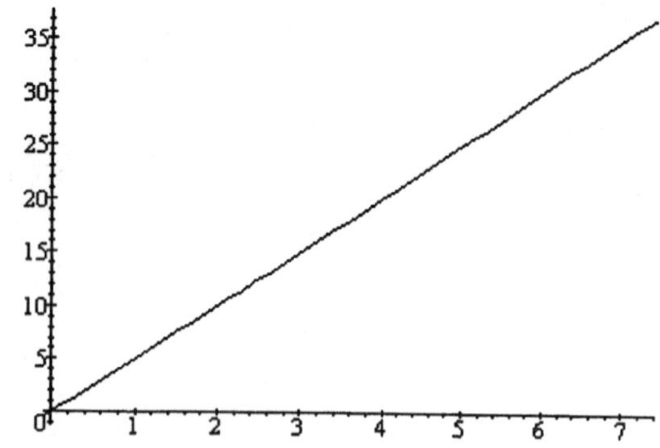

a. The truck can go about 35 miles on 7 gallons of fuel.
b. About 4 gallons of fuel are needed to travel 20 miles.
c. On 6.5 gallons of fuel the truck can travel about 32 miles.

43. ROAD MAPS

| Location | (number, letter) |
| --- | --- |
| Rockford | (5, B) |
| Mt. Carroll | (1, C) |
| Harvard | (7, A) |
| 251/30 | (5, E) |

**Writing**

45. Answers will vary.
47. Answers will vary.

**Review**

49. $-3 - 3(-5) = -3 + 15 = 12$
51. The opposite of $-8$ is $8$.

53. $-4x + 7 = -21$
$-4x + 7 - 7 = -21 - 7$
$-4x = -28$
$\frac{-4x}{-4} = \frac{-28}{-4}$
$x = 7$

55. $(x+1)(x+y)^2$ for $x = -2, y = -5$
$(-2+1)(-2+(-5))^2$
$= (-1)(-7)^2$
$= -49$

**Vocabulary**

1. The equation $y = x + 1$ is an equation in **two** variables.
3. In equations containing the variables $x$ and $y$, $x$ is called the **independent** variable and $y$ is called the **dependent** variable.

**Concepts**

5. a. $y = -2x + 6$ has two variables.
   b. $(4, -2)$ does satisfy the equation.
   $$-2 \overset{?}{=} -2(4) + 6$$
   $$-2 \overset{?}{=} -8 + 6$$
   $$-2 = -2$$
   c. $x = -3, y = 12$ is a solution.
   $$12 \overset{?}{=} -2(-3) + 6$$
   $$12 \overset{?}{=} 6 + 6$$
   $$12 = 12$$
   d. There are an infinite number of solutions to this equation.

7. Answers may vary. Three other solutions are $(1, 4), (3, 2)$ and $(5, 0)$.

9.

| $x$ | $y = x^3$ |
|---|---|
| 0 | $0^3 = 1$ |
| $-1$ | $(-1)^3 = -1$ |
| $-2$ | $(-2)^3 = -8$ |
| 1 | $1^3 = 1$ |
| 2 | $2^3 = 8$ |

11. Instead of the crooked line the student should have first re-considered the calculations of ordered pairs.

13. A smooth curve should have been drawn through the plotted points.

**Notation**

15. $y = -x + 4$
    substitute $(-2, 6)$ into this equation,
    $$6 \overset{?}{=} -(-2) + 4$$
    $$6 \overset{?}{=} 2 + 4$$
    $$6 = 6$$

**Practice**

17. $y = 2x - 4$

$4 \overset{?}{=} 2(4) - 4$

$4 \overset{?}{=} 8 - 4$

$4 = 4$

$(4, 4)$ satisfies this equation.

19. $y = |x - 2|$

$-3 \overset{?}{=} |4 - 2|$

$-3 \overset{?}{=} |2|$

$-3 \neq 2$

$(4, -3)$ does not satisfy this equation.

21.

| $x$ | $y = x - 3$ |
|---|---|
| 0 | $0 - 3 = -3$ |
| 1 | $1 - 3 = -2$ |
| $-2$ | $-2 - 3 = -5$ |

23.

| $x$ | $y = x^2 - 3$ |
|---|---|
| 0 | $0^2 - 3 = -3$ |
| 2 | $2^2 - 3 = 1$ |
| $-2$ | $(-2)^2 - 3 = 1$ |

25.

| $x$ | $y = 2x - 3$ |
|---|---|
| $-1$ | $2(-1) - 3 = -5$ |
| 0 | $2(0) - 3 = -3$ |
| 1 | $2(1) - 3 = -1$ |

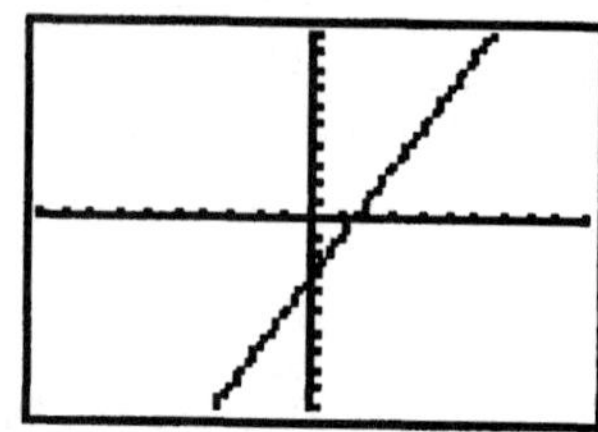

27.

| $x$ | $y = -2x + 1$ |
|---|---|
| $-1$ | $-2(-1) + 1 = 3$ |
| 0 | $-2(0) + 1 = 1$ |
| 1 | $-2(1) + 1 = -1$ |

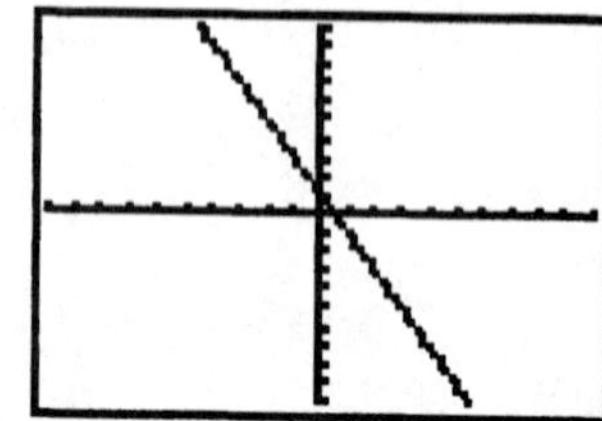

29.

| $x$ | $y = x^2 + 1$ |
|---|---|
| $-1$ | $(-1)^2 + 1 = 2$ |
| $0$ | $0^2 + 1 = 1$ |
| $1$ | $1^2 + 1 = 2$ |

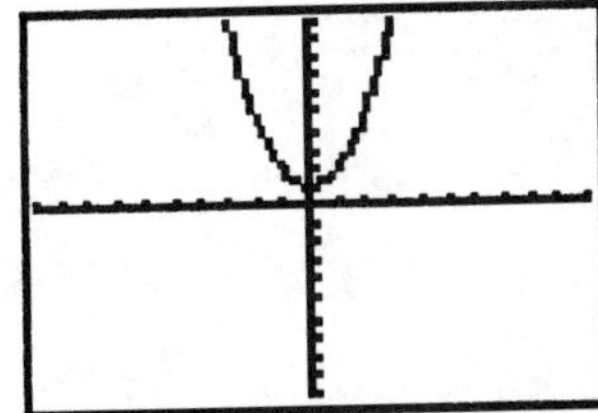

This graph is raised one unit higher on the $y$-axis than $y = x^2$.

31.

| $x$ | $y = (x - 2)^2$ |
|---|---|
| $-1$ | $(-1 - 2)^2 = 9$ |
| $0$ | $(0 - 2)^2 = 4$ |
| $1$ | $(1 - 2)^2 = 1$ |

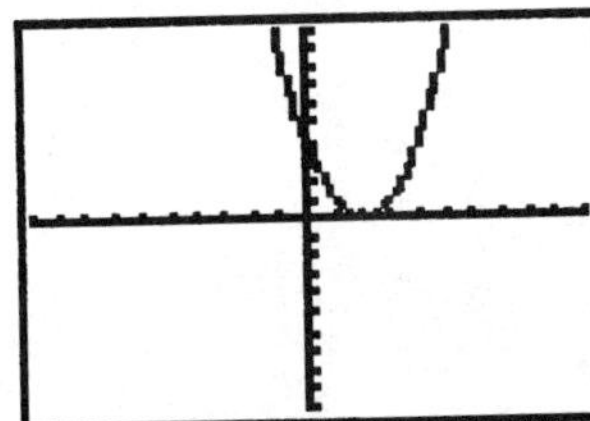

This graph is shifted two units to the right compared to $y = x^2$.

33.

| $x$ | $y = -|x|$ |
|---|---|
| $-1$ | $-|-1| = -1$ |
| $0$ | $-|0| = 0$ |
| $1$ | $-|1| = -1$ |

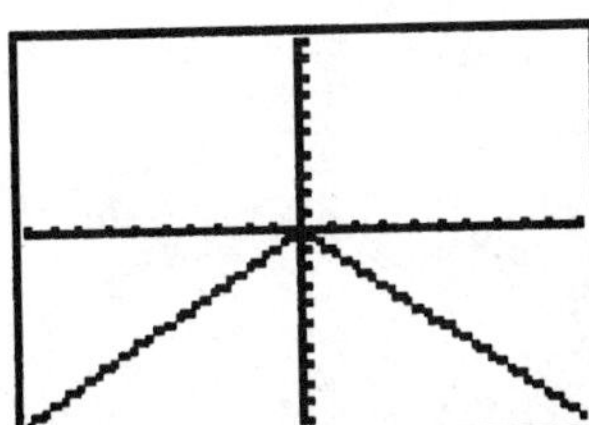

This graph is upside down compared to $y = |x|$.

35.

| $x$ | $y = \lvert x + 2 \rvert$ |
| --- | --- |
| $-1$ | $\lvert -1 + 2 \rvert = 1$ |
| $0$ | $\lvert 0 + 2 \rvert = 2$ |
| $1$ | $\lvert 1 + 2 \rvert = 3$ |

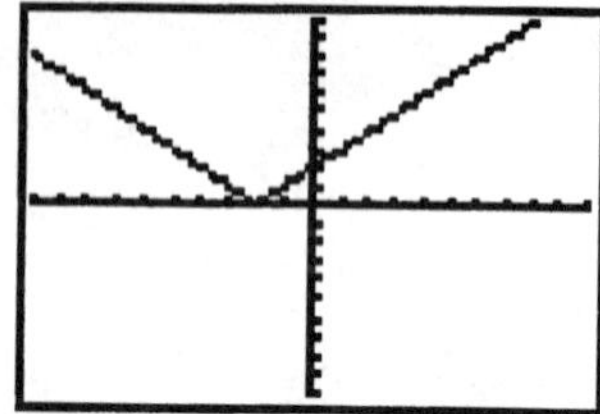

This graph is shifted two units to the left compared to $y = \lvert x \rvert$.

37.

| $x$ | $y = -x^3$ |
| --- | --- |
| $-1$ | $-(-1)^3 = 1$ |
| $0$ | $0^3 = 0$ |
| $1$ | $-1^3 = -1$ |

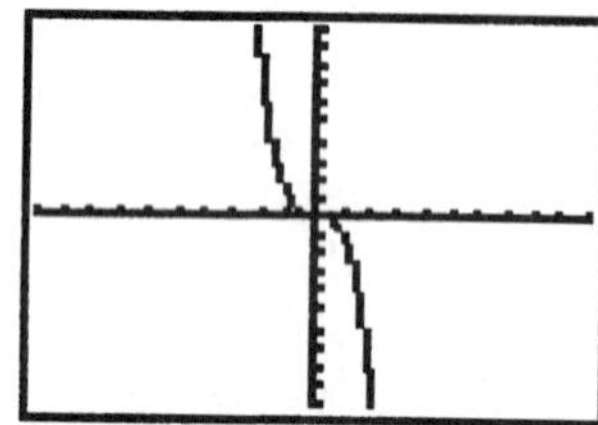

This graph is upside down compared to $y = x^3$.

39.

| $x$ | $y = x^3 - 2$ |
| --- | --- |
| $-1$ | $(-1)^3 - 2 = -3$ |
| $0$ | $0^3 - 2 = -2$ |
| $1$ | $1^3 - 2 = -1$ |

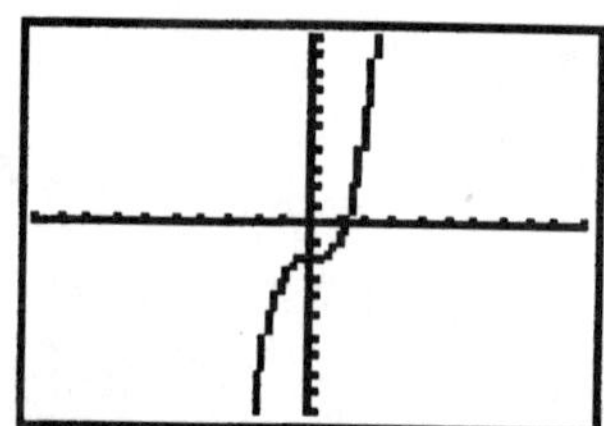

This graph is shifted down two units compared to $y = x^3$.

## Applications

41. 8 BALL

| $x$ | $-1$ | $2$ | $5$ | $8$ |
| --- | --- | --- | --- | --- |
| $y$ | $0$ | $4$ | $0$ | $-4$ |

43. SUSPENSION BRIDGES

| $x$ | $0$ | $2$ | $4$ | $-2$ | $-4$ |
| --- | --- | --- | --- | --- | --- |
| $y$ | $0$ | $1$ | $4$ | $1$ | $4$ |

45. MANUFACTURING
   a. $(2, 8)$ tells us that a 2 inch bolt costs 8 cents to manufacture.
   b. It costs 12 cents to manufacture a 7 inch bolt.
   c. A 4 inch bolt is the least expensive to manufacture.
   d. As the length of the bolt increases beyond 1 inch the cost decreases until a 4 inch bolt.
      As the length of the bolt increases beyond 4 inches the manufacturing cost also increases.

47. HOUSE MARKET VALUE
   a. This house was purchased for $90,000.
   b. 3 years after the purchase this house reached its lowest value.
   c. In year 6 the market value of the house began to surpass the purchase price.
   d. The market value of this home declined in the first three years but then increased for
      the next five years.

**Writing**

49. Answers will vary
51. Answers will vary
53. Answers will vary

**Review**

55. $\frac{x}{8} = -12$
    $8\left(\frac{x}{8}\right) = 8(-12)$
    $x = -96$

57. $\frac{x+5}{6}$ is an expression

59. $x = 0.005 \cdot 250$
    $x = 1.25$

61. $-2.5 - (-2.6)$
    $= -2.5 + 2.6$
    $= 0.1$

# Section 9.3 Graphing Linear Equations

**Vocabulary**

1. An equation whose graph is a line and whose variables are to the first power is called a **linear** equation.
3. The **$y$-intercept** of a line is the point $(0, b)$ where the line intersects the $y$-axis.
5. Lines parallel to the $y$-axis are **vertical** lines.

**Concepts**

7. a. $y = x^3$ is nonlinear
   b. $2x + 3y = 6$ is linear
   c. $y = |x + 2|$ is nonlinear
   d. $x = -2$ is linear
   e. $y = -x^2$ is nonlinear

9. a. The $y$ variable is raised to the first power and the variable is raised to the first power.
   b. The $y$ variable is raised to the first power and the variable is raised to the second power.
   c. The $y$ variable is raised to the first power and the variable is raised to the third power.

11. $5y = 2x + 10$
    $y = \frac{2}{5}x + 2$

| $x$ | $y$ |
|-----|-----|
| 10 | 6 |
| $-5$ | 0 |
| 5 | 4 |

13. $x - 2y = 4$
    $-2y = -x + 4$
    $y = \frac{1}{2}x - 2$

| $x$ | $y$ |
|-----|-----|
| 0 | $-2$ |
| 4 | 0 |
| 1 | $-\frac{3}{2}$ |

15. The coordinates of point A will satisfy this equation since A is on the line.

17. These three points do not correspond to a linear equation since they do not lie in a straight line. There must have been an error in finding these solutions.

19. The $x$-intercept is $(-3, 0)$ and the $y$-intercept is $(0, -1)$.

21. a. To find the $y$-intercept of the graph of a linear equation, we let $x = 0$ and solve for $y$.
    b. To find the $x$-intercept of the graph of a linear equation, we let $y = 0$ and solve for $x$.

**Notation**

23. a. $-4x = -y - 6$
    $-4x + y = -6$ or $4x - y = 6$
  b. $y = \frac{1}{2}x$
    $2(y) = 2(\frac{1}{2}x)$
    $2y = x$
    $x - 2y = 0$
  c. $3 = \frac{x}{3} + y$
    $3(3) = 3(\frac{x}{3} + y)$
    $9 = x + 3y$
  d. $x = 12$
    $x + 0y = 12$

**Practice**

25. $y = -x + 2$

| $x$ | $y$ |
| --- | --- |
| $-1$ | 3 |
| 0 | 2 |
| 1 | 1 |

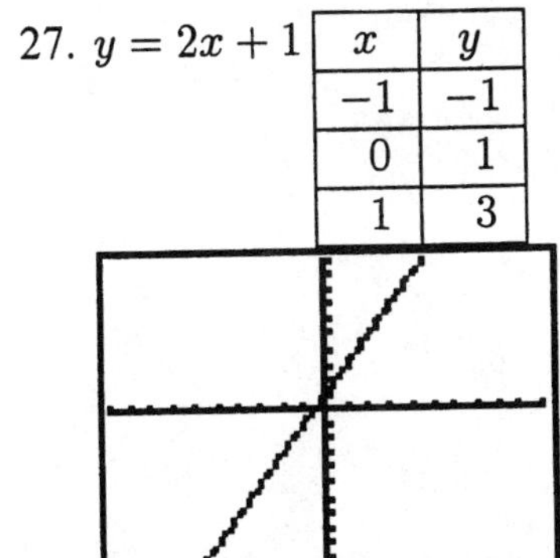

27. $y = 2x + 1$

| $x$ | $y$ |
| --- | --- |
| $-1$ | $-1$ |
| 0 | 1 |
| 1 | 3 |

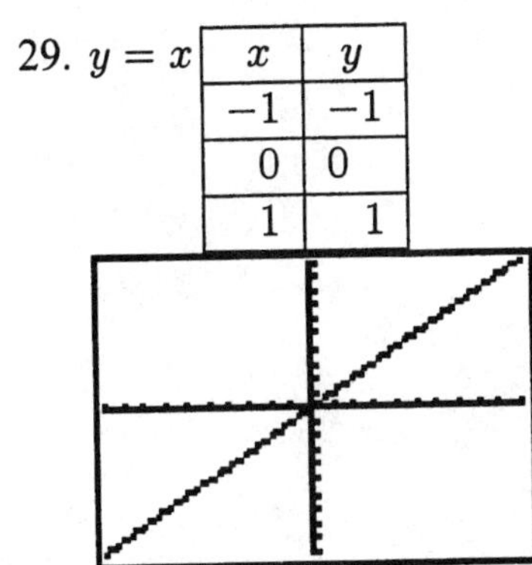

29. $y = x$

| $x$ | $y$ |
| --- | --- |
| $-1$ | $-1$ |
| 0 | 0 |
| 1 | 1 |

31. $y = -3x$

| $x$ | $y$ |
|---|---|
| $-1$ | 3 |
| 0 | 0 |
| 1 | $-3$ |

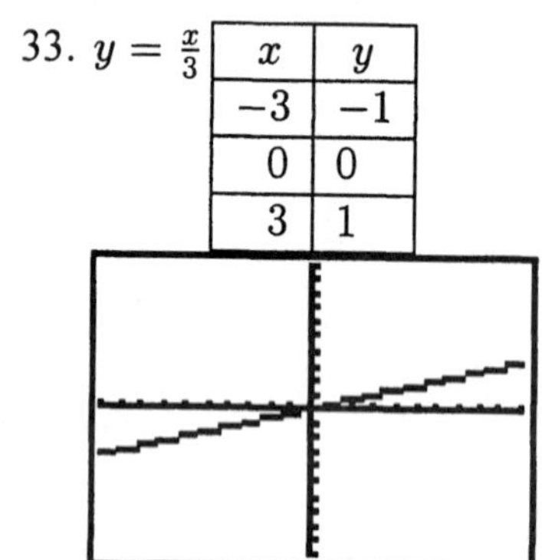

33. $y = \frac{x}{3}$

| $x$ | $y$ |
|---|---|
| $-3$ | $-1$ |
| 0 | 0 |
| 3 | 1 |

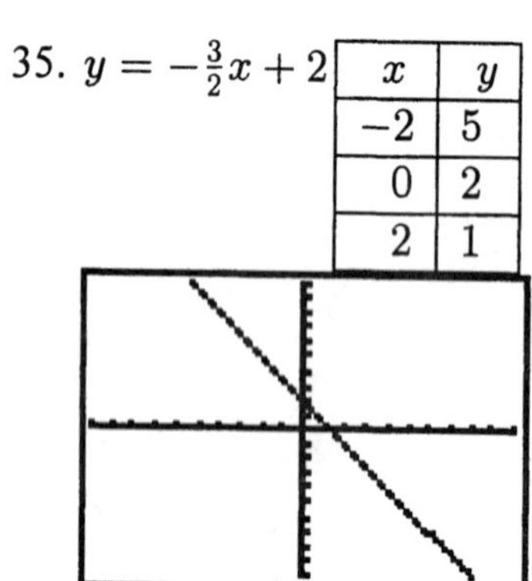

35. $y = -\frac{3}{2}x + 2$

| $x$ | $y$ |
|---|---|
| $-2$ | 5 |
| 0 | 2 |
| 2 | 1 |

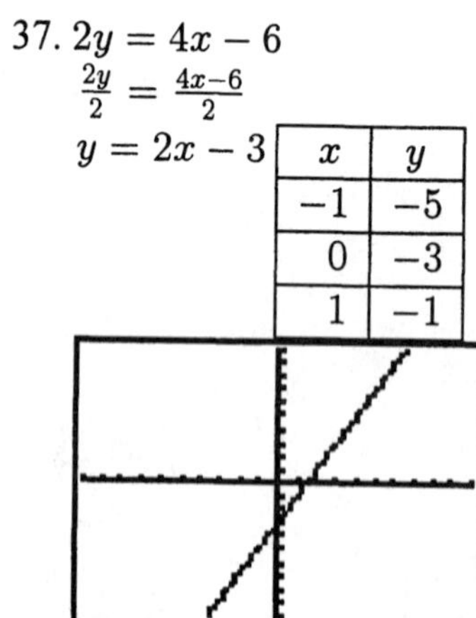

37. $2y = 4x - 6$

$$\frac{2y}{2} = \frac{4x-6}{2}$$

$y = 2x - 3$

| $x$ | $y$ |
|---|---|
| $-1$ | $-5$ |
| 0 | $-3$ |
| 1 | $-1$ |

39. $2y = x - 4$

$\frac{2y}{2} = \frac{x-4}{2}$

$y = \frac{1}{2}x - 2$

| $x$ | $y$ |
| --- | --- |
| $-1$ | $-\frac{5}{2}$ |
| $0$ | $-2$ |
| $1$ | $-\frac{3}{2}$ |

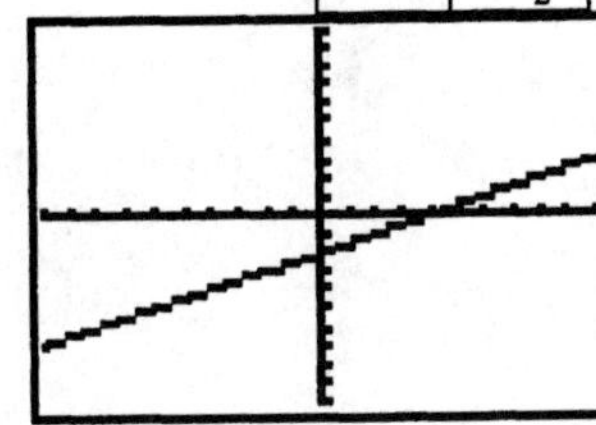

41. $2y + x = -2$

$2y = -x - 2$

$y = -\frac{1}{2}x - 1$

| $x$ | $y$ |
| --- | --- |
| $-2$ | $0$ |
| $0$ | $-1$ |
| $2$ | $-2$ |

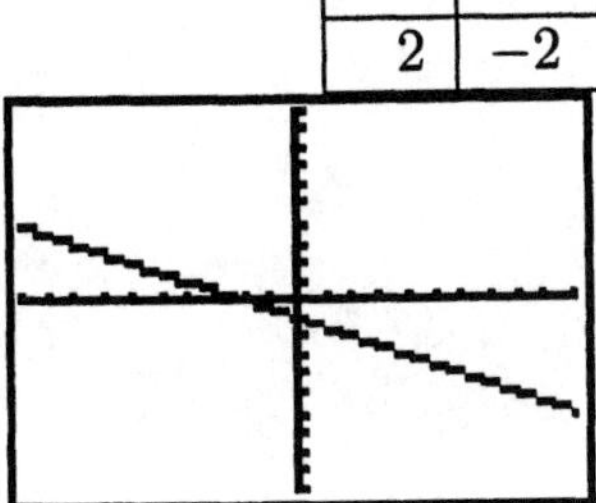

43. $y = 4$

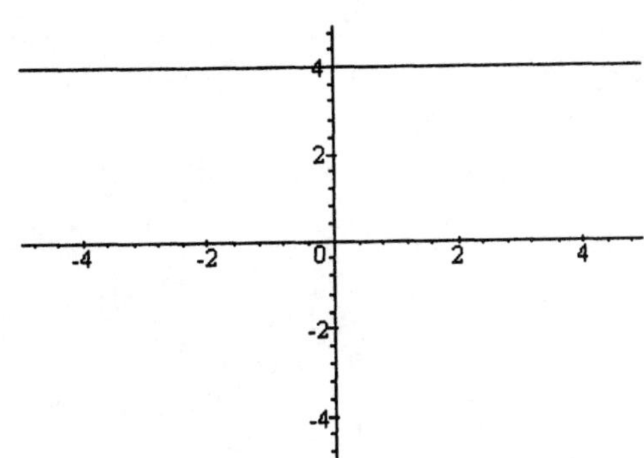

45. $x = -2$

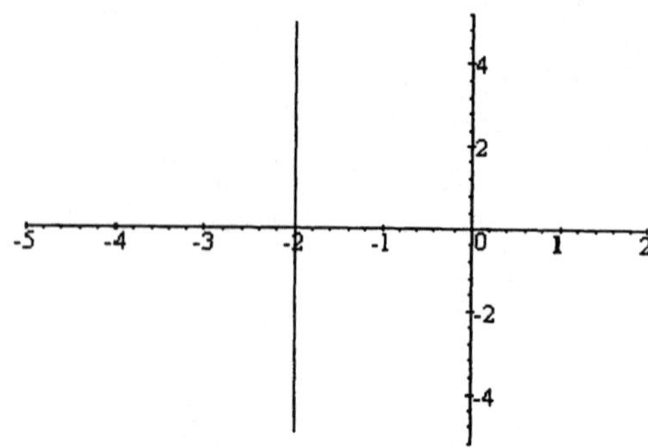

47. $y = -\frac{1}{2}$

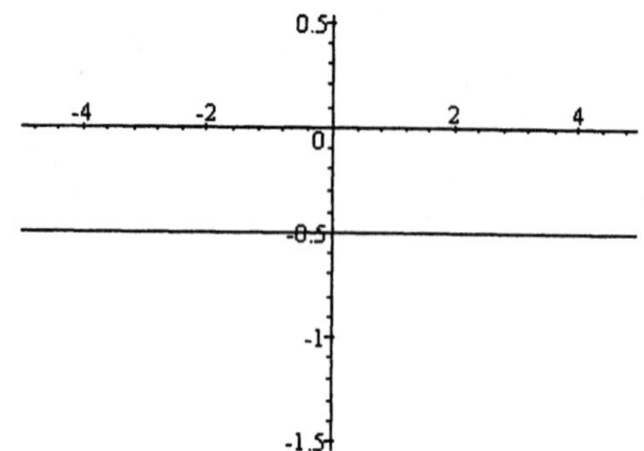

49. $x = \frac{4}{3}$

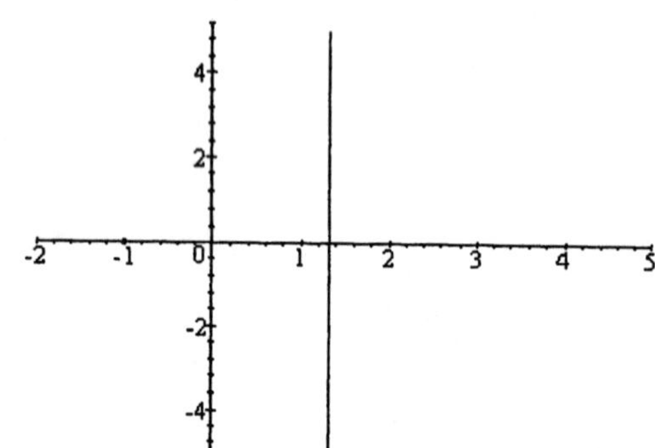

51. $2y - 2x = 6$

| If $x = 0$, | If $y = 0$, |
|---|---|
| $2y - 2(0) = 6$ | $2(0) - 2x = 6$ |
| $y = 0$ | $x = -3$ |

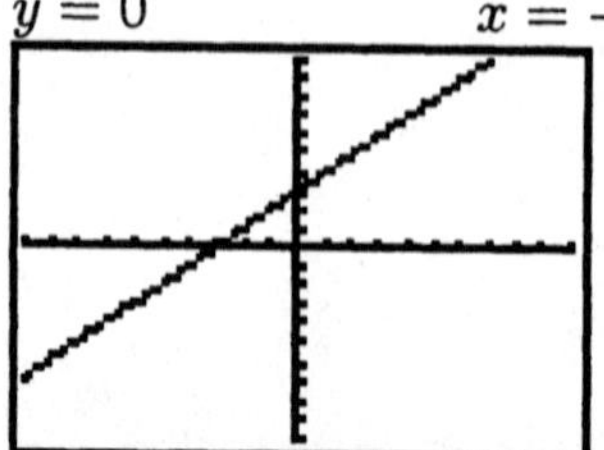

53. $-4y + 9x = -9$

If $x = 0,$  
$-4y + 9(0) = -9$  
$y = \frac{9}{4}$

If $y = 0,$  
$-4(0) + 9x = -9$  
$x = -1$

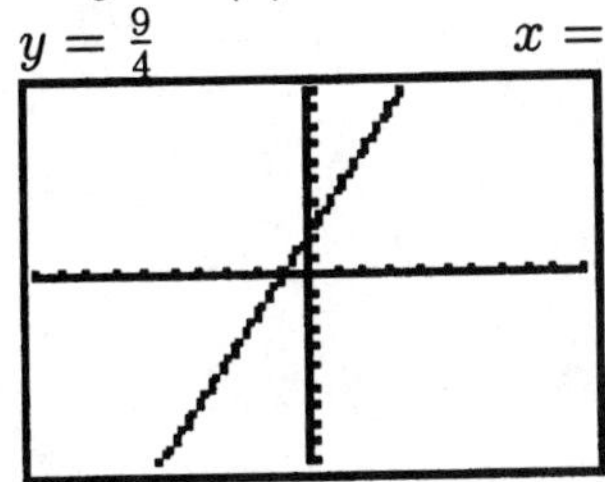

55. $4x + 5y = 20$

If $x = 0,$  
$4(0) + 5y = 20$  
$y = 4$

If $y = 0,$  
$4x + 5(0) = 20$  
$x = 5$

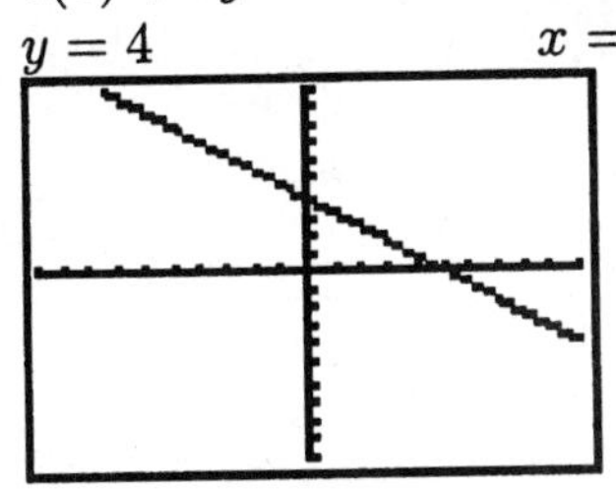

57. $3x + 4y = 12$

If $x = 0,$  
$3(0) + 4y = 12$  
$y = 3$

If $y = 0,$  
$3x + 4(0) = 12$  
$x = 4$

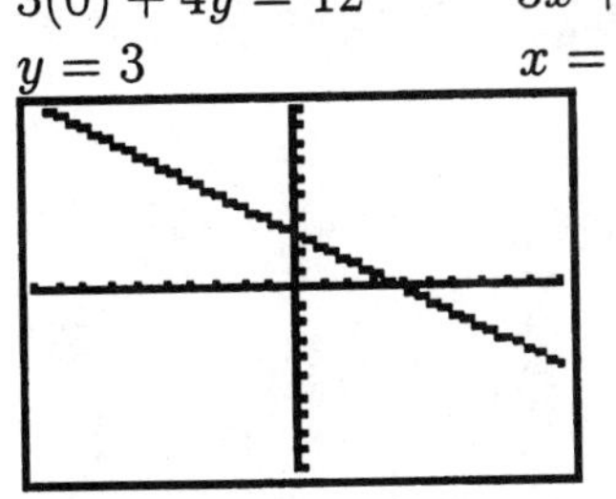

59. $15y + 5x = -15$

If $x = 0,$  
$15y + 5(0) = -15$  
$y = -1$

If $y = 0,$  
$15(0) + 5x = -15$  
$x = -3$

61. $3x + 4y = 8$

If $x = 0$,

$3(0) + 4y = 8$

$y = 2$

If $y = 0$,

$3x + 4(0) = 8$

$x = \frac{8}{3}$

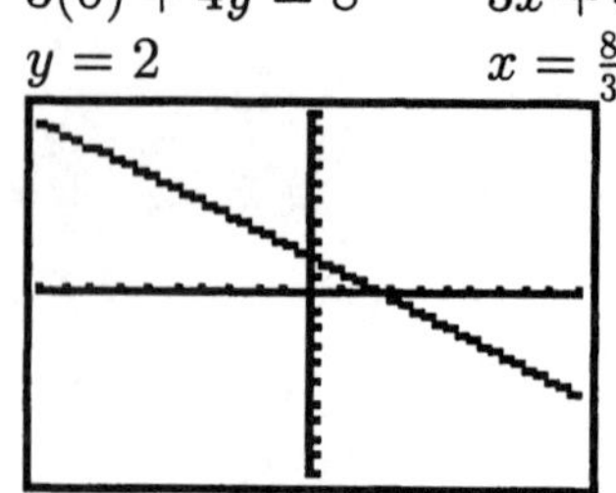

**Applications**

63. EDUCATIONAL COSTS

a. $c = \$50 + \$25u$

b.

| $u$ | $c$ |
| --- | --- |
| 4 | 150 |
| 8 | 250 |
| 14 | 400 |

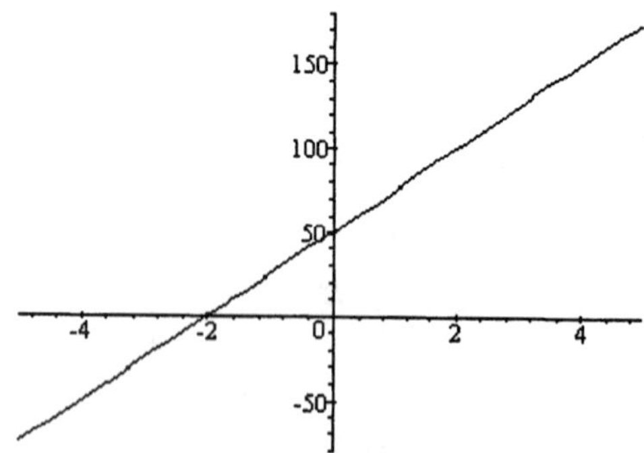

c. A student taking 18 units the first semester will pay $500 and the cost for 12 units the next semester is $350 for a total of $850.

d. The $y$-intercept tells us that before a student takes any units there is a $50 fee.

65. PHYSIOLOGY

$h = 3.9r + 28.9$

a.
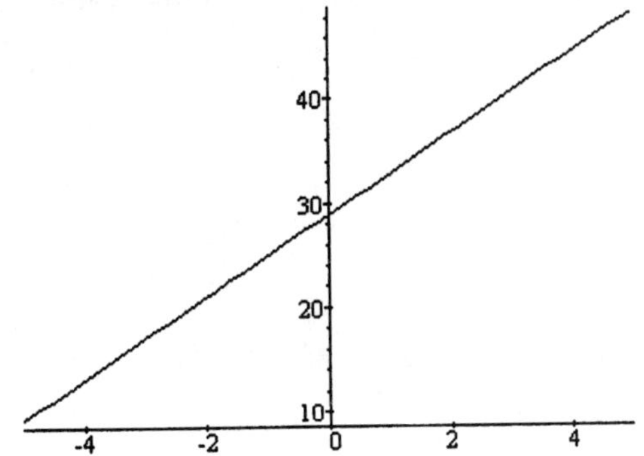

| $r$ | $h$ |
| --- | --- |
| 7 | 56.2 |
| 8.5 | 62.1 |
| 9 | 64.0 |

b. From the graph, we see that the longer the radius bone the taller the woman.

c. A woman with a radius bone of 7.5 inches will be about 58 inches tall.

## Writing

67. Answers will vary

69. Answers will vary

## Review

71. $-(-5 - 4c) = 5 + 4c$

73. $\frac{x+6}{2} = 1$

$2\left(\frac{x+6}{2}\right) = 2 \cdot 1$

$x + 6 = 2$

$x + 6 - 6 = 2 - 6$

$x = -4$

75. $P = r - c$ where $P$ is profit, $r$ is revenue and $c$ is cost.

77. $1 + 2[-3 - 4(2 - 8^2)]$

$= 1 + 2[-3 - 4(2 - 64)]$

$= 1 + 2[-3 - 4(-62)]$

$= 1 + 2[-3 + 248]$

$= 1 + 2(245)$

$= 1 + 490$

$= 491$

**Vocabulary**

1. A **ratio** is the quotient of two numbers.
3. The **slope** of a line is defined to be the ratio of the change in $y$ to the change in $x$.
5. The rate of **change** of a linear relationship can be found by finding the slope of the graph of the line.

**Concepts**

7. a. Graph $l_2$ has a positive slope.
   b. Graph $l_1$ has a negative slope.
   c. Graph $l_4$ has a zero slope.
   d. Graph $l_3$ has an undefined slope.

9.

| $x$ | $y$ |
|-----|-----|
| $-4$ | $2$ |
| $5$ | $-7$ |

Using $m = \frac{y_2 - y_1}{x_2 - x_1}$

$m = \frac{-7-2}{5-(-4)}$

$m = \frac{-9}{9}$

$m = -1$

11. GROWTH RATE

The rate of change will be the slope of the line. Using the ordered pairs of $(4, 37)$ and $(5, 40)$ the rate of change of the boy's height is

$m = \frac{40-37}{5-4}$

$m = \frac{3}{1}$

The rate of change is 3 inches per year for this boy.

13. DEPRECIATION

The rate of change will be the slope of the line. (Note the units of the $y$-axis are in 1,000 dollar increments.) Using the ordered pairs of $(0, 25000)$ and $(10, 0)$ the rate of change is

$m = \frac{0-25000}{10-0}$

$m = \frac{-25000}{10}$

$m = -2,500$

During this time the sound equipment's value declined by $2,500$ per year.

15. UNCOLA

a. Using the given ordered pairs $(90, 200)$ and $(97, 200)$ the slope of this line is

$m = \frac{200-200}{97-90}$

$m = \frac{0}{7}$

The slope of the line for this period of time is 0. A line with a slope of 0 is horizontal, which tells us that 7-Up sales did not change during this time frame.

b. Mountain Dew had the greatest rate of change from 1998 to 1999 as observed in the steepness of the graph.

**Notation**

17. The formula for finding the slope through these points is $m = \frac{y_2 - y_1}{x_2 - x_1}$ or $m = \frac{y_1 - y_2}{x_1 - x_2}$

19. $m = \frac{y_2 - y_1}{x_2 - x_1}$ : $m = \frac{4-3}{2-1} = \frac{1}{1} = 1$

21. $m = \frac{y_2 - y_1}{x_2 - x_1}$ : $m = \frac{4-7}{3-2} = \frac{-3}{1} = -3$

23. $m = \frac{y_2 - y_1}{x_2 - x_1}$ : $m = \frac{0-5}{0-4} = \frac{-5}{-4} = \frac{5}{4}$

25. $m = \frac{y_2 - y_1}{x_2 - x_1}$ : $m = \frac{5-6}{-3-(-5)} = \frac{-1}{2} = -\frac{1}{2}$

27. $m = \frac{y_2 - y_1}{x_2 - x_1}$ : $m = \frac{-2-(-8)}{-2-(-12)} = \frac{6}{10} = \frac{3}{5}$

29. $m = \frac{y_2 - y_1}{x_2 - x_1}$ : $m = \frac{7-7}{5-(-4)} = \frac{0}{9} = 0$

31. $m = \frac{y_2 - y_1}{x_2 - x_1}$ : $m = \frac{-4-(-3)}{8-8} = \frac{-1}{0}$ undefined

33. $m = \frac{y_2 - y_1}{x_2 - x_1}$ : $m = \frac{0-(-4)}{-6-0} = \frac{4}{-6} = -\frac{2}{3}$

35. $m = \frac{y_2 - y_1}{x_2 - x_1}$ : $m = \frac{1.75-(-7.75)}{-2.5-(-0.5)} = \frac{9.5}{-2} = -4.75$

37. $m = \frac{y_2 - y_1}{x_2 - x_1}$ : $m = \frac{1-(-3)}{2-(-4)} = \frac{4}{6} = \frac{2}{3}$

39. $m = \frac{y_2 - y_1}{x_2 - x_1}$ : $m = \frac{0-(-4)}{1-(-2)} = \frac{4}{3}$

41. $m = \frac{y_2 - y_1}{x_2 - x_1}$ : $m = \frac{4-(-3)}{-4-4} = \frac{7}{-8} = -\frac{7}{8}$

43. $m = \frac{y_2 - y_1}{x_2 - x_1}$ : $m = \frac{3-2}{-1-4} = \frac{1}{-5} = -\frac{1}{5}$

45. $(0,1), m = 2$

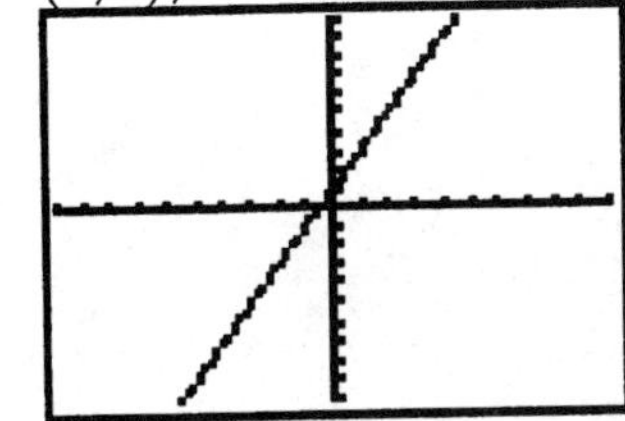

47. $(-3,-3), m = -\frac{3}{2}$

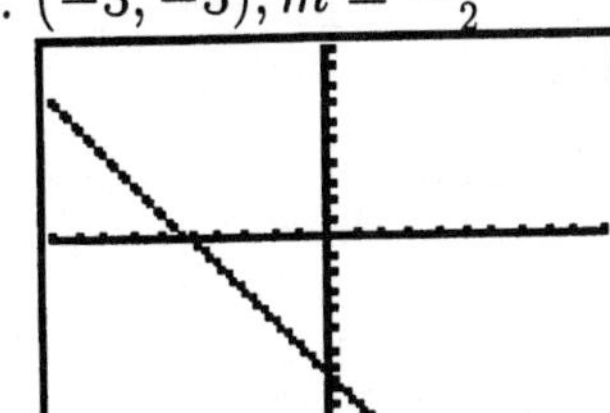

49. $(5, -3), m = \frac{3}{4}$

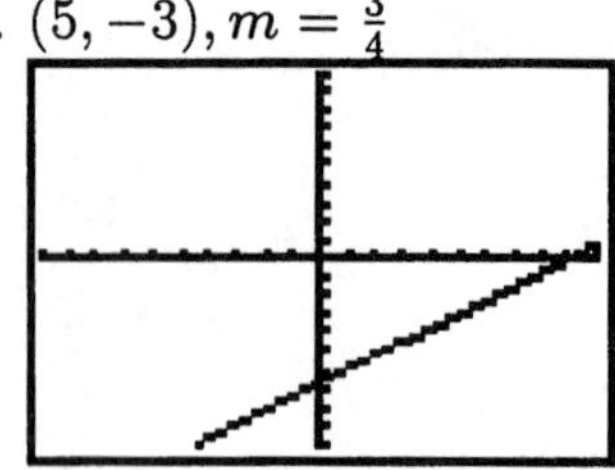

51. $(0, 0), m = -4$

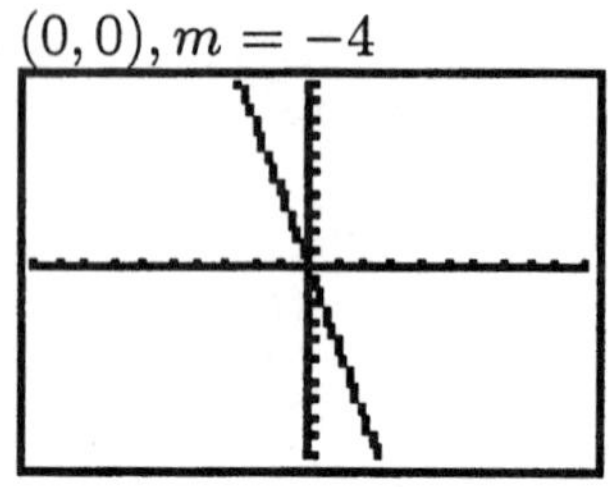

53. $(-5, 1), m = 0$

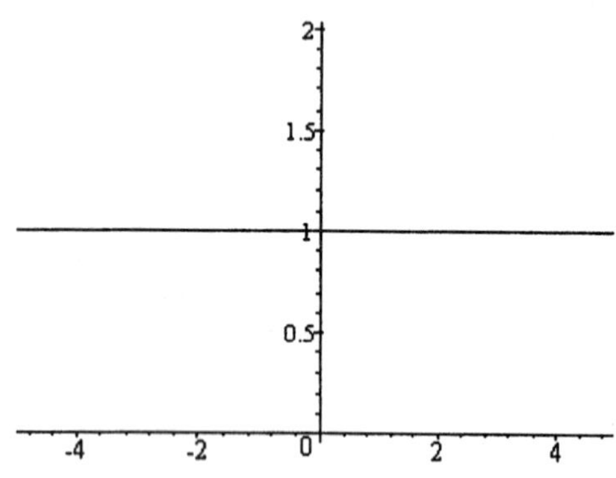

55. $(-1, -4)$, undefined slope.

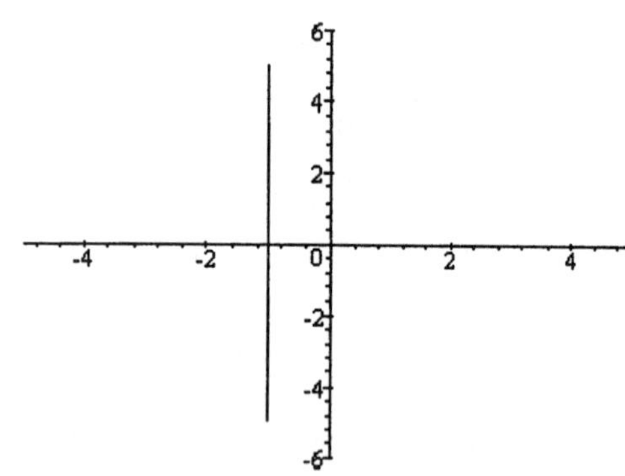

## Applications

57. POOL DESIGN

The ordered pairs for the drop off are $(5, -3)$ and $(20, -9)$

$$m = \frac{y_2 - y_1}{x_2 - x_1}$$

$$m = \frac{-3 - (-9)}{5 - 20}$$

$$m = \frac{6}{-15}$$

$$m = -\frac{2}{5}$$

59. GRADE OF ROAD

Using the ordered pairs $(0, 264)$ and $(5280, 0)$ the slope of this road is

$$m = \frac{0-264}{5280-0}$$

$$m = \frac{-264}{5280}$$

$$m = \frac{-1}{20}$$

Since $\frac{-1}{20}$ is equivalent to $-5\%$, the road sign for truckers should read 5% DECLINE AHEAD.

61. ACCESSIBILITY

a. $m = \frac{rise}{run}$

$$m = \frac{2}{16}$$

$$m = \frac{1}{8}$$

b. $m = \frac{rise}{run}$

$$m = \frac{1}{16-4}$$

$$m = \frac{1}{12}$$

c. An advantage for design 1 is there is only one ramp to construct which may result in less expense, but the disadvantage is the steepness compared to design 2. An advantage for design 2 is it is less steep compared to design 1, but with two ramps the cost will probably exceed design 1.

63. ENGINE OUTPUT

Using the ordered pairs $(2400, 150)$ and $(4800, 330)$, the rate of change of horsepower per revolutions per minute is given by

$$m = \frac{330-150}{4800-2400}$$

$$m = \frac{180}{2400}$$

$$m = \frac{3}{40}$$

The rate of change is $\frac{3 \text{ hp}}{40 \text{ rpm}}$.

**Writing**

65. Answers will vary
67. Answers will vary

**Review**

69. $(-3, 6)$ lies in quadrant II.

71. Substituting $(-1, -2)$ into $y = x^2 + 1$ yields,

$$-2 \overset{?}{=} (-1)^2 + 1$$

$$-2 \overset{?}{=} 1 + 1$$

$$-2 \neq 2$$

$(-1, -2)$ is not a solution to this equation.

73. $y = 2x + 2$ is a linear equation.

**Vocabulary**

1.  The equation $y = mx + b$ is called the **slope-intercept** form for the equation of a line.
3.  **Parallel** lines do not intersect.
5.  The numbers $\frac{5}{6}$ and $-\frac{6}{5}$ are called negative **reciprocals**. Their product is $-1$.

**Concepts**

7.  TREE GROWTH

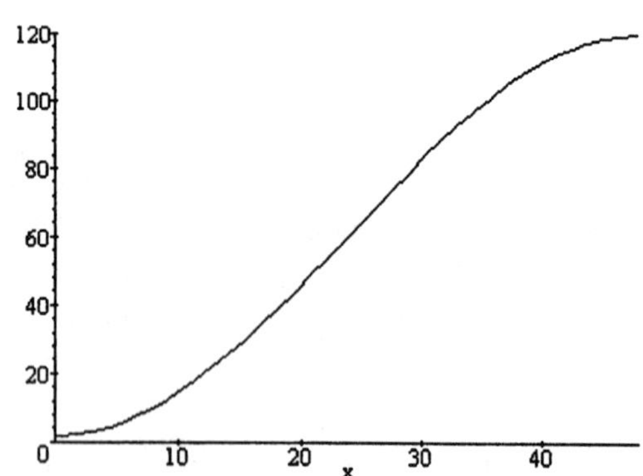

The graph does not indicate a linear relationship between the age of the tree and its height as the graph is curved.

9.  NAVIGATION
    a. The $y$-intercept shows that the ship may travel at 18 knots with no head waves.
    b. The rate of change as wave height increases is $m = \frac{18-14}{0-8} = \frac{4}{-8} = -\frac{1}{2}$. This says that for every foot of wave increase the boat must slow down $\frac{1}{2}$ knot in speed.
    c. $y = -\frac{1}{2}x + 18$

11. a. The $y$-intercept of $l_1$ is $(0,0)$.
    b. $l_1$ and $l_2$ appear to have the same slope but have different $y$-intercepts.

13. a. $m = -\frac{2}{3}$
    b. $m = \frac{1}{4}$
    c. $m = -8$
    d. $m = 3$
    e. $m = 1$
    f. $m = -1$

15. $\frac{-2y}{-2} = \frac{6x}{-2} - \frac{12}{-2}$
    $y = -3x + 6$

**Notation**

17. $6x - 2y = 10$
$6x - 6x - 2y = -6x + 10$
$-2y = -6x + 10$
$\frac{-2y}{-2} = \frac{-6x}{-2} + \frac{10}{-2}$
$y = 3x - 5$
The slope is 3 and the $y$-intercept is $(0, -5)$

**Practice**

19. $y = 4x + 2$
$m = 4, (0, 2)$

21. $y = \frac{x}{4} - \frac{1}{2}$
$m = \frac{1}{4}, (0, -\frac{1}{2})$

23. $y = \frac{1}{2}x + 6$
$m = \frac{1}{2}, (0, 6)$

25. $6y = x - 6$
$y = \frac{x}{6} - 1$
$m = \frac{1}{6}, (0, -1)$

27. $x + y = 8$
$y = -x + 8$
$m = -1, (0, 8)$

29. $2x + 3y = 6$
$3y = -2x + 6$
$y = -\frac{2}{3}x + 2$
$m = -\frac{2}{3}, (0, 2)$

31. $3y - 13 = 0$
$3y = 13$
$y = \frac{13}{3}$
$m = 0, (0, \frac{13}{3})$

33. $y = -5x$
$m = -5, (0, 0)$

35. $y = 5x - 3$

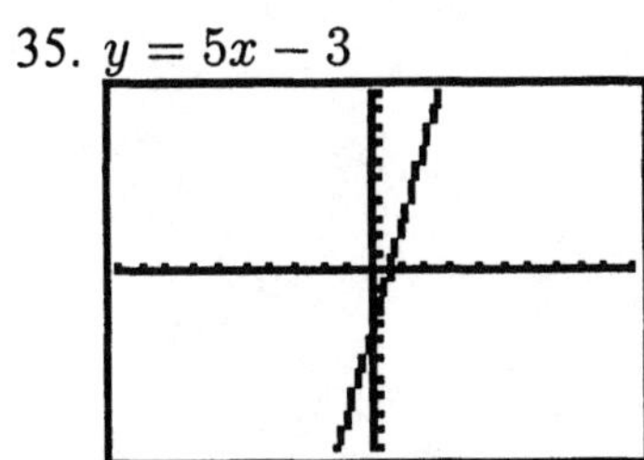

37. $y = \frac{1}{4}x - 2$

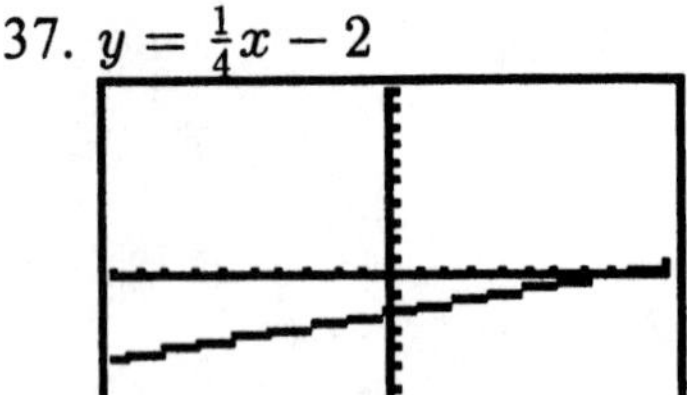

39. $y = -3x + 6$

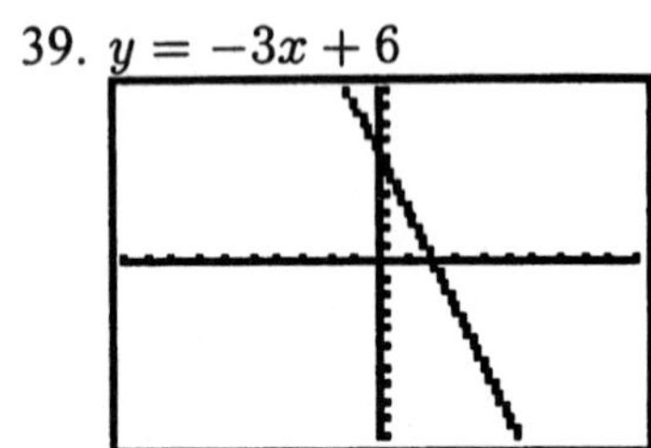

41. $y = -\frac{8}{3}x + 5$

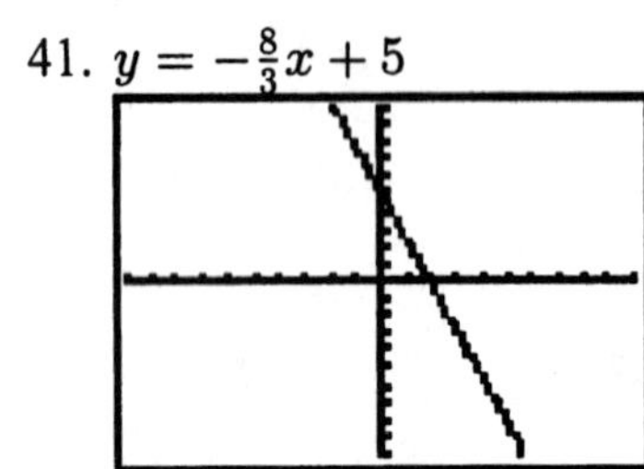

43. $y = 3x + 3$

$m = 3, (0, 3)$

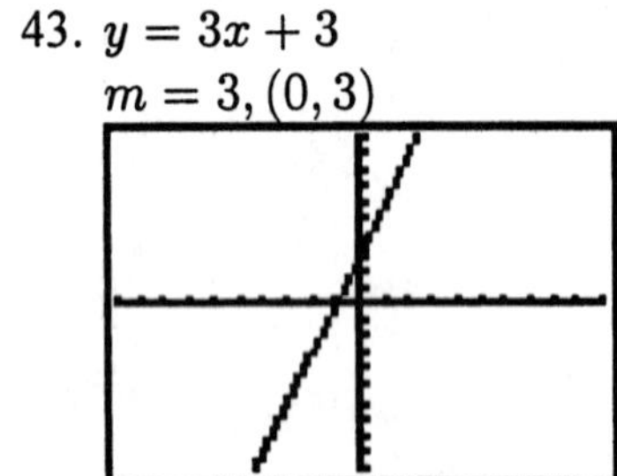

45. $y = -\frac{x}{2} + 2$

$m = -\frac{1}{2}, (0, 2)$

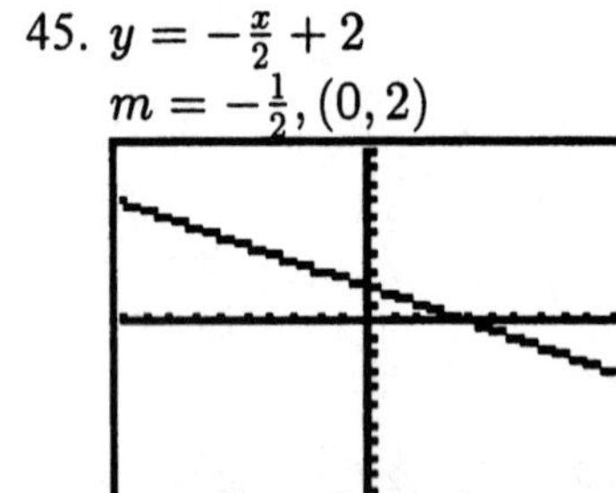

47. $3x + 4y = 16$

$3x - 3x + 4y = 16 - 3x$

$4y = 16 - 3x$

$\frac{4y}{4} = \frac{16 - 3x}{4}$

$y = 4 - \frac{3}{4}x$

$m = -\frac{3}{4}$

$b = (0, 4)$

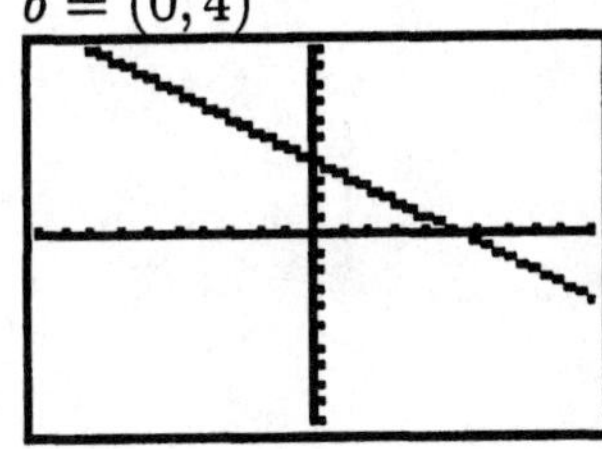

49. $10x - 5y = 5$

$10x - 10x - 5y = 5 - 10x$

$\frac{-5y}{-5} = \frac{5 - 10x}{-5}$

$y = -1 + 2x$

$m = 2$

$b = (0, -1)$

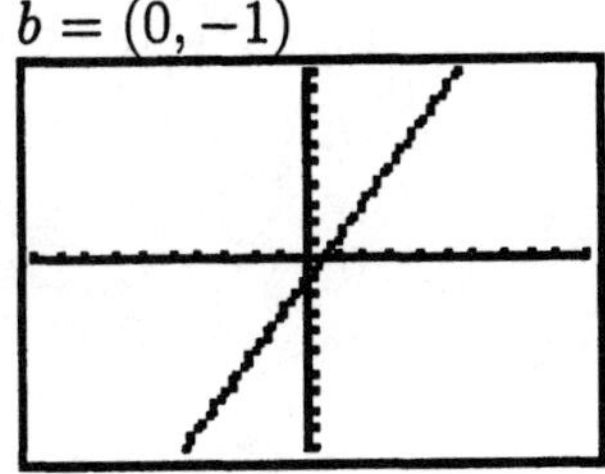

## Applications

51. PRODUCTION COSTS

a. $y = 5000 + 2000x$

b. If $x = 8$ then

$y = 5000 + 2000 \cdot 8$

$y = \$21,000$

53. CHEMISTRY EXPERIMENT

$y = 5x - 10$

55. **SALAD BAR**
   a. $y = 0.20x + 1.00$
   b.

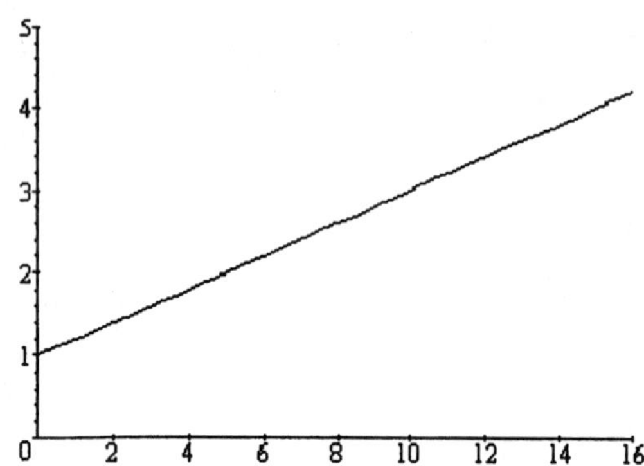

   c. The $y$-intercept would change to $(0, 2)$ but the slope would remain the same.
   d. The slope would change from $0.2$ to $0.3$ but the $y$-intercept would be the same.

57. **EMPLOYMENT SERVICE**
   $y = -20x + 500$

**Writing**

59. Answers will vary
61. Answers will vary

**Review**

63. $m = \frac{1-(-2)}{-6-6}$
   $m = \frac{3}{-12}$
   $m = -\frac{1}{4}$

65. $-4 - (-4) = -4 + 4 = 0$

67. Subtraction should be performed first in $[-2(4 - 8) + 4^2]$.

69. Let $p$ represent the percentage,
   $6p = 1.5$
   $\frac{6p}{6} = \frac{1.5}{6}$
   $p = 0.25$
   Therefore, 25% of 6 is 1.5

**Vocabulary**

1.  $y - y_1 = m(x - x_1)$ is called the **point-slope** form of the equation of a line.
3.  In illustration 1, point P has an **x-coordinate** of 2 and a **y-coordinate** of $-1$.

**Concepts**

5.  a. The linear equation $y = 2x - 3$ has slope 2 and $y$-intercept of $(0, -3)$.
    b. The linear equation $y - 4 = 6(x - 5)$ passes through the point $(5, 4)$ and has slope 6.

7.  a. $(-4, -2)$ and $(3, 2)$
    b. $m = \frac{2-(-2)}{3-(-4)} = \frac{4}{7}$
    c. $y - 2 = \frac{4}{7}(x - 3)$ or
    $y - (-2) = \frac{4}{7}(x - (-4))$
    $y + 2 = \frac{4}{7}(x + 4)$

9.  a. It is not sufficient to have only one point a linepasses through to write an equation of the line.
    b. It is not sufficient to have only the slope of a line to write an equation of the line.
    c. This table of values is sufficient as the slope can be determined as well as points the line passes through.

**Notation**

11. In $y - y_1 = m(x - x_1)$, we read $x_1$ as "**x sub one**."

13. $y - 2 = -3(x - 4)$
    $y - 2 = -3x + 12$
    $y - 2 + 2 = -3x + 12 + 2$
    $y = -3x + 14$

15. $y - y_1 = m(x - x_1)$
    where $(x_1, y_1) = (-1, 5)$ and $m = -2$
    $y - 5 = -2(x - (-1))$
    $y - 5 = -2(x + 1)$
    $y - 5 = -2x - 2$
    $y - 5 + 5 = -2x - 2 + 5$
    $y = -2x + 3$

**Practice**

17. $y - y_1 = m(x - x_1)$
    where $(x_1, y_1) = (2, 1)$ and $m = 3$
    $y - 1 = 3(x - 2)$

19. $y - y_1 = m(x - x_1)$
    where $(x_1, y_1) = (-5, -1)$ and $m = -\frac{4}{5}$
    $y - (-1) = -\frac{4}{5}(x - (-5))$
    $y + 1 = -\frac{4}{5}(x + 5)$

21. $y - y_1 = m(x - x_1)$
where $(x_1, y_1) = (10, 1)$ and $m = \frac{1}{5}$
$y - 1 = \frac{1}{5}(x - 10)$
$y - 1 = \frac{1}{5}x - 2$
$y - 1 + 1 = \frac{1}{5}x - 2 + 1$
$y = \frac{1}{5}x - 1$

23. $y - y_1 = m(x - x_1)$
where $(x_1, y_1) = (-9, 8)$ and $m = -5$
$y - 8 = -5(x - (-9))$
$y - 8 = -5(x + 9)$
$y - 8 = -5x - 45$
$y - 8 + 8 = -5x - 45 + 8$
$y = -5x - 37$

25. $y - y_1 = m(x - x_1)$
where $(x_1, y_1) = (6, -4)$ and $m = -\frac{4}{3}$
$y - (-4) = -\frac{4}{3}(x - 6)$
$y + 4 = -\frac{4}{3}x + 8$
$y + 4 - 4 = -\frac{4}{3}x + 8 - 4$
$y = -\frac{4}{3}x + 4$

27. $y - y_1 = m(x - x_1)$
where $(x_1, y_1) = (3, 0)$ and $m = -\frac{2}{3}$
$y - 0 = -\frac{2}{3}(x - 3)$
$y = -\frac{2}{3}x + 2$

29. $y - y_1 = m(x - x_1)$
where $(x_1, y_1) = (0, 4)$ and $m = 8$
$y - 4 = 8(x - 0)$
$y - 4 = 8x$
$y - 4 + 4 = 8x + 4$
$y = 8x + 4$

31. $y - y_1 = m(x - x_1)$
where $(x_1, y_1) = (0, 0)$ and $m = -3$
$y - 0 = -3(x - 0)$
$y = -3x$

33. Passes through $(1, 7)$ and $(-2, 1)$
$m = \frac{7-1}{1-(-2)} = \frac{6}{3} = 2$
$y - 7 = 2(x - 1)$
$y - 7 = 2x - 2$
$y - 7 + 7 = 2x - 2 + 7$
$y = 2x + 5$

35. Passes through $(-4, 3)$ and $(2, 0)$

$m = \frac{3-0}{-4-2} = \frac{3}{-6} = -\frac{1}{2}$

$y - 0 = -\frac{1}{2}(x - 2)$

$y = -\frac{1}{2}x + 1$

37. Passes through $(5, 5)$ and $(7, 5)$

$m = \frac{5-5}{5-7} = \frac{0}{-2} = 0$

$y - 5 = 0(x - 5)$

$y - 5 = 0$

$y = 5$

39. Passes through $(5, 1)$ and $(-5, 0)$

$m = \frac{0-1}{-5-5} = \frac{-1}{-10} = \frac{1}{10}$

$y - 0 = \frac{1}{10}(x - (-5))$

$y = \frac{1}{10}x + \frac{1}{2}$

41. Passes through $(-8, 2)$ and $(-8, 17)$

$m = \frac{17-2}{-8-(-8)} = \frac{15}{0} = $ undefined

This is a vertical line, $x = -8$.

43. Vertical line through $(4, 5)$ is $x = 4$

45. Horizontal line through $(4, 5)$ is $y = 5$

**Applications**

47. POLE VAULT

a.

| Position | Coordinates |
|---|---|
| 1 | $(0, 0)$ and $(-5, 2)$ |
| 2 | $(0, 0)$ and $(-3, 6)$ |
| 3 | $(0, 0)$ and $(-1, 7)$ |
| 4 | $(0, 0)$ and $(0, 10)$ |

b.

| Position | Slope |
|---|---|
| 1 | $m = \frac{2-0}{-5-0} = -\frac{2}{5}$ |
| 3 | $m = \frac{7-0}{-1-0} = -7$ |
| 4 | $m = \frac{10-0}{0-0}$ undefined |

Position 1

$y - 0 = -\frac{2}{5}(x - 0)$

$y = -\frac{2}{5}x$

Position 3

$y - 0 = -7(x - 0)$

$y = -7x$

Position 4

$x = 0$

c. The pole is curved and the slope-intercept form is for straight lines.

49. TOXIC CLEANUP
   a. Using the information given to form ordered pairs, we have 800 yd$^3$ after 3 months and 720 yd$^3$ after 5 months which gives $(3, 800)$ and $(4, 720)$.
   $$m = \frac{800-720}{3-5} = \frac{80}{-2} = -40$$
   $$y - 800 = -40(x - 3)$$
   $$y - 800 = -40x + 120$$
   $$y = -40x + 920$$
   b. One year after the project began is 12 months later,
   $$y = -40 \cdot 12 + 920$$
   $$y = -480 + 920$$
   $$y = 440 \text{ yd}^3 \text{ left after 1 year.}$$

51. COUNSELING
   The rate of change for patient visits is $m = \frac{105-75}{2-1} = 30$.
   If we assume this is a linear equation that passes through the points $(1, 75)$ and $(2, 105)$ then the equation becomes
   $$c - 75 = 30(t - 1)$$
   $$c - 75 = 30t - 30$$
   $$c - 75 + 75 = 30t - 30 + 75$$
   $$c = 30t + 45$$

53. CONVERTING TEMPERATURES
   a. Two ordered pairs of the form $(C, F)$ are $(0, 32)$ and $(100, 212)$
   b. $m = \frac{212-32}{100-0} = \frac{180}{100} = \frac{9}{5}$
   $$F - 32 = \frac{9}{5}(C - 0)$$
   $$F - 32 = \frac{9}{5}C$$
   $$F - 32 + 32 = \frac{9}{5}C + 32$$
   $$F = \frac{9}{5}C + 32$$

55. AIR CONDITIONING
   Using the information given to form ordered pairs of (minute, temperature), we have $(30, 75)$ and $(45, 71)$. The second point is found using the fact that the air is being cooled $4°$ every 15 minutes.
   $$m = \frac{75-71}{30-45} = \frac{4}{-15} = -\frac{4}{15}$$
   $$y - 75 = -\frac{4}{15}(x - 30)$$
   $$y - 75 = -\frac{4}{15}x + 8$$
   $$y = -\frac{4}{15}x + 83$$

**Writing**

57. Answers will vary
59. Answers will vary

**Review**

61. $m = \frac{8-4}{-6-2} = \frac{4}{-8} = -\frac{1}{2}$

63. $A = \pi r^2$ where $r = \frac{12}{2} = 6$
    $A = \pi 6^2$
    $A = 36\pi$
    $A \approx 113.0973355$
    To the nearest tenth $A = 113.1 \text{ ft}^2$

65. $(-1)^5 = (-1)(-1)(-1)(-1)(-1) = -1$

67. The coefficient of the second term is 6.

**Section 9.7 Functions**

**Vocabulary**

1.  A **function** is a rule that assigns to each value of the input a single value of another variable.
3.  For $y = 2x + 8$, $x$ is called the **independent** variable, and $y$ is called the **dependent** variable.

**Concepts**

5.  a. If positive real numbers are substituted for $x$ in $f(x) = x^2$ then positive real numbers will result.
    b. If negative real numbers are substituted for $x$ in $f(x) = x^2$ then positive real numbers will result.
    c. If zero is substituted for $x$ then zero will result.
    d. The domain is all real numbers and the range is all positive real numbers and zero, $y \geq 0$.

7.  1. In the equation $y = -5x + 1$, find the value of $y$ when $x = -1$.
    2. In the equation $f(x) = -5x + 1$ find $f(-1)$.

9.  a. $(-2, 4), (-2, -4)$
    b. This is not the graph of a function as $-2$ yields two output values.

**Notation**

11. The function notation $f(4) = -5$ states that when 4 is substituted for $x$ in function $f$, the result is $-5$.
    This fact can be illustrated graphically be plotting the point $(4, -5)$.

13. If $f(x) = 6 - 5x$, then $f(0) = 6$ is read as "$f$ **of** zero **is** 6."

**Practice**

15. $y = 2x + 10$ defines a function

17. $y = x^2$ defines a function

19. $y^2 = x$ does not define a function
    If $x = 1$ then
    $y^2 = 1$ and $y$ can equal 1 or $-1$

21. $y = x^3$ defines a function
23. $x = 3$ is not a function, this is a vertical line.
25. This data represents a function.

27. This data does not represent a function.
    If $x = -1$ then $y = 0$ and $y = 2$.

29. This data does not represent a function.
    If $t = 3$ then $d = 3$ and $y = 4$.

31. This graph represents a function.

33. This graph does not represent a function.
    $(3, 4), (3, -1)$

35. This graph does not represent a function.
   $(0, 2), (0, -4)$

37. $f(x) = x + 1$
   Domain: all reals  Range: all reals

39. $f(x) = x^2$
   Domain: all reals  Range: $y \geq 0$

41. $f(x) = x^3$
   Domain: all reals  Range: all reals

43. $f(x) = 4x - 1$
   a. $f(1) = 4 \cdot 1 - 1 = 3$
   b. $f(-2) = 4(-2) - 1 = -9$
   c. $f(\frac{1}{4}) = 4 \cdot \frac{1}{4} - 1 = 0$
   d. $f(50) = 4 \cdot 50 - 1 = 199$

45. $h(t) = 2t^2$
   a. $h(0.4) = 2(0.4)^2 = 0.32$
   b. $h(-3) = 2(-3)^2 = 18$
   c. $h(1000) = 2(1000)^2 = 2000000$
   d. $h(\frac{1}{8}) = 2(\frac{1}{8})^2 = \frac{2}{64} = \frac{1}{32}$

47. $s(x) = |x - 7|$
   a. $s(0) = |0 - 7| = 7$
   b. $s(-7) = |-7 - 7| = 14$
   c. $s(7) = |7 - 7| = 0$
   d. $s(8) = |8 - 7| = 1$

49. $f(x) = x^3 - x$
   a. $f(1) = 1^3 - 1 = 0$
   b. $f(10) = 10^3 - 10 = 990$
   c. $f(-3) = (-3)^3 - (-3) = -24$
   d. $f(6) = 6^3 - 6 = 210$

51. $f(x) = 3.4x^2 - 1.2x + 0.5$
   $f(-0.3) = 3.4(-0.3)^2 - 1.2(-0.3) + 0.5$
   $f(-0.3) = 1.166$

53. $f(x) = -2 - 3x$

| $x$ | $f(x)$ |
|---|---|
| 0 | $-2 - 3(0) = -2$ |
| 1 | $-2 - 3(1) = -5$ |
| $-1$ | $-2 - 3(-1) = 1$ |
| $-2$ | $-2 - 3(-2) = 4$ |

Domain: all reals, Range: all reals

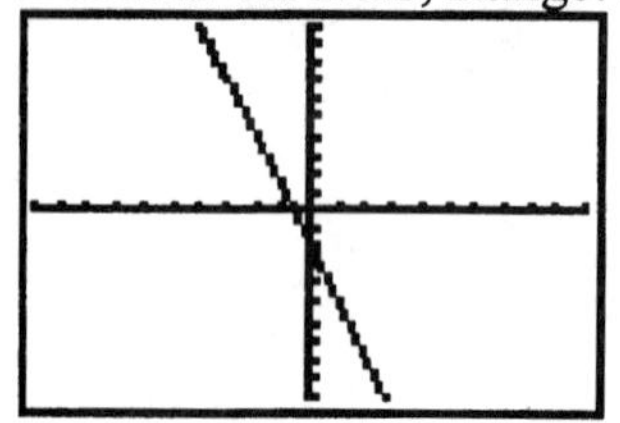

55. $f(x) = \frac{1}{2}x - 2$

| $x$ | $y$ |
|---|---|
| $-2$ | $\frac{1}{2}(-2) - 2 = -3$ |
| 0 | $\frac{1}{2}(0) - 2 = -2$ |
| 2 | $\frac{1}{2}(2) - 2 = -1$ |

Domain: all reals, Range: all reals

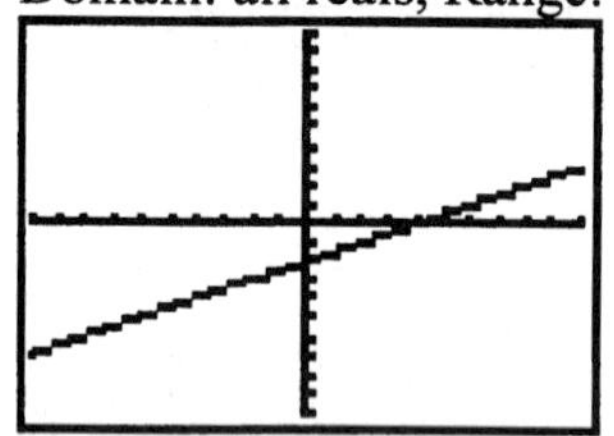

57. $s(x) = 2 - x^2$

| $x$ | $s(x)$ |
|---|---|
| 0 | $2 - 0^2 = 2$ |
| 1 | $2 - 1^2 = 1$ |
| 2 | $2 - 2^2 = -2$ |
| $-1$ | $2 - (-1)^2 = 1$ |
| $-2$ | $2 - (-2)^2 = -2$ |

Domain: all reals, Range: $y \leq 2$

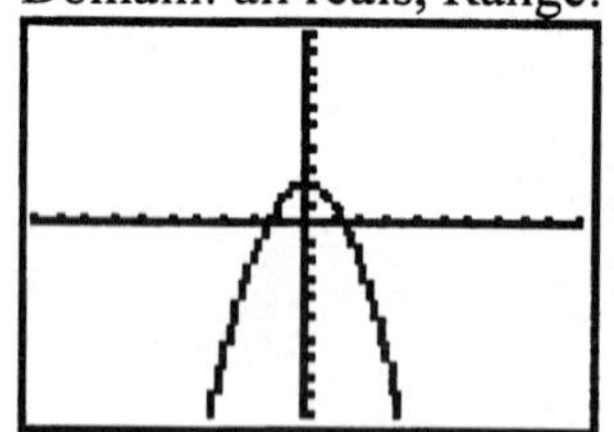

**Applications**

59. REFLECTIONS
    An absolute value function could serve as a model of this, $f(x) = |x|$

61. TIDES
    a. The domain is $0 \le x \le 24$
    b. $f(3) = 0.5$
    c. $f(6) = 1.5$
    d. $f(15) = -1.5$
    e. $f(12)$ tells us the low tide at noon was $-2.5$ meters.
    f. $f(21) \approx 1.6$

63. LAWN SPRINKLERS
    $A(r) = \pi r^2$
    $A(5) = \pi 5^2 = 25\pi \approx 78.5 \text{ ft}^2$
    $A(10) = \pi 10^2 = 100\pi \approx 314.2 \text{ft}^2$
    $A(20) = \pi 20^2 = 400\pi \approx 1256.6 \text{ ft}^2$

**Writing**

65. Answers will vary

**Review**

67. A horizontal line passing through $(-3, 6)$ is $y = 6$

69. Profit is the difference of revenue and cost
    $P = r - c$

71. $-3(2x - 4) = -6x + 12$

73. There are $12d$ eggs in $d$ dozen.

**Chapter 9 Key Concepts**

1. $m = -3$ and $b = -4$
   $y = -3x - 4$

3. $2x - 4y = 8$
   $2x - 4y + 4y = 8 + 4y$
   $2x = 8 + 4y$
   $2x - 8 = 8 - 8 + 4y$
   $2x - 8 = 4y$
   $\frac{2x-8}{4} = \frac{4y}{4}$
   $\frac{1}{2}x - 2 = y$

| $x$ | $y$ |
|---|---|
| 0 | $-2$ |
| 4 | 0 |
| $-2$ | $-3$ |

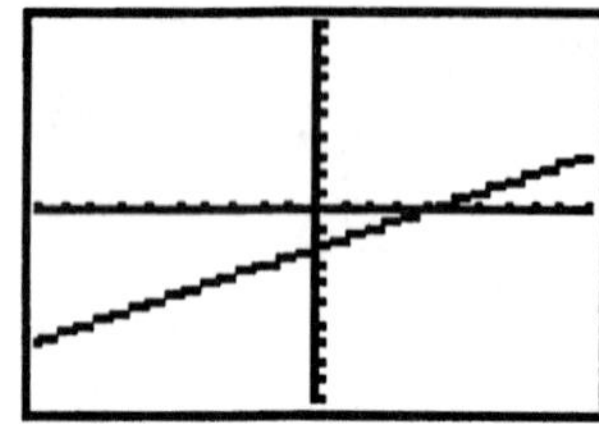

5. a. $(-2, -2)$ is a point on this line.
   b. The slope is $m = -2$.
   c. The equation of this line is $y = 2x - 6$.

7. $f(x) = 35x + 25$
   $f(3) = 35 \cdot 3 + 25$
   $f(3) = 130$
   $f(3)$ tells us that renting the cement mixer for 3 days will cost \$130.

**Chapter 9 Review**

**Section 9.1**

1.  a.

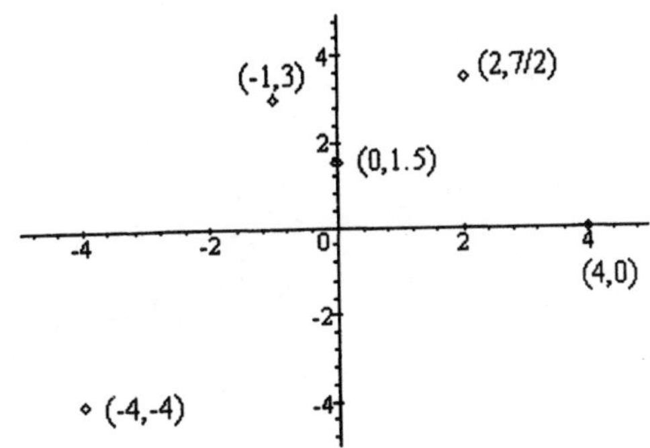

b.

| $x$ | $y$ |
|----|----|
| 3 | $-1$ |
| 0 | 0 |
| $-3$ | 1 |

3.  SNOWFALL
    a. 2 feet of snow was on the ground on day 1.
    b. It snowed 2 feet between days 2 and 3.
    c. 6 feet of snow was on the ground on day 6.

**Section 9.2**

5.  Is $(-3, 5)$ a solution of $y = |2 + x|$?
    $$5 \overset{?}{=} |2 + (-3)|$$
    $$5 \overset{?}{=} |-1|$$
    $$5 \neq 1$$
    $(-3, 5)$ is not a solution

7.  a. If $t = 70$ then $O = 9000$
    b. $(40, 18)$ tells us that 40 orange trees per acre will produce the highest yield of 18,000 oranges.

**Section 9.3**

9.  $5x + 2y = 10$
    $A = 5$
    $B = 2$
    $C = 10$

11. $x + 2y = 6$

$x - x + 2y = 6 - x$

$2y = 6 - x$

$\frac{2y}{2} = \frac{6-x}{2}$

$y = -\frac{1}{2}x + 3$

| $x$ | $y$ | $(x, y)$ |
|---|---|---|
| $-1$ | $\frac{7}{2}$ | $(-1, \frac{7}{2})$ |
| $0$ | $3$ | $(0, 3)$ |
| $1$ | $\frac{5}{2}$ | $(1, \frac{5}{2})$ |

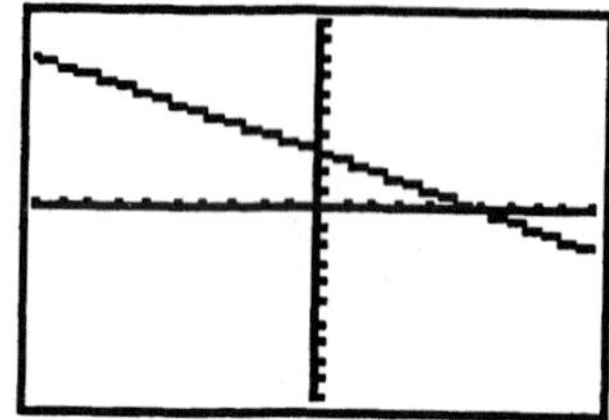

13. a. $y = 4$

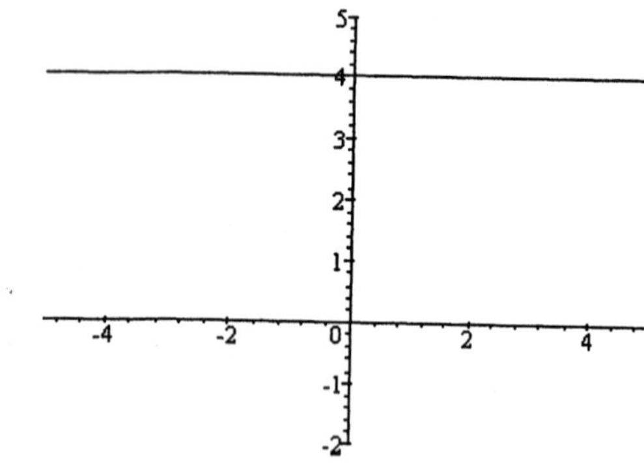

b. $x = -1$

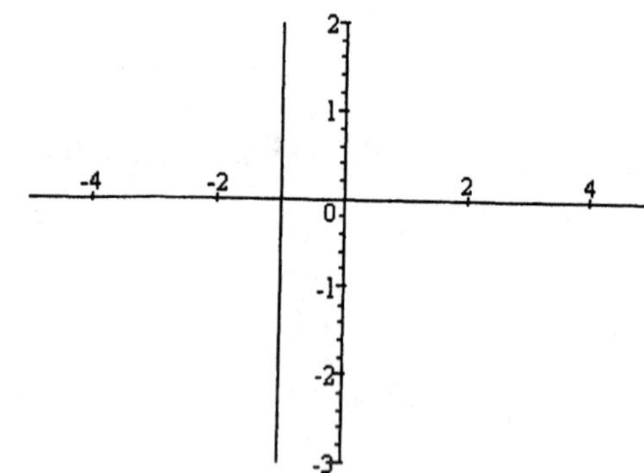

**Section 9.4**

15. a. $m = \frac{-2-(-3)}{4-0} = \frac{1}{4}$

b. $m = \frac{-3-(-17)}{2-4} = \frac{14}{-2} = -7$

c. $m = \frac{-5-(-5)}{5-2} = \frac{0}{3} = 0$

d. $m = \frac{-7-(-4)}{3-1} = \frac{-3}{2} = -\frac{3}{2}$

17. TOURISM
   a. The largest decline in tourism was from 1992 to 1994, 47.3 million to 44.8 million.
   $\frac{44.8-47.3}{1994-1992} = \frac{-2.5}{2} = -1.25$ million tourists per year.
   b. The largest increase was from 1986 to 1988, 26 million to 34.1 million.
   $\frac{34.1-26}{1988-1986} = \frac{8.1}{2} = 4.05$ million tourists per year.

**Section 9.5**

19. $9x - 3y = 15$
   $9x - 3y + 3y = 15 + 3y$
   $9x = 15 + 3y$
   $9x - 15 = 15 - 15 + 3y$
   $9x - 15 = 3y$
   $\frac{9x-15}{3} = \frac{3y}{3}$
   $3x - 5 = y$
   slope $m = 3$ and $y$-intercept $(0, -5)$

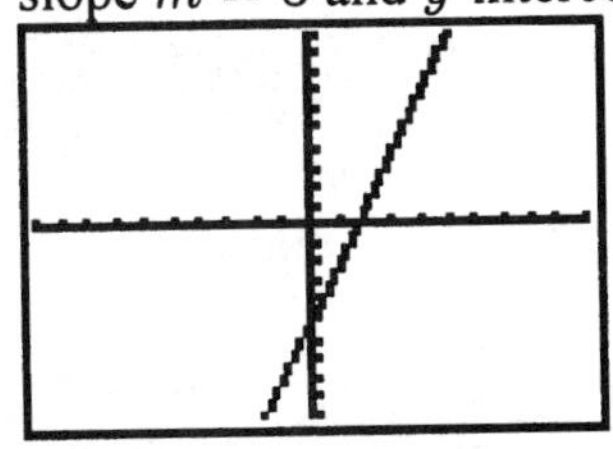

21. a. These lines would be parallel since the slopes are identical.
   b. These lines would be perpendicular since the slopes are negative reciprocals.

**Section 9.6**

23. a. $m = \frac{7-1}{3-(-6)} = \frac{6}{9} = \frac{2}{3}$
   $y - 7 = \frac{2}{3}(x - 3)$
   $y - 7 = \frac{2}{3}x - 2$
   $y - 7 + 7 = \frac{2}{3}x - 2 + 7$
   $y = \frac{2}{3}x + 5$
   b. Since a horizontal line has slope zero, $y = -8$

**Section 9.7**

25. a. $y = 3x - 2$ defines a function
   b. $y^2 = x$ does not define a function.
   c. This data does not define a function, observe $x = 2$.

27. $g(x) = 1 - 6x$
   a. $g(1) = 1 - 6 \cdot 1 = -5$
   b. $g(-6) = 1 - 6(-6) = 37$
   c. $g(0.5) = 1 - 6(0.5) = -2$
   d. $g\left(\frac{3}{2}\right) = 1 - 6 \cdot \frac{3}{2} = -8$

29. a. This is not a function, since it does not pass the vertical line test.
   b. This is a function, since it passes the vertical line test.

1. There were 10 dogs in the kennel two days before the holiday.
3. There were 30 dogs in the kennel 1 day before the holiday and on day 3 of the holiday weekend.

5. $y = x^2 - 4$

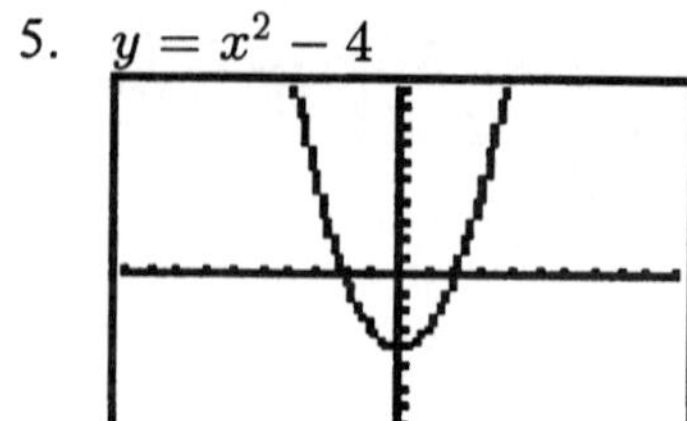

7. $\frac{-6+0}{2} = -3$ and $\frac{2+(-10)}{2} = -4$ therefore $(-3, -4)$ is the midpoint.

9. $2x - 3y = 6$
$x$-intercept: let $y = 0$
$2x = 6$
$\frac{2x}{2} = \frac{6}{2}$
$x = 3$
$x$-intercept is $(3, 0)$
$y$-intercept: let $x = 0$
$-3y = 6$
$\frac{-3y}{-3} = \frac{6}{-3}$
$y = -2$
$y$-intercept is $(0, -2)$

11. $x = -4$

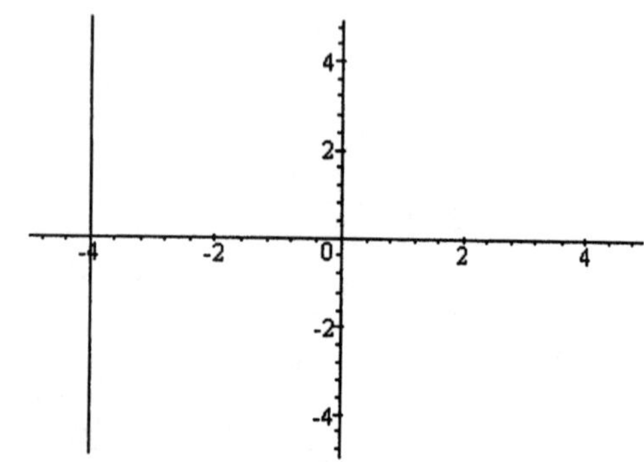

13. $m = \frac{3-(-1)}{-1-3} = \frac{4}{-4} = -1$

15. The slope of a line perpendicular to a line with slope $-\frac{7}{8}$ is $\frac{8}{7}$ since $\frac{8}{7}$ is the negative reciprocal of $\frac{7}{8}$.

17. The steepest incline is from mile 15 to mile 20.
$(15, 150), (20, 300)$
$m = \frac{300-150}{20-15}$
$m = \frac{150}{5}$
$m = 30$ feet per mile

19. DEPRECIATION

$v = 15{,}000 - 1500x$

21. This is not a graph of a function since it fails the vertical line test.

23. $y = 2x - 8$ defines a function.

25. $g(x) = 3.5x^3$

$g(6) = 3.5(6)^3$

$g(6) = 756$

**Chapters 1 - 9 Cumulative Review**

1. TWA
   a. In quarter two the loss was the least, approximately $5 million.
   b. In quarter for the loss was the greatest, approximately $270 million.

3. $\frac{1}{250} = 0.004$

5. AUTO SALES

| Rank | Vehicle | 99 Rank | % |
|---|---|---|---|
| 1 | F-series | 1 | $+0.9$ |
| 2 | Silverado | 2 | $+0.9$ |
| 3 | Explorer | 5 | $+3.8$ |
| 4 | Camry | 3 | $-5.6$ |
| 5 | Accord | 6 | $+0.1$ |

7. $b^2 - 4ac$
   $(-8)^2 - 4(2)(4)$
   $= 64 - 32$
   $= 32$

9. There are three terms and the coefficient of the second term is $-2$.

11. $5a + 10 - a = 4a + 10$

13. $(a + 2) - (a - 2) = a + 2 - a + 2 = 4$

15. $3x - 5 = 13$
    $3x = 18$
    $x = 6$

17. $\frac{2x}{3} - 2 = 4$
    $\frac{2}{3}x = 6$
    $x = \frac{3}{2}(6)$
    $x = 9$

19. $-3(2y - 2) - y = 5$
    $-6y + 6 - y = 5$
    $-7y = -1$
    $y = \frac{1}{7}$

21. $\frac{1}{3} + \frac{c}{5} = -\frac{3}{2}$
    $30\left(\frac{1}{3} + \frac{c}{5}\right) = 30\left(-\frac{3}{2}\right)$
    $10 + 6c = -45$
    $6c = -55$
    $c = -\frac{55}{6}$

23. $y = mx + b$
    $y - b = mx$
    $\frac{y - b}{m} = x$

25. The measures of the base angles are equal, the sum of the measures of the interior angles of a triangle is $180^0$.
$$x + x + 22 = 180$$
$$2x = 158$$
$$x = 79^0$$

27. ROAD TRIP

|       | $r$ | $t$ | $d = rt$ |
|-------|-----|-----|----------|
| Bus   | 60  | $t$ | $60t$    |
| Truck | 50  | $t$ | $50t$    |

We are asked to find the time that the bus and truck will be 75 miles apart,
$$60t - 50t = 75$$
$$10t = 75$$
$$\frac{10t}{10} = \frac{75}{10}$$
$$t = 7.5$$
In 7.5 hours the bus and truck will be 75 miles apart.

29. $-\frac{3}{16}x \geq -9$
$$3x \leq 144$$
$$x \leq 48$$
$$(-\infty, 48]$$

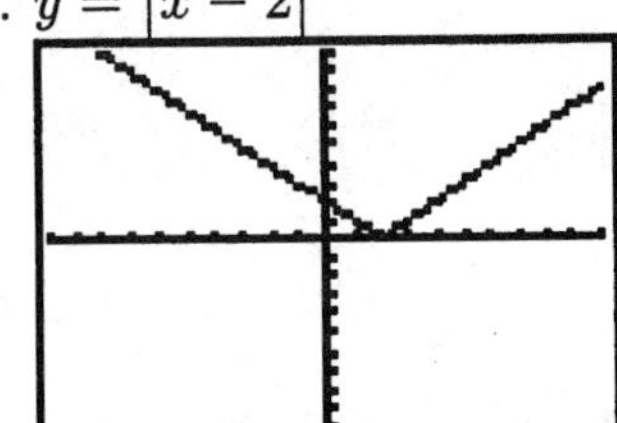

31. MEDICATION
    a. A 5-year old child should receive 1 teaspoon.
    b. A 9-year old child should receive 3 teaspoons.

33. $y = |x - 2|$

35. The slope of $y = 5$ is zero.

37. $4x - 6y = -12$
$$-6y = -4x - 12$$
$$y = \frac{2}{3}x + 2$$
$$m = \frac{2}{3}, \; y\text{-intercept is } (0, 2)$$

39. $m = -\frac{7}{8}, (2, -9)$
$$y - (-9) = -\frac{7}{8}(x - 2)$$
$$y + 9 = -\frac{7}{8}(x - 2)$$

## Section 10.1 Natural-Number Exponents

### Vocabulary

1. The **base** of the exponential expression $(-5)^3$ is $-5$. The **exponent** is 3.
3. The entire expression $x^n$ is called a **power** of $x$.

### Concepts

5. $(3x)^4 = (3x)(3x)(3x)(3x)$

7. $x^m x^n = x^{m+n}$

9. $\left(\frac{a}{b}\right)^n = \frac{a^n}{b^n}$

11. $\frac{x^m}{x^n} = x^{m-n}$

13. $(xy) = (xy)^1$

15. a. Answers may vary, $(3x^2)^{22}$
    b. Answers may vary, $\left(\frac{3a^3}{b}\right)^{11}$

17. a. $x^2 + x^2 = 2x^2$
    b. $x^2 - x^2 = 0$
    c. $x^2 \cdot x^2 = x^{2+2} = x^4$
    d. $\frac{x^2}{1} = x^2$

19. a. $x^3 + x^2$ is simplified
    b. $x^3 - x^2$ is simplified
    c. $x^3 \cdot x^2 = x^{3+2} = x^5$
    d. $\frac{x^3}{x^2} = x^{3-2} = x$

21. Using the formula for area of a square, $A = s^2$ where $s = a^5$,
$$A = a^5 \cdot a^5$$
$$A = a^{5+5}$$
$$A = a^{10} \text{ mi}^2$$

23. Using the formula for area of a rectangular solid, $V = lwh$, where $l = x^4, w = x^3, h = x^2$
$$V = x^4 \cdot x^3 \cdot x^2$$
$$V = x^{4+3+2}$$
$$V = x^9 \text{ m}^3$$

### Notation

25. $(x^4 x^2)^3 = (x^{4+2})^3 = (x^6)^3 = x^{18}$

27. $4^3$ base 4, exponent 3

29. $x^5$ base $x$, exponent 5

31. $(-3x)^2$ base $-3x$, exponent 2

33. $-\frac{1}{3}y^6$ base $y$, exponent 6

35. $(-4)^2 = (-4)(-4) = 16$

37. $-4^2 = -4 \cdot 4 = -16$

39. $x^5 = x \cdot x \cdot x \cdot x \cdot x$

41. $\left(\frac{t^2}{2}\right)^3 = \left(\frac{t^2}{2}\right)\left(\frac{t^2}{2}\right)\left(\frac{t^2}{2}\right)$

43. $4t(4t)(4t)(4t) = (4t)^4$

45. $-4 \cdot t \cdot t \cdot t = -4t^3$

**Practice**

47. $12^3 \cdot 12^4 = 12^{3+4} = 12^7$

49. $2(2^3)(2^2) = 2^{1+3+2} = 2^6$

51. $a^3 \cdot a^3 = a^{3+3} = a^6$

53. $x^4 x^3 = x^{4+3} = x^7$

55. $a^3 a a^5 = a^{3+1+5} = a^9$

57. $y^3(y^2 y^4)$
$\quad = y^3(y^{2+4})$
$\quad = y^3(y^6)$
$\quad = y^{3+6}$
$\quad = y^9$

59. $\frac{8^{12}}{8^4} = 8^{12-4} = 8^8$

61. $\frac{x^{15}}{x^3} = x^{15-3} = x^{12}$

63. $\frac{c^{10}}{c^9} = c^{10-9} = c$

65. $(3^2)^4 = 3^8$

67. $(y^5)^3 = y^{15}$

69. $(m^{50})^{10} = m^{500}$

71. $(a^2 b^3)(a^3 b^3) = a^{2+3} b^{3+3} = a^5 b^6$

73. $(cd^4)(cd) = c^{1+1} d^{4+1} = c^2 d^5$

75. $xy^2 \cdot x^2 y = x^{1+2} y^{2+1} = x^3 y^3$

77. $\dfrac{y^3 y^4}{y y^2} = \dfrac{y^{3+4}}{y^{1+2}} = \dfrac{y^7}{y^3} = y^{7-3} = y^4$

79. $\dfrac{c^3 d^7}{cd} = c^{3-1} d^{7-1} = c^2 d^6$

81. $(x^2 x^3)^5 = (x^{2+3})^5 = (x^5)^5 = x^{5(5)} = x^{25}$

83. $(3zz^2 z^3)^5$
$\quad = (3z^{1+2+3})^5$
$\quad = (3z^6)^5$
$\quad = 3^5 z^{6(5)}$
$\quad = 243z^{30}$

85. $(x^5)^2 (x^7)^3$
$\quad = x^{5(2)} x^{7(3)}$
$\quad = x^{10} x^{21}$
$\quad = x^{10+21}$
$\quad = x^{31}$

87. $(uv)^4 = u^4 v^4$

89. $(a^3 b^2)^3 = a^{3(3)} b^{2(3)} = a^9 b^6$

91. $(-2r^2 s^3)^3 = (-2)^3 r^{2(3)} s^{3(3)} = -8r^6 s^9$

93. $\left(\dfrac{a}{b}\right)^3 = \dfrac{a^3}{b^3}$

95. $\left(\dfrac{x^2}{y^3}\right)^5 = \dfrac{x^{2(5)}}{y^{3(5)}} = \dfrac{x^{10}}{y^{15}}$

97. $\left(\dfrac{-2a}{b}\right)^5 = \dfrac{(-2)^5 a^5}{b^5} = \dfrac{-32a^5}{b^5}$

99. $\dfrac{(6k)^7}{(6k)^4} = (6k)^{7-4} = (6k)^3 = 216k^3$

101. $\dfrac{(a^2 b)^{15}}{(a^2 b)^9} = (a^2 b)^{15-9} = (a^2 b)^6 = a^{12} b^6$

103. $\dfrac{a^2 a^3 a^4}{(a^4)^2} = \dfrac{a^{2+3+4}}{a^{4(2)}} = \dfrac{a^9}{a^8} = a^{9-8} = a$

105. $\dfrac{(ab^2)^3}{(ab)^2}$
$\quad = \dfrac{a^3 b^{2(3)}}{a^2 b^2}$
$\quad = \dfrac{a^3 b^6}{a^2 b^2}$
$\quad = a^{3-2} b^{6-2}$
$\quad = ab^4$

107. $\dfrac{(r^4 s^3)^4}{(rs^3)^3}$

$= \dfrac{r^{4(4)} s^{3(4)}}{r^3 s^{3(3)}}$

$= \dfrac{r^{16} s^{12}}{r^3 s^9}$

$= r^{16-3} s^{12-9}$

$= r^{13} s^3$

109. $\left(\dfrac{y^3 y}{2yy^2}\right)^3$

$= \dfrac{y^{3(3)} y^3}{2^3 y^3 y^{2(3)}}$

$= \dfrac{y^9 y^3}{8 y^3 y^6}$

$= \dfrac{y^{9+3}}{8 y^{3+6}}$

$= \dfrac{y^{12}}{8 y^9}$

$= \dfrac{1}{8} y^{12-9}$

$= \dfrac{1}{8} y^3$

111. $\left(\dfrac{3t^3 t^4 t^5}{4t^2 t^6}\right)^3$

$= \dfrac{3^3 (t^{3+4+5})^3}{4^3 (t^{2+6})^3}$

$= \dfrac{27 (t^{12})^3}{64 (t^8)^3}$

$= \dfrac{27 t^{12(3)}}{64 t^{8(3)}}$

$= \dfrac{27 t^{36}}{64 t^{24}}$

$= \dfrac{27}{64} t^{36-24}$

$= \dfrac{27}{64} t^{12}$

## Applications

113. **ART HISTORY**

a. The man's height corresponds to the side length of the square, using the formula for the area of a square, $A = s^2$ where

$s = 5x,$

$A = (5x)^2$

$A = 5^2 x^2$

$A = 25x^2 \text{ ft}^2$

b. $3x$ represents the radius of the circle, using the formula for the area of a circle,

$A = \pi r^2,$

$A = \pi (3x)^2$

$A = 9\pi x^2 \text{ ft}^2$

115. **BOUNCING BALL**

a.

b. $32(\frac{1}{2}) = 16$ ft

$32(\frac{1}{2})^2 = 32(\frac{1}{4}) = 8$ ft

$32(\frac{1}{2})^3 = 32(\frac{1}{8}) = 4$ ft

$32(\frac{1}{2})^4 = 32(\frac{1}{16}) = 2$ ft

Since each rebound is one-half of the previous height the above expressions represent the first four bounces of the ball measured in feet.

117. **COMPUTERS**

a. $11111111, 11111101, 11111001, 00000011$

b. Since there are eight total digits comprised of zeros or ones, the total numbers of bytes can be represented with $2^8$.

**Writing**

119. Answers will vary.

121. Answers will vary.

**Review**

123. $y = 2x - 1$

| $x$ | $y$ |
|-----|-----|
| $-2$ | $-5$ |
| $-1$ | $-3$ |
| $0$ | $-1$ |
| $1$ | $1$ |

These ordered pairs match graph c.

125. $y = 3$ is a horizontal line, graph d.

## Section 10.2 Zero and Negative Integer Exponents

### Vocabulary

1. In the exponential expression $8^{-3}$, 8 is the **base** and $-3$ is the **exponent**.
3. Another way to write $2^{-3}$ is to write its **reciprocal** and to change the sign of the exponent: $2^{-3} = \frac{1}{2^3}$.

### Concepts

5. a. $\frac{6^4}{6^4}$

$$= 6^{4-4}$$
$$= 6^0$$

   b. $\frac{6^4}{6^4}$

$$= \frac{6 \cdot 6 \cdot 6 \cdot 6}{6 \cdot 6 \cdot 6 \cdot 6}$$
$$= 1$$

   c. So, we define $6^0$ to be 1, and in general, if $x$ is any nonzero real number, then $x^0 = 1$.

7.

| $x$ | $3^x$ |
|---|---|
| 2 | $3^2 = 9$ |
| 1 | $3^1 = 3$ |
| 0 | $3^0 = 1$ |
| $-1$ | $3^{-1} = \frac{1}{3}$ |
| $-2$ | $3^{-2} = \frac{1}{3^2} = \frac{1}{9}$ |

9.

| $x$ | $(-9)^x$ |
|---|---|
| 2 | $(-9)^2 = 81$ |
| 1 | $(-9)^1 = -9$ |
| 0 | $(-9)^0 = 1$ |
| $-1$ | $(-9)^{-1} = -\frac{1}{9}$ |
| $-2$ | $(-9)^{-2} = \frac{1}{(-9)^2} = \frac{1}{81}$ |

11.

| $x$ | $y$ | Power |
|---|---|---|
| 2 | 4 | $2^2$ |
| 1 | 2 | $2^1$ |
| 0 | 1 | $2^0$ |
| $-1$ | $\frac{1}{2}$ | $2^{-1}$ |
| $-2$ | $\frac{1}{4}$ | $2^{-2}$ |

### Notation

13. $(y^5 y^3)^{-5}$
$$= (y^8)^{-5}$$
$$= y^{-40}$$
$$= \frac{1}{y^{40}}$$

15. In the expression $3x^{-2}$, $x$ is the base and $-2$ is the exponent.

17. a. $-4^2$, 4 is the base and 2 is the exponent.
   $$-4^2 = -16$$
   b. $4^{-2}$, 4 is the base and $-2$ is the exponent.
   $$4^{-2} = \frac{1}{4^2} = \frac{1}{16}$$
   c. $-4^{-2}$, 4 is the base and $-2$ is the exponent.
   $$-4^{-2} = \frac{1}{-4^2} = -\frac{1}{16}$$

**Practice**

19. $7^0 = 1$

21. $\left(\frac{1}{4}\right)^0 = 1$

23. $2x^0 = 2 \cdot 1 = 2$

25. $(-x)^0 = 1$

27. $\left(\frac{a^2 b^3}{ab^4}\right)^0 = 1$

29. $\frac{5}{2x^0} = \frac{5}{2(1)} = \frac{5}{2}$

31. $12^{-2} = \frac{1}{12^2} = \frac{1}{144}$

33. $(-4)^{-1} = \frac{1}{-4} = -\frac{1}{4}$

35. $\frac{1}{5^{-3}} = \frac{1}{\frac{1}{5^3}} = 5^3 = 125$

37. $\frac{2^{-4}}{3^{-1}} = \frac{\frac{1}{2^4}}{\frac{1}{3}} = \frac{1}{2^4}\left(\frac{3}{1}\right) = \frac{3}{2^4} = \frac{3}{16}$

39. $-4^{-3} = -\frac{1}{4^3} = -\frac{1}{64}$

41. $-(-4)^{-3} = -\frac{1}{(-4)^3} = \frac{1}{64}$

43. $x^{-2} = \frac{1}{x^2}$

45. $-b^{-5} = -\frac{1}{b^5}$

47. $(2y)^{-4} = \frac{1}{(2y)^4} = \frac{1}{16y^4}$

49. $(ab^2)^{-3} = \frac{1}{(ab^2)^3} = \frac{1}{a^3 b^6}$

51. $2^5 \cdot 2^{-2}$
$$= 2^5 \cdot \frac{1}{2^2}$$
$$= \frac{2^5}{2^2}$$
$$= 2^{5-2}$$
$$= 2^3$$
$$= 8$$

53. $4^{-3} \cdot 4^{-2} \cdot 4^5$

$\quad = 4^{-3+(-2)} \cdot 4^5$

$\quad = 4^{-5} \cdot 4^5$

$\quad = \frac{1}{4^5} \cdot 4^5$

$\quad = \frac{4^5}{4^5}$

$\quad = 4^{5-5}$

$\quad = 4^0$

$\quad = 1$

55. $\left(\frac{7}{8}\right)^{-1} = \frac{7^{-1}}{8^{-1}} = \frac{\frac{1}{7}}{\frac{1}{8}} = \frac{1}{7} \cdot \frac{8}{1} = \frac{8}{7}$

57. $\frac{3^5 \cdot 3^{-2}}{3^3} = \frac{3^5}{3^3} \cdot \frac{1}{3^2} = \frac{3^5}{3^5} = 1$

59. $\frac{y^4}{y^5} = y^{4-5} = y^{-1} = \frac{1}{y}$

61. $\frac{(r^2)^3}{(r^3)^4} = \frac{r^6}{r^{12}} = r^{6-12} = r^{-6} = \frac{1}{r^6}$

63. $\frac{y^4 y^3}{y^4 y^{-2}} = \frac{y^{4+3}}{y^{4-2}} = \frac{y^7}{y^2} = y^{7-2} = y^5$

65. $\frac{10a^4 a^{-2}}{5a^2 a^0}$

$\quad = \frac{10}{5} \cdot \frac{a^4}{a^2(1)} \cdot \frac{1}{a^2}$

$\quad = 2a^{4-2} \cdot \frac{1}{a^2}$

$\quad = 2a^2 \cdot \frac{1}{a^2}$

$\quad = 2a^{2-2}$

$\quad = 2a^0$

$\quad = 2$

67. $(ab^2)^{-2}$

$\quad = \frac{1}{(ab^2)^2}$

$\quad = \frac{1}{a^2 b^{2 \cdot 2}}$

$\quad = \frac{1}{a^2 b^4}$

69. $(x^2 y)^{-3}$

$\quad = \frac{1}{(x^2 y)^3}$

$\quad = \frac{1}{x^{2 \cdot 3} y^3}$

$\quad = \frac{1}{x^6 y^3}$

71. $(x^{-4} x^3)^3$

$\quad = (x^{-4+3})^3$

$\quad = (x^{-1})^3$

$\quad = x^{-1 \cdot 3}$

$\quad = x^{-3}$

$\quad = \frac{1}{x^3}$

73. $(a^{-2}b^3)^{-4}$

$$= a^{-2(-4)}b^{3(-4)}$$
$$= a^8 b^{-12}$$
$$= a^8 \cdot \frac{1}{b^{12}}$$
$$= \frac{a^8}{b^{12}}$$

75. $(-2x^3 y^{-2})^{-5}$

$$= (-2)^{-5} x^{3 \cdot (-5)} y^{-2 \cdot (-5)}$$
$$= \frac{1}{(-2)^5} x^{-15} y^{10}$$
$$= -\frac{1}{32} \cdot \frac{1}{x^{15}} \cdot y^{10}$$
$$= -\frac{y^{10}}{32x^{15}}$$

77. $\left(\frac{a^3}{a^{-4}}\right)^2 = (a^{3-(-4)})^2 = (a^7)^2 = a^{14}$

79. $\left(\frac{b^5}{b^{-2}}\right)^{-2}$

$$= (b^{5-(-2)})^{-2}$$
$$= (b^7)^{-2}$$
$$= \frac{1}{(b^7)^2}$$
$$= \frac{1}{b^{14}}$$

81. $\left(\frac{4x^2}{3x^{-5}}\right)^4$

$$= \left(\frac{4}{3}\right)^4 (x^{2-(-5)})^4$$
$$= \frac{256}{81}(x^7)^4$$
$$= \frac{256x^{28}}{81}$$

83. $\left(\frac{12y^3 z^{-2}}{3y^{-4}z^3}\right)^2$

$$= (4y^{3-(-4)} z^{-2-3})^2$$
$$= (4y^7 z^{-5})^2$$
$$= 4^2 y^{7 \cdot 2} z^{-5 \cdot 2}$$
$$= 16y^{14} z^{-10}$$
$$= \frac{16y^{14}}{z^{10}}$$

85. $x^{2m} x^m = x^{2m+m} = x^{3m}$

87. $u^{2m} u^{-3m} = u^{2m+(-3m)} = u^{-m} = \frac{1}{u^m}$

89. $\frac{y^{3m}}{y^{2m}} = y^{3m-2m} = y^m$

91. $\frac{x^{3n}}{x^{6n}} = x^{3n-6n} = x^{-3n} = \frac{1}{x^{3n}}$

**Applications**

93. DECIMAL NUMERATION SYSTEM
Hundreds, $10^2$
Tens, $10^1$
Ones, $10^0$
Tenths, $10^{-1}$
Hundredths, $10^{-2}$
Thousandths, $10^{-3}$
Ten-Thousandths, $10^{-4}$

95. RETIREMENT YEARS
Using the formula for present value, $P = A(1 + i)^{-n}$ and substituting appropriate given values,
$i = 0.08, A = 100,000, n = 40$
$P = 100,000(1 + (0.08))^{-40}$
$P = \frac{100,000}{(1+(0.08))^{40}}$
$P = \frac{100,000}{(1.08)^{40}}$
Use a calculator to and round to the nearest penny, $P \approx \$4603.09$
To have \$100,000 in 40 years this couple should invest \$4603.09 in an account yielding 8%.

**Writing**

97. Answers will vary.

**Review**

99. IQ TEST
Using the formula $IQ = \frac{\text{mental age}}{\text{chronological age}} \cdot 100$ where the chronological age is 10 and her $IQ$ is 135.
Solve the equation for mental age.
$IQ(\text{chronological age}) = (\text{mental age}) \cdot 100$
$\frac{IQ(\text{chronological age})}{100} = \text{mental age}$
Now substitute appropriate values,
$\text{mental age} = \frac{135(10)}{100}$
$\text{mental age} = \frac{1350}{100}$
$\text{mental age} = 13.5$
This girl's mental age is 13.5 years.

101. Using the point slope form of an equation, $y = mx + b$, where $m$ is slope and $b$ is the
$y$-intercept, we are told $m = \frac{3}{4}$ and $b = -5$
$y = \frac{3}{4}x + (-5) = \frac{3}{4}x - 5$

**Section 10.3 Scientific Notation**

**Vocabulary**

1.  A number is written in **scientific** notation when it is written as the product of a number between 1 (including 1) and 10.

**Concepts**

3.  $2.5 \times 10^2 = 2.5 \times 100 = 250$

5.  $2.5 \times 10^{-5} = 2.5 \times \frac{1}{10,000}$
    $= \frac{2.5}{10,000} = 0.000025$

7.  $387,000 = 3.87 \times 10^5$

9.  $0.00387 = 3.87 \times 10^{-3}$

11. When multiplying a decimal by $10^5$, the decimal point moves 5 places to the **right**.

13. Dividing a decimal by $10^4$ is equivalent to multiplying it by $10^{-4}$.

15. When a real number greater than 1 is written in scientific notation, the exponent on 10 is a **positive** number.

**Notation**

17. $63.7 \times 10^5$
    $= 6.37 \times 10^1 \times 10^5$
    $= 6.37 \times 10^{1+5}$
    $= 6.37 \times 10^6$

**Practice**

19. $23,000 = 2.3 \times 10^4$

21. $1,700,000 = 1.7 \times 10^6$

23. $0.062 = 6.2 \times 10^{-2}$

25. $0.0000051 = 5.1 \times 10^{-6}$

27. $42.5 \times 10^2$
    $= 4.25 \times 10^1 \times 10^2$
    $= 4.25 \times 10^{1+2}$
    $= 4.25 \times 10^3$

29. $0.25 \times 10^{-2}$
    $= 2.5 \times 10^{-1} \times 10^{-2}$
    $= 2.5 \times 10^{-1+(-2)}$
    $= 2.5 \times 10^{-3}$

31. $2.3 \times 10^2 = 2.3 \times 100 = 230$

33. $8.12 \times 10^5 = 8.12 \times 100,000 = 812,000$

35. $1.15 \times 10^{-3} = 1.15 \times \frac{1}{1000} = 0.00115$

37. $9.76 \times 10^{-4} = 9.76 \times \frac{1}{10,000} = 0.000976$

39. $25 \times 10^6 = 25 \times 1,000,000 = 25,000,000$

41. $0.51 \times 10^{-3} = 0.51 \times \frac{1}{1000} = 0.00051$

43. ASTRONOMY
$25,700,000,000,000 \text{ miles} = 2.57 \times 10^{13} \text{ miles}$

45. GEOGRAPHY
$6.38 \times 10^7$ square miles
$= 6.38 \times 10,000,000$
$= 63,800,000 \text{ mi}^2$

47. LENGTH OF A METER
$0.00622 = 6.22 \times 10^{-3}$ mile

49. $(3.4 \times 10^2)(2.1 \times 10^3)$
$= 3.4 \times 10^2 \times 2.1 \times 10^3$
$= 3.4 \times 2.1 \times 10^2 \times 10^3$
$= 7.14 \times 10^{2+3}$
$= 7.14 \times 10^5$
$= 714,000$

51. $\frac{9.3 \times 10^2}{3.1 \times 10^{-2}}$
$= \frac{9.3}{3.1} \times \frac{10^2}{10^{-2}}$
$= 3 \times 10^{2-(-2)}$
$= 3 \times 10^4$
$= 3 \times 10,000$
$= 30,000$

53. $\frac{96,000}{(12,000)(0.00004)}$
$= \frac{9.6 \times 10^4}{1.2 \times 10^4 \times 4 \times 10^{-5}}$
$= \frac{9.6 \times 10^4}{1.2 \times 4 \times 10^4 \times 10^{-5}}$
$= \frac{9.6 \times 10^4}{4.8 \times 10^{4+(-5)}}$
$= \frac{9.6 \times 10^4}{4.8 \times 10^{-1}}$
$= \frac{9.6}{4.8} \times \frac{10^4}{10^{-1}}$
$= 2 \times 10^{4-(-1)}$
$= 2 \times 10^5$
$= 2 \times 100,000$
$= 200,000$

55. $(456.4)^6 = 9.038030748 \times 10^{15}$

57. $(0.009)^{-6} = 1.881676423 \times 10^{12}$

59. $\left(\frac{1}{3}\right)^{-55} = \frac{1^{-55}}{3^{-55}} = 3^{55} = 1.7449211 \times 10^{26}$

**Applications**

61. WAVELENGTH
Shortest to longest

| Type | Wavelength (m) |
| --- | --- |
| gamma | $8.9 \times 10^{-14}$ |
| x-ray | $2.3 \times 10^{-11}$ |
| ultraviolet | $6.1 \times 10^{-8}$ |
| visible light | $9.3 \times 10^{-6}$ |
| infrared | $3.7 \times 10^{-5}$ |
| microwave | $1.1 \times 10^{-2}$ |
| radio | $3.0 \times 10^{2}$ |

63. PROTON
Since we are told the mass of one proton is $1.7 \times 10^{-24}$ gram, the mass of one million protons is the product of the mass of one and one million.
$$= (1.7 \times 10^{-24} \text{gram})(1,000,000)$$
$$= (1.7 \times 10^{-24} \text{gram})(1 \times 10^6)$$
$$= 1.7 \times 10^{-24} \times 10^6 \text{gram}$$
$$= 1.7 \times 10^{-24+6} \text{gram}$$
$$= 1.7 \times 10^{-18} \text{gram}$$

65. LIGHT YEAR
Since we are told there are $5,280$ feet/mile,
$$(5.87 \times 10^{12} \text{ miles})(5,280 \tfrac{\text{feet}}{\text{mile}})$$
$$= 5.87 \times 10^{12} \times 5.280 \times 10^3 \text{ feet}$$
$$= (5.87)(5.28) \times 10^{12}10^3$$
$$= 30.9936 \times 10^{12+3} \text{ feet}$$
$$= 3.09936 \times 10^1 \times 10^{15} \text{ feet}$$
$$= 3.09936 \times 10^{1+15} \text{feet}$$
$$= 3.09936 \times 10^{16} \text{feet}$$

67. INTEREST EARNED
Using the simple interest formula, $I = Prt$, where $I$ = interest
$P$ = principal = $\$5.09 \times 10^{12}$
$r$ = interest rate = $4\%$
$t$ = time = 1 year
$I = (5.09 \times 10^{12})(0.04)(1)$
$I = 0.2036 \times 10^{12}$
$I = 2.036 \times 10^{-1} \times 10^{12}$
$I = 2.036 \times 10^{-1+12}$
$I = \$\,2.036 \times 10^{11}$

69. SIZE OF THE MILITARY
   a. The number of troops in 1993 was $1.7 \times 10^6$ or $1,700,000$.
   b. The smallest number of troops was $1.4 \times 10^6$ or $1,400,000$ in 1999.
      The largest number of troops was $2.05 \times 10^6$ or $2,050,000$ in 1986.

**Writing**

71. Answers will vary.
73. Answers will vary.

**Review**

75. If $y = -1$ then $-5y^{55} = -5(-1)^{55} = (-5)(-1) = 5$
    Remember a negative number raised to an odd power is negative.

77. $5 + z = z + 5$
    Commutative Property of Addition

79. $3(x - 4) - 6 = 0$
    $3x - 3 \cdot 4 - 6 = 0$
    $3x - 12 - 6 = 0$
    $3x - 18 = 0$
    $3x - 18 + 18 = 0 + 18$
    $3x = 18$
    $\frac{3x}{3} = \frac{18}{3}$
    $x = 6$

## Vocabulary

1. A **polynomial** is a term or a sum of terms in which all variables have whole-number exponents.
3. The degree of a polynomial is the same as the degree of its **term** with the largest degree.
5. The **degree** of the monomial $3x^7$ is 7.
7. $-x^3 - 6x^2 + 9x - 2$ is a polynomial **in** $x$ and is written in **decreasing** powers of $x$.
9. The notation $f(x)$ is read as f **of** $x$.

## Concepts

11. $x^3 - 5x^2 - 2$ is a polynomial

13. $\frac{1}{2x} + 3$ is not a polynomial

15. $x^2 - y^2$ is a polynomial

17. $3x + 7$ is a binomial

19. $y^2 + 4y + 3$ is a trinomial

21. $3z^2$ is a monomial

23. $t - 32$ is a binomial

25. $s^2 - 23s + 31$ is a trinomial

27. $3x^5 - x^4 - 3x^3 + 7$ is none of these

29. $2a^2 - 3ab + b^2$ is a trinomial

31. $3x^4$ fourth degree

33. $-2x^2 + 3x + 1$ second degree

35. $3x - 5$ first degree

37. $-5r^2s^2 - r^3s + 3$ fourth degree

39. $x^{12} + 3x^2y^3$ twelfth degree

41. 38 zero degree

**Notation**

43. $f(x) = -2x^2 + 3x - 1$
$f(2) = -2(2)^2 + 3(2) - 1$
$f(2) = -2(4) + 6 - 1$
$f(2) = -8 + 6 - 1$
$f(2) = -2 - 1$
$f(2) = -3$

45. $f(x) = x^3 + 2x^2 - 1$ is a polynomial function since the right hand side is the sum of one or more terms containing whole number exponents, which defines a polynomial.

**Practice**

47. $f(x) = 5x - 3$
$f(2) = 5(2) - 3 = 10 - 3 = 7$

49. $f(x) = 5x - 3$
$f(-1) = 5(-1) - 3 = -5 - 3 = -8$

51. $f(x) = 5x - 3$
$f(\frac{1}{5}) = 5(\frac{1}{5}) - 3 = 1 - 3 = -2$

53. $f(x) = 5x - 3$
$f(-0.9) = 5(-0.9) - 3 = -4.5 - 3 = -7.5$

55. $g(x) = -x^2 - 4$
$g(0) = -(0)^2 - 4 = -4$

57. $g(x) = -x^2 - 4$
$g(-1) = -(-1)^2 - 4 = -1 - 4 = -5$

59. $g(x) = -x^2 - 4$
$g(1.3) = -(1.3)^2 - 4 = -1.69 - 4 = -5.69$

61. $g(x) = -x^2 - 4$
$g(-13.6) = -(-13.6)^2 - 4$
$g(-13.6) = -184.96 - 4$
$g(-13.6) = -188.96$

63. $h(x) = x^3 - 2x + 3$
$h(0) = 0^3 - 2(0) + 3 = 0 - 0 + 3 = 3$

65. $h(x) = x^3 - 2x + 3$
$h(-2) = (-2)^3 - 2(-2) + 3$
$h(-2) = -8 + 4 + 3$
$h(-2) = -1$

67. $h(x) = x^3 - 2x + 3$
$h(0.9) = 0.9^3 - 2(0.9) + 3$
$h(0.9) = 0.729 - 1.8 + 3$
$h(0.9) = 1.929$

69. $h(x) = x^3 - 2x + 3$
$h(-8.1) = (-8.1)^3 - 2(-8.1) + 3$
$h(-8.1) = -531.441 + 16.2 + 3$
$h(-8.1) = -512.241$

71. $f(x) = -x^4 - x^3 + x^2 + x - 1$
$f(1) = -(1)^4 - (1)^3 + (1)^2 + (1) - 1$
$f(1) = -1$

73. $f(x) = -x^4 - x^3 + x^2 + x - 1$
$f(-2) = -(-2)^4 - (-2)^3 + (-2)^2 + (-2) - 1$
$f(-2) = -7$

**Applications**

75. PACKAGING
If 3 inch squares are cut out then the volume of the resulting box is
$f(3) = 4(3)^3 - 44(3)^2 + 120(3)$
$f(3) = 72 \text{ in}^3$

77. WATER BALLOONS
$f(t) = -16t^2 + 12t + 20$
After 0.5 seconds the height of the balloon was 22 feet.
$f(0.5) = -16(0.5)^2 + 12(0.5) + 20$
$f(0.5) = -16(0.25) + 6 + 20$
$f(0.5) = -4 + 26$
$f(0.5) = 22$

After 1.5 seconds the height of the balloon was 2 feet.
$f(1.5) = -16(1.5)^2 + 12(1.5) + 20$
$f(1.5) = -16(2.25) + 18 + 20$
$f(1.5) = -36 + 38$
$f(1.5) = 2$

79. SUSPENSION BRIDGES
$f(s) = 400 + 0.0066667s^2 - 0.0000001s^4$
If $s$ is 24.6 feet then the approximate length of the cable is 404 feet.
$f(24.6) = 400 + 0.0066667(24.6)^2 - 0.0000001(24.6)^4$
$f(24.6) \approx 400 + 4.034420172 - 0.03662186256$
$f(24.6) \approx 403.9977983$

81. DOLPHINS

$f(x) = -0.05x^2 + 2x$

$f(15)_{dolphin\,1} = -0.05(15)^2 + 2 \cdot 15$

$f(15)_{dolphin\,1} = -0.05(225) + 30$

$f(15)_{dolphin\,1} = -11.25 + 30$

$f(15)_{dolphin\,1} = 18.75 \text{ feet}$

$f(20)_{dolphin\,2} = -0.05(20)^2 + 2 \cdot 20$

$f(20)_{dolphin\,2} = -0.05(400) + 40$

$f(20)_{dolphin\,2} = -20 + 40$

$f(20)_{dolphin\,2} = 20 \text{ feet}$

$f(30)_{dolphin\,3} = -0.05(30)^2 + 2 \cdot 30$

$f(30)_{dolphin\,3} = -0.05(900) + 60$

$f(30)_{dolphin\,3} = -45 + 60$

$f(30)_{dolphin\,3} = 15 \text{ feet}$

**Writing**

83. Answers will vary.

**Review**

85. $-4(3y + 2) \le 28$

$-4(3y) - 4(2) \le 28$

$-12y - 8 \le 28$

$-12y - 8 + 8 \le 28 + 8$

$-12y \le 36$

$\dfrac{-12y}{-12} \ge \dfrac{36}{-12}$

$y \ge -3$

87. $(x^2 x^4)^3 = x^{2 \cdot 3} x^{4 \cdot 3} = x^6 x^{12} = x^{6+12} = x^{18}$

89. $\left(\dfrac{y^2 y^5}{y^4}\right)^3$

$= \left(\dfrac{y^{2+5}}{y^4}\right)^3$

$= \left(\dfrac{y^7}{y^4}\right)^3$

$= (y^{7-4})^3$

$= (y^3)^3$

$= y^{3 \cdot 3}$

$= y^9$

**Section 10.5 Adding and Subtracting Polynomials**

**Vocabulary**

1. The expression $(x^2 - 3x + 2) + (x^2 - 4x)$ is the sum of two **polynomials**.
3. "To add or subtract like terms" means to combine their **coefficients** and keep the same variables with the same exponents.

**Concepts**

5. To add like monomials, combine like **terms**.
7. To add two polynomials, combine any **like** terms contained in the polynomials.
9. When the sign preceding parentheses is a $-$ sign, we can remove the parentheses by dropping the sign and the parentheses, and **changing** the sign of every term within the parentheses.

11. $-(-2x^2 - 3x + 4)$
$= (-1)(-2x^2) - (-1)(3x) + (-1)(4)$
$= 2x^2 + 3x - 4$

13. JETS
The polynomial representing the length of the passenger jet is
$(9x - 15) + (2x + 3)$ feet
$= 9x - 15 + 2x + 3$ feet
$= (11x - 12)$ feet

**Notation**

15. $(5x^2 + 3x) - (7x^2 - 2x)$
$= 5x^2 + 3x - 7x^2 + 2x$
$= 5x^2 - 7x^2 + 3x + 2x$
$= -2x^2 + 5x$

**Practice**

17. $4y + 5y = 9y$

19. $8t^2 + 4t^2 = 12t^2$

21. $-32u^3 - 16u^3 = -32u^3 + (-16u^3) = -48u^3$

23. $1.8x - 1.9x = 1.8x + (-1.9x) = -0.1x$

25. $\frac{1}{2}st + \frac{3}{2}st = \frac{4}{2}st = 2st$

27. $3r - 4r + 7r = -1r + 7r = 6r$

29. $-4ab + 4ab - ab = 0 + (-ab) = -ab$

31. $(3x)^2 - 4x^2 + 10x^2$
$= 9x^2 + (-4x^2) + 10x^2$
$= 15x^2$

33. $(3x + 7) + (4x - 3)$
$\quad = 3x + 7 + 4x - 3$
$\quad = 7x + 4$

35. $(4a + 3) - (2a - 4)$
$\quad = 4a + 3 - 2a + 4$
$\quad = 2a + 7$

37. $(2x + 3y) + (5x - 10y)$
$\quad = 2x + 3y + 5x - 10y$
$\quad = 7x - 7y$

39. $(-8x - 3y) - (-11x + y)$
$\quad = -8x - 3y + 11x - y$
$\quad = 3x - 4y$

41. $(3x^2 - 3x - 2) + (3x^2 + 4x - 3)$
$\quad = 3x^2 - 3x - 2 + 3x^2 + 4x - 3$
$\quad = 6x^2 + x - 5$

43. $(2b^2 + 3b - 5) - (2b^2 - 4b - 9)$
$\quad = 2b^2 + 3b - 5 - 2b^2 + 4b + 9$
$\quad = 7b + 4$

45. $(2x^2 - 3x + 1) - (4x^2 - 3x + 2) + (2x^2 + 3x + 2)$
$\quad = 2x^2 - 3x + 1 - 4x^2 + 3x - 2 + 2x^2 + 3x + 2$
$\quad = 3x + 1$

47. $(4.52x^2 + 1.13x - 0.89) + (9.02x^2 - 7.68x + 7.04)$
$\quad = 13.54x^2 - 6.55x + 6.15$

49.
$$\begin{array}{r} 3x^2 + 4x + 5 \\ + \ 2x^2 - 3x + 6 \\ \hline 5x^2 + x + 11 \end{array}$$

51.
$$\begin{array}{r} 2x^3 - 3x^2 + 4x - 7 \\ + \ -9x^3 - 4x^2 - 5x + 6 \\ \hline -7x^3 - 7x^2 - x - 1 \end{array}$$

53.
$$\begin{array}{r} -3x^2 + 4x + \ 25 \\ + \ 5x^2 \qquad - 12 \\ \hline 2x^2 + 4x + 13 \end{array}$$

55.
$$\begin{array}{r} 3x^2 + 4x - 5 \\ - \ -2x^2 - 2x + 3 \\ \hline \end{array}$$
becomes
$$\begin{array}{r} 3x^2 + 4x - 5 \\ + \ 2x^2 + 2x - 3 \\ \hline 5x^2 + 6x - 8 \end{array}$$

57. $$4x^3 + 4x^2 - 3x + 10$$
$$-\ \ 5x^3 - 2x^2 - 4x - 4$$
becomes
$$4x^3 + 4x^2 - 3x + 10$$
$$+\ -5x^3 + 2x^2 + 4x + 4$$
$$-x^3 + 6x^2 + x + 14$$

59. $$-2x^2y^2 \qquad\quad + 12y^2$$
$$-\ \ 10x^2y^2 + 9xy - 24y^2$$
becomes
$$-2x^2y^2 \qquad\quad + 12y^2$$
$$+\ -10x^2y^2 - 9xy + 24y^2$$
$$-12x^2y^2 - 9xy + 36y^2$$

61. $[(3t^3 + t^2) + (-t^3 + 6t - 3)] - [t^3 - 2t^2 + 2]$
$\quad = [3t^3 + t^2 - t^3 + 6t - 3] + [-t^3 + 2t^2 - 2]$
$\quad = 2t^3 + t^2 + 6t - 3 - t^3 + 2t^2 - 2$
$\quad = t^3 + 3t^2 + 6t - 5$

63. $[3x^2 + 4x - 7] + [-2x^2 - 7x + 1 + (-4x^2 + 8x - 1)]$
$\quad = 3x^2 + 4x - 7 - 2x^2 - 7x + 1 - 4x^2 + 8x - 1$
$\quad = -3x^2 + 5x - 7$

65. $2(x + 3) + 4(x - 2)$
$\quad = 2x + 6 + 4x - 8$
$\quad = 6x - 2$

67. $-2(x^2 + 7x - 1) - 3(x^2 - 2x + 2)$
$\quad = -2x^2 - 14x + 2 - 3x^2 + 6x - 6$
$\quad = -5x^2 - 8x - 4$

69. $2(2y^2 - 2y + 2) - 4(3y^2 - 4y - 1) + 4(y^2 - y - 1)$
$\quad = 4y^2 - 4y + 4 - 12y^2 + 16y + 4 + 4y^2 - 4y - 4$
$\quad = -4y^2 + 8y + 8$

71. $2(ab^2 - b) - 3(a + 2ab) + (b - a + a^2b)$
$\quad = 2ab^2 - 2b - 3a - 6ab + b - a + a^2b$
$\quad = a^2b + 2ab^2 - 6ab - 4a - b$

73. This is an isoceles triangle therefore two sides have the same measure.
$\quad 2(x^2 + 3x + 1) + (x^2 - 4)$
$\quad = 2x^2 + 6x + 2 + x^2 - 4$
$\quad = (3x^2 + 6x - 2)$ yards

**Applications**

75. GREEK ARCHITECTURE
$\quad (x^2 - 3x + 2) - (5x - 10)$ feet
$\quad = x^2 - 3x + 2 - 5x + 10$ feet
$\quad = (x^2 - 8x + 12)$ feet

77. **AUTO MECHANICS**

$\pi + (x^2 - 3x) + 1.5\pi + (x^2 + 9x) + 2\pi + (x^2 + 5x)$
$= 4.5\pi + x^2 - 3x + x^2 + 9x + x^2 + 5x$
$= (4.5\pi + 3x^2 + 11x)$ inches
$= (3x^2 + 11x + 4.5\pi)$ inches

79. **VALUE OF A HOUSE**

Using the given polynomial function $y = 900x + 105,000$ and substituting $x = 10$ years gives
$y = 900(10) + 105,000$
$y = 9000 + 105,000$
$y = \$114,000$

81. **VALUE OF TWO HOUSES**

The polynomial for the value of both houses is found by summing the polynomials for each house.
$y = 900x + 105,000 + 1000x + 120,000$
$y = 1900x + 225,000$

83. **VALUE OF A COMPUTER**

Let $x$ represent years of depreciation, therefore $-1100x$ represents depreciation for $x$ years.
Since the first computer cost $6600 we can write a polynomial for the value after $x$ years.
$y = -1100x + 6600$
(take a moment, when $x = 0$ the computer has not depreciated yet, it was just purchased, so $y = 6600$)

85. **VALUE OF TWO COMPUTERS**

The combined value of both computers after $x$ years is found by first combining their depreciations
and then combining their intial costs.
Depreciation: $-1100 + (-1700) = -2800$
Initial Cost: $6600 + 9200 = 15,800$
$y = -2800x + 15,800$
(when $x = 0$ these computers have not depreciated yet, they were just purchased, so $y = 15,800$)

**Writing**

87. Answers will vary
89. Answers will vary

**Review**

91. The sum of the measures of the interior angles of a triangle is $180^0$.

93. $-4(3x - 3) \geq -12$
$-12x + 12 \geq -12$
$-12x + 12 - 12 \geq -12 - 12$
$-12x \geq -24$
$\frac{-12x}{-12} \leq \frac{-24}{-12}$
$x \leq 2$

**Section 10.6 Multiplying Polynomials**

**Vocabulary**

1. The expression $(2a - 4)(3a + 5)$ is the product of two **binomials**.
3. In the acronym FOIL, F stands for **first** terms, O for **outer** terms, I for **inner** terms, and L for **last** terms.

**Concepts**

5. $(2x + 5)(3x - 4)$
   The product of the first terms is $2x \cdot 3x = 6x^2$.

7. $(2x + 5)(3x - 4)$
   The product of the inner terms is $5 \cdot 3x = 15x$

9. STAMPS
   The area of this stamp is the product of its length and width or,
   $(3x - 1)(2x + 1)$
   $= (3x - 1)(2x) + (3x - 1)(1)$
   $= 6x^2 - 2x + 3x - 1$
   $= (6x^2 + x - 1) \text{ cm}^2$

**Notation**

11. $7x(3x^2 - 2x + 5)$
    $= 7x \cdot 3x^2 - 7x \cdot (2x) + 7x \cdot 5$
    $= 21x^3 - 14x^2 + 35x$

**Practice**

13. $(3x^2)(4x^3) = 12x^5$

15. $(3b^2)(-2b)(4b^3) = (-6b^3)(4b^3) = -24b^6$

17. $(2x^2 y^3)(3x^3 y^2) = 6x^5 y^5$

19. $(x^2 y^5)(x^2 z^5)(-3z^3)$
    $= (x^4 y^5 z^5)(-3z^3)$
    $= -3x^4 y^5 z^8$

21. $3(x + 4) = 3x + 12$

23. $-4(t + 7) = -4t - 28$

25. $3x(x - 2) = 3x^2 - 6x$

27. $-2x^2(3x^2 - x) = -6x^4 + 2x^3$

29. $3xy(x + y) = 3x^2 y + 3xy^2$

31. $2x^2(3x^2 + 4x - 7) = 6x^4 + 8x^3 - 14x^2$

33. $(3x)(-2x^2)(x + 4)$
$\quad = -6x^3(x + 4)$
$\quad = -6x^4 - 24x^3$

35. $(a + 4)(a + 5)$
$\quad = a \cdot a + a \cdot 5 + 4 \cdot a + 4 \cdot 5$
$\quad = a^2 + 5a + 4a + 20$
$\quad = a^2 + 9a + 20$

37. $(3x - 2)(x + 4)$
$\quad = 3x \cdot x + 3x \cdot 4 - 2 \cdot x - 2 \cdot 4$
$\quad = 3x^2 + 12x - 2x - 8$
$\quad = 3x^2 + 10x - 8$

39. $(2a + 4)(3a - 5)$
$\quad = 2a \cdot 3a + 2a \cdot (-5) + 4 \cdot 3a + 4 \cdot (-5)$
$\quad = 6a^2 - 10a + 12a - 20$
$\quad = 6a^2 + 2a - 20$

41. $(3x - 5)(2x + 1)$
$\quad = 3x \cdot 2x + 3x \cdot 1 - 5 \cdot 2x - 5 \cdot 1$
$\quad = 6x^2 + 3x - 10x - 5$
$\quad = 6x^2 - 7x - 5$

43. $(x + 3)(2x - 3)$
$\quad = x \cdot 2x + x \cdot (-3) + 3 \cdot 2x + 3 \cdot (-3)$
$\quad = 2x^2 - 3x + 6x - 9$
$\quad = 2x^2 + 3x - 9$

45. $(2t + 3s)(3t - s)$
$\quad = 2t \cdot 3t + 2t \cdot (-s) + 3s \cdot 3t + 3s \cdot (-s)$
$\quad = 6t^2 - 2st + 9st - 3s^2$
$\quad = 6t^2 + 7st - 3s^2$

47. $(x + y)(x + z)$
$\quad = x \cdot x + x \cdot z + y \cdot x + y \cdot z$
$\quad = x^2 + xz + xy + yz$

49. $(4t - u)(-3t + u)$
$\quad = 4t \cdot (-3t) + 4t \cdot u - u \cdot (-3t) - u \cdot u$
$\quad = -12t^2 + 4tu + 3tu - u^2$
$\quad = -12t^2 + 7tu - u^2$

51. $4(2x + 1)(x - 2)$
$\quad = (8x + 4)(x - 2)$
$\quad = 8x^2 - 16x + 4x - 8$
$\quad = 8x^2 - 12x - 8$

53. $3a(a+b)(a-b)$
   $= 3a[a^2 - ab + ab - b^2]$
   $= 3a(a^2 - b^2)$
   $= 3a^3 - 3ab^2$

55. $2t(t+2) + 3t(t-5)$
   $= 2t^2 + 4t + 3t^2 - 15t$
   $= 5t^2 - 11t$

57. $(x+y)(x-y) + x(x+y)$
   $= x^2 - xy + xy - y^2 + x^2 + xy$
   $= 2x^2 + xy - y^2$

59. $(x+4)(x+4)$
   $= x^2 + 8x + 16$

61. $(t-3)(t-3)$
   $= t^2 - 6t + 9$

63. $(r+4)(r-4)$
   $= r^2 - 4r + 4r - 16$
   $= r^2 - 16$

65. $(4x+5)(4x-5)$
   $= 16x^2 - 20x + 20x - 25$
   $= 16x^2 - 25$

67. $(2s+1)(2s+1)$
   $= 4s^2 + 4s + 1$

69. $(x+5)^2$
   $= (x+5)(x+5)$
   $= x^2 + 10x + 25$

71. $(x-2y)^2$
   $= (x-2y)(x-2y)$
   $= x^2 - 4xy + 4y^2$

73. $(2a-3b)^2$
   $= (2a-3b)(2a-3b)$
   $= 4a^2 - 12ab + 9b^2$

75. $(4x+5y)^2$
   $= (4x+5y)(4x+5y)$
   $= 16x^2 + 40xy + 25y^2$

77. $A_{triangle} = \frac{1}{2}bh$
   $A_{triangle} = \frac{1}{2}(4x-2)(2x-2)$
   $A_{triangle} = \frac{1}{2}(8x^2 - 8x - 4x + 4)$
   $A_{triangle} = \frac{1}{2}(8x^2 - 12x + 4)$
   $A_{triangle} = (4x^2 - 6x + 2)\,\text{cm}^2$

79. $A_{circle} = \pi r^2$
$A_{circle} = \pi(x+3)^2$
$A_{circle} = \pi(x+3)(x+3)$
$A_{circle} = \pi(x^2 + 3x + 3x + 9)$
$A_{circle} = \pi(x^2 + 6x + 9)\,\text{in}^2$

81. $(x+2)(x^2 - 2x + 3)$
$= x^3 - 2x^2 + 3x + 2x^2 - 4x + 6$
$= x^3 - x + 6$

83. $(4t+3)(t^2 + 2t + 3)$
$= 4t^3 + 8t^2 + 12t + 3t^2 + 6t + 9$
$= 4t^3 + 11t^2 + 18t + 9$

85. $(-3x + y)(x^2 - 8xy + 16y^2)$
$= -3x^3 + 24x^2y - 48xy^2 + x^2y - 8xy^2 + 16y^3$
$= -3x^3 + 25x^2y - 56xy^2 + 16y^3$

87. $x^2 - 2x + 1$
$$\begin{array}{r} x + 2 \\ \hline x^3 - 2x^2 + x \\ 2x^2 - 4x + 2 \\ \hline x^3 - 3x + 2 \end{array}$$

89. $4x^2 + 3x - 4$
$$\begin{array}{r} 3x + 2 \\ \hline 12x^3 + 9x^2 - 12x \\ 8x^2 + 6x - 8 \\ \hline 12x^3 + 17x^2 - 6x - 8 \end{array}$$

91. $(s-4)(s+1) = s^2 + 5$
$s^2 - 3s - 4 = s^2 + 5$
$s^2 - s^2 - 3s - 4 = s^2 - s^2 + 5$
$-3s - 4 = 5$
$-3s - 4 + 4 = 5 + 4$
$-3s = 9$
$\frac{-3s}{-3} = \frac{9}{-3}$
$s = -3$

93. $z(z+2) = (z+4)(z-4)$
$z^2 + 2z = z^2 - 16$
$z^2 - z^2 + 2z = z^2 - z^2 - 16$
$2z = -16$
$\frac{2z}{2} = \frac{-16}{2}$
$z = -8$

95. $(x+4)(x-4) = (x-2)(x+6)$
$x^2 - 16 = x^2 + 6x - 2x - 12$
$x^2 - 16 = x^2 + 4x - 12$
$x^2 - x^2 - 16 = x^2 - x^2 + 4x - 12$
$-16 = 4x - 12$
$-16 + 12 = 4x - 12 + 12$
$-4 = 4x$
$\frac{-4}{4} = \frac{4x}{4}$
$-1 = x$

97. $(a-3)^2 = (a+3)^2$
$(a-3)(a-3) = (a+3)(a+3)$
$a^2 - 6a + 9 = a^2 + 6a + 9$
$a^2 - a^2 - 6a + 9 = a^2 - a^2 + 6a + 9$
$-6a + 9 = 6a + 9$
$-6a + 6a + 9 = 6a + 6a + 9$
$9 = 12a + 9$
$9 - 9 = 12a + 9 - 9$
$0 = 12a$
$0 = a$

**Applications**

99. TOYS

The perimeter of the toy screen is, $P = 2l + 2w$, where $l = 7x + 3$ and $w = (7x + 4) - 2x$, where $2x$ is the total width of the frame. Note: The question asks about the dimensions of the *screen* only.
$P = 2(7x + 3) + 2((7x + 4) - 2x)$
$P = 14x + 6 + 2(5x + 4)$
$P = 14x + 6 + 10x + 8$
$P = (24x + 14)\,\text{cm}$

The area of the toy screen is, $A = lw$
$A = (7x + 3)((7x + 4) - 2x)$
$A = (7x + 3)(5x + 4)$
$A = 35x^2 + 28x + 15x + 12$
$A = (35x^2 + 43x + 12)\,\text{cm}^2$

101. GARDENING

a.

| | Corn | Tomatoes | Beans | Carrots |
|---|---|---|---|---|
| $Area$ | $x^2$ ft$^2$ | $6x$ ft$^2$ | $5x$ ft$^2$ | $30$ ft$^2$ |

$Area = l \cdot w$

The total area of the garden is the sum of these four areas,

$A = x^2 + 6x + 5x + 30$

$A = (x^2 + 11x + 30) \text{ ft}^2$

b. Depending on perspective, the length of this garden is $(6 + x)$ feet and the width is $(5 + x)$ feet.

$A = lw$

$A = (6 + x)(5 + x)$

$A = 30 + 6x + 5x + x^2$

$A = (x^2 + 11x + 30) \text{ ft}^2$

c. The area calculated in part (a) is the same as the area calculated in part (b).

103. INTEGER PROBLEM

Let $a$ and $a + 1$ represent two consecutive integers, then the difference of the squares may be represented as $(a + 1)^2 - a^2$ and we are told this difference is 11.

$(a + 1)^2 - a^2 = 11$

$(a + 1)(a + 1) - a^2 = 11$

$a^2 + 2a + 1 - a^2 = 11$

$2a + 1 = 11$

$2a + 1 - 1 = 11 - 1$

$2a = 10$

$\frac{2a}{2} = \frac{10}{2}$

$a = 5$

Therefore $a + 1 = 5 + 1 = 6$, so the two consecutive integers are 5 and 6.

105. STONE-GROUND FLOUR

Let $r$ be the radius of the smaller mill stone, then the radius of the larger stone is $(r + 3)$.

The area of the smaller stone is $A_{small} = \pi r^2$ and the area of the larger stone is $A_{larger} = \pi(r + 3)^2$.

We are told that the difference of these two areas is $15\pi$ m$^2$, so the equation to represent this difference is

$A_{larger} - A_{small} = 15\pi$

$\pi(r + 3)^2 - \pi r^2 = 15\pi$

$\pi(r + 3)(r + 3) - \pi r^2 = 15\pi$

$\pi(r^2 + 6r + 9) - \pi r^2 = 15\pi$

$\pi r^2 + 6\pi r + 9\pi - \pi r^2 = 15\pi$

$6\pi r + 9\pi = 15\pi$

$6\pi r + 9\pi - 9\pi = 15\pi - 9\pi$

$6\pi r = 6\pi$

$\frac{6\pi r}{6\pi} = \frac{6\pi}{6\pi}$

$r = 1 \text{ m}$

The radius of the smaller stone is one meter therefore the radius of the larger mill stone is

$r + 3 = 1 + 3 = 4 \text{ m}$

107. **BASEBALL**

Let $s$ represent the distance between bases in softball, then $s + 30$ is the distance between bases in the major league.
$$A_{major\,league} = (s + 30)^2$$
$$A_{softball} = s^2$$
The difference of these two areas, we are told, is 4500 ft$^2$.
$$A_{major\,league} - A_{softball} = 4500$$
$$(s + 30)^2 - s^2 = 4500$$
$$(s + 30)(s + 30) - s^2 = 4500$$
$$s^2 + 60s + 900 - s^2 = 4500$$
$$60s + 900 = 4500$$
$$60s + 900 - 900 = 4500 - 900$$
$$60s = 3600$$
$$\frac{60s}{60} = \frac{3600}{60}$$
$$s = 60$$
The distance between the bases in softball is 60 feet therefore the distance between the bases in the major league is $(s + 30) = 60 + 30 = 90$ feet.

## Writing

109.　Answers will vary

## Review

111. The slope of line AB is found using
$$m = \frac{y_2 - y_1}{x_2 - x_1}$$
where the ordered pairs of A and B are $(-4, -2)$ and $(2, 4)$ respectively.
Substituting into the formula for slope
$$m = \frac{4 - (-2)}{2 - (-4)}, \; m = \frac{6}{6} = 1.$$

113. The slope of line CD is found using $m = \frac{y_2 - y_1}{x_2 - x_1}$ where the ordered pairs of C and D are $(2, -2)$ and $(-1, 0)$ respectively. Substituting into the formula for slope,
$$m = \frac{-2 - (0)}{2 - (-1)},$$
$$m = \frac{-2}{3} = -\frac{2}{3}.$$

115. The $y$-intercept of AB is $(0, 2)$ as read from the graph.

**Section 10.7 Dividing Polynomials by Monomials**

**Vocabulary**

1. A **polynomial** is an algebraic expression that is the sum of one or more terms containing whole-number exponents.
3. A binomial is a polynomial with **two** terms.
5. $\frac{x^m}{x^n} = x^{m-n}$ is a rule for **exponents**.

**Concepts**

7. $\frac{18x+9}{9} = \frac{18x}{9} + \frac{9}{9}$

9. Slashes denote common factors being divided out from the denominator and numerator.

11. a. $d = rt$
$$\frac{d}{r} = \frac{rt}{r}$$
$$\frac{d}{r} = t$$

b.

|            | $r$  | $t$    | $d$    |
|------------|------|--------|--------|
| Motorcycle | $2x$ | $3x^2$ | $6x^3$ |

13. $(10x + 35)$ has a value in nickels of
$$\tfrac{1}{5}(10x + 35)$$
$$= \tfrac{1}{5}(10x) + \tfrac{1}{5}(35)$$
$$= (2x + 7)$$

**Notation**

15. $\frac{a^2 b^3}{a^3 b^2} = \frac{a \cdot a \cdot b \cdot b \cdot b}{a \cdot a \cdot a \cdot b \cdot b}$
$$= \frac{b}{a}$$

**Practice**

17. $\frac{5}{15} = \frac{1}{3}$

19. $\frac{-125}{75} = -\frac{5}{3}$

21. $\frac{120}{160} = \frac{3}{4}$

23. $\frac{-3612}{-3612} = 1$

25. $\frac{x^5}{x^2} = x^3$

27. $\frac{r^3 s^2}{r s^3} = \frac{r^2}{s}$

29. $\frac{8x^3 y^2}{4x y^3} = \frac{2x^2}{y}$

31. $\frac{12u^5 v}{-4u^2 v^3} = -\frac{3u^3}{v^2}$

33. $\dfrac{-16r^3y^2}{-4r^2y^4} = \dfrac{4r}{y^2}$

35. $\dfrac{-65rs^2t}{15r^2s^3t} = -\dfrac{13}{3rs}$

37. $\dfrac{x^2x^3}{xy^6} = \dfrac{x^5}{xy^6} = \dfrac{x^4}{y^6}$

39. $\dfrac{(a^3b^4)^3}{ab^4} = \dfrac{a^9b^{12}}{ab^4} = a^8b^8$

41. $\dfrac{15(r^2s^3)^2}{-5(rs^5)^3} = -\dfrac{3r^4s^6}{r^3s^{15}} = -\dfrac{3r}{s^9}$

43. $\dfrac{-32(x^3y)^3}{128(x^2y^2)^3} = -\dfrac{x^9y^3}{4x^6y^6} = -\dfrac{x^3}{4y^3}$

45. $\dfrac{-(4x^3y^3)^2}{(x^2y^4)^3} = -\dfrac{16x^6y^6}{x^6y^{12}} = -\dfrac{16}{y^6}$

47. $\dfrac{(a^2a^3)^4}{(a^4)^3} = \dfrac{(a^5)^4}{(a^4)^3} = \dfrac{a^{20}}{a^{12}} = a^8$

49. $\dfrac{6x+9}{3} = \dfrac{1}{3}(6x+9) = \dfrac{6x}{3} + \dfrac{9}{3} = 2x+3$

51. $\dfrac{5x-10y}{25xy} = \dfrac{1}{25xy}(5x-10y) = \dfrac{5x}{25xy} - \dfrac{10y}{25xy}$
$= \dfrac{1}{5y} - \dfrac{2}{5x}$

53. $\dfrac{3x^2+6y^3}{3x^2y^2} = \dfrac{1}{3x^2y^2}(3x^2+6y^3) = \dfrac{3x^2}{3x^2y^2} + \dfrac{6y^3}{3x^2y^2}$
$= \dfrac{1}{y^2} + \dfrac{2y}{x^2}$

55. $\dfrac{15a^3b^2-10a^2b^3}{5a^2b^2} = \dfrac{1}{5a^2b^2}(15a^3b^2 - 10a^2b^3)$
$= \dfrac{15a^3b^2}{5a^2b^2} - \dfrac{10a^2b^3}{5a^2b^2}$
$= 3a - 2b$

57. $\dfrac{4x-2y+8z}{4xy} = \dfrac{1}{4xy}(4x-2y+8z)$
$= \dfrac{4x}{4xy} - \dfrac{2y}{4xy} + \dfrac{8z}{4xy}$
$= \dfrac{1}{y} - \dfrac{1}{2x} + \dfrac{2z}{xy}$

59. $\dfrac{12x^3y^2-8x^2y-4x}{4xy}$
$= \dfrac{1}{4xy}(12x^3y^2 - 8x^2y - 4x)$
$= \dfrac{12x^3y^2}{4xy} - \dfrac{8x^2y}{4xy} - \dfrac{4x}{4xy}$
$= 3x^2y - 2x - \dfrac{1}{y}$

61. $\dfrac{-25x^2y+30xy^2-5xy}{-5xy}$
$= \dfrac{1}{-5xy}(-25x^2y + 30xy^2 - 5xy)$
$= \dfrac{-25x^2y}{-5xy} + \dfrac{30xy^2}{-5xy} - \dfrac{5xy}{-5xy}$
$= 5x - 6y + 1$

63. $\dfrac{5x(4x-2y)}{2y}$

$= \dfrac{20x^2 - 10xy}{2y}$

$= \dfrac{1}{2y}(20x^2 - 10xy)$

$= \dfrac{20x^2}{2y} - \dfrac{10xy}{2y}$

$= \dfrac{10x^2}{y} - 5x$

65. $\dfrac{(-2x)^3 + (3x^2)^2}{6x^2}$

$= \dfrac{-8x^3 + 9x^4}{6x^2}$

$= \dfrac{1}{6x^2}(-8x^3 + 9x^4)$

$= \dfrac{-8x^3}{6x^2} + \dfrac{9x^4}{6x^2}$

$= -\dfrac{4x}{3} + \dfrac{3x^2}{2}$

67. $\dfrac{4x^2y^2 - 2(x^2y^2 + xy)}{2xy}$

$= \dfrac{4x^2y^2 - 2x^2y^2 - 2xy}{2xy}$

$= \dfrac{2x^2y^2 - 2xy}{2xy}$

$= \dfrac{1}{2xy}(2x^2y^2 - 2xy)$

$= \dfrac{2x^2y^2}{2xy} - \dfrac{2xy}{2xy}$

$= xy - 1$

69. $\dfrac{(3x-y)(2x-3y)}{6xy}$

$= \dfrac{6x^2 - 9xy - 2xy + 3y^2}{6xy}$

$= \dfrac{6x^2 - 11xy + 3y^2}{6xy}$

$= \dfrac{1}{6xy}(6x^2 - 11xy + 3y^2)$

$= \dfrac{6x^2}{6xy} - \dfrac{11xy}{6xy} + \dfrac{3y^2}{6xy}$

$= \dfrac{x}{y} - \dfrac{11}{6} + \dfrac{y}{2x}$

71. $\dfrac{(a+b)^2 - (a-b)^2}{2ab}$

$= \dfrac{(a+b)(a+b) - (a-b)(a-b)}{2ab}$

$= \dfrac{a^2 + 2ab + b^2 - (a^2 - 2ab + b^2)}{2ab}$

$= \dfrac{a^2 + 2ab + b^2 - a^2 + 2ab - b^2}{2ab}$

$= \dfrac{4ab}{2ab}$

$= 2$

## Applications

73. POOL

Assuming that each side of the rack has the same measure, each side has measure $\dfrac{(6x^2 - 3x + 9)}{3}$ where the denominator represents the three sided rack.

$\dfrac{(6x^2 - 3x + 9)}{3}$

$= \dfrac{1}{3}(6x^2 - 3x + 9)$

$= (2x^2 - x + 3)$ inches

75. AIR CONDITIONING

Since volume is found using the formula $V = lwh$, we can solve this for $h$ which is,

$$\frac{V}{lw} = \frac{lwh}{lw}$$
$$\frac{V}{lw} = h$$

Now substituting the given amounts,

$$\frac{(36x^3 - 24x^2)}{4x \cdot 3x} = h$$
$$\frac{(36x^3 - 24x^2)}{12x^2} = h$$
$$\frac{1}{12x^2}(36x^3 - 24x^2) = h$$
$$\frac{36x^3}{12x^2} - \frac{24x^2}{12x^2} = h$$
$$(3x - 2) \text{ feet} = h$$

77. CONFIRMING FORMULAS

A. $l = \frac{P - 2w}{2}$

B. $l = \frac{P}{2} - w$

Leaving formula A alone and finding a common denominator on the right side of formula B yields,

$$l = \frac{P}{2} - w\left(\frac{2}{2}\right)$$
$$l = \frac{P}{2} - \left(\frac{2w}{2}\right)$$
$$l = \frac{P}{2} - \frac{2w}{2}$$
$$l = \frac{1}{2}(P - 2w)$$
$$l = \frac{P - 2w}{2} \text{ which is formula A, therefore the two formulas are the same.}$$

79. ELECTRIC BILLS

E. $\frac{0.08x + 5}{x}$

F. $0.08x + \frac{5}{x}$

Leaving formula E alone and finding a common denominator for formula F yields,

$$0.08x\left(\frac{x}{x}\right) + \frac{5}{x}$$
$$= \left(\frac{0.08x^2}{x}\right) + \frac{5}{x}$$
$$= \frac{0.08x^2}{x} + \frac{5}{x}$$
$$= \frac{1}{x}(0.08x^2 + 5)$$
$$= \frac{0.08x^2 + 5}{x}, \text{ which is not formula E, therefore the two formulas are not the same.}$$

**Writing**

81. Answers will vary

**Review**

83. $5a^2b + 2ab^2$ is a binomial
85. $-2x^3 + 3x^2 - 4x + 12$ is none of the above
87. The degree of $3x^2 - 2x + 4$ is two.

**Vocabulary**

1. In the division $x + 1 \overline{)x^2 + 2x + 1}$ , the expression $x + 1$ is called the **divisor** and $x^2 + 2x + 1$ is called the **dividend**.

3. If a division does not come out even, the leftover part is called a **remainder**.

**Concepts**

5. $4x^3 + 7x - 2x^2 + 6$ in descending order is $4x^3 - 2x^2 + 7x + 6$

7. $9x + 2x^2 - x^3 + 6x^4$ in descending order is $6x^4 - x^3 + 2x^2 + 9x$

9. The missing term(s) in $5x^4 + 2x^2 - 1$ are the $x^3$ and $x^1$ terms, think of them as having $0$ as their coefficients.

11. The second quotient $(x^3 + 3x^2 + 9x + 27)$ seems reasonable since $x$ divided into $x^4$ is $x^3$.

13. a. If $d = rt$ then
$$\frac{d}{t} = \frac{rt}{t}$$
$$\frac{d}{t} = r$$

b.

|  | $r$ | $t$ | $d$ |
|---|---|---|---|
| Subway | $x - 3$ | $x + 4$ | $x^2 + x - 12$ |

$$r = \frac{x^2 + x - 12}{x + 4}$$

$$\begin{array}{r} x - 3 \\ x + 4 \overline{)x^2 + x - 12} \\ \underline{x^2 + 4x} \\ -3x - 12 \\ \underline{-3x - 12} \end{array}$$

15. To check $\frac{3x^2 + 8x + 4}{3x + 2} = x + 2$ using multiplication,

$$(3x + 2)(x + 2) \stackrel{?}{=} 3x^2 + 8x + 4$$
$$3x^2 + 6x + 2x + 4 \stackrel{?}{=} 3x^2 + 8x + 4$$
$$3x^2 + 8x + 4 = 3x^2 + 8x + 4$$

The division is correct.

**Notation**

17.
$$\begin{array}{r} x + 2 \\ x + 2 \overline{)x^2 + 4x + 4} \\ \underline{x^2 + 2x} \\ 2x + 4 \\ \underline{2x + 4} \end{array}$$

**Practice**

19. $\dfrac{x^2+4x-12}{x-2}$

$$
\begin{array}{r}
x + 6 \\
x - 2 \overline{)x^2 + 4x - 12} \\
\underline{x^2 - 2x} \phantom{xxxxx} \\
6x - 12 \\
\underline{6x - 12}
\end{array}
$$

21. $\dfrac{y^2+13y+12}{y+1}$

$$
\begin{array}{r}
y + 12 \\
y + 1 \overline{)y^2 + 13y + 12} \\
\underline{y^2 + \phantom{1}y} \phantom{xxxxxx} \\
12y + 12 \\
\underline{12y + 12}
\end{array}
$$

23. $\dfrac{6a^2+5a-6}{2a+3}$

$$
\begin{array}{r}
3a - 2 \\
2a + 3 \overline{)6a^2 + 5a - 6} \\
\underline{6a^2 + 9a} \phantom{xxxx} \\
-4a - 6 \\
\underline{-4a - 6}
\end{array}
$$

25. $\dfrac{3b^2+11b+6}{3b+2}$

$$
\begin{array}{r}
b + 3 \\
3b + 2 \overline{)3b^2 + 11b + 6} \\
\underline{3b^2 + \phantom{1}2b} \phantom{xxxx} \\
9b + 6 \\
\underline{9b + 6}
\end{array}
$$

27.

$$
\begin{array}{r}
2x + 1 \\
5x + 3 \overline{)10x^2 + 11x + 3} \\
\underline{10x^2 + \phantom{1}6x} \phantom{xxx} \\
5x + 3 \\
\underline{5x + 3}
\end{array}
$$

29.

$$
\begin{array}{r}
x - 7 \\
2x + 4 \overline{)2x^2 - 10x - 28} \\
\underline{2x^2 + \phantom{1}4x} \phantom{xxxx} \\
-14x - 28 \\
\underline{-14x - 28}
\end{array}
$$

31.

$$
\begin{array}{r}
3x + 2 \\
2x - 1 \overline{)6x^2 + x - 2} \\
\underline{6x^2 - 3x} \phantom{xxx} \\
4x - 2 \\
\underline{4x - 2}
\end{array}
$$

33.
$$
\begin{array}{r}
2x - 1 \phantom{xxxxxx} \\
x + 3 \overline{) 2x^2 + 5x - 3} \\
\underline{2x^2 + 6x} \phantom{xxx} \\
-x - 3 \\
\underline{-x - 3}
\end{array}
$$

35.
$$
\begin{array}{r}
x^2 + 2x - 1 \phantom{xxxxxx} \\
2x + 3 \overline{) 2x^3 + 7x^2 + 4x - 3} \\
\underline{2x^3 + 3x^2} \phantom{xxxxxx} \\
4x^2 + 4x \phantom{xxx} \\
\underline{4x^2 + 6x} \phantom{xxx} \\
-2x - 3 \\
\underline{-2x - 3}
\end{array}
$$

37.
$$
\begin{array}{r}
2x^2 + 2x + 1 \phantom{xxxxxx} \\
3x + 2 \overline{) 6x^3 + 10x^2 + 7x + 2} \\
\underline{6x^3 + \phantom{x} 4x^2} \phantom{xxxxxx} \\
6x^2 + 7x \phantom{xxx} \\
\underline{6x^2 + 4x} \phantom{xxx} \\
3x + 2 \\
\underline{3x + 2}
\end{array}
$$

39.
$$
\begin{array}{r}
x^2 + x + 1 \phantom{xxxxxx} \\
2x + 1 \overline{) 2x^3 + 3x^2 + 3x + 1} \\
\underline{2x^3 + x^2} \phantom{xxxxxx} \\
2x^2 + 3x \phantom{xxx} \\
\underline{2x^2 + x} \phantom{xxx} \\
2x + 1 \\
\underline{2x + 1}
\end{array}
$$

41. $\dfrac{2x^2 + 5x + 2}{2x + 3}$
$$
\begin{array}{r}
x + 1 + \frac{-1}{2x+3} \phantom{xxx} \\
2x + 3 \overline{) 2x^2 + 5x + 2} \\
\underline{2x^2 + 3x} \phantom{xxx} \\
2x + 2 \\
\underline{2x + 3} \\
-1
\end{array}
$$

43. $\dfrac{4x^2 + 6x - 1}{2x + 1}$
$$
\begin{array}{r}
2x + 2 + \frac{-3}{2x+1} \phantom{xxx} \\
2x + 1 \overline{) 4x^2 + 6x - 1} \\
\underline{4x^2 + 2x} \phantom{xxx} \\
4x - 1 \\
\underline{4x + 2} \\
-3
\end{array}
$$

45. $\dfrac{x^3+3x^2+3x+1}{x+1}$

$$\begin{array}{r}
x^2 + 2x + 1 \phantom{xxxx} \\[2pt]
\hline
x+1{\overline{\smash{\big)}\,x^3 + 3x^2 + 3x + 1\phantom{)}}} \\[-2pt]
\underline{x^3 + \phantom{x}x^2}\phantom{xxxxxxxxxx} \\[-2pt]
2x^2 + 3x\phantom{xxxx} \\[-2pt]
\underline{2x^2 + 2x}\phantom{xxxx} \\[-2pt]
x + 1 \\[-2pt]
\underline{x + 1}
\end{array}$$

47. $\dfrac{2x^3+7x^2+4x+3}{2x+3}$

$$\begin{array}{r}
x^2 + 2x - 1 + \dfrac{6}{2x+3} \\[2pt]
\hline
2x+3{\overline{\smash{\big)}\,2x^3 + 7x^2 + 4x + 3}} \\[-2pt]
\underline{2x^3 + 3x^2}\phantom{xxxxxxxxx} \\[-2pt]
4x^2 + 4x\phantom{xxxx} \\[-2pt]
\underline{4x^2 + 6x}\phantom{xxxx} \\[-2pt]
-2x + 3 \\[-2pt]
\underline{-2x - 3} \\[-2pt]
6
\end{array}$$

49. $\dfrac{2x^3+4x^2-2x+3}{x-2}$

$$\begin{array}{r}
2x^2 + 8x + 14 + \dfrac{31}{x-2} \\[2pt]
\hline
x-2{\overline{\smash{\big)}\,2x^3 + 4x^2 - 2x + 3}} \\[-2pt]
\underline{2x^3 - 4x^2}\phantom{xxxxxxxxx} \\[-2pt]
8x^2 - \phantom{1}2x\phantom{xxxx} \\[-2pt]
\underline{8x^2 - 16x}\phantom{xxxx} \\[-2pt]
14x + \phantom{1}3 \\[-2pt]
\underline{14x - 28} \\[-2pt]
31
\end{array}$$

51. $\dfrac{x^2-1}{x-1}$

$$\begin{array}{r}
x + 1\phantom{xxx} \\[2pt]
\hline
x-1{\overline{\smash{\big)}\,x^2 + 0x - 1}} \\[-2pt]
\underline{x^2 - \phantom{1}x}\phantom{xxxx} \\[-2pt]
x - 1 \\[-2pt]
\underline{x - 1}
\end{array}$$

53. $\dfrac{4x^2-9}{2x+3}$

$$\begin{array}{r}
2x - 3\phantom{xxx} \\[2pt]
\hline
2x+3{\overline{\smash{\big)}\,4x^2 + 0x - 9}} \\[-2pt]
\underline{4x^2 + 6x}\phantom{xxxx} \\[-2pt]
-6x - 9 \\[-2pt]
\underline{-6x - 9}
\end{array}$$

55. $\frac{x^3+1}{x+1}$

$$
\begin{array}{r}
x^2 - x + 1 \\
x + 1 \overline{)x^3 + 0x^2 + 0x + 1} \\
\underline{x^3 + x^2} \\
-x^2 + 0x \\
\underline{-x^2 - x} \\
x + 1 \\
\underline{x + 1}
\end{array}
$$

57. $\frac{a^3+a}{a+3}$

$$
\begin{array}{r}
a^2 - 3a + 10 + \frac{-30}{a+3} \\
a + 3 \overline{)a^3 + 0a^2 + a} \\
\underline{a^3 + 3a^2} \\
-3a^2 + a \\
\underline{-3a^2 - 9a} \\
10a \\
\underline{10a + 30} \\
-30
\end{array}
$$

59.
$$
\begin{array}{r}
5x^2 - x + 4 + \frac{16}{3x-4} \\
3x - 4 \overline{)15x^3 - 23x^2 + 16x} \\
\underline{15x^3 - 20x^2} \\
-3x^2 + 16x \\
\underline{-3x^2 + 4x} \\
12x \\
\underline{12x - 16} \\
16
\end{array}
$$

**Applications**

61. FURNACE FILTER

   a. The length is the quotient of the width and area, $l = \frac{A}{w}$

$$
\begin{array}{r}
x - 6 \\
x + 4 \overline{)x^2 - 2x - 24} \\
\underline{x^2 + 4x} \\
-6x - 24 \\
\underline{-6x - 24}
\end{array}
$$

   The length of this furnace filter is $(x - 6)$ inches.

   b. The perimeter of the filter is $P = 2l + 2w$

$$P = 2(x - 6) + 2(x + 4)$$
$$P = 2x - 12 + 2x + 8$$
$$P = (4x - 4) \text{ inches}$$

63. **COMMUNICATION**

The number of poles used is the quotient of $\frac{8x^3-6x^2+5x-21}{2x-3}$

$$
\begin{array}{r}
4x^2 + 3x + 7 \\
2x - 3 \overline{) 8x^3 - 6x^2 + 5x - 21} \\
\underline{8x^3 - 12x^2} \\
6x^2 + 5x \\
\underline{6x^2 - 9x} \\
14x - 21 \\
\underline{14x - 21}
\end{array}
$$

There were $(4x^2 + 3x + 7)$ poles used on this line.

**Writing**

65. Answers will vary

**Review**

67. $(x^5 x^6)^2 = (x^{11})^2 = x^{22}$

69. $3(2x^2 - 4x + 5) + 2(x^2 + 3x - 7)$
$= 6x^2 - 12x + 15 + 2x^2 + 6x - 14$
$= 8x^2 - 6x + 1$

71. The slopes of two parallel lines are the same.

**Chapter 10 Key Concepts**

1.  a. This is a polynomial in $x$. It is written in **descending** powers of $x$.
    b. This polynomial has four terms.
    c. The degrees of each term from left to right are $3, 2, 1, 0$.
    d. This polynomial has degree three.
    e. The coefficients from left to right are $1, -2, 6, -8$.

3.  a. binomial
    b. none of these
    c. trinomial
    d. monomial

5.  To add two polynomials, remove the parentheses and **combine** any like terms.

7.  To multiply two polynomials, multiply **each** term of one polynomial by **each** term of the other polynomial and combine like terms.

9.  $(2x + 3) + (x - 8) = 3x - 5$

11. $(2x + 3)(x - 8)$
    $$= 2x^2 - 16x + 3x - 24$$
    $$= 2x^2 - 13x - 24$$

13. $(y^2 + y - 6) + (y + 3)$
    $$= y^2 + 2y - 3$$

15. $(y^2 + y - 6)(y + 3)$
    $$= y^3 + 3y^2 + y^2 + 3y - 6y - 18$$
    $$= y^3 + 4y^2 - 3y - 18$$

17. $f(x) = x^3 - 2x + 5$
    $$f(-2) = (-2)^3 - 2(-2) + 5$$
    $$f(-2) = 1$$

## Section 10.1

1.  a. $-3x^4 = -3 \cdot x \cdot x \cdot x \cdot x$
    b. $\left(\frac{1}{2}pq\right)^3 = \frac{1}{2}pq \cdot \frac{1}{2}pq \cdot \frac{1}{2}pq$

3.  a. $x^3 x^2 = x^5$
    b. $-3y(y^5) = -3y^6$
    c. $(y^7)^3 = y^{21}$
    d. $(3x)^4 = 81x^4$
    e. $b^3 b^4 b^5 = b^{12}$
    f. $-z^2(z^3 y^2) = -y^2 z^5$
    g. $(-16s)^2 s = 256s^3$
    h. $(2x^2 y)^2 = 4x^4 y^2$
    i. $(x^2 x^3)^3 = (x^5)^3 = x^{15}$
    j. $\left(\frac{x^2 y}{xy^2}\right)^2 = \left(\frac{x}{y}\right)^2 = \frac{x^2}{y^2}$
    k. $\frac{x^7}{x^3} = x^4$
    l. $\frac{(5y^2 z^3)^3}{(yz)^5} = \frac{125 y^6 z^9}{y^5 z^5} = 125 y z^4$

## Section 10.2

5.  a. $x^0 = 1$
    b. $(3x^2 y^2)^0 = 1$
    c. $(3x^0)^2 = (3 \cdot 1)^2 = 9$
    d. $10^{-3} = \frac{1}{10^3} = \frac{1}{1000}$
    e. $\left(\frac{3}{4}\right)^{-1} = \frac{1}{\left(\frac{3}{4}\right)} = \frac{4}{3}$
    f. $-5^{-2} = \frac{1}{-5^2} = -\frac{1}{25}$
    g. $x^{-5} = \frac{1}{x^5}$
    h. $-6y^4 y^{-5} = -6y^{-1} = -6\left(\frac{1}{y}\right) = -\frac{6}{y}$
    i. $\frac{x^{-3}}{x^7} = x^{-10} = \frac{1}{x^{10}}$
    j. $(x^{-3} x^{-4})^{-2} = (x^{-7})^{-2} = x^{14}$
    k. $\left(\frac{x^2}{x}\right)^{-5} = x^{-5} = \frac{1}{x^5}$
    l. $\left(\frac{3z^4}{z^3}\right)^{-2} = (3z)^{-2} = \frac{1}{(3z)^2} = \frac{1}{9z^2}$

## Section 10.3

7.  a. $728 = 7.28 \times 10^2$
    b. $9{,}370{,}000 = 9.37 \times 10^6$
    c. $0.0136 = 1.36 \times 10^{-2}$
    d. $0.00942 = 9.42 \times 10^{-3}$
    e. $0.018 \times 10^{-2} = 1.8 \times 10^{-2} \times 10^{-2} = 1.8 \times 10^{-4}$
    f. $753 \times 10^3 = 7.53 \times 10^2 \times 10^3 = 7.53 \times 10^5$

9.  a. $\dfrac{(0.00012)(0.00004)}{0.00000016}$

$= \dfrac{(1.2\times10^{-4})(4\times10^{-5})}{1.6\times10^{-7}}$

$= \dfrac{4.8\times10^{-9}}{1.6\times10^{-7}}$

$= 3 \times 10^{-2}$

$= 0.03$

b. $\dfrac{(4800)(20,000)}{600,000}$

$= \dfrac{(4.8\times10^{3})(2\times10^{4})}{6\times10^{5}}$

$= \dfrac{9.6\times10^{7}}{6\times10^{5}}$

$= 1.6 \times 10^{2}$

$= 160$

11. ATOMS

Since the atom is $1.0 \times 10^{-8}$ cm wide and a nucleus is $1.0 \times 10^{-13}$ cm, the number of nuclei placed end-to-end is the quotient of the width of the atom to the width of the nucleus.

$\dfrac{1.0\times10^{-8}\text{ cm}}{1.0\times10^{-13}\text{ cm}} = 1.0 \times 10^{5} = 100,000$

**Section 10.4**

13. a. This polynomial has four terms.
    b. The leading term is $3x^3$.
    c. The coefficient of the second term is $-1$.
    d. The constant term is 10.

15. Letting $f(x) = 3x^2 + 2x + 1$
    a. $f(3) = 3(3)^2 + 2(3) + 1 = 34$
    b. $f(0) = 3(0)^2 + 2(0) + 1 = 1$
    c. $f(-2) = 3(-2)^2 + 2(-2) + 1 = 9$
    d. $f(-0.2) = 3(-0.2)^2 + 2(-0.2) + 1 = 0.72$

**Section 10.5**

17. a. $3x^6 + 5x^5 - x^6 = 2x^6 + 5x^5$
    b. $x^2y^2 - 3x^2y^2 = -2x^2y^2$
    c. $(3x^2 + 2x) + (5x^2 - 8x) = 8x^2 - 6x$
    d. $3(9x^2 + 3x + 7) - 2(11x^2 - 5x + 9)$
    $= 27x^2 + 9x + 21 - 22x^2 + 10x - 18$
    $= 5x^2 + 19x + 3$

**Section 10.6**

19. a. $(2x^2)(5x) = 10x^3$
    b. $(-6x^4z^3)(x^6z^2) = -6x^{10}z^5$
    c. $(2rst)(-3r^2s^3t^4) = -6r^3s^4t^5$
    d. $5b^3 \cdot 6b^2 \cdot 4b^6 = 120b^{11}$

21. a. $(x + 3)(x + 2) = x^2 + 5x + 6$
 b. $(2x + 1)(x - 1) = 2x^2 - x - 1$
 c. $(3a - 3)(2a + 2) = 6a^2 - 6$
 d. $6(a - 1)(a + 1) = 6(a^2 - 1) = 6a^2 - 6$
 e. $(a - b)(2a + b) = 2a^2 - ab - b^2$
 f. $(-3x - y)(2x + y) = -6x^2 - 5xy - y^2$

23. a. $(3x + 1)(x^2 + 2x + 1)$
 $= 3x^3 + 6x^2 + 3x + x^2 + 2x + 1$
 $= 3x^3 + 7x^2 + 5x + 1$
 b. $(2a - 3)(4a^2 + 6a + 9)$
 $= 8a^3 + 12a^2 + 18a - 12a^2 - 18a - 27$
 $= 8a^3 - 27$

25. APPLIANCES

$\text{Base}_{perimeter} = 2(x + 6) + 2(2x - 1)$
$\text{Base}_{perimeter} = 2x + 12 + 4x - 2$
$\text{Base}_{perimeter} = (6x + 10) \text{ inches}$

$\text{Base}_{area} = (x + 6)(2x - 1)$
$\text{Base}_{area} = (2x^2 + 11x - 6) \text{ inches}^2$

$\text{Volume} = (x + 6)(2x - 1)(3x)$
$\text{Volume} = (2x^2 + 11x - 6)(3x)$
$\text{Volume} = (6x^3 + 33x^2 - 18x) \text{ inches}^3$

**Section 10.7**

27. a. $\frac{8x+6}{2} = \frac{1}{2}(8x + 6) = 4x + 3$
 b. $\frac{14xy-21x}{7xy} = \frac{1}{7xy}(14xy - 21x) = \frac{14xy}{7xy} - \frac{21x}{7xy} = 2 - \frac{3}{y}$
 c. $\frac{15a^2b+20ab^2-25ab}{5ab}$
  $= \frac{1}{5ab}(15a^2b + 20ab^2 - 25ab)$
  $= \frac{15a^2b}{5ab} + \frac{20ab^2}{5ab} - \frac{25ab}{5ab}$
  $= 3a + 4b - 5$
 d. $\frac{(x+y)^2+(x-y)^2}{-2xy}$
  $= -\frac{1}{2xy}[(x^2 + 2xy + y^2) + (x^2 - 2xy + y^2)]$
  $= -\frac{1}{2xy}(2x^2 + 2y^2)$
  $= \frac{2x^2}{-2xy} + \frac{2y^2}{-2xy}$
  $= -\frac{x}{y} - \frac{y}{x}$

29. a.
$$x + 2 \overline{)\phantom{x^2}} \quad \begin{array}{l} x + 1 + \frac{3}{x+2} \\ \overline{x^2 + 3x + 5} \\ \underline{x^2 + 2x} \\ \phantom{x^2+2}x + 5 \\ \phantom{x^2+2}\underline{x + 2} \\ \phantom{x^2+2x+}3 \end{array}$$

b.
$$x - 1 \overline{)\phantom{x^2}} \quad \begin{array}{l} x - 5 \\ \overline{x^2 - 6x + 5} \\ \underline{x^2 - x} \\ \phantom{x^2}-5x + 5 \\ \phantom{x^2}\underline{-5x + 5} \end{array}$$

c.
$$x + 3 \overline{)\phantom{2x^2}} \quad \begin{array}{l} 2x + 1 \\ \overline{2x^2 + 7x + 3} \\ \underline{2x^2 + 6x} \\ \phantom{2x^2}x + 3 \\ \phantom{2x^2}\underline{x + 3} \end{array}$$

d.
$$3x - 1 \overline{)\phantom{3x^2}} \quad \begin{array}{l} x + 5 + \frac{3}{3x-1} \\ \overline{3x^2 + 14x - 2} \\ \underline{3x^2 - x} \\ \phantom{3x^2}15x - 2 \\ \phantom{3x^2}\underline{15x - 5} \\ \phantom{3x^2+15x}3 \end{array}$$

e.
$$2x - 1 \overline{)\phantom{6x^3}} \quad \begin{array}{l} 3x^2 + 2x + 1 + \frac{2}{2x-1} \\ \overline{6x^3 + x^2 + 0x + 1} \\ \underline{6x^3 - 3x^2} \\ \phantom{6x^3}4x^2 + 0x \\ \phantom{6x^3}\underline{4x^2 - 2x} \\ \phantom{6x^3+4x^2}2x + 1 \\ \phantom{6x^3+4x^2}\underline{2x - 1} \\ \phantom{6x^3+4x^2+2x}2 \end{array}$$

f.
$$3x + 1 \overline{)\phantom{9x^3}} \quad \begin{array}{l} 3x^2 - x - 4 \\ \overline{9x^3 + 0x^2 - 13x - 4} \\ \underline{9x^3 + 3x^2} \\ \phantom{9x^3}-3x^2 - 13x \\ \phantom{9x^3}\underline{-3x^2 - x} \\ \phantom{9x^3+3x^2}-12x - 4 \\ \phantom{9x^3+3x^2}\underline{-12x - 4} \end{array}$$

31. ZOOLOGY

$d = 8x^2 + 2x - 3$

$t = 2x - 1$

First solve $d = rt$ for $r$,

$d = rt$

$\frac{d}{t} = \frac{rt}{t}$

$\frac{d}{t} = r$

now substitute

$\frac{8x^2 + 2x - 3}{2x - 1} = r$

$$
\begin{array}{r}
4x + 3 \\
2x - 1 \overline{)\,8x^2 + 2x - 3} \\
\underline{8x^2 - 4x}\phantom{00000} \\
6x - 3 \\
\underline{6x - 3}
\end{array}
$$

This snail was traveling at a rate of $(4x + 3)\ \frac{\text{inches}}{\text{minute}}$.

1. $2xxxyyyy = 2x^3y^4$

3. $y^2(yy^3) = y^2y^4 = y^6$

5. $3x^0 = 3 \cdot 1 = 3$

7. $\frac{y^2}{yy^{-2}} = \frac{y^2}{y^{-1}} = y^3$

9. $V = s^3$ for cube,
$V = (10y^4)^3$
$V = 1000y^{12}$ inches$^3$

11. ELECTRICITY
$6,250,000,000,000,000,000 = \quad 6.25 \times 10^{18}$

13. $3x^2 + 2$ is a binomial

15. If $f(x) = x^2 + x - 2$ then $f(-2) = (-2)^2 + (-2) - 2$
$f(-2) = 4 - 2 - 2$
$f(-2) = 0$

17. $-6(x - y) + 2(x + y) - 3(x + 2y)$
$= -6x + 6y + 2x + 2y - 3x - 6y$
$= -7x + 2y$

19. $(-2x^3)(2x^2y) = -4x^5y$

21. $(x - 9)(x + 9) = x^2 - 81$

23. $(2x - 5)(3x + 4)$
$= 6x^2 + 8x - 15x - 20$
$= 6x^2 - 7x - 20$

25. $(a + 2)^2 = (a - 3)^2$
$(a + 2)(a + 2) = (a - 3)(a - 3)$
$a^2 + 4a + 4 = a^2 - 6a + 9$
$a^2 - a^2 + 4a + 4 = a^2 - a^2 - 6a + 9$
$4a + 4 = -6a + 9$
$4a + 6a + 4 = -6a + 6a + 9$
$10a + 4 = 9$
$10a + 4 - 4 = 9 - 4$
$10a = 5$
$\frac{10a}{10} = \frac{5}{10}$
$a = \frac{1}{2}$

27. $\frac{6a^2 - 12b^2}{24ab}$

$= \frac{1}{24ab}(6a^2 - 12b^2)$

$= \frac{1}{24ab}(6a^2) - \frac{1}{24ab}(12b^2)$

$= \frac{a}{4b} - \frac{b}{2a}$

29. Answers will vary

**Chapters 1-10 Cumulative Review**

1. PERSONAL SAVINGS RATE
   According to this illustration Americans in September of 1998 were spending more than they were earning for this month.

3. $\frac{7}{10} - \frac{1}{14} = \frac{49}{70} - \frac{5}{70} = \frac{44}{70} = \frac{22}{35}$

5. RACING
   This driver has $(250 - x)$ laps to finish.

7. The coefficient of the second term is 5.

9. $3x - 5x + 2y = -2x + 2y$

11. $2x^2 y^3 - xy(xy^2)$
    $= 2x^2 y^3 - x^2 y^3$
    $= x^2 y^3$

13. $3(x - 5) + 2 = 2x$
    $3x - 15 + 2 = 2x$
    $3x - 13 = 2x$
    $3x - 2x - 13 = 2x - 2x$
    $x - 13 = 0$
    $x - 13 + 13 = 0 + 13$
    $x = 13$

15. $A = \frac{1}{2}h(b + B)$
    $2 \cdot A = 2\left(\frac{1}{2}h(b + B)\right)$
    $2A = h(b + B)$
    $\frac{2A}{(b+B)} = \frac{h(b+B)}{(b+B)}$
    $\frac{2A}{(b+B)} = h$

17. $4^2 - 5^2 = 16 - 25 = -9$

19. $\frac{-3-(-7)}{2^2-3} = \frac{-3+7}{4-3} = \frac{4}{1} = 4$

21. $8(4 + x) > 10(6 + x)$
    $32 + 8x > 60 + 10x$
    $32 + 8x - 8x > 60 + 10x - 8x$
    $32 > 60 + 2x$
    $32 - 60 > 60 - 60 + 2x$
    $-28 > 2x$
    $\frac{-28}{2} > \frac{2x}{2}$
    $-14 > x$ or $x < -14$
    $(-\infty, -14)$

23. $y = x^2$

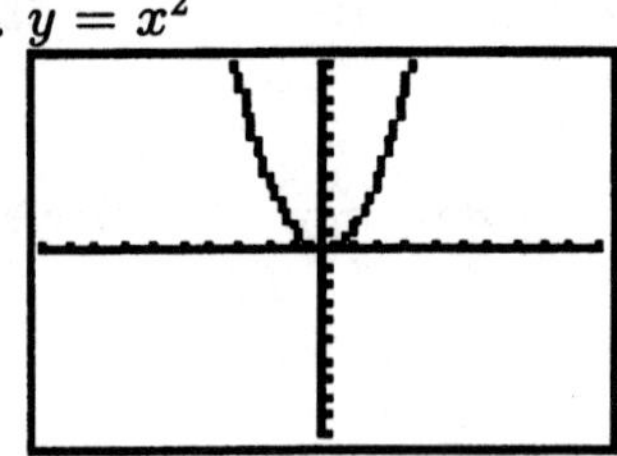

25. $4x - 3y = 12$ rewritten in slope intercept form is
$$-3y = -4x + 12$$
$$y = \tfrac{4}{3}x - 4$$

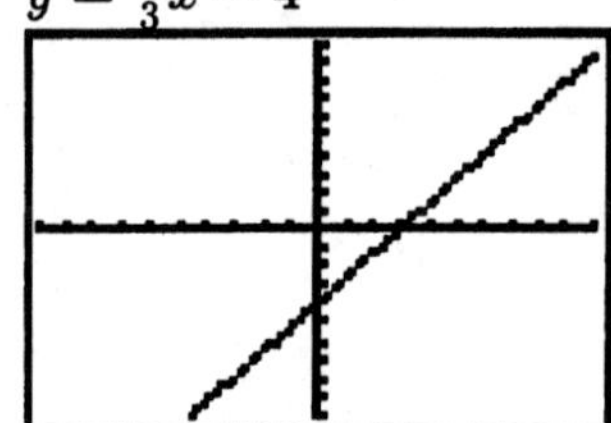

27. $m = \dfrac{4-8}{-2-6}$
$$m = \dfrac{-4}{-8}$$
$$m = \tfrac{1}{2}$$

29. $y = -4x + 3$ has slope $m = -4$.

31. $m = \tfrac{2}{3}$ and $b = 5$
$$y = \tfrac{2}{3}x + 5$$

33. Remember a horizontal line has slope 0, and using the point slope form
$$y - y_1 = m(x - x_1)$$
$$y - 4 = 0(x - 2)$$
$$y - 4 = 0$$
$$y = 4$$

35. $y = -\tfrac{3}{4}x + \tfrac{15}{4}$
$$4x - 3y = 25$$
Re-write the second equation in slope-intercept form,
$$-3y = -4x + 25$$
$$y = \tfrac{4}{3}x - \tfrac{25}{3}$$
The slopes are negative reciprocals of one another, therefore the lines are perpendicular.

37. $y = x^3 - 4$ is a function.

39. If $f(x) = 2x^2 - 3$ then,
$$f(0) = 2(0)^2 - 3$$
$$f(0) = -3$$

41. If $f(x) = 2x^2 - 3$ then,
$$f(-2) = 2(-2)^2 - 3$$
$$f(-2) = 8 - 3$$
$$f(-2) = 5$$

43. The domain of this function is all real numbers and the range is all non-negative real numbers, or $y \geq 0$.

45. A positive rate of change could be associated with water heating up at a rate of 10 degrees per minute and a negative rate of change could be associated with water in an ice tray cooling at a rate of 2 degrees per minute.

47. $(y^3 y^5)y^6 = y^8 y^6 = y^{14}$

49. $\frac{x^3 x^4}{x^2 x^3} = \frac{x^7}{x^5} = x^2$

51. $x^{-5} = \frac{1}{x^5}$

53. $(x^{-4})^2 = \frac{1}{(x^4)^2} = \frac{1}{x^8}$

55. $615,000 = 6.15 \times 10^5$

57. $5.25 \times 10^{-4} = 0.000525$

59. The degree of $3x^2 + 2x - 5$ is 2.

61. MUSICAL INSTRUMENTS
$$f(x) = 0.01875x^4 - 0.15x^3 + 1.2x$$
If the gong is hung in the middle of the support the deflection is found by substituting $x = \frac{1}{2}(4) = 2$ in the above equation.
$$f(2) = 0.01875(2)^4 - 0.15(2)^3 + 1.2(2)$$
$$f(2) = 1.5$$
Therefore the amount of deflection is 1.5 inches.

63. $(3x^2 + 2x - 7) - (2x^2 - 2x + 7)$
$$= 3x^2 + 2x - 7 - 2x^2 + 2x - 7$$
$$= x^2 + 4x - 14$$

65. $-5x^2(7x^3 - 2x^2 - 2)$
$$= -35x^5 + 10x^4 + 10x^2$$

67. $(3x - 7)(2x + 8)$
$$= 6x^2 + 24x - 14x - 56$$
$$= 6x^2 + 10x - 56$$

69. $(3x + 1)^2$
$$= (3x + 1)(3x + 1)$$
$$= 9x^2 + 6x + 1$$

71. $\frac{6x^2-8x}{2x}$

$\quad = \frac{1}{2x}(6x^2 - 8x)$

$\quad = \frac{6x^2}{2x} - \frac{8x}{2x}$

$\quad = 3x - 4$

**Section 11.1 Factoring Out the Greatest Common Factor and Factoring by Grouping**

**Vocabulary**

1.  A natural number greater than 1 whose only factors are 1 and itself is called a **prime** number.
3.  The GCF of several natural numbers is the **largest** number that divides each of the numbers exactly.
5.  The process of finding the individual factors of a known product is called **factoring**.

**Concepts**

7.  $6a + 9b + 3 = 3(2a + 3b + 1)$
    This is the correct method, notice there is a 1 instead of 0 in this solution.

9.  $30a^3 - 12a^2 = 6a^2(5a - 2)$
    $6a^2$ is the GCF not $6a$

11. This illustrates factoring out the greatest common factor.

13. $3x$ is the GCF of these three since it is common to each.

15. For this particular problem the FOIL method could be used to check the factoring done.
    $(j + 2)(3j^2 + 2)$ should equal $3j^2 + 6j^2 + 2j + 4$

**Notation**

17. $b^3 - 6b^2 + 2b - 12$
    $= b^2(b - 6) + 2(b-6)$
    $= (b - 6)(b^2 + 2)$

19. The implied coefficient of $xy$ is 1.

**Practice**

21. $12 = 4 \cdot 3 = 2^2 \cdot 3$

23. $15 = 3 \cdot 5$

25. $40 = 8 \cdot 5 = 2^3 \cdot 5$

27. $98 = 2 \cdot 49 = 2 \cdot 7^2$

29. $225 = 25 \cdot 9 = 5^2 \cdot 3^2$

31. $288 = 144 \cdot 2 = 16 \cdot 9 \cdot 2 = 2^5 \cdot 3^2$

33. $4a + 12 = 4(a + 3)$

35. $4y^2 + 8y - 2xy = 2y(2y + 4 - x)$

37. $3x + 6 = 3(x + 2)$

39. $12x^2 - 6x - 24 = 6(2x^2 - x - 4)$

41. $t^3 + 2t^2 = t^2(t + 2)$

43. $a^3 - a^2 = a^2(a - 1)$

45. $24x^2y^3 + 8xy^2 = 8xy^2(3xy + 1)$

47. $12uvw^3 - 18uv^2w^2 = 6uvw^2(2w - 3v)$

49. $3x + 3y - 6z = 3(x + y - 2z)$

51. $ab + ac - ad = a(b + c - d)$

53. $12r^2 - 3rs + 9r^2s^2 = 3r(4r - s + 3rs^2)$

55. $\pi R^2 - \pi ab = \pi(R^2 - ab)$

57. $3(x + 2) - x(x + 2) = (x + 2)(3 - x)$

59. $h^2(14 + r) + (14 + r) = (h^2 + 1)(14 + r)$

61. $-a - b = -1(a + b) = -(a + b)$

63. $-2x + 5y = -1(2x - 5y) = -(2x - 5y)$

65. $-3m - 4n + 1 = -1(3m + 4n - 1)$
$= -(3m + 4n - 1)$

67. $-3ab - 5ac + 9bc = -1(3ab + 5ab - 9bc)$
$= -(3ab + 5ab - 9bc)$

69. $-3x^2 - 6x = -3x(x + 2)$

71. $-4a^2b^3 + 12a^3b^2 = -4a^2b^2(b - 3a)$

73. $-4a^2b^2c^2 + 14a^2b^2c - 10ab^2c^2$
$= -2ab^2c(2ac - 7a + 5c)$

75. $2x + 2y + ax + ay$
$= 2(x + y) + a(x + y)$
$= (2 + a)(x + y)$

77. $7r + 7s - kr - ks$
$= 7(r + s) - k(r + s)$
$= (7 - k)(r + s)$

79. $xr + xs + yr + ys$
$= x(r + s) + y(r + s)$
$= (x + y)(r + s)$

81. $2ax + 2bx + 3a + 3b$
$= 2x(a + b) + 3(a + b)$
$= (2x + 3)(a + b)$

83. $2ab + 2ac + 3b + 3c$
$= 2a(b + c) + 3(b + c)$
$= (2a + 3)(b + c)$

85. $6x^2 - 2x - 15x + 5$
$= 2x(3x - 1) - 5(3x - 1)$
$= (2x - 5)(3x - 1)$

87. $9mp + 3mq - 3np - nq$
$= 3m(3p + q) - n(3p + q)$
$= (3m - n)(3p + q)$

89. $2xy + y^2 - 2x - y$
$= y(2x + y) - (2x + y)$
$= (y - 1)(2x + y)$

91. $8z^5 + 12z^2 - 10z^3 - 15$
$= 4z^2(2z^3 + 3) - 5(2z^3 + 3)$
$= (4z^2 - 5)(2z^3 + 3)$

93. $ax^3 + bx^3 + 2ax^2y + 2bx^2y$
$= x^2(ax + bx + 2ay + 2by)$
$= x^2[x(a + b) + 2y(a + b)]$
$= x^2[(x + 2y)(a + b)]$

95. $4a^2b + 12a^2 - 8ab - 24a$
$= 4a(ab + 3a - 2b - 6)$
$= 4a[a(b + 3) - 2(b + 3)]$
$= 4a[(a - 2)(b + 3)]$

97. $x^3y - x^2y - xy^2 + y^2$
$= y(x^3 - x^2 - xy + y)$
$= y[x^2(x - 1) - y(x - 1)]$
$= y[(x^2 - y)(x - 1)]$

**Applications**

99. PICTURE FRAMING
   a. The area of the picture frame is $6x \cdot 2x^2 = 12x^3$ in$^2$
   b. The area of the picture is $4x \cdot 5x = 20x^2$ in$^2$
   c. The area of the picture mat is
$$(6x \cdot 2x^2) - (4x \cdot 5x)$$
$$= (12x^3 - 20x^2) \text{ in}^2$$
$$= 4x^2(3x - 5) \text{ in}^2$$

101. COOKING
   a. The length of the side of the griddle is $4r$ inches. The area of the griddle is $(4r)^2 = 16r^2$ in$^2$
   b. Each pancake covers $\pi r^2$ in$^2$, so all four cover $4\pi r^2$ in$^2$
   c. The amount of surface area not covered is the difference of $16r^2 - 4\pi r^2 = 4r^2(4 - \pi)$ in$^2$

**Writing**

103.  Answers will vary
105.  Answers will vary

**Review**

107.  $\left(\dfrac{y^3 y}{2yy^2}\right)^3 = \left(\dfrac{y^4}{2y^3}\right)^3 = \left(\dfrac{y}{2}\right)^3 = \dfrac{y^3}{8}$

109.  $(3, 5)$ does lie on the graph of $4x - y = 7$ since $4(3) - 5 = 12 - 5 = 7$

## Section 11.2 Factoring Trinomials of the Form $x^2 + bx + c$

### Vocabulary

1. A polynomial, such as $x^2 - x - 6$, that has exactly three terms is called a **trinomial**. A polynomial, such as $x - 3$, that has exactly two terms is called a **binomial**.
3. Since $10 = (-5)(-2)$, we say $-5$ and $-2$ are **factors** of 10.
5. The **leading** coefficient of the trinomial $x^2 - 3x + 2$ is 1, the **coefficient** of the middle term is $-3$, and the last **term** is 2.

### Concepts

7. Two factorizations of 4 that involve only positive numbers are $4 \cdot 1$ and $2 \cdot 2$. Two factorizations of 4 that involve only negative numbers are $-4(-1)$ and $-2(-2)$.

9. Before attempting to factor a trinomial into two binomials, always factor out any **common** factors first.

11. Two factors of 18 whose sum is $-9$ are $-6$ and $-3$.

13.

| Factors of 8 | Sum of factors of 8 |
|---|---|
| $1(8)$ | $1 + 8 = 9$ |
| $2(4)$ | $2 + 4 = 6$ |
| $-1(-8)$ | $-1 + -8 = -9$ |
| $-2(-4)$ | $-2 + -4 = -6$ |

15. $x^2 - 2x - 15$
    a. The coefficient of the $x^2$ term is 1.
    b. The last term is $-15$.
       The product of $-5$ and 3 is $-15$.
    c. The coefficient of the middle term is $-2$.
       The sum of $-5$ and 3 is $-2$.

17. The sum of two negatives could not equal 7.

19. a. If $c$ is positive then the two chosen integers have the same sign.
    b. If $c$ is negative then the two chosen integers have different signs.

### Notation

21. $6 + 5x + x^2 = x^2 + 5x + 6 = (x + 3)(x + 2)$

### Practice

23. $x^2 + 3x + 2 = (x + 2)(x + 1)$

25. $t^2 - 9t + 14 = (t - 7)(t - 2)$

27. $a^2 + 6a - 16 = (a + 8)(a - 2)$

29. $z^2 + 12z + 11 = (z + 11)(z + 1)$

31. $m^2 - 5m + 6 = (m - 3)(m - 2)$

33. $a^2 - 4a - 5 = (a - 5)(a + 1)$

35. $x^2 + 5x - 24 = (x + 8)(x - 3)$

37. $a^2 - 10a - 39 = (a + 3)(a - 13)$

39. $u^2 + 10u + 15$ prime

41. $s^2 + 11s - 26 = (s + 13)(s - 2)$

43. $r^2 - 2r + 4$ prime

45. $m^2 - m - 12 = (m - 4)(m + 3)$

47. $x^2 + 4xy + 4y^2 = (x + 2y)^2$

49. $m^2 + 3mn - 10n^2 = (m + 5n)(m + 2n)$

51. $a^2 - 4ab - 12b^2 = (a - 6b)(a + 2b)$

53. $r^2 - 2rs + 4s^2$ is prime

55. $-x^2 - 7x - 10$
$$= -1(x^2 + 7x + 10)$$
$$= -1(x + 5)(x + 2)$$

57. $-t^2 - 15t + 34$
$$= -1(t^2 + 15t - 34)$$
$$= -1(t + 17)(t - 2)$$

59. $-r^2 + 14r - 40$
$$= -1(r^2 - 14r + 40)$$
$$= -1(r - 10)(r - 4)$$

61. $-a^2 - 4ab - 3b^2 = -(a + 3b)(a - b)$

63. $-x^2 + 6xy + 7y^2 = -(x - 7y)(x + y)$

65. $4 - 5x + x^2 = (x - 4)(x - 1)$

67. $10y + 9 + y^2 = (y + 1)(y + 9)$

69. $-r^2 + 2 + r$
$$= -1(r^2 - r - 2)$$
$$= -1(r - 2)(r + 1)$$

71. $4rx + r^2 + 3x^2$
$= r^2 + 4rx + 3x^2$
$= (r + 3x)(r + x)$

73. $-3ab + a^2 + 2b^2$
$= a^2 - 3ab + 2b^2$
$= (a - 2b)(a - b)$

75. $2x^2 + 10x + 12$
$= 2(x^2 + 5x + 6)$
$= 2(x + 3)(x + 2)$

77. $-5a^2 + 25a - 30$
$= -5(a^2 - 5a + 6)$
$= -5(a - 3)(a - 2)$

79. $3z^2 - 15z + 12$
$= 3(z^2 - 5z + 4)$
$= 3(z - 4)(z - 1)$

81. $12xy + 4x^2y - 72y$
$= 4y(x^2 + 3x - 18)$
$= 4y(x + 6)(x - 3)$

83. $-4x^2y - 4x^3 + 24xy^2$
$= -4x(x^2 + xy - 6y^2)$
$= -4x(x + 3y)(x - 2y)$

**Applications**

85. PETS
$x^3 + 12x^2 + 27x$
$= x(x^2 + 12x + 27)$
$= x(x + 9)(x + 3)$
The cage has length $(x + 9)$ inches, width $x$ inches and height $(x + 3)$ inches.

**Writing**

87. Answers will vary
89. Answers will vary
91. Answers will vary

**Review**

93. $x - 3 > 5$
$x > 8$

95. $-3x - 5 \geq 4$

$\quad\quad -3x \geq 9$

$\quad\quad x \leq -3$

$$\longleftarrow\!\!\!\rule{2cm}{0.4pt}]\overset{-3}{\phantom{x}}\!\!\!\rule{1cm}{0.4pt}\longrightarrow$$

# Section 11.3 Factoring Trinomials of the Form $ax^2 + bx + c$

## Vocabulary

1. The trinomial $3x^2 - x - 12$ has a **leading** coefficient of 3. The **last** term is $-12$.
3. $(x - 2)(5x - 1)$
   The product of the **outer** terms is $-x$ and the product of the **inner** terms is $-10x$.
5. The **middle** term of $4x^2 - 7x + 13$ is $-7x$.
7. The **sum** of the middle terms of the polynomial $4a^2 - 12a - a + 3$ is $-13a$.

## Concepts

9. The first coefficients in the binomials must be factors of 5. The last numbers of the binomials must be factors of $-8$.

11. Since the last term of the trinomial is **positive** and the middle term is **negative**, the integers must be **negative** factors of 6.

13. Since the last term of the trinomial is **negative** the signs of the integers will be **different**.

15. We need to find two integers whose product is 24 and whose sum is $-11$.

## Notation

17. a. Answers will vary, $x^2 - 5x + 2$
    b. Answers will vary, $-99x^2 - 56x + 12$

## Practice

19. $3a^2 + 13a + 4 = (3a + 1)(a + 4)$

21. $4z^2 - 13z + 3 = (z - 3)(4z - 1)$

23. $2m^2 + 5m - 12 = (2m - 3)(m + 4)$

25. $12t^2 + 17t + 6$
$$= 12t^2 + 9t + 8t + 6$$
$$= 3t(4t + 3) + 2(4t + 3)$$
$$= (4t + 3)(3t + 2)$$

27. $2x^2 - 3x + 1 = (2x - 1)(x - 1)$

29. $3a^2 + 13a + 4 = (3a + 1)(a + 4)$

31. $4z^2 + 13z + 3 = (z + 3)(4z + 1)$

33. $6y^2 + 7y + 2 = (3y + 2)(2y + 1)$

35. $6x^2 - 7x + 2 = (3x - 2)(2x - 1)$

37. $3a^2 - 4a - 4 = (3a + 2)(a - 2)$

39. $2x^2 - 3x - 2 = (x - 2)(2x + 1)$

41. $2m^2 + 5m - 10$  prime

43. $10y^2 - 3y - 1 = (5y + 1)(2y - 1)$

45. $12y^2 - 5y - 2 = (4y + 1)(3y - 2)$

47. $-5t^2 - 13t - 6$
$= -1(5t^2 + 13t + 6)$
$= -1(5t + 3)(t + 2)$
$= -(5t + 3)(t + 2)$

49. $-16m^2 + 14m - 3$
$= -1(16m^2 - 14m + 3)$
$= -1(2m - 1)(8m - 3)$
$= -(2m - 1)(8m - 3)$

51. $4a^2 - 4ab + b^2 = (2a - b)(2a - b)$

53. $6r^2 + rs - 2s^2 = (2r - s)(3r + 2s)$

55. $4x^2 + 8xy + 3y^2 = (2x + 3y)(2x + y)$

57. $4a^2 - 15ab + 9b^2 = (4a - 3b)(a - 3b)$

59. $3x^2 - 13x - 10 = (3x + 2)(x - 5)$

61. $8a^2 - 26a + 15 = (2a - 5)(4a - 3)$

63. $12y^2 - 25y + 12 = (4y - 3)(3y - 4)$

65. $3x^2 + x + 6$ prime

67. $2a^2 + 5ab + 3b^2 = (2a + 3b)(a + b)$

69. $6p^2 + pq - q^2 = (3p - q)(2p + q)$

71. $4x^2 + 10x - 6$
$= 2(2x^2 + 5x - 3)$
$= 2(2x - 1)(x + 3)$

73. $-y^3 - 13y^2 - 12y$
$= -y(y^2 + 13y + 12)$
$= -y(y + 12)(y + 1)$

75. $6x^3 - 15x^2 - 9x$
   $= 3x(2x^2 - 5x - 3)$
   $= 3x(2x + 1)(x - 3)$

77. $30r^5 + 63r^4 - 30r^3$
   $= 3r^3(10r^2 + 21r - 10)$
   $= 3r^3(5r - 2)(2r + 5)$

79. $-16m^3n - 20m^2n^2 - 6mn^3$
   $= -2mn(8m^2 + 10mn + 3n^2)$
   $= -2mn(4m + 3n)(2m + n)$

81. $-28u^3v^3 + 26u^2v^4 - 6uv^5$
   $= -2uv^3(14u^2 - 13uv + 3v^2)$
   $= -2uv^3(7u - 3v)(2u - v)$

## Applications

83. OFFICE FURNITURE
   $4x^2 + 20x - 11 = (2x - 1)(2x + 11)$
   The length is $2x + 11$ and the width is $2x - 1$.
   The difference between the length and the width is
   $2x + 11 - (2x - 1)$
   $= 2x + 11 - 2x + 1$
   $= 12$ inches

## Writing

85. Answers will vary
87. Answers will vary

## Review

89. $(x^2x^5)^2 = (x^7)^2 = x^{14}$

91. $\frac{1}{2^{-3}} = 2^3 = 8$

**Section 11.4 Special Factorizations and a Factoring Strategy**

**Vocabulary**

1. The binomial $x^2 - 25$ is called a **difference** of two squares.
3. The binomial $x^3 + 27$ is  called a sum of two **cubes**.  The binomial $x^3 - 8$ is called a **difference** of two cubes.

**Concepts**

5. $25x^2 + 30x + 9$
   a. The first term is the square of $5x$.
   b. The last term is the square of $3$.
   c. The middle term is twice the product of $5x$ and $3$.

7. To factor the square of a First quantity minus the square of the Last quantity, we multiply the **First** plus the **Last** by the **First** minus the **Last**.

9. a. $36x^2 = (6x)^2$
   b. $100x^4 = (10x^2)^2$
   c. $27m^3 = (3m)^3$
   d. $a^6 = (a^2)^3$

11. The first ten perfect square integers are $1, 4, 9, 16, 25, 36, 49, 64, 81, 100$

13. a. $9h^2 - 6h + 7$ is not a perfect square trinomial since $7$ is not a perfect square.
   b. $j^2 - 8j - 16$ is not a perfect square trinomial since the sign of the last term is negative.
   c. $25r^2 + 20r + 16$ is not a perfect square trinomial since $20r$ is not twice the product of $5r$ and $4$.

**Notation**

15. $(6x)^2 - (5y)^2 = 36x^2 - 25y^2$

17. $(3a)^2 - 2(3a)(5b) + (5b)^2 = 9a^2 - 30ab + 25b^2$

19. $(x + 8)(x + 8) = (x + 8)^2$

**Practice**

21. $a^2 - 6a + 9 = (a - 3)^2$

23. $4x^2 + 4x + 1 = (2x + 1)^2$

25. $x^2 + 6x + 9 = (x + 3)(x + 3) = (x + 3)^2$

27. $y^2 - 8y + 16 = (y - 4)(y - 4) = (y - 4)^2$

29. $t^2 + 20t + 100 = (t + 10)(t + 10) = (t + 10)^2$

31. $u^2 - 18u + 81 = (u - 9)(u - 9) = (u - 9)^2$

33. $4x^2 + 12x + 9 = (2x + 3)(2x + 3) = (2x + 3)^2$

35. $36x^2 + 12x + 1 = (6x + 1)(6x + 1) = (6x + 1)^2$

37. $a^2 + 2ab + b^2 = (a + b)(a + b) = (a + b)^2$

39. $16x^2 - 8xy + y^2 = (4 - y)(4 - y) = (4 - y)^2$

41. $y^2 - 49 = (y + 7)(y - 7)$

43. $t^2 - w^2 = (t + w)(t - w)$

45. $x^2 - 16 = (x - 4)(x + 4)$

47. $4y^2 - 1 = (2y + 1)(2y - 1)$

49. $9x^2 - y^2 = (3x - y)(3x + y)$

51. $16a^2 - 25b^2 = (4a - 5b)(4a + 5b)$

53. $a^2 + b^2 = a^2 + b^2$ prime

55. $a^4 - 144b^2 = (a^2 + 12b)(a^2 - 12b)$

57. $t^2 z^2 - 64 = (tz - 8)(tz + 8)$

59. $8x^2 - 32y^2$
$\quad = 8(x^2 - 4y^2)$
$\quad = 8(x - 2y)(x + 2y)$

61. $7a^2 - 7$
$\quad = 7(a^2 - 1)$
$\quad = 7(a - 1)(a + 1)$

63. $6x^4 - 6x^2 y^2$
$\quad = 6x^2(x^2 - y^2)$
$\quad = 6x^2(x - y)(x + y)$

65. $x^4 - 81$
$\quad = (x^2 - 9)(x^2 + 9)$
$\quad = (x + 3)(x - 3)(x^2 + 9)$

67. $a^4 - 16$
$\quad = (a^2 - 4)(a^2 + 4)$
$\quad = (a - 2)(a + 2)(a^2 + 4)$

69. $81r^4 - 256s^4$
$\quad = (9r^2 - 16s^2)(9r^2 + 16s^2)$
$\quad = (3r - 4s)(3r + 4s)(9r^2 + 16s^2)$

71. $a^3 + 8 = (a + 2)(a^2 - 2a + 4)$

73. $b^3 + 27 = (b + 3)(b^2 - 3b + 9)$

75. $y^3 + 1 = (y + 1)(y^2 - y + 1)$

77. $a^3 - 27 = (a - 3)(a^2 + 3a + 9)$

79. $8 + x^3 = (2 + x)(4 - 2x + x^2)$

81. $s^3 - t^3 = (s - t)(s^2 + st + t^2)$

83. $a^3 + 8b^3 = (a + 2b)(a^2 - 2ab + 4b^2)$

85. $64x^3 - 27 = (4x - 3)(16x^2 + 12x + 9)$

87. $a^6 - b^3 = (a^2 - b)(a^4 + a^2b + b^2)$

89. $x^9 + y^6 = (x^3 + y^2)(x^6 - x^3y^2 + y^4)$

91. $2x^3 + 54$
$$= 2(x^3 + 27)$$
$$= 2(x + 3)(x^2 - 3x + 9)$$

93. $-x^3 + 216$
$$= -1(x^3 - 216)$$
$$= -1(x - 6)(x^2 + 6x + 36)$$

95. $64m^3x - 8n^3x$
$$= 8x(8m^3 - n^3)$$
$$= 8x(2m - n)(4m^2 + 2mn + n^2)$$

97. $x^4y + 216xy^4$
$$= xy(x^3 + 216y^3)$$
$$= xy(x + 6y)(x^2 - 6xy + 36y^2)$$

99. $81r^4s^2 - 24rs^5$
$$= 3rs^2(27r^3 - 8s^3)$$
$$= 3rs^2(3r - 2s)(9r^2 + 6rs + 4s^2)$$

**Applications**

101.  GENETICS
$$p^2 + 2pq + q^2 = 1$$
$$(p + q)(p + q) = 1$$
$$(p + q)^2 = 1$$

103.  PHYSICS
$$0.5gt_1^2 - 0.5gt_2^2$$
$$0.5g(t_1^2 - t_2^2)$$
$$0.5g(t_1 - t_2)(t_1 + t_2)$$

**Writing**

105. Answers will vary

107. Answers will vary

**Review**

109. $\dfrac{5x^2+10y^2-15xy}{5xy}$

$= \dfrac{5x^2}{5xy} + \dfrac{10y^2}{5xy} - \dfrac{15xy}{5xy}$

$= \dfrac{x}{y} + \dfrac{2y}{x} - 3$

111.
$$\begin{array}{r} 3a + 2 \\ \hline 2a - 1 \overline{)\, 6a^2 + a - 2} \\ \underline{6a^2 - 3a\phantom{aaaaa}} \\ 4a - 2 \\ \underline{4a - 2} \end{array}$$

**Additional Factoring Problems**

113. $a^2(x - a) - b^2(x - a)$
$= (a^2 - b^2)(x - a)$
$= (a - b)(a + b)(x - a)$

115. $70p^4q^3 - 35p^4q^2 + 49p^5q^2$
$= 7p^4q^2(10q - 5 + 7p)$

117. $2ab^2 + 8ab - 24a$
$= 2a(b^2 + 4b - 12)$
$= 2a(b + 6)(b - 2)$

119. $-8p^3q^7 - 4p^2q^3$
$= -4p^2q^3(2pq^4 + 1)$

121. $20m^2 + 100m + 125$
$= 5(4m^2 + 20m + 25)$
$= 5(2m + 5)(2m + 5)$
$= 5(2m + 5)^2$

123. $x^2 + 7x + 1$ prime

125. $-2x^5 + 128x^2$
$= -2x^2(x^3 - 64)$
$= -2x^2(x - 4)(x^2 + 4x + 16)$

127. $6t^4 + 14t^3 - 40t^2$
$= 2t^2(3t^2 + 7t - 20)$
$= 2t^2(3t - 5)(t + 4)$

129. $x^2y^2 - 2x^2 - y^2 + 2$
$= x^2(y^2 - 2) - (y^2 - 2)$
$= (x^2 - 1)(y^2 - 2)$
$= (x + 1)(x - 1)(y^2 - 2)$

131. $8p^6 - 27q^6$
$= (2p^2 - 3q^2)(4p^4 + 6p^2q^2 + 9q^4)$

133. $125p^3 - 64y^3$
$= (5p - 4y)(25p^2 + 20py + 16y^2)$

135. $-16x^4y^2z + 24x^5y^3z^4 - 15x^2y^3z^7$
$= -x^2y^2z(16x^2 - 24x^3yz^3 + 15yz^6)$

137. $81p^4 - 16q^4$
$= (9p^2 - 4q^2)(9p^2 + 4q^2)$
$= (3p - 2q)(3p + 2q)(9p^2 + 4q^2)$

139. $4x^2 + 9y^2$ prime

141. $54x^3 + 250y^6$
$= 2(27x^3 + 125y^6)$
$= 2(3x + 5y^2)(9x^2 - 15xy^2 + 25y^4)$

143. $10r^2 - 13r - 4$ prime

145. $49p^2 + 28pq + 4q^2$
$= (7p + 2q)(7p + 2q)$
$= (7p + 2q)^2$

**Section 11.5**

**Vocabulary**

1. Any equation that can be written in the form $ax^2 + bx + c = 0$ is called a **quadratic** equation.

**Concepts**

3. When the products of two numbers is zero, at least one of them is **zero**. Symbolically, we can state this: If $ab = 0$, then $a = 0$ or $b = 0$.

5. To write a quadratic equation in *quadratic form* means that one side of the equation must be **zero** and the other side must be in the form $ax^2 + bx + c$.

7. a. $3x^2 + 4x + 2 = 0$ quadratic
   b. $3x + 7 = 0$ linear
   c. $2 = -16 - 4x$ linear
   d. $-6x + 2 = x^2$ quadratic

9. a. $x^2 + 6x - 16$ for $x = 0$
   $= 0^2 + 6 \cdot 0 - 16$
   $= -16$
   b. $x^2 + 6x - 16 = (x + 8)(x - 2)$
   c. $x^2 + 6x - 16 = 0$
   $(x + 8)(x - 2) = 0$
   $(x + 8) = 0 \quad$ or $\quad (x - 2) = 0$
   $x = -8 \qquad$ or $\quad x = 2$

11. a. The first step for $x^2 + 7x = -6$ is to add 6 to each side.
    b. The first step for $x(x + 7) = -3$ is to distribute the multiplication by $x$.

**Notation**

13. $7y^2 + 14y = 0$
    $7y(y + 2) = 0$
    $7y = 0 \qquad$ or $\qquad y + 2 = 0$
    $y = 0 \qquad$ or $\qquad y = -2$

**Practice**

15. $(x - 2)(x + 3) = 0$
    $x - 2 = 0 \quad$ or $\qquad x + 3 = 0$
    $x = 2 \qquad$ or $\qquad x = -3$

17. $(2s - 5)(s + 6) = 0$
    $2s - 5 = 0 \quad$ or $\qquad s + 6 = 0$
    $2s = 5 \qquad$ or $\qquad s = -6$
    $s = \frac{5}{2}$

19. $(x-1)(x+2)(x-3)=0$
$x-1=0$ or $x+2=0$ or $x-3=0$
$x=1$ or $x=-2$ or $x=3$

21. $x(x-3)=0$
$x=0$ or $x-3=0$
$x=3$

23. $x(2x-5)=0$
$x=0$ or $2x-5=0$
$2x=5$
$x=\frac{5}{2}$

25. $w^2-7w=0$
$w(w-7)=0$
$w=0$ or $w-7=0$
$w=7$

27. $3x^2+8x=0$
$x(3x+8)=0$
$x=0$ or $3x+8=0$
$3x=-8$
$x=-\frac{8}{3}$

29. $8s^2-16s=0$
$8s(s-2)=0$
$8s=0$ or $s-2=0$
$s=0$ $s=2$

31. $x^2-25=0$
$(x+5)(x-5)=0$
$x+5=0$ or $x-5=0$
$x=-5$ $x=5$

33. $4x^2-1=0$
$(2x+1)(2x-1)=0$
$2x+1=0$ or $2x-1=0$
$2x=-1$ $2x=1$
$x=-\frac{1}{2}$ $x=\frac{1}{2}$

35. $9y^2-4=0$
$(3y+2)(3y-2)=0$
$3y+2=0$ or $3y-2=0$
$3y=-2$ $3y=2$
$y=-\frac{2}{3}$ $y=\frac{2}{3}$

37. $x^2=100$
$x^2-100=0$
$(x+10)(x-10)=0$
$x+10=0$ or $x-10=0$
$x=-10$ $x=10$

39. $4x^2 = 81$
$$4x^2 - 81 = 0$$
$$(2x + 9)(2x - 9) = 0$$
$$2x + 9 = 0 \quad \text{or} \quad 2x - 9 = 0$$
$$2x = -9 \qquad\qquad 2x = 9$$
$$x = -\tfrac{9}{2} \qquad\qquad x = \tfrac{9}{2}$$

41. $x^2 - 13x + 12 = 0$
$$(x - 12)(x - 1) = 0$$
$$x - 12 = 0 \quad \text{or} \quad x - 1 = 0$$
$$x = 12 \qquad\qquad x = 1$$

43. $x^2 - 4x - 21 = 0$
$$(x - 7)(x + 3) = 0$$
$$x - 7 = 0 \quad \text{or} \quad x + 3 = 0$$
$$x = 7 \qquad\qquad x = -3$$

45. $x^2 - 9x + 8 = 0$
$$(x - 8)(x - 1) = 0$$
$$x - 8 = 0 \quad \text{or} \quad x - 1 = 0$$
$$x = 8 \qquad\qquad x = 1$$

47. $a^2 + 8a = -15$
$$a^2 + 8a + 15 = 0$$
$$(a + 3)(a + 5) = 0$$
$$a + 3 = 0 \quad \text{or} \quad a + 5 = 0$$
$$a = -3 \qquad\qquad a = -5$$

49. $2y - 8 = -y^2$
$$y^2 + 2y - 8 = 0$$
$$(y + 4)(y - 2) = 0$$
$$y + 4 = 0 \quad \text{or} \quad y - 2 = 0$$
$$y = -4 \qquad\qquad y = 2$$

51. $x^3 + 3x^2 + 2x = 0$
$$x(x^2 + 3x + 2) = 0$$
$$x(x + 2)(x + 1) = 0$$
$$x = 0 \quad \text{or} \quad x + 2 = 0 \quad \text{or} \quad x + 1 = 0$$
$$x = -2 \qquad\qquad x = -1$$

53. $k^3 - 6k^2 - 27k = 0$
$$k(k^2 - 6k - 27) = 0$$
$$k(k - 9)(k + 3) = 0$$
$$k = 0 \quad \text{or} \quad k - 9 = 0 \quad \text{or} \quad k + 3 = 0$$
$$k = 9 \qquad\qquad k = -3$$

55. $(x - 1)(x^2 + 5x + 6) = 0$
$$(x - 1)(x + 3)(x + 2) = 0$$
$$x - 1 = 0 \quad \text{or} \quad x + 3 = 0 \quad \text{or} \quad x + 2 = 0$$
$$x = 1 \qquad\qquad x = -3 \qquad\qquad x = -2$$

57. $2x^2 - 5x + 2 = 0$
$(2x - 1)(x - 2) = 0$
$2x - 1 = 0 \quad \text{or} \quad x - 2 = 0$
$2x = 1 \qquad\qquad x = 2$
$x = \frac{1}{2}$

59. $5x^2 - 6x + 1 = 0$
$(5x - 1)(x - 1) = 0$
$5x - 1 = 0 \quad \text{or} \quad x - 1 = 0$
$5x = 1 \qquad\qquad x = 1$
$x = \frac{1}{5}$

61. $4r^2 + 4r = -1$
$4r^2 + 4r + 1 = 0$
$(2r + 1)(2r + 1) = 0$
$2r + 1 = 0 \quad \text{or} \quad 2r + 1 = 0$
$2r = -1 \qquad\qquad 2r = -1$
$r = -\frac{1}{2} \qquad\qquad r = -\frac{1}{2}$

63. $-15x^2 + 2 = -7x$
$15x^2 - 7x - 2 = 0$
$(3x - 2)(5x + 1) = 0$
$3x - 2 = 0 \quad \text{or} \quad 5x + 1 = 0$
$3x = 2 \qquad\qquad 5x = -1$
$x = \frac{2}{3} \qquad\qquad x = -\frac{1}{5}$

65. $x(2x - 3) = 20$
$2x^2 - 3x - 20 = 0$
$(2x + 5)(x - 4) = 0$
$2x + 5 = 0 \quad \text{or} \quad x - 4 = 0$
$2x = -5 \qquad\qquad x = 4$
$x = -\frac{5}{2}$

67. $(d + 1)(8d + 1) = 18d$
$8d^2 + 9d + 1 - 18d = 0$
$8d^2 - 9d + 1 = 0$
$(8d - 1)(d - 1) = 0$
$8d - 1 = 0 \quad \text{or} \quad d - 1 = 0$
$8d = 1 \qquad\qquad d = 1$
$d = \frac{1}{8}$

69. $2x(3x^2 + 10x) = -6x$
$2x(3x^2 + 10x) + 2x(3) = 0$
$2x(3x^2 + 10x + 3) = 0$
$2x(3x + 1)(x + 3) = 0$
$2x = 0 \quad \text{or} \quad 3x + 1 = 0 \quad \text{or} \quad x + 3 = 0$
$x = 0 \qquad\qquad 3x = -1 \qquad\qquad x = -3$
$\qquad\qquad\qquad x = -\frac{1}{3}$

71. $x^3 + 7x^2 = x^2 - 9x$

$\quad x^3 + 6x^2 + 9x = 0$

$\quad x(x^2 + 6x + 9) = 0$

$\quad x(x + 3)(x + 3) = 0$

$\quad x = 0 \quad \text{or} \quad x + 3 = 0 \quad \text{or} \quad x + 3 = 0$

$\qquad\qquad\qquad\qquad x = -3 \qquad\qquad x = -3$

**Applications**

73. TIME OF FLIGHT

$\quad h = vt - 16t^2$

If the object hits the ground $h = 0$,

$\quad 0 = 144t - 16t^2$

$\quad 16t^2 - 144t = 0$

$\quad t(16t - 144) = 0$

$\quad t = 0 \quad \text{or} \quad 16t - 144 = 0$

$\qquad\qquad\qquad\quad 16t = 144$

$\qquad\qquad\qquad\quad t = 9$

After 9 seconds the object will strike the ground.

75. OFFICIATING

$\quad h = -16t^2 + 22t + 3$

The coin is in the air until $h = 0$,

$\quad 0 = -16t^2 + 22t + 3$

$\quad 16t^2 - 22t - 3 = 0$

$\quad (8t + 1)(2t - 3) = 0$

$\quad 8t + 1 = 0 \quad \text{or} \quad 2t - 3 = 0$

$\quad 8t = -1 \qquad\quad 2t = 3$

$\quad t = -\frac{1}{8} \qquad\quad t = \frac{3}{2}$

The negative time is inappropriate, therefore the captain has $\frac{3}{2}$ or 1.5 seconds to call heads or tails.

77. EXHIBITION DIVING

$\quad h = -16t^2 + 64$

When $h = 0$, the diver hits the water

$\quad 0 = -16t^2 + 64$

$\quad 16t^2 - 64 = 0$

$\quad (4t - 8)(4t + 8) = 0$

$\quad 4t - 8 = 0 \quad \text{or} \quad 4t + 8 = 0$

$\quad 4t = 8 \qquad\qquad 4t = -8$

$\quad t = 2 \qquad\qquad\quad t = -2$

The dive lasts 2 seconds.

79. **CHOREOGRAPHY**

$d = \frac{1}{2}r(r+1)$
If $d = 36$, then,
$36 = \frac{1}{2}r(r+1)$
$72 = r(r+1)$
$72 = r^2 + r$
$0 = r^2 + r - 72$
$0 = (r+9)(r-8)$
$r + 9 = 0$ or $r - 8 = 0$
$r = -9 \qquad r = 8$
There are 8 rows of dancers.

81. **INSULATION**

$A = lw$
$36 = (2w+1)w$
$36 = 2w^2 + w$
$0 = 2w^2 + w - 36$
$0 = (2w+9)(w-4)$
$2w + 9 = 0$ or $w - 4 = 0$
$2w = -9 \qquad w = 4$
$w = -\frac{9}{2}$
The width of this piece of insulation is 4 meters and the length is $2w + 1 = 2 \cdot 4 + 1 = 9$ meters.

83. **BOATING**

Using the pythagorean theorem, $c^2 = a^2 + b^2$ where $c$ represents the ramp and $a$ represents the rise and $b$ represents the run.
We are told that $c = a + 8, b = a + 7$
$c^2 = a^2 + b^2$
$(a+8)^2 = a^2 + (a+7)^2$
$a^2 + 16a + 64 = a^2 + a^2 + 14a + 49$
$0 = a^2 - 2a - 15$
$0 = (a-5)(a+3)$
$a - 5 = 0$ or $a + 3 = 0$
$a = 5 \qquad a = -3$
The rise of the ramp is 5 meters.
The run of the ramp is $5 + 7 = 12$ meters.
The ramp is $5 + 8 = 13$ meters.

85. **GARDENING TOOLS**

The cutting edge and the back form a $90^0$ angle, so   using the pythagorean theorem, $c^2 = a^2 + b^2$,
gives
$x^2 + (x+1)^2 = (x+2)^2$
$x^2 + x^2 + 2x + 1 = x^2 + 4x + 4$
Simplifying gives $x^2 - 2x - 3 = 0$
$(x-3)(x+1) = 0$
$x - 3 = 0$   or   $x + 1 = 0$
$x = 3 \qquad x = -1$
The back is 3 mm.
The cutting edge is $x + 1 = 3 + 1 = 4$ mm.
The span is $x + 2 = 3 + 2 = 5$ mm.

87. DESIGNING A TENT

The area of a triangle is found using $A = \frac{1}{2}bh$ where $b = 2h + 2$ and $A = 30$.

$30 = \frac{1}{2} \cdot (2h + 2)h$

$60 = 2h^2 + 2h$

$0 = 2h^2 + 2h - 60$

$0 = (2h - 10)(h + 6)$

$2h - 10 = 0$ or $h + 6 = 0$

$2h = 10 \qquad\quad h = -6$

$h = 5$

The height of the triangle is 5 feet and the base is $2 \cdot 5 + 2 = 12$ feet.

89. TUBING

$A = bh$ where we are told that $A = 60, h = b + 7$

$60 = b(b + 7)$

$60 = b^2 + 7b$

$0 = b^2 + 7b - 60$

$0 = (b - 5)(b + 12)$

$b - 5 = 0$ or $b + 12 = 0$

$b = 5 \qquad\quad b = -12$

The circumference of this tube is 5 inches.

91. HOUSE CONSTRUCTION

$A = \frac{h(B+b)}{2}$

$A = 24$

$B = 8$

$b = h$

$24 = \frac{h(8+h)}{2}$

$48 = 8h + h^2$

$0 = h^2 + 8h - 48$

$0 = (h + 12)(h - 4)$

$h + 12 = 0$ or $h - 4 = 0$

$h = -12 \qquad\quad h = 4$

The height of the truss is 4 meters.

**Writing**

93. Answers will vary

**Review**

95. EXERCISE

Let $t$ represent length of exercise, 15 minutes $\leq t < 30$ minutes

97. $l = 3w$ and $P = 120$

$P = 2l + 2w$

$120 = 2 \cdot 3w + 2w$

$120 = 8w$

$15 = w$

$l = 3 \cdot 15 = 45$

The area of this rectangle is $A = 45 \cdot 15 = 675 \text{ cm}^2$

**Chapter 11 Key Concepts**

1. Factor $3x + 27$
   $3(x + 9)$

3. $-3a^2 + 21a - 36$
   $= -3(a^2 - 7a + 12)$
   $= -3(a - 4)(a - 3)$

5. $rt + 2r + st + 2s$
   $= r(t + 2) + s(t + 2)$
   $= (r + s)(t + 2)$

7. $6t^2 - 19t + 15 = (3t - 5)(2t - 3)$

9. $2r^3 - 50r = 2r(r^2 - 25) = 2r(r - 5)(r + 5)$

**Section 11.1**

1. a. $35 = 7 \cdot 5$
   b. $45 = 5 \cdot 9 = 5 \cdot 3 \cdot 3 = 5 \cdot 3^2$
   c. $96 = 2 \cdot 48$
      $= 2 \cdot 16 \cdot 3$
      $= 2 \cdot 2^4 \cdot 3$
      $= 2^5 \cdot 3$
   d. $99 = 11 \cdot 9 = 11 \cdot 3^2$
   e. $2050 = 2 \cdot 1025$
      $= 2 \cdot 5 \cdot 205$
      $= 2 \cdot 5 \cdot 5 \cdot 41$
      $= 2 \cdot 5^2 \cdot 41$
   f. $4096 = 2 \cdot 2048$
      $= 2 \cdot 32 \cdot 64$
      $= 2 \cdot 2^5 \cdot 2 \cdot 32$
      $= 2^7 \cdot 2^5$
      $= 2^{12}$

3. a. $-a - 7 = -1(a + 7)$
   b. $-4t^2 + 3t - 1 = -1(4t^2 - 3t + 1)$

**Section 11.2**

5.

| Product of factors of 6 | Sum of factors of 6 |
| --- | --- |
| 1(6) | $1 + 6 = 7$ |
| 2(3) | $2 + 3 = 5$ |
| $-1(-6)$ | $-1 + (-6) = -7$ |
| $-2(-3)$ | $-2 + (-3) = -5$ |

7. Use the FOIL method to check this factorization,
   $(x - 4)(x + 5)$
   $= x^2 - 4x + 5x - 20$
   $= x^2 + x - 20$

**Section 11.3**

9. a. $2x^2 - 5x - 3 = (2x + 1)(x - 3)$
   b. $10y^2 + 21y - 10 = (5y - 2)(2y + 5)$
   c. $-3x^2 + 14x + 5 = -(3x + 1)(x - 5)$
   d. $6p^3 - 9p^2 - 6p$
      $= 3p(2p^2 - 3p - 2)$
      $= 3p(2p + 1)(p - 2)$
   e. $4b^2 - 17bc - 4c^2 = (4b - c)(b - 4c)$
   f. $3y^2 + 7y - 11$ is prime

**Section 11.4**

11. a. $x^2 + 10x + 25 = (x+5)(x+5) = (x+5)^2$
    b. $9y^2 - 24y + 16 = (3y-4)(3y-4) = (3y-4)^2$
    c. $-z^2 + 2z - 1 = -(z-1)(z-1) = -(z-1)^2$
    d. $25a^2 + 20ab + 4b^2 = (5a+2b)(5a+2b) = (5a+2b)^2$

13. a. $h^3 + 1 = (h+1)(h^2 - h + 1)$
    b. $125p^3 + q^3 = (5p+q)(25p^2 - 5pq + q^2)$
    c. $x^3 - 27 = (x-3)(x^2 + 3x + 9)$
    d. $16x^5 - 54x^2y^3$
       $= 2x^2(8x^3 - 27y^3)$
       $= 2x^2(2x - 3y)(4x^2 + 6xy + 9y^2)$

**Section 11.5**

15. a. $x^2 + 2x = 0$
    $x(x+2) = 0$
    $x = 0$     or     $x + 2 = 0$
                             $x = -2$
    b. $x(x-6) = 0$
    $x = 0$     or     $x - 6 = 0$
                             $x = 6$
    c. $x^2 - 9 = 0$
    $(x-3)(x+3) = 0$
    $x - 3 = 0$   or     $x + 3 = 0$
    $x = 3$                $x = -3$
    d. $a^2 - 7a + 12 = 0$
    $(a-3)(a-4) = 0$
    $a - 3 = 0$   or     $a - 4 = 0$
    $a = 3$               $a = 4$
    e. $t^2 + 4t + 4 = 0$
    $(t+2)(t+2) = 0$
    $t + 2 = 0$   or     $t + 2 = 0$
    $t = -2$             $t = -2$
    f. $-x^2 + 2x + 24 = 0$
    $-(x^2 - 2x - 24) = 0$
    $x^2 - 2x - 24 = 0$
    $(x-6)(x+4) = 0$
    $x - 6 = 0$   or     $x + 4 = 0$
    $x = 6$              $x = -4$
    g. $5a^2 - 6a + 1 = 0$
    $(5a-1)(a-1) = 0$
    $5a - 1 = 0$   or     $a - 1 = 0$
    $5a = 1$            $a = 1$
    $a = \frac{1}{5}$

j. $2p^3 = 2p(p+2)$
$2p^3 - 2p(p+2) = 0$
$2p(p^2 - (p+2)) = 0$
$2p(p^2 - p - 2) = 0$
$2p(p-2)(p+1) = 0$
$2p = 0$ or $p - 2 = 0$ or $p + 1 = 0$
$p = 0 \qquad p = 2 \qquad p = -1$

17. GARDENING

$A = 27$ feet$^2$

$l = 2w + 3$

Remembering the area of a rectangle is $A = lw$

$27 = (2w + 3)w$

$27 = 2w^2 + 3w$

$0 = 2w^2 + 3w - 27$

$0 = (2w + 9)(w - 3)$

$2w + 9 = 0$ or $\qquad w - 3 = 0$

$w = -\frac{9}{2} \qquad\qquad w = 3$

The width of the garden is 3 feet and its length is

$l = 2w + 3$

$l = 2 \cdot 3 + 3$

$l = 9$ feet

1. $196 = 4 \cdot 49 = 2^2 \cdot 7^2$

3. $4x + 16 = 4(x + 4)$

5. $q^2 - 81 = (q - 9)(q + 9)$

7. $16x^4 - 81$
$= (4x^2 - 9)(4x^2 + 9)$
$= (2x - 3)(2x + 3)(4x^2 + 9)$

9. $-x^2 + 9x + 22$
$= -1(x^2 - 9x - 22)$
$= -1(x - 11)(x + 2)$

11. $2a^2 + 5a - 12 = (2a - 3)(a + 4)$

13. $x^3 + 8 = (x + 2)(x^2 - 2x + 4)$

15. LANDSCAPING
$4r^2 - \pi r^2 = r^2(4 - \pi)$

17. The greatest common factor of $4a^3b^2$ and $18ab^2$ is GCF $= 2ab^2$

19. $(x + 3)(x - 2) = 0$
$x + 3 = 0$ or $x - 2 = 0$
$x = -3 \qquad\quad x = 2$

21. $6x^2 - x = 0$
$x(6x - 1) = 0$
$x = 0$ or $6x - 1 = 0$
$\qquad\qquad 6x = 1$
$\qquad\qquad x = \frac{1}{6}$

23. $6x^2 + x - 1 = 0$
$(3x - 1)(2x + 1) = 0$
$3x - 1 = 0$ or $2x + 1 = 0$
$3x = 1 \qquad\quad 2x = -1$
$x = \frac{1}{3} \qquad\quad x = -\frac{1}{2}$

25. DRIVING SAFETY

$$A = lw$$
$$54 = w(w + 3)$$
$$54 = w^2 + 3w$$
$$0 = w^2 + 3w - 54$$
$$(w + 9)(w - 6) = 0$$
$$w + 9 = 0 \quad \text{or} \quad w - 6 = 0$$
$$w = -9 \qquad w = 6$$

The width is 6 feet so the length is $6 + 3 = 9$ feet.

27. Using $a^2 + b^2 = c$,

$$(x - 4)^2 + (x - 2)^2 = x^2$$
$$x^2 - 8x + 16 + x^2 - 4x + 4 = x^2$$
$$2x^2 - 12x + 20 = x^2$$
$$x^2 - 12x + 20 = 0$$
$$(x - 10)(x - 2) = 0$$
$$x - 10 = 0 \quad \text{or} \quad x - 2 = 0$$
$$x = 10 \qquad x = 2$$

The hypotenuse is 10 and not 2 since $x - 4$ would be negative if the hypotenuse was 2.

**Chapters 1-11 Cumulative Review**

1. HEART RATE

   The difference in the heart rate of a 70 year old and a 35 year old is approximately $185 - 150 = 35$ beats per minute.

3. $\frac{124}{125} = 0.992$

5. $\frac{16}{5} \div \frac{10}{3} = \frac{16}{5} \cdot \frac{3}{10} = \frac{24}{25}$

7. $3 + 2[-1 - 4(5)]$
   $= 3 + 2[-1 - 20]$
   $= 3 + 2[-21]$
   $= 3 + (-42)$
   $= -39$

9. $\frac{-(-2)-5}{1-2} = \frac{2-5}{-1} = 3$

11. $-8y^2 - 5y^2 + 6 = -13y^2 + 6$

13. $-(3a + 1) + a = 2$
    $-3a - 1 + a = 2$
    $-2a = 3$
    $a = -\frac{3}{2}$

15. $\frac{3t-21}{2} = t - 6$
    $3t - 21 = 2t - 12$
    $t = 9$

17. $A = P + Prt$
    $A - P = Prt$
    $\frac{A-P}{Pr} = t$

19. GEOMETRY TOOL

    The question is asking us to find the distance around half of a circle. Using $C = 2\pi r$ where $r = 3$, then $C \approx \frac{2\pi(3)}{2} \approx 9.4$ inches.

21. $x$ twentry dollar bills have $\$20x$ value.

23. $y = (x + 2)^2$

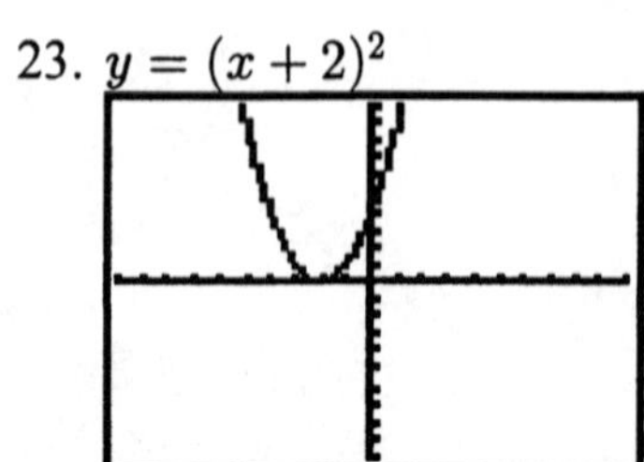

25. $3x - 3y = 6$
$-3y = -3x + 6$
$y = x - 2$
Slope is 1 and the $y$-intercept is $(0, -2)$.

27. $m = -3, (-4, 1)$
$y - y_1 = m(x - x_1)$
$y - 1 = -3(x - (-4))$
$y - 1 = -3x - 12$
$y = -3x - 11$

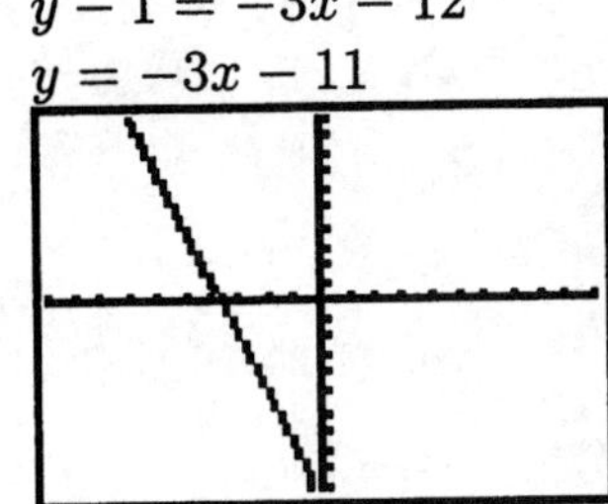

29. If two lines are parallel then the slopes are the same.

31. $g(x) = 3x - x^3$
$g(-2) = 3(-2) - (-2)^3$
$g(-2) = -6 - (-8)$
$g(-2) = 2$

33. $-y^2(4y^3) = -4y^5$

35. $\left(\frac{b^5}{b^{-2}}\right)^{-2} = (b^7)^{-2} = b^{-14} = \frac{1}{b^{14}}$

37. $(x^2 - 3x + 8) - (3x^2 + x + 3)$
$= x^2 - 3x + 8 - 3x^2 - x - 3$
$= -2x^2 - 4x + 5$

39. $(y - 6)^2 = y^2 - 12y + 36$

41. $\dfrac{12a^2b^2 - 8a^2b - 4ab}{4ab}$
$= \dfrac{12a^2b^2}{4ab} - \dfrac{8a^2b}{4ab} - \dfrac{4ab}{4ab}$
$= 3ab - 2a - 1$

43. PLAYPEN
a. The perimeter is $P = 2(x + 3) + 2(x + 1)$
$P = 2x + 6 + 2x + 2$
$P = (4x + 8)$ inches
b. The area of the floor is $A = (x + 3)(x + 1)$
$A = (x^2 + 4x + 3)$ inches$^2$
c. The volume is $V = (x + 1)(x + 3)x$
$V = x(x^2 + 4x + 3)$
$V = (x^3 + 4x^2 + 3x)$ inches$^3$

45. $b^3 - 3b^2 = b^2(b - 3)$

47. $2x^2 - 3x - 2 = (2x+1)(x-2)$

49. $-5a^2 + 25a - 30$
$\quad = -5(a^2 - 5a + 6)$
$\quad = -5(a-3)(a-2)$

51. $t^3 - 8 = (t-2)(t^2 + 2t + 4)$

53. $15s^2 - 20s = 0$
$\quad 5s(3s - 4) = 0$
$\quad 5s = 0 \ \text{ or } \ 3s - 4 = 0$
$\quad s = 0 \ \ \text{ or } \ 3s = 4$
$\quad\quad\quad\quad\quad s = \frac{4}{3}$

## Section 12.1 Simplifying Rational Expressions

### Vocabulary

1. In a fraction, the part above the fraction bar is called the **numerator** and the part below the fraction bar is called the **denominator**.
3. Division by 0 is **undefined**.
5. To **simplify** a rational expression, means to factor the numerator and denominator completely and divide out common factors.

### Concepts

7. a. $x = 0$
   b. $x = 6$ will make the denominator zero.
   c. $x = -6$ will make the denominator zero.

9. The common factor divided out is $x + 1$.

11. Error is that $x$ is not a common factor so it cannot be divided out.

### Notation

13. $\dfrac{x^2+5x-6}{x^2-1}$
    $= \dfrac{(x+6)(x-1)}{(x+1)(x-1)}$
    $= \dfrac{(x+6)}{(x+1)}$

### Practice

15. $\dfrac{x-2}{x-5} = \dfrac{6-2}{6-5} = \dfrac{4}{1} = 4$

17. $\dfrac{-2x-3}{x^2-1} = \dfrac{-2(6)-3}{6^2-1} = \dfrac{-12-3}{35} = \dfrac{-15}{35} = -\dfrac{3}{7}$

19. $\dfrac{x^2-4x-12}{x^2+x-2} = \dfrac{6^2-4(6)-12}{6^2+6-2} = \dfrac{36-24-12}{36+6-2} = \dfrac{0}{40} = 0$

21. $x - 2 = 0$ when $x = 2$

23. This rational expression is never undefined.

25. $2x - 1 = 0$
    $2x = 1$
    $x = \frac{1}{2}$

27. $x^2 - 36 = 0$
    $x^2 = 36$
    $x = 6$ or $x = -6$

29. $x^2 + x - 2 = 0$
$(x + 2)(x - 1) = 0$
$x + 2 = 0$ or $x - 1 = 0$
$x = -2 \qquad x = 1$

31. $\frac{28}{35} = \frac{7 \cdot 4}{7 \cdot 5} = \frac{4}{5}$

33. $\frac{9}{27} = \frac{9}{9 \cdot 3} = \frac{1}{3}$

35. $-\frac{36}{48} = -\frac{2 \cdot 3 \cdot 6}{2 \cdot 4 \cdot 6} = -\frac{3}{4}$

37. $\frac{45}{9a} = \frac{9 \cdot 5}{9a} = \frac{5}{a}$

39. $\frac{5+5}{5z} = \frac{10}{5z} = \frac{2}{z}$

41. $\frac{(3+4)a}{24-3} = \frac{7a}{21} = \frac{7a}{7 \cdot 3} = \frac{a}{3}$

43. $\frac{2x}{3x} = \frac{2}{3}$

45. $\frac{6x^2}{4x^2} = \frac{6}{4} = \frac{3}{2}$

47. $\frac{2x^2}{3y}$ already in lowest terms

49. $\frac{15x^2y}{5xy^2} = \frac{15x}{5y} = \frac{3x}{y}$

51. $\frac{6x+3}{3y} = \frac{3(2x+1)}{3y} = \frac{2x+1}{y}$

53. $\frac{x+3}{3x+9} = \frac{(x+3)}{3(x+3)} = \frac{1}{3}$

55. $\frac{x-7}{7-x} = \frac{x-7}{-1(-7+x)} = \frac{x-7}{-1(x-7)} = \frac{1}{-1} = -1$

57. $\frac{6x-30}{5-x} = \frac{6(x-5)}{-1(-5+x)} = \frac{6(x-5)}{-1(x-5)} = \frac{6}{-1} = -6$

59. $\frac{12-3x^2}{x^2-x-2} = \frac{-3(-4+x^2)}{(x-2)(x+1)} = \frac{-3(x^2-4)}{(x-2)(x+1)} = \frac{-3(x-2)(x+2)}{(x-2)(x+1)} = \frac{-3(x+2)}{(x+1)}$

61. $\frac{x^2+3x+2}{x^2+x-2} = \frac{(x+1)(x+2)}{(x-1)(x+2)} = \frac{x+1}{x-1}$

63. $\frac{x^2-8x+15}{x^2-x-6} = \frac{(x-3)(x-5)}{(x-3)(x+2)} = \frac{(x-5)}{(x+2)}$

65. $\frac{2x^2-8x}{x^2-6x+8} = \frac{2x(x-4)}{(x-4)(x-2)} = \frac{2x}{x-2}$

67. $\frac{2-a}{a^2-a-2} = \frac{-1(-2+a)}{(a+1)(a-2)} = \frac{-1(a-2)}{(a+1)(a-2)} = \frac{-1}{a+1}$

69. $\frac{x^2+3x+2}{x^3+x^2} = \frac{(x+2)(x+1)}{x^2(x+1)} = \frac{x+2}{x^2}$

71. $\dfrac{x^2-8x+16}{x^2-16} = \dfrac{(x-4)(x-4)}{(x+4)(x-4)} = \dfrac{x-4}{x+4}$

73. $\dfrac{2x^2-8}{x^2-3x+2}$
$= \dfrac{2(x^2-4)}{(x-2)(x-1)}$
$= \dfrac{2(x-2)(x+2)}{(x-2)(x-1)}$
$= \dfrac{2(x+2)}{(x-1)}$

75. $\dfrac{5x^2+2x-3}{x^2+2x-15}$ in lowest terms

77. $\dfrac{x^2-3(2x-3)}{9-x^2}$
$= \dfrac{x^2-6x+9}{-1(x^2-9)}$
$= \dfrac{(x-3)(x-3)}{-1(x+3)(x-3)}$
$= -\dfrac{x-3}{x+3}$

79. $\dfrac{4(x+3)+4}{3(x+2)+6}$
$= \dfrac{4x+12+4}{3x+6+6}$
$= \dfrac{4x+16}{3x+12}$
$= \dfrac{4(x+4)}{3(x+4)}$
$= \dfrac{4}{3}$

81. $\dfrac{x^2-9}{(2x+3)-(x+6)}$
$= \dfrac{(x+3)(x-3)}{2x+3-x-6}$
$= \dfrac{(x+3)(x-3)}{(x-3)}$
$= x+3$

83. $\dfrac{y-xy}{xy-x}$ in lowest terms

85. $\dfrac{6a-6b+6c}{9a-9b+9c} = \dfrac{6(a-b+c)}{9(a-b+c)} = \dfrac{6}{9} = \dfrac{2}{3}$

87. $\dfrac{15x-3x^2}{25y-5xy} = \dfrac{3x(5-x)}{5y(5-x)} = \dfrac{3x}{5y}$

89. $\dfrac{a+b-c}{c-a-b} = \dfrac{a+b-c}{-1(-c+a+b)} = \dfrac{a+b-c}{-1(a+b-c)} = \dfrac{1}{-1} = -1$

## Applications

91. ROOFING
$pitch = \dfrac{rise}{run}$
$pitch = \dfrac{x^2+4x+4}{(x^2-4)}$
$pitch = \dfrac{(x+2)(x+2)}{(x+2)(x-2)}$
$pitch = \dfrac{x+2}{x-2}$

93. **WORD PROCESSORS**

$f(x) = \frac{8000}{x}$

| $x$ (font) | $f(x)$ |
|---|---|
| 8 | $\frac{8000}{8} = 1000$ |
| 10 | $\frac{8000}{10} = 800$ |
| 12 | $\frac{8000}{12} \approx 667$ |
| 16 | $\frac{8000}{16} = 500$ |
| 24 | $\frac{8000}{24} \approx 333$ |
| 36 | $\frac{8000}{36} \approx 222$ |

## Writing

95. Answers will vary

97. Answers will vary.

## Review

99. Associative property of addition $a + (b + c) = (a + b) + c$.

101. If $ab = 0$ then $a = 0$ or $b = 0$ or they both equal zero.

103. The opposite of $-\frac{5}{3}$ is $\frac{5}{3}$.

## Section 12.2 Multiplying and Dividing Rational Expressions

### Vocabulary

1.  In a fraction, the part above the fraction bar is called the **numerator**.

### Concepts

3.  To multiply fractions, we multiply their **numerators** and multiply their **denominators**.

5.  To write a polynomial in fraction form, we insert a denominator of 1.

7.  To divide fractions, we invert the **divisor** and **multiply**.

### Notation

9.  $\dfrac{x^2+x}{3x-6} \cdot \dfrac{x-2}{x+1} = \dfrac{(x^2+x)(x-2)}{(3x-6)(x+1)}$

$\qquad = \dfrac{x(x+1)(x-2)}{3(x-2)(x+1)}$

$\qquad = \dfrac{x}{3}$

### Practice

11. $\dfrac{3}{y} \cdot \dfrac{y}{2} = \dfrac{3}{2}$

13. $\dfrac{5y}{7} \cdot \dfrac{7}{5} = y$

15. $\dfrac{7z}{9z} \cdot \dfrac{4z}{2z} = \dfrac{7}{9} \cdot \dfrac{2}{1} = \dfrac{14}{9}$

17. $\dfrac{2x^2 y}{3xy} \cdot \dfrac{3xy^2}{2} = \dfrac{2xxy \cdot 3xyy}{3xy \cdot 2} = x^2 y^2$

19. $\dfrac{8x^2 y^2}{4x^2} \cdot \dfrac{2xy}{2y} = \dfrac{2 \cdot 2 \cdot 2xxyy \cdot 2xy}{2 \cdot 2xx \cdot 2y} = 2xy^2$

21. $\dfrac{-2xy}{x^2} \cdot \dfrac{3xy}{2} = \dfrac{-2xy \cdot 3xy}{xx \cdot 2} = -3y^2$

23. $\dfrac{ab^2}{a^2 b} \cdot \dfrac{b^2 c^2}{abc} \cdot \dfrac{abc^2}{a^3 c^2} = \dfrac{abb \cdot bbcc \cdot abcc}{aab \cdot abc \cdot aaacc} = \dfrac{b^3 c}{a^4}$

25. $\dfrac{10r^2 st^3}{6rs^2} \cdot \dfrac{3r^3 t}{2rst} \cdot \dfrac{2s^3 t^4}{5s^2 t^3}$

$\qquad = \dfrac{5rt^3}{3s} \cdot \dfrac{3r^2}{2s} \cdot \dfrac{2st}{5}$

$\qquad = \dfrac{5rttt \cdot 3rr \cdot 2st}{3s \cdot 2s \cdot 5}$

$\qquad = \dfrac{t^4 r^3}{s}$

27. $\dfrac{z+7}{7} \cdot \dfrac{z+2}{z}$

$\qquad = \dfrac{(z+7)(z+2)}{7z}$

$\qquad = \dfrac{z^2 + 9z + 14}{7z}$

29. $\frac{x-2}{2} \cdot \frac{2x}{x-2}$

$= \frac{(x-2)2x}{2(x-2)}$

$= \frac{2x}{2}$

$= x$

31. $\frac{x+5}{5} \cdot \frac{x}{x+5}$

$= \frac{(x+5)x}{5(x+5)}$

$= \frac{x}{5}$

33. $\frac{5}{m} \cdot m = 5$

35. $4d \cdot \frac{3}{2d} = 2 \cdot 3 = 6$

37. $15x\left(\frac{x+1}{15x}\right) = x + 1$

39. $12y\left(\frac{y+8}{6y}\right) = 2(y + 8) = 2y + 16$

41. $(x + 8)\frac{x+5}{x+8} = x + 5$

43. $10(h + 9)\frac{h-3}{h+9} = 10(h - 3) = 10h - 30$

45. $\frac{(x+1)^2}{x+1} \cdot \frac{x+2}{x+1} = \frac{(x+1)(x+1)(x+2)}{(x+1)(x+1)} = x + 2$

47. $\frac{2x+6}{x+3} \cdot \frac{3}{4x} = \frac{2(x+3)\cdot 3}{(x+3)\cdot 2\cdot 2x} = \frac{3}{2x}$

49. $\frac{x^2-x}{x} \cdot \frac{3x-6}{3x-3} = \frac{x(x-1)}{x} \cdot \frac{3(x-2)}{3(x-1)} = \frac{(x-1)(x-2)}{(x-1)} = x - 2$

51. $\frac{7y-14}{y-2} \cdot \frac{x^2}{7x} = \frac{7(y-2)}{(y-2)} \cdot \frac{x\cdot x}{7x} = \frac{7x}{7} = x$

53. $\frac{x^2+x-6}{5x} \cdot \frac{5x-10}{x+3} = \frac{(x+3)(x-2)\cdot 5(x-2)}{5x(x+3)} = \frac{(x-2)^2}{x}$

55. $\frac{m^2-2m-3}{2m+4} \cdot \frac{m^2-4}{m^2+3m+2}$

$= \frac{(m-3)(m+1)(m+2)(m-2)}{2(m+2)(m+2)(m+1)}$

$= \frac{(m-3)(m-2)}{2(m+2)}$

$= \frac{m^2-5m+6}{2m+4}$

57. $\frac{3x^2+5x+2}{x^2-9} \cdot \frac{x-3}{x^2-4} \cdot \frac{x^2+5x+6}{6x+4}$

$= \frac{(3x+2)(x+1)(x-3)(x+3)(x+2)}{(x-3)(x+3)(x-2)(x+2)2(3x+2)}$

$= \frac{(x+1)}{(x-2)2}$

$= \frac{x+1}{2x-4}$

59. $\frac{2}{y} \div \frac{4}{3} = \frac{2}{y} \cdot \frac{3}{4} = \frac{2}{y} \cdot \frac{3}{2\cdot 2} = \frac{3}{2y}$

61. $\frac{3x}{2} \div \frac{x}{2} = \frac{3x}{2} \cdot \frac{2}{x} = 3$

63. $\frac{3x}{y} \div \frac{2x}{4} = \frac{3x}{y} \cdot \frac{4}{2x} = \frac{3x}{y} \cdot \frac{2 \cdot 2}{2x} = \frac{6}{y}$

65. $\frac{4x}{3x} \div \frac{2y}{9y} = \frac{4}{3} \div \frac{2}{9} = \frac{4}{3} \cdot \frac{9}{2} = 6$

67. $\frac{x^2}{3} \div \frac{2x}{4} = \frac{x^2}{3} \div \frac{x}{2} = \frac{x^2}{3} \cdot \frac{2}{x} = \frac{2x}{3}$

69. $\frac{x^2 y}{3xy} \div \frac{xy^2}{6y} = \frac{x}{3} \div \frac{xy}{6} = \frac{x}{3} \cdot \frac{6}{xy} = \frac{2}{y}$

71. $\frac{x+2}{3x} \div \frac{x+2}{2} = \frac{x+2}{3x} \cdot \frac{2}{x+2} = \frac{2}{3x}$

73. $\frac{(z-2)^2}{3z^2} \div \frac{z-2}{6z} = \frac{(z-2)^2}{3z^2} \cdot \frac{6z}{(z-2)} = \frac{2(z-2)}{z}$

75. $\frac{(z-7)^2}{z+2} \div \frac{z(z-7)}{5z^2} = \frac{(z-7)^2}{(z+2)} \cdot \frac{5z^2}{z(z-7)} = \frac{5z(z-7)}{(z+2)}$

77. $\frac{x^2-4}{3x+6} \div \frac{x-2}{x+2} = \frac{(x-2)(x+2)}{3(x+2)} \cdot \frac{(x+2)}{(x-2)} = \frac{x+2}{3}$

79. $\frac{x^2-1}{3x-3} \div \frac{x+1}{3} = \frac{(x-1)(x+1)}{3(x-1)} \cdot \frac{3}{(x+1)} = 1$

81. $\frac{x^2-2x-35}{3x^2+27x} \div \frac{x^2+7x+10}{6x^2+12x} = \frac{(x-7)(x+5)}{3x(x+9)} \cdot \frac{6x(x+2)}{(x+5)(x+2)} = \frac{2(x-7)}{(x+9)}$

83. $\frac{2d^2+8d-42}{d-3} \div \frac{2d^2+14d}{d^2+5d} = \frac{2(d-3)(d+7)}{(d-3)} \cdot \frac{d(d+5)}{2d(d+7)} = d + 5$

85. $\frac{x}{3} \cdot \frac{9}{4} \div \frac{x^2}{6} = \frac{3x}{4} \cdot \frac{6}{x^2} = \frac{9}{2x}$

87. $\frac{x^2}{18} \div \frac{x^3}{6} \div \frac{12}{x^2} = \quad \frac{x^2}{18} \cdot \frac{6}{x^3} \div \frac{12}{x^2} = \frac{1}{3x} \cdot \frac{x^2}{12} \quad = \frac{x}{36}$

89. $\frac{z^2-4}{2z+6} \div \frac{z+2}{4} \cdot \frac{z+3}{z-2} = \frac{(z-2)(z+2)}{2(z+3)} \cdot \frac{4}{(z+2)} \cdot \frac{(z+3)}{(z-2)} = 2$

91. $\frac{x-x^2}{x^2-4}\left(\frac{2x+4}{x+2} \div \frac{5}{x+2}\right)$

$\quad = \frac{x(1-x)}{(x-2)(x+2)}\left(\frac{2(x+2)}{(x+2)} \cdot \frac{(x+2)}{5}\right)$

$\quad = \frac{x(1-x)}{(x-2)(x+2)} \cdot \frac{2(x+2)}{5}$

$\quad = \frac{2x(1-x)}{5(x-2)}$

93. $\frac{y^2}{x+1} \cdot \frac{x^2+2x+1}{x^2-1} \div \frac{3y}{xy-y}$

$\quad = \frac{y^2}{(x+1)} \cdot \frac{(x+1)(x+1)}{(x+1)(x-1)} \div \frac{3y}{y(x-1)}$

$\quad = \frac{y^2}{(x-1)} \cdot \frac{y(x-1)}{3y}$

$\quad = \frac{y^2}{3}$

95. $\frac{x^2+x-6}{x^2-4} \cdot \frac{x^2+2x}{x-2} \div \frac{x^2+3x}{x+2}$

$= \frac{(x+3)(x-2)}{(x-2)(x+2)} \cdot \frac{x(x+2)}{(x-2)} \div \frac{x(x+3)}{(x+2)}$

$= \frac{x(x+3)}{(x-2)} \div \frac{x(x+3)}{(x+2)}$

$= \frac{x(x+3)}{(x-2)} \cdot \frac{(x+2)}{x(x+3)}$

$= \frac{x+2}{x-2}$

## Applications

97. INTERNATIONAL ALPHABET

$A = 6 \cdot \left(\frac{2x+1}{2}\right)^2$

$A = 6 \cdot \left(\frac{(2x+1)^2}{2^2}\right)$

$A = \frac{6}{4} \cdot (4x^2 + 4x + 1)$

$A = \frac{3}{2} \cdot (4x^2 + 4x + 1)$

$A = \frac{12x^2+12x+3}{2}$ in$^2$

## Writing

99. Answers will vary

101. Answers will vary

## Review

103. $2x^3y^2(-3x^2y^4) = -6x^5y^6$

105. $(3y)^{-4} = \frac{1}{(3y)^4} = \frac{1}{81y^4}$

107. $-4(y^3 - 4y^2 + 3y - 2) - 4(-2y^3 - y)$

$\quad = -4y^3 + 16y^2 - 12y + 8 + 8y^3 + 4y$

$\quad = 4y^3 + 16y^2 - 8y + 8$

# Section 12.3 Adding and Subtracting Rational Expressions

## Vocabulary

1.  The **LCD** for a set of fractions is the smallest number that each denominator divides exactly.

## Concepts

3.  To add two fractions with like denominators, we add their **numerators** and keep the **common denominator**.

## Notation

5.  $\dfrac{6a-1}{4a+1} + \dfrac{2a+3}{4a+1} = \dfrac{6a-1+2a+3}{4a+1}$
$$= \dfrac{8a+2}{4a+1}$$
$$= \dfrac{2(4a+1)}{4a+1}$$
$$= 2$$

## Practice

7.  $\dfrac{x}{9} + \dfrac{2x}{9} = \dfrac{3x}{9} = \dfrac{x}{3}$

9.  $\dfrac{2x}{y} + \dfrac{2x}{y} = \dfrac{4x}{y}$

11. $\dfrac{4}{7y} + \dfrac{10}{7y} = \dfrac{14}{7y} = \dfrac{2}{y}$

13. $\dfrac{y+2}{10z} + \dfrac{y+4}{10z} = \dfrac{2y+6}{10z} = \dfrac{2(y+3)}{10z} = \dfrac{y+3}{5z}$

15. $\dfrac{3x-5}{x-2} + \dfrac{6x-13}{x-2} = \dfrac{9x-18}{x-2} = \dfrac{9(x-2)}{(x-2)} = 9$

17. $\dfrac{a}{a^2+5a+6} + \dfrac{3}{a^2+5a+6} = \dfrac{a+3}{(a+3)(a+2)} = \dfrac{1}{a+2}$

19. $\dfrac{35y}{72} - \dfrac{44y}{72} = -\dfrac{9y}{72} = -\dfrac{y}{8}$

21. $\dfrac{2x}{y} - \dfrac{x}{y} = \dfrac{x}{y}$

23. $\dfrac{9y}{3x} - \dfrac{6y}{3x} = \dfrac{3y}{3x} = \dfrac{y}{x}$

25. $\dfrac{6x-5}{3xy} - \dfrac{3x-5}{3xy} = \dfrac{6x-5-3x+5}{3xy} = \dfrac{3x}{3xy} = \dfrac{1}{y}$

27. $\dfrac{3y-2}{2y+6} - \dfrac{2y-5}{2y+6} = \dfrac{3y-2-2y+5}{2y+6} = \dfrac{y+3}{2(y+3)} = \dfrac{1}{2}$

29. $\dfrac{2c}{c^2-d^2} - \dfrac{2d}{c^2-d^2}$
$$= \dfrac{2c-2d}{c^2-d^2}$$
$$= \dfrac{2(c-d)}{(c-d)(c+d)}$$
$$= \dfrac{2}{c+d}$$

31. $\dfrac{13x}{15} + \dfrac{12x}{15} - \dfrac{5x}{15}$

$= \dfrac{13x+12x-5x}{15}$

$= \dfrac{20x}{15}$

$= \dfrac{4x}{3}$

33. $\dfrac{x}{y} + \dfrac{2x}{y} - \dfrac{3x}{y}$

$= \dfrac{x+2x-3x}{y}$

$= \dfrac{0}{y}$

$= 0$

35. $\dfrac{3x}{y+2} - \dfrac{3y}{y+2} + \dfrac{x+y}{y+2}$

$= \dfrac{3x-3y+x+y}{y+2}$

$= \dfrac{4x-2y}{y+2}$

$= \dfrac{2(2x-y)}{y+2}$

37. $\dfrac{x+1}{x-2} - \dfrac{2(x-3)}{x-2} + \dfrac{3(x+1)}{x-2}$

$= \dfrac{x+1-2x+6+3x+3}{x-2}$

$= \dfrac{2x+10}{x-2}$

$= \dfrac{2(x+5)}{x-2}$

39. $\dfrac{25}{4} \cdot \dfrac{5x}{5x} = \dfrac{125x}{20x}$

41. $\dfrac{8}{x} \cdot \dfrac{xy}{xy} = \dfrac{8xy}{x^2 y}$

43. $\dfrac{3x}{x+1} \cdot \dfrac{x+1}{x+1} = \dfrac{3x^2+3x}{(x+1)^2}$

45. $\dfrac{2y}{x} \cdot \dfrac{x+1}{x+1} = \dfrac{2xy+2y}{x^2+x}$

47. $\dfrac{z}{z-1} \cdot \dfrac{z+1}{z+1} = \dfrac{z^2+z}{z^2-1}$

49. $\dfrac{2}{x+1} \cdot \dfrac{x+2}{x+2} = \dfrac{2x+4}{x^2+3x+2}$

51. The LCD of $2x$ and $6x$ is $6x$.

53. The LCD of $6y$ and $9xy^2$ is $18xy^2$.

55. The LCD of $x^2 - 1$ and $x + 1$ is $x^2 - 1$.

57. The LCD of $x^2 + 6x$, $x + 6$ and $x$ is $x^2 + 6x$.

59. The LCD of $x^2 - 4x - 5$ and $x^2 - 25$ is $(x + 1)(x + 5)(x - 5)$.

61. $\dfrac{2y}{9} + \dfrac{y}{3} = \dfrac{2y}{9} + \dfrac{3y}{9} = \dfrac{5y}{9}$

63. $\dfrac{21x}{14} - \dfrac{5x}{21} = \dfrac{63x}{42} - \dfrac{10x}{42} = \dfrac{53x}{42}$

65. $\frac{4x}{3} + \frac{2x}{y} = \frac{4xy}{3y} + \frac{6x}{3y} = \frac{4xy+6x}{3y}$

67. $\frac{2}{x} - 3x = \frac{2}{x} - \frac{3x^2}{x} = \frac{2-3x^2}{x}$

69. $\frac{y+2}{5y^2} + \frac{y+4}{15y}$
$= \frac{3(y+2)}{15y^2} + \frac{9(y+4)}{15y^2}$
$= \frac{3y+6+y^2+4y}{15y^2}$
$= \frac{y^2+7y+6}{15y^2}$

71. $\frac{x+5}{xy} - \frac{x-1}{x^2 y}$
$= \frac{x(x+5)}{x^2 y} - \frac{x-1}{x^2 y}$
$= \frac{x^2+5x-x+1}{x^2 y}$
$= \frac{x^2+4x+1}{x^2 y}$

73. $\frac{x}{x+1} + \frac{x-1}{x}$
$= \frac{x^2}{x(x+1)} + \frac{(x-1)(x+1)}{x(x+1)}$
$= \frac{x^2+x^2-1}{x^2+x}$
$= \frac{2x^2-1}{x^2+x}$

75. $\frac{x-1}{x} + \frac{y+1}{y}$
$= \frac{y(x-1)}{xy} + \frac{x(y+1)}{xy}$
$= \frac{xy-y+xy+x}{xy}$
$= \frac{2xy+x-y}{xy}$

77. $\frac{x}{x-2} + \frac{4+2x}{x^2-4}$
$= \frac{x}{x-2} + \frac{4+2x}{(x-2)(x+2)}$
$= \frac{x(x+2)}{(x-2)(x+2)} + \frac{4+2x}{(x-2)(x+2)}$
$= \frac{x^2+2x+4+2x}{(x-2)(x+2)}$
$= \frac{x^2+4x+4}{(x-2)(x+2)}$
$= \frac{(x+2)(x+2)}{(x-2)(x+2)}$
$= \frac{(x+2)}{(x-2)}$

79. $\frac{x+1}{x-1} + \frac{x-1}{x+1}$
$= \frac{(x+1)(x+1)}{(x-1)(x+1)} + \frac{(x-1)(x-1)}{(x+1)(x-1)}$
$= \frac{x^2+2x+1+x^2-2x+1}{(x-1)(x+1)}$
$= \frac{2x^2+2}{(x-1)(x+1)}$

81. $\dfrac{5}{a-4} + \dfrac{7}{4-a}$

$= \dfrac{5}{a-4} + \dfrac{7}{-1(a-4)}$

$= \dfrac{5(-1)}{(-1)(a-4)} + \dfrac{7}{(-1)(a-4)}$

$= -\dfrac{-5+7}{a-4}$

$= -\dfrac{2}{a-4}$

83. $\dfrac{t+1}{t-7} - \dfrac{t+1}{7-t}$

$= \dfrac{t+1}{t-7} - \dfrac{t+1}{(-1)(t-7)}$

$= \dfrac{t+1}{t-7} + \dfrac{t+1}{t-7}$

$= \dfrac{2t+2}{t-7}$

$= \dfrac{2(t+1)}{t-7}$

85. $\dfrac{2x+2}{x-2} - \dfrac{2x}{2-x}$

$= \dfrac{2x+2}{(x-2)} - \dfrac{2x}{-1(x-2)}$

$= \dfrac{2x+2}{(x-2)} + \dfrac{2x}{(x-2)}$

$= \dfrac{4x+2}{x-2}$

$= \dfrac{2(2x+1)}{x-2}$

87. $\dfrac{b}{b+1} - \dfrac{b+1}{2(b+1)}$

$= \dfrac{2b}{2(b+1)} - \dfrac{b+1}{2(b+1)}$

$= \dfrac{2b-b-1}{2(b+1)}$

$= \dfrac{b-1}{2(b+1)}$

89. $\dfrac{2}{(a+3)(a+1)} + \dfrac{1}{a+3}$

$= \dfrac{2}{(a+3)(a+1)} + \dfrac{a+1}{(a+3)(a+1)}$

$= \dfrac{3+a}{(a+3)(a+1)}$

$= \dfrac{1}{a+1}$

91. $\dfrac{x+1}{2x+4} - \dfrac{x^2}{2x^2-8}$

$= \dfrac{x+1}{2(x+2)} - \dfrac{x^2}{2(x^2-4)}$

$= \dfrac{(x+1)}{2(x+2)} - \dfrac{x^2}{2(x-2)(x+2)}$

$= \dfrac{(x+1)(x-2)}{2(x+2)(x-2)} - \dfrac{x^2}{2(x-2)(x+2)}$

$= \dfrac{-x^2+x^2-x-2}{2(x-2)(x+2)}$

$= \dfrac{-1(x+2)}{2(x-2)(x+2)}$

$= -\dfrac{1}{2(x-2)}$

93. $\dfrac{2x}{x^2-3x+2} + \dfrac{2x}{x-1} - \dfrac{x}{x-2}$

$\quad = \dfrac{2x}{(x-1)(x-2)} + \dfrac{2x}{x-1} - \dfrac{x}{x-2}$

$\quad = \dfrac{2x}{(x-1)(x-2)} + \dfrac{2x(x-2)}{(x-1)(x-2)} - \dfrac{x(x-1)}{(x-2)(x-1)}$

$\quad = \dfrac{2x+2x^2-4x-x^2+x}{(x-1)(x-2)}$

$\quad = \dfrac{x^2-x}{(x-1)(x-2)}$

$\quad = \dfrac{x(x-1)}{(x-1)(x-2)}$

$\quad = \dfrac{x}{(x-2)}$

95. $\dfrac{2x}{x-1} + \dfrac{3x}{x+1} - \dfrac{x+3}{x^2-1}$

$\quad = \dfrac{2x}{x-1} + \dfrac{3x}{x+1} - \dfrac{x+3}{(x-1)(x+1)}$

$\quad = \dfrac{2x(x+1)}{(x-1)(x+1)} + \dfrac{3x(x-1)}{(x+1)(x-1)} - \dfrac{x+3}{(x-1)(x+1)}$

$\quad = \dfrac{-x-3+2x^2+2x+3x^2-3x}{(x-1)(x+1)}$

$\quad = \dfrac{5x^2-2x-3}{(x-1)(x+1)}$

$\quad = \dfrac{(x-1)(5x+3)}{(x-1)(x+1)}$

$\quad = \dfrac{(5x+3)}{(x+1)}$

## Applications

97. The total height of this funnel is $\dfrac{3}{2x^2} + \dfrac{10}{3x} = \dfrac{3\cdot3}{3\cdot2x^2} + \dfrac{2x\cdot10}{2x\cdot3x} = \dfrac{9+20x}{6x^2}$ cm

## Writing

99. Answers will vary
101.  Answers will vary

## Review

103.  $49 = 7 \cdot 7 = 7^2$

105.  $136 = 2 \cdot 2 \cdot 2 \cdot 17 = 2^3 \cdot 17$

## Section 12.4 Complex Fractions

### Vocabulary

1.  If a fraction has a fraction in its numerator or denominator, it is called **complex fraction**.

### Concepts

3.  To simplify a complex fraction using method 1, we write the numerator and denominator of a complex fraction as **single** fractions and then **divide**.

### Notation

5.  $$\frac{\frac{2}{a}-\frac{1}{b}}{\frac{1}{a}+\frac{2}{b}}$$
$$=\frac{\frac{2b-a}{ab}}{\frac{b+2a}{ab}}$$
$$=\frac{2b-a}{ab}\div\frac{b+2a}{ab}$$
$$=\frac{2b-a}{ab}\cdot\frac{ab}{b+2a}$$
$$=\frac{(2b-a)ab}{ab(b+2a)}$$
$$=\frac{(2b-a)}{(b+2a)}$$

### Practice

7.  $\dfrac{\frac{2}{3}}{\frac{3}{4}}=\dfrac{2}{3}\div\dfrac{3}{4}=\dfrac{2}{3}\cdot\dfrac{4}{3}=\dfrac{8}{9}$

9.  $\dfrac{\frac{4}{5}}{\frac{32}{15}}=\dfrac{4}{5}\div\dfrac{32}{15}=\dfrac{4}{5}\cdot\dfrac{5\cdot3}{4\cdot8}=\dfrac{3}{8}$

11.  $\dfrac{\frac{2}{3}+1}{\frac{1}{3}+1}=\dfrac{3\left(\frac{2}{3}+1\right)}{3\left(\frac{1}{3}+1\right)}=\dfrac{2+3}{1+3}=\dfrac{5}{4}$

13.  $\dfrac{\frac{1}{2}+\frac{3}{4}}{\frac{3}{2}+\frac{1}{4}}=\dfrac{4\left(\frac{1}{2}+\frac{3}{4}\right)}{4\left(\frac{3}{2}+\frac{1}{4}\right)}=\dfrac{2+3}{6+1}=\dfrac{5}{7}$

15.  $\dfrac{\frac{x}{y}}{\frac{1}{x}}=\dfrac{x}{y}\div\dfrac{1}{x}=\dfrac{x}{y}\cdot\dfrac{x}{1}=\dfrac{x^2}{y}$

17.  $\dfrac{\frac{5t^2}{9x^2}}{\frac{3t}{x^2t}}=\dfrac{5t^2}{9x^2}\div\dfrac{3t}{x^2t}=\dfrac{5t^2}{9x^2}\cdot\dfrac{x^2}{3}=\dfrac{5t^2}{27}$

19.  $\dfrac{\frac{1}{x}-3}{\frac{5}{x}+2}=\dfrac{x\left(\frac{1}{x}-3\right)}{x\left(\frac{5}{x}+2\right)}=\dfrac{1-3x}{5+2x}$

21.  $\dfrac{\frac{2}{x}+2}{\frac{4}{x}+2}=\dfrac{x\left(\frac{2}{x}+2\right)}{x\left(\frac{4}{x}+2\right)}=\dfrac{2+2x}{4+2x}=\dfrac{2(1+x)}{2(2+x)}=\dfrac{1+x}{2+x}$

23. $\dfrac{\frac{3y}{x}-y}{y-\frac{y}{x}}$

$= \dfrac{x\left(\frac{3y}{x}-y\right)}{x\left(y-\frac{y}{x}\right)}$

$= \dfrac{3y-xy}{xy-y}$

$= \dfrac{y(3-x)}{y(x-1)}$

$= \dfrac{(3-x)}{(x-1)}$

25. $\dfrac{\frac{1}{x+1}}{1+\frac{1}{x+1}}$

$= \dfrac{(x+1)\left(\frac{1}{x+1}\right)}{(x+1)\left(1+\frac{1}{x+1}\right)}$

$= \dfrac{1}{x+1+1}$

$= \dfrac{1}{x+2}$

27. $\dfrac{\frac{x}{x+2}}{\frac{x}{x+2}+x}$

$= \dfrac{(x+2)\left(\frac{x}{x+2}\right)}{(x+2)\left(\frac{x}{x+2}+x\right)}$

$= \dfrac{x}{x+x^2+2x}$

$= \dfrac{x}{x^2+3x}$

$= \dfrac{x}{x(x+3)}$

$= \dfrac{1}{x+3}$

29. $\dfrac{1}{\frac{1}{x}+\frac{1}{y}}$

$= \dfrac{xy(1)}{xy\left(\frac{1}{x}+\frac{1}{y}\right)}$

$= \dfrac{xy}{y+x}$

31. $\dfrac{\frac{2}{x}}{\frac{2}{y}-\frac{4}{x}}$

$= \dfrac{xy\left(\frac{2}{x}\right)}{xy\left(\frac{2}{y}-\frac{4}{x}\right)}$

$= \dfrac{2y}{2x-4y}$

$= \dfrac{2y}{2(x-2y)}$

$= \dfrac{y}{x-2y}$

33. $\dfrac{3+\frac{3}{x-1}}{3-\frac{3}{x}}$

$= \dfrac{\left(\frac{3(x-1)}{x-1}+\frac{3}{x-1}\right)}{\left(\frac{3x}{x}-\frac{3}{x}\right)}$

$= \dfrac{3x-3+3}{x-1} \div \dfrac{3x-3}{x}$

$= \dfrac{3x}{x-1} \cdot \dfrac{x}{3x-3}$

$= \dfrac{3x^2}{3x^2-3x-3x+3}$

$= \dfrac{3x^2}{3x^2-6x+3}$

$= \dfrac{3x^2}{3(x^2-2x+1)}$

$= \dfrac{x^2}{x^2-2x+1}$

$= \dfrac{x^2}{(x-1)(x-1)}$

35. $\dfrac{\frac{3}{x}+\frac{4}{x+1}}{\frac{2}{x+1}-\frac{3}{x}}$

$$= \dfrac{\frac{3(x+1)}{x(x+1)}+\frac{4x}{x(x+1)}}{\frac{2x}{x(x+1)}-\frac{3(x+1)}{x(x+1)}}$$

$$= \dfrac{3x+3+4x}{x(x+1)} \div \dfrac{2x-3x-3}{x(x+1)}$$

$$= \dfrac{7x+3}{x(x+1)} \cdot \dfrac{x(x+1)}{-x-3}$$

$$= \dfrac{7x+3}{-x-3}$$

37. $\dfrac{\frac{2}{x}-\frac{3}{x+1}}{\frac{2}{x+1}-\frac{3}{x}}$

$$= \dfrac{\frac{2(x+1)}{x(x+1)}-\frac{3x}{x(x+1)}}{\frac{2x}{x(x+1)}-\frac{3(x+1)}{x(x+1)}}$$

$$= \dfrac{2x+2-3x}{x(x+1)} \div \dfrac{2x-3x-3}{x(x+1)}$$

$$= \dfrac{2-x}{x(x+1)} \cdot \dfrac{x(x+1)}{-x-3}$$

$$= \dfrac{2-x}{-x-3}$$

$$= \dfrac{-1(x-2)}{-1(x+3)}$$

$$= \dfrac{(x-2)}{(x+3)}$$

39. $\dfrac{\frac{1}{y^2+y}-\frac{1}{xy+x}}{\frac{1}{xy+x}-\frac{1}{y^2+y}}$

$$= \dfrac{\frac{1}{y^2+y}-\frac{1}{xy+x}}{-1\left(-\frac{1}{xy+x}+\frac{1}{y^2+y}\right)}$$

$$= -\dfrac{\frac{1}{y^2+y}-\frac{1}{xy+x}}{\frac{1}{y^2+y}-\frac{1}{xy+x}}$$

$$= -1$$

41. $\dfrac{x^{-2}}{y^{-1}}$

$$= \dfrac{\frac{1}{x^2}}{\frac{1}{y}}$$

$$= \dfrac{1}{x^2} \div \dfrac{1}{y}$$

$$= \dfrac{1}{x^2} \cdot \dfrac{y}{1}$$

$$= \dfrac{y}{x^2}$$

43. $\dfrac{1+x^{-1}}{x^{-1}-1}$

$$= \dfrac{1+\frac{1}{x}}{\frac{1}{x}-1}$$

$$= \dfrac{x\left(1+\frac{1}{x}\right)}{x\left(\frac{1}{x}-1\right)}$$

$$= \dfrac{(x+1)}{(1-x)}$$

45. $\dfrac{a^{-2}+a}{a}$

$$= \dfrac{\frac{1}{a^2}+a}{a}$$

$$= \dfrac{a^2\left(\frac{1}{a^2}+a\right)}{a^2 \cdot a}$$

$$= \dfrac{1+a^3}{a^3}$$

47. $\dfrac{2x^{-1}+4x^{-2}}{2x^{-2}+x^{-1}}$

$\quad = \dfrac{\frac{2}{x}+\frac{4}{x^2}}{\frac{2}{x^2}+\frac{1}{x}}$

$\quad = \dfrac{x^2\left(\frac{2}{x}+\frac{4}{x^2}\right)}{x^2\left(\frac{2}{x^2}+\frac{1}{x}\right)}$

$\quad = \dfrac{2x+4}{2+x}$

$\quad = \dfrac{2(x+2)}{2+x}$

$\quad = 2$

49. $\dfrac{1-25y^{-2}}{1+10y^{-1}+25y^{-2}}$

$\quad = \dfrac{1-\frac{25}{y^2}}{1+\frac{10}{y}+\frac{25}{y^2}}$

$\quad = \dfrac{y^2\left(1-\frac{25}{y^2}\right)}{y^2\left(1+\frac{10}{y}+\frac{25}{y^2}\right)}$

$\quad = \dfrac{y^2-25}{y^2+10y+25}$

$\quad = \dfrac{(y-5)(y+5)}{(y+5)(y+5)}$

$\quad = \dfrac{(y-5)}{(y+5)}$

## Applications

51. GARDENING TOOLS

$\quad \dfrac{\text{Opening of the cutting blades}}{\text{Opening of the handles}}$

$\quad = \dfrac{\frac{x}{2}}{\frac{7x}{3}}$

$\quad = \dfrac{x}{2} \div \dfrac{7x}{3}$

$\quad = \dfrac{x}{2} \cdot \dfrac{3}{7x}$

$\quad = \dfrac{3}{14}$

53. ELECTRONICS

$\quad \text{Total Resistance} = \dfrac{1}{\frac{1}{R_1}+\frac{1}{R_2}}$

$\quad \text{Total Resistance} = \dfrac{R_1 \cdot R_2 \cdot 1}{R_1 \cdot R_2\left(\frac{1}{R_1}+\frac{1}{R_2}\right)}$

$\quad \text{Total Resistance} = \dfrac{R_1 R_2}{R_2+R_1}$

## Writing

55. Answers will vary

## Review

57. $t^3 t^4 t^2 = t^{3+4+2} = t^9$

59. $-2r\left(r^3\right)^2 = -2r \cdot r^6 = -2r^7$

61. $\left(\dfrac{3r}{4r^3}\right)^4 = \dfrac{81r^4}{256r^{12}} = \dfrac{81}{256r^8}$

63. $\left(\dfrac{6r^{-2}}{2r^3}\right)^{-2}$

$= \left(\dfrac{3r^{-2}}{r^3}\right)^{-2}$

$= \left(\dfrac{\frac{3}{r^2}}{r^3}\right)^{-2}$

$= \left(\dfrac{3}{r^2} \div r^3\right)^{-2}$

$= \left(\dfrac{3}{r^2} \cdot \dfrac{1}{r^3}\right)^{-2}$

$= \left(\dfrac{3}{r^5}\right)^{-2}$

$= \dfrac{1}{\left(\frac{3}{r^5}\right)^2}$

$= 1 \div \dfrac{9}{r^{10}}$

$= 1 \cdot \dfrac{r^{10}}{9}$

$= \dfrac{r^{10}}{9}$

**Vocabulary**

1. Equations that contain one or more rational expressions, such as $\frac{x+2}{x+3} + \frac{1}{x^2+2x-3} = 1$ area called **rational equations**.

3. If you multiply both sides of an equation by an expression that involves a variable, you must **check** the solution.

5. In the formula $I = Pr$, $I$ stands for the amount of **interest** earned in one year, $P$ stands for the **principal**, and $r$ stands for the annual interest **rate**.

**Concepts**

7. a. $x = 5$ is a solution since
$$\frac{1}{5-1} \overset{?}{=} 1 - \frac{3}{5-1}$$
$$\frac{1}{4} = 1 - \frac{3}{4}$$
$$\frac{1}{4} = \frac{1}{4}$$
b. $x = 5$ is not a solution since
$$\frac{5}{5-5} = \frac{5}{0}$$ is an undefined fraction.

9. a.

| | Time | Amount Assembled |
|---|---|---|
| Marvin | 6 hours | $\frac{1}{6}$ |
| Kyla | 5 hours | $\frac{1}{5}$ |

b. If they worked together they would assemble $\frac{1}{6} + \frac{1}{5} = \frac{5}{30} + \frac{6}{30} = \frac{11}{30}$ in 1 hour.

11. If the complete theater is emptied in 6 minutes then $\frac{1}{6}$ of the theatre is emptied in 1 minute.

13. $I = Pr$
a. $\frac{I}{P} = \frac{Pr}{P}$
$\frac{I}{P} = r$
b. $\frac{I}{r} = \frac{Pr}{r}$
$\frac{I}{r} = P$

**Notation**

15. $\frac{2}{a} + \frac{1}{2} = \frac{7}{2a}$
$$2a\left(\frac{2}{a} + \frac{1}{2}\right) = 2a\left(\frac{7}{2a}\right)$$
$$\frac{4a}{a} + \frac{2a}{2} = \frac{14a}{2a}$$
$$4 + a = 7$$
$$4 + a - 4 = 7 - 4$$
$$a = 3$$

17. The original equation was $\frac{3}{5} + \frac{7}{x+2} = 2$.

**Practice**

19. $\frac{x}{2} + 4 = \frac{3x}{2}$

$2\left(\frac{x}{2} + 4\right) = 2\left(\frac{3x}{2}\right)$

$x + 8 = 3x$

$8 = 2x$

$4 = x$

Check

$\frac{4}{2} + 4 \stackrel{?}{=} \frac{3(4)}{2}$

$6 = 6$

21. $\frac{x+1}{3} + \frac{x-1}{5} = \frac{2}{15}$

$15\left(\frac{x+1}{3} + \frac{x-1}{5}\right) = 15\left(\frac{2}{15}\right)$

$5x + 5 + 3x - 3 = 2$

$2x = 0$

$x = 0$

Check

$\frac{0+1}{3} + \frac{0-1}{5} \stackrel{?}{=} \frac{2}{15}$

$\frac{1}{3} - \frac{1}{5} \stackrel{?}{=} \frac{2}{15}$

$\frac{5}{15} - \frac{3}{15} \stackrel{?}{=} \frac{2}{15}$

$\frac{2}{15} = \frac{2}{15}$

23. $\frac{3}{x} + 2 = 3$

$x\left(\frac{3}{x} + 2\right) = 3x$

$3 + 2x = 3x$

$3 = x$

Check

$\frac{3}{3} + 2 \stackrel{?}{=} 3$

$3 = 3$

25. $\frac{5}{a} - \frac{4}{a} = 8 + \frac{1}{a}$

$\frac{1}{a} = 8 + \frac{1}{a}$

$a\left(\frac{1}{a}\right) = a\left(8 + \frac{1}{a}\right)$

$1 = 8a + 1$

$0 = 8a$

$0 = a$

Check

$\frac{5}{0} - \frac{4}{0} \stackrel{?}{=} 8 + \frac{1}{0}$

This is an extraneous solution.

27. $\frac{3}{4h} + \frac{2}{h} = 1$

$4h\left(\frac{3}{4h} + \frac{2}{h}\right) = 4h \cdot 1$

$3 + 8 = 4h$

$11 = 4h$

$\frac{11}{4} = h$

Check

$\frac{3}{4\left(\frac{11}{4}\right)} + \frac{2}{\frac{11}{4}} \stackrel{?}{=} 1$

$\frac{3}{11} + \frac{8}{11} \stackrel{?}{=} 1$

$\frac{11}{11} = 1$

$1 = 1$

29. $\frac{a}{4} - \frac{4}{a} = 0$

$4a\left(\frac{a}{4} - \frac{4}{a}\right) = 4a \cdot 0$

$a^2 - 16 = 0$

$a^2 = 16$

$a = \pm 4$

Check

$a = 4$ $\qquad\qquad$ $a = -4$

$\frac{4}{4} - \frac{4}{4} \stackrel{?}{=} 0$ $\qquad$ $\frac{-4}{4} - \frac{4}{-4} \stackrel{?}{=} 0$

$1 - 1 \stackrel{?}{=} 0$ $\qquad\qquad$ $-1 + 1 \stackrel{?}{=} 0$

$0 = 0$ $\qquad\qquad$ $0 = 0$

31. $\frac{2}{y+1} + 5 = \frac{12}{y+1}$

$5 = \frac{10}{y+1}$

$(y+1)(5) = (y+1)\left(\frac{10}{y+1}\right)$

$5y + 5 = 10$

$5y = 5$

$y = 1$

Check

$\frac{2}{1+1} + 5 \stackrel{?}{=} \frac{12}{1+1}$

$1 + 5 = 6$

33. $\frac{x}{x-5} - \frac{5}{x-5} = 3$

$\frac{x-5}{x-5} = 3$

$1 \neq 3$

No solution

35. $\frac{3r}{2} - \frac{3}{r} = \frac{3r}{2} + 3$

$2r\left(\frac{3r}{2} - \frac{3}{r}\right) = 2r\left(\frac{3r}{2} + 3\right)$

$3r^2 - 6 = 3r^2 + 6r$

$-6 = 6r$

$-1 = r$

Check

$\frac{3(-1)}{2} - \frac{3}{(-1)} \stackrel{?}{=} \frac{3(-1)}{2} + 3$

$-\frac{3}{2} + 3 = -\frac{3}{2} + 3$

37. $\frac{1}{3} + \frac{2}{x-3} = 1$

$\frac{2}{x-3} = \frac{2}{3}$

$3(x-3)\left(\frac{2}{x-3}\right) = \left(\frac{2}{3}\right)3(x-3)$

$6 = 2x - 6$

$12 = 2x$

$6 = x$

Check

$\frac{1}{3} + \frac{2}{6-3} \stackrel{?}{=} 1$

$\frac{1}{3} + \frac{2}{3} \stackrel{?}{=} 1$

$1 = 1$

39. $\frac{z-4}{z-3} = \frac{z+2}{z+1}$

$(z+1)(z-3)\left(\frac{z-4}{z-3}\right) = (z+1)(z-3)\left(\frac{z+2}{z+1}\right)$

$(z+1)(z-4) = (z-3)(z+2)$

$z^2 - 3z - 4 = z^2 - z - 6$

$0 = 2z - 2$

$2z = 2$

$z = 1$

Check

$\frac{1-4}{1-3} \stackrel{?}{=} \frac{1+2}{1+1}$

$\frac{-3}{-2} = \frac{3}{2}$

$\frac{3}{2} = \frac{3}{2}$

41. $\frac{v}{v+2} + \frac{1}{v-1} = 1$

$(v+2)(v-1)\left(\frac{v}{v+2} + \frac{1}{v-1}\right) = 1(v+2)(v-1)$

$v^2 - v + v + 2 = v^2 + v - 2$

$v^2 + 2 = v^2 + v - 2$

$2 = v - 2$

$4 = v$

Check

$\frac{4}{4+2} + \frac{1}{4-1} \stackrel{?}{=} 1$

$\frac{4}{6} + \frac{1}{3} \stackrel{?}{=} 1$

$\frac{2}{3} + \frac{1}{3} \stackrel{?}{=} 1$

$1 = 1$

43. $\frac{a^2}{a+2} - \frac{4}{a+2} = a$

$\frac{a^2-4}{a+2} = a$

$(a+2)\left(\frac{a^2-4}{a+2}\right) = a(a+2)$

$a^2 - 4 = a^2 + 2a$

$-4 = 2a$

$-2 = a$

Check

$\frac{(-2)^2}{-2+2} - \frac{4}{-2+2} \stackrel{?}{=} -2$

$\frac{4}{0} - \frac{4}{0} \stackrel{?}{=} -2$

No solution, extraneous solution

45. $\frac{7}{q^2-q-2} + \frac{1}{q+1} = \frac{3}{q-2}$

$\frac{7}{(q-2)(q+1)} + \frac{1}{q+1} = \frac{3}{q-2}$

$(q-2)(q+1)\left(\frac{7}{(q-2)(q+1)} + \frac{1}{q+1}\right) = (q-2)(q+1)\left(\frac{3}{q-2}\right)$

$7 + q - 2 = 3q + 3$

$q + 5 = 3q + 3$

$2 = 2q$

$1 = q$

Check

$\frac{7}{1^2-1-2} + \frac{1}{1+1} \stackrel{?}{=} \frac{3}{1-2}$

$\frac{7}{-2} + \frac{1}{2} \stackrel{?}{=} \frac{3}{-1}$

$-\frac{6}{2} \stackrel{?}{=} -3$

$-3 = -3$

47. $\frac{u}{u-1} + \frac{1}{u} = \frac{u^2+1}{u^2-u}$

$\frac{u}{u-1} + \frac{1}{u} = \frac{u^2+1}{u(u-1)}$

$u(u-1)\left(\frac{u}{u-1} + \frac{1}{u}\right) = u(u-1)\left(\frac{u^2+1}{u(u-1)}\right)$

$u^2 + u - 1 = u^2 + 1$

$u = 2$

Check

$\frac{2}{2-1} + \frac{1}{2} \stackrel{?}{=} \frac{2^2+1}{2^2-2}$

$2 + \frac{1}{2} \stackrel{?}{=} \frac{5}{2}$

$\frac{4}{2} + \frac{1}{2} = \frac{5}{2}$

49. $\frac{n}{n^2-9} + \frac{n+8}{n+3} = \frac{n-8}{n-3}$

$\frac{n}{(n-3)(n+3)} + \frac{n+8}{n+3} = \frac{n-8}{n-3}$

$(n-3)(n+3)\left(\frac{n}{(n-3)(n+3)} + \frac{n+8}{n+3}\right) = (n-3)(n+3)\left(\frac{n-8}{n-3}\right)$

$n + n^2 + 5n - 24 = n^2 - 5n - 24$

$6n - 24 = -5n - 24$

$11n = 0$

$n = 0$

Check

$\frac{0}{0^2-9} + \frac{0+8}{0+3} \stackrel{?}{=} \frac{0-8}{0-3}$

$0 + \frac{8}{3} \stackrel{?}{=} \frac{-8}{-3}$

$\frac{8}{3} = \frac{8}{3}$

51. $\frac{5}{x+4} + \frac{1}{x+4} = x - 1$

$\frac{6}{x+4} = x - 1$

$(x+4)\left(\frac{6}{x+4}\right) = (x+4)(x-1)$

$6 = x^2 + 3x - 4$

$x^2 + 3x - 10 = 0$

$(x+5)(x-2) = 0$

$x + 5 = 0 \quad$ or $\quad x - 2 = 0$

$x = -5 \qquad\qquad x = 2$

Check

$x = -5;$

$\frac{6}{-5+4} \stackrel{?}{=} -5 - 1$

$-6 = -6$

$x = 2;$

$\frac{6}{2+4} \stackrel{?}{=} 2 - 1$

$1 = 1$

53. $\frac{3}{x+1} - \frac{x-2}{2} = \frac{x-2}{x+1}$

$2(x+1)\left(\frac{3}{x+1} - \frac{x-2}{2}\right) = 2(x+1)\left(\frac{x-2}{x+1}\right)$

$2(3) - (x+1)(x-2) = 2(x-2)$

$6 - (x^2 - x - 2) = 2x - 4$

$6 - x^2 + x + 2 = 2x - 4$

$x^2 + x - 12 = 0$

$(x+4)(x-3) = 0$

$x + 4 = 0 \quad$ or $\quad x - 3 = 0$

$x = -4 \qquad\qquad x = 3$

Check

$x = -4;$

$(-4)^2 + (-4) - 12 \stackrel{?}{=} 0$

$16 - 4 - 12 \stackrel{?}{=} 0$

$0 = 0$

$x = 3;$

$(3)^2 + 3 - 12 \stackrel{?}{=} 0$

$9 + 3 - 12 \stackrel{?}{=} 0$

$0 = 0$

55. $\frac{b+2}{b+3} + 1 = \frac{-7}{b-5}$

$(b+3)(b-5)\left(\frac{b+2}{b+3} + 1\right) = (b+3)(b-5)\left(\frac{-7}{b-5}\right)$

$b^2 - 3b - 10 + b^2 - 2b - 15 = -7b - 21$

$2b^2 - 5b - 25 = -7b - 21$

$2b^2 + 2b - 4 = 0$

$2(b^2 + b - 2) = 0$

$b^2 + b - 2 = 0$

$(b+2)(b-1) = 0$

$b + 2 = 0 \qquad \text{or} \qquad b - 1 = 0$

$b = -2 \qquad\qquad\qquad b = 1$

Check both solutions:

$b = -2$

$\frac{-2+2}{-2+3} + 1 \stackrel{?}{=} \frac{-7}{-2-5}$

$\frac{0}{1} + 1 \stackrel{?}{=} \frac{-7}{-7}$

$1 = 1$

Check

$b = 1$

$\frac{1+2}{1+3} + 1 \stackrel{?}{=} \frac{-7}{1-5}$

$\frac{3}{4} + 1 \stackrel{?}{=} \frac{-7}{-4}$

$\frac{7}{4} = \frac{7}{4}$

57. $\frac{x}{x-1} - \frac{12}{x^2-x} = \frac{-1}{x-1}$

$x(x-1)\left(\frac{x}{x-1} - \frac{12}{x(x-1)}\right) = x(x-1)\left(\frac{-1}{x-1}\right)$

$x^2 - 12 = -x$

$x^2 + x - 12 = 0$

$(x+4)(x-3) = 0$

$x + 4 = 0 \qquad \text{or} \qquad x - 3 = 0$

$x = -4 \qquad\qquad\qquad x = 3$

Check both solutions:

$x = -4$

$\frac{-4}{-4-1} - \frac{12}{(-4)^2-(-4)} \stackrel{?}{=} \frac{-1}{(-4)-1}$

$\frac{4}{5} - \frac{12}{20} \stackrel{?}{=} \frac{1}{5}$

$\frac{4}{5} - \frac{3}{5} \stackrel{?}{=} \frac{1}{5}$

$\frac{1}{5} = \frac{1}{5}$

Check

$x = 3$

$\frac{3}{3-1} - \frac{12}{3^2-3} \stackrel{?}{=} \frac{-1}{3-1}$

$\frac{3}{2} - \frac{12}{6} \stackrel{?}{=} -\frac{1}{2}$

$\frac{3}{2} - \frac{4}{2} \stackrel{?}{=} -\frac{1}{2}$

$-\frac{1}{2} = -\frac{1}{2}$

59. $1 - \frac{3}{b} = \frac{-8b}{b^2+3b}$

$1 - \frac{3}{b} = \frac{-8b}{b(b+3)}$

$b(b+3)\left(1 - \frac{3}{b}\right) = b(b+3)\left(\frac{-8b}{b(b+3)}\right)$

$b^2 + 3b - 3b - 9 = -8b$

$b^2 + 8b - 9 = 0$

$(b+9)(b-1) = 0$

$b + 9 = 0 \qquad \text{or} \qquad b - 1 = 0$

$b = -9 \qquad\qquad\qquad b = 1$

Check both solutions:

$b = -9$

$1 - \frac{3}{-9} \overset{?}{=} \frac{-8(-9)}{(-9)^2+3(-9)}$

$1 + \frac{1}{3} \overset{?}{=} \frac{72}{54}$

$\frac{4}{3} = \frac{4}{3}$

Check

$b = 1$

$1 - \frac{3}{1} \overset{?}{=} \frac{-8(1)}{1^2+3(1)}$

$-2 \overset{?}{=} \frac{-8}{4}$

$-2 = -2$

61. $\frac{1}{a} + \frac{1}{b} = 1$

$ab\left(\frac{1}{a} + \frac{1}{b}\right) = ab(1)$

$b + a = ab$

$a - ab = -b$

$a(1 - b) = -b$

$a = -\frac{b}{1-b}$

Factor out $-1$ from the denominator

$a = \frac{b}{b-1}$

63. $I = \frac{E}{R+r}$

$(R+r)I = \frac{E}{R+r}(R+r)$

$IR + rI = E$

$rI = E - IR$

$r = \frac{E-IR}{I}$

65. $\frac{a}{b} = \frac{c}{d}$

$d\left(\frac{a}{b}\right) = d\left(\frac{c}{d}\right)$

$d\left(\frac{a}{b}\right) = c$

$d = \frac{c}{\left(\frac{a}{b}\right)}$

$d = c\left(\frac{b}{a}\right)$

$d = \frac{bc}{a}$

67. The number is 2 since $\frac{3(2)}{4+2} = \frac{6}{6} = 1$.

69. The number is 5 since $\frac{3}{4} = \frac{3+5}{4+2(5)} = \frac{8}{14} = \frac{4}{7}$.

71. The numbers are $\frac{2}{3}$ and $\frac{3}{2}$ since $\frac{2}{3} + \frac{3}{2} = \frac{4}{6} + \frac{9}{6} = \frac{13}{6}$.

**Applications**

73. OPTICS

$$\frac{1}{f} = \frac{1}{d_1} + \frac{1}{d_2}$$
$$fd_1d_2\left(\frac{1}{f}\right) = fd_1d_2\left(\frac{1}{d_1} + \frac{1}{d_2}\right)$$
$$d_1d_2 = fd_2 + fd_1$$
$$d_1d_2 = f(d_2 + d_1)$$
$$\frac{d_1d_2}{(d_2+d_1)} = f$$

75. MEDICINE

$$H = \frac{RB}{R+B}$$
$$H(R + B) = RB$$
$$HR + HB = RB$$
$$HR - RB = -HB$$
$$R(H - B) = -HB$$
$$R = \frac{-HB}{(H-B)}$$

Factoring out $-1$ in the denominator yields

$$R = \frac{HB}{B-H}$$

77. FILLING A POOL

Pipe One(1hour) + Pipe Two(1hour) = Full Pool

$$\frac{1}{5} + \frac{1}{4} = \frac{1}{F}$$
$$20F\left(\frac{1}{5} + \frac{1}{4}\right) = 20F\left(\frac{1}{F}\right)$$
$$4F + 5F = 20$$
$$9F = 20$$
$$F = \frac{20}{9}$$
$$F = 2\frac{2}{9}$$

Using both pipes it will take $2\frac{2}{9}$ hours to fill this pool.

79. ROOFING A HOUSE

Homeowner(1 day) + Professional(1 day) = Roof

$$\frac{1}{7} + \frac{1}{4} = \frac{1}{r}$$
$$28r\left(\frac{1}{7} + \frac{1}{4}\right) = 28r\left(\frac{1}{r}\right)$$
$$4r + 7r = 28$$
$$11r = 28$$
$$r = \frac{28}{11}$$
$$r = 2\frac{6}{11}$$

If these two worked together it would take $2\frac{6}{11}$ days.

81. TOURING

|  | $r$ | $t$ | $d$ |
|---|---|---|---|
| Riding | $r + 10$ | $t$ | 28 |
| Walking | $r$ | $t$ | 8 |

Time to ride 28 miles $=$ Time to walk 8 miles

$\frac{28}{r+10} = \frac{8}{r}$

$r(r + 10)\left(\frac{28}{r+10}\right) = r(r + 10)\left(\frac{8}{r}\right)$

$28r = 8r + 80$

$20r = 80$

$r = 4 \, \text{mph}$

This man walks at a rate of 4 mph. If he wants to walk a 30 mile trail he should allow 7.5 hours.

$d = rt$

$\frac{d}{r} = t$

$\frac{30}{4} = t$

$7.5 = t$

83. BOATING

$r$ represents the rate of the river

|  | $r$ | $t$ | $d$ |
|---|---|---|---|
| Boat (downstream) | $18 + r$ | $t$ | 22 |
| Boat (upstream) | $18 - r$ | $t$ | 14 |

$\frac{22}{18+r} = \frac{14}{18-r}$

$(18 + r)(18 - r)\left(\frac{22}{18+r}\right) = (18 + r)(18 - r)\left(\frac{14}{18-r}\right)$

$22(18 - r) = 14(18 + r)$

$396 - 22r = 252 + 14r$

$144 = 36r$

$4 = r$

The river flows at 4 mph.

85. COMPARING INVESTMENTS

|  | Principal | Rate | Interest |
|---|---|---|---|
| CD One | $P$ | $r - 1$ | 175 |
| CD Two | $P$ | $r$ | 200 |

$\frac{175}{r-1} = \frac{200}{r}$

$r(r - 1)\left(\frac{175}{r-1}\right) = r(r - 1)\left(\frac{200}{r}\right)$

$175r = 200(r - 1)$

$175r = 200r - 200$

$200 = 25r$

$8 = r$

$r - 1 = 7$

The two interest rates are 7% and 8%.

87. SHARING COSTS

$n$ people contributed $d$ dollars for a total of \$35.

$nd = 35, d = \frac{35}{n}$

If $n + 2$ had contributed $d - 2$ for a total of \$35.

$(n + 2)(d - 2) = 35$

$nd - 2n + 2d - 4 = 35$

$nd + 2d = 35 + 2n + 4$

$d(n + 2) = 35 + 2n + 4$

$d = \frac{35 + 2n + 4}{(n + 2)}$

Since $d = d$

$\frac{35}{n} = \frac{35 + 2n + 4}{(n + 2)}$

$n(n + 2)\left(\frac{35}{n}\right) = n(n + 2)\left(\frac{39 + 2n}{(n + 2)}\right)$

$35n + 70 = 39n + 2n^2$

$0 = 2n^2 + 4n - 70$

$0 = n^2 + 2n - 35$

$0 = (n + 7)(n - 5)$

$n + 7 = 0$ or $\quad n - 5 = 0$

$n = -7 \qquad\qquad n = 5$

5 people contributed to the \$35 gift.


89. SALES

| | no. calculators | price |
|---|---|---|
| regular price | $c$ | $\frac{120}{c}$ |
| discount price | $c + 10$ | $\frac{120}{c} - 1$ |

At the discount price the bookstore is spending \$120, but receiving 10 more calculators since the per item cost is decreased by \$1.

$120 = (c + 10)\left(\frac{120}{c} - 1\right)$

$120 = 120 - c + \frac{1200}{c} - 10$

$0 = -c + \frac{1200}{c} - 10$

$0 \cdot c = c\left(-c + \frac{1200}{c} - 10\right)$

$0 = -c^2 - 10c + 1200$

$0 = c^2 + 10c - 1200$

$0 = (c - 30)(c + 40)$

$c - 30 = 0$ or $\quad c + 40 = 0$

$c = 30 \qquad\qquad c = -40$

30 calculators can be purchased at the regular price.

91. RIVER TOUR

Using $d = rt, t = \frac{d}{r}$

|  | rate | time | distance |
|---|---|---|---|
| upstream | $r - 5$ | $t$ | 60 |
| downstream | $r + 5$ | $t$ | 60 |

The time upstream plus the time downstream will equal 5.

$\frac{60}{r-5} + \frac{60}{r+5} = 5$

$(r - 5)(r + 5)\left(\frac{60}{r-5} + \frac{60}{r+5}\right) = 5(r - 5)(r + 5)$

$60(r + 5) + 60(r - 5) = 5(r^2 - 25)$

$60r + 300 + 60r - 300 = 5r^2 - 125$

$0 = 5r^2 - 120r - 125$

$0 = 5(r^2 - 24r - 25)$

$0 = (r - 25)(r + 1)$

$r - 25 = 0 \quad$ or $\quad r + 1 = 0$

$r = 25 \qquad\qquad r = -1$

This boat has a still water speed of 25 mph.

## Writing

93. Answers will vary
95. Answers will vary

## Review

97. $x^2 + 4x = x(x + 4)$

99. $2x^2 + x - 3 = (2x + 3)(x - 1)$

101. $x^4 - 16 = (x^2 - 4)(x^2 + 4) = (x - 2)(x + 2)(x^2 + 4)$

## Section 12.6 Proportions and Similar Triangles

**Vocabulary**

1.  A **ratio** is the quotient of two numbers or the quotient of two quantities with the same units.  A **rate** is a quotient of two quantities that have different units.

3.  In the proportion $\frac{a}{b} = \frac{c}{d}$, $a$ and $d$ are called the **extremes** of the proportion. The second and third terms of a proportion are called the **means** of the proportion.

5.  If two triangles have the same **shape**, they are said to be similar.

**Concepts**

7.  The equation $\frac{a}{b} = \frac{c}{d}$ is a proportion if the product $ad$ is equal to the product $bc$.

9.  $\frac{5}{3} \overset{?}{=} \frac{75}{45}$

$\frac{5}{3} \overset{?}{=} \frac{5 \cdot 15}{3 \cdot 15}$

$\frac{5}{3} = \frac{5}{3}$

Therefore, $x = 45$ is a solution.

11. MINIATURES

The ratio on the right hand side should be $\frac{h}{48}$.

**Notation**

13. $\frac{12}{18} = \frac{x}{24}$

$12 \cdot 24 = 18 \cdot x$

$288 = 18x$

$\frac{288}{18} = \frac{18x}{18}$

$16 = x$

15. Read $\triangle ABC$ as **triangle** ABC.

**Practice**

17. $\frac{9}{7} \overset{?}{=} \frac{81}{70}$

$9(70) \overset{?}{=} 7(81)$

$630 \neq 567$

Not a proportion

19. $\frac{7}{3} \overset{?}{=} \frac{14}{6}$

$7(6) \overset{?}{=} 3(14)$

$42 = 42$

A proportion

21. $\frac{9}{19} \overset{?}{=} \frac{38}{80}$

    $9(80) \overset{?}{=} 19(38)$

    $720 \neq 722$

    Not a proportion

23. $\frac{10.4}{3.6} \overset{?}{=} \frac{41.6}{14.4}$

    $10.4(14.4) \overset{?}{=} 3.6(41.6)$

    $149.76 = 149.76$

    A proportion

25. $\frac{2}{3} = \frac{x}{6}$

    $2 \cdot 6 = 3x$

    $\frac{2 \cdot 6}{3} = \frac{3x}{3}$

    $\frac{12}{3} = x$

    $4 = x$

27. $\frac{5}{10} = \frac{3}{c}$

    $5c = 3 \cdot 10$

    $\frac{5c}{5} = \frac{3 \cdot 10}{5}$

    $c = \frac{30}{5}$

    $c = 6$

29. $\frac{6}{x} = \frac{8}{4}$

    $6 \cdot 4 = 8x$

    $\frac{6 \cdot 4}{8} = \frac{8x}{8}$

    $\frac{24}{8} = x$

    $3 = x$

31. $\frac{x}{3} = \frac{9}{3}$

    $3x = 3 \cdot 9$

    $\frac{3x}{3} = \frac{3 \cdot 9}{3}$

    $x = \frac{27}{3}$

    $x = 9$

33. $\frac{x+1}{5} = \frac{3}{15}$

    $15(x + 1) = 3 \cdot 5$

    $15x + 15 = 15$

    $15x + 15 - 15 = 15 - 15$

    $15x = 0$

    $\frac{15x}{15} = \frac{0}{15}$

    $x = 0$

35. $\frac{x+3}{12} = \frac{-7}{6}$

$6(x+3) = -7 \cdot 12$

$6x + 18 = -84$

$6x + 18 - 18 = -84 - 18$

$6x = -102$

$\frac{6x}{6} = -\frac{102}{6}$

$x = -17$

37. $\frac{4-x}{13} = \frac{11}{26}$

$26(4-x) = 13 \cdot 11$

$104 - 26x = 143$

$104 - 104 - 26x = 143 - 104$

$-26x = 39$

$\frac{-26x}{-26} = \frac{39}{-26}$

$x = -\frac{3}{2}$

39. $\frac{2x+1}{18} = \frac{14}{3}$

$3(2x+1) = 18 \cdot 14$

$6x + 3 = 252$

$6x + 3 - 3 = 252 - 3$

$6x = 249$

$\frac{6x}{6} = \frac{249}{6}$

$x = \frac{83}{2}$ or $41.5$

41. $\frac{y}{4} = \frac{4}{y}$

$y(y) = 4(4)$

$y^2 = 16$

$y = 4$  or  $y = -4$

43. $\frac{2}{c} = \frac{c-3}{2}$

$2(2) = c(c-3)$

$4 = c^2 - 3c$

$c^2 - 3c - 4 = 0$

$(c-4)(c+1) = 0$

$c - 4 = 0$  or  $c + 1 = 0$

$c = 4$ $\qquad\qquad$ $c = -1$

45. $\frac{2}{x+6} = \frac{-2x}{5}$

$2(5) = (x+6)(-2x)$

$10 = -2x^2 - 12x$

$2x^2 + 12x + 10 = 0$

$x^2 + 6x + 5 = 0$

$(x+5)(x+1) = 0$

$x + 5 = 0$  or  $x + 1 = 0$

$x = -5$ $\qquad\qquad$ $x = -1$

**Applications**

47. GROCERY SHOPPING

$$\frac{3 \text{ pints}}{\$1} = \frac{51 \text{ pints}}{\$y}$$
$$3y = 1 \cdot 51$$
$$3y = 51$$
$$\frac{3y}{3} = \frac{51}{3}$$
$$y = 17$$

51 pints of yogurt will cost $17.

49. ADVERTISING

$$\frac{30 \text{ seconds}}{\$1.2 \text{ million}} = \frac{45 \text{ seconds}}{x}$$
$$30x = 1.2 \cdot 45$$
$$30x = 54$$
$$\frac{30x}{30} = \frac{54}{30}$$
$$x = 1.8$$

At this rate a 45 second advertisement costs $1.8 million.

51. MIXING PERFUME

$$\frac{3 \text{ drops essence}}{7 \text{ drops alcohol}} = \frac{d \text{ drops essence}}{56 \text{ drops alcohol}}$$
$$3 \cdot 56 = 7d$$
$$168 = 7d$$
$$\frac{168}{7} = \frac{7d}{7}$$
$$24 = d$$

24 drops of essence are required for 56 drops of alcohol.

53. COOKING

First recognize that 15 servings requires that the recipe be increased by two and a half or $\frac{5}{2}$.
Check: $6 \cdot \frac{5}{2} = 15$ servings.

| broth | $3\left(\frac{5}{2}\right) = \frac{15}{2} = 7\frac{1}{2}$ cups |
| --- | --- |
| rice | $\frac{2}{3}\left(\frac{5}{2}\right) = \frac{5}{3} = 1\frac{2}{3}$ cups |
| onions | $\frac{1}{4}\left(\frac{5}{2}\right) = \frac{5}{8}$ cups |
| carrots | $\left(\frac{1}{2}\right)\left(\frac{5}{2}\right) = \frac{5}{4}$ cups |
| cream | $1\left(\frac{5}{2}\right) = \frac{5}{2} = 2\frac{1}{2}$ cups |
| flour | $2\left(\frac{5}{2}\right) = 5$ tablespoons |
| pepper | $\left(\frac{1}{8}\right)\left(\frac{5}{2}\right) = \frac{5}{16}$ teaspoon |

55. QUALITY CONTROL

$$\frac{500 \text{ shirts}}{17 \text{ rejects}} = \frac{15{,}000 \text{ shirts}}{x \text{ rejects}}$$
$$500x = 17 \cdot 15{,}000$$
$$500x = 255{,}000$$
$$\frac{500x}{500} = \frac{255{,}000}{500}$$
$$x = 510$$

You would expect to find 510 crooked collars in a run of 15,000 shirts.

57. HIP-HOP

$$\frac{674 \text{ syllables}}{54.9 \text{ seconds}} = \frac{s \text{ syllables}}{60 \text{ seconds}}$$

$$674 \cdot 60 = 54.9s$$

$$40440 = 54.9s$$

$$\frac{40440}{54.9} = \frac{54.9s}{54.9}$$

$$736.61202 \approx s$$

At the world record rate 737 syllables could be rapped in 1 minute.

59. COMPUTING A PAYCHECK

$$\frac{\$412}{40 \text{ hours}} = \frac{\$d}{30 \text{ hours}}$$

$$412 \cdot 30 = 40d$$

$$12{,}360 = 40d$$

$$\frac{12{,}360}{40} = \frac{40d}{40}$$

$$309 = d$$

Bill would earn \$309 for a 30 hour work week.

61. MODEL RAILROADING

Let $c$ represent the actual caboose length and convert 3.5 inches to feet.

$3\frac{1}{2}$ inches is equivalent to $\frac{7}{2}$ inches.

$$\frac{12 \text{ inches}}{1 \text{ foot}} = \frac{\frac{7}{2} \text{ inches}}{x \text{ feet}}$$

$$12x = \frac{7}{2}$$

$$x = \frac{\frac{7}{2}}{12} = \frac{7}{24} \text{ feet}$$

Now determine how long a real caboose is,

$$\frac{169 \text{ feet}}{1 \text{ foot}} = \frac{c \text{ feet}}{\frac{7}{24} \text{ feet}}$$

$$c = \frac{7}{24} \cdot 169$$

$$c = 49 \text{ feet } 3\frac{1}{2} \text{ inches}$$

A real caboose is 49 feet $3\frac{1}{2}$ inches long.

63. DRIVER'S LICENSE

Let $d$ represent the number of people in Oregon with a driver's lincense.

$$\frac{824}{1000} = \frac{d}{3{,}282{,}000}$$

$$1000d = 824 \cdot 3{,}282{,}000$$

$$d = 2{,}704{,}368$$

2,704,368 people in Oregon have a driver's license.

65. PHOTO ENLARGEMENTS

$$\frac{5 \text{ inches}}{6\frac{1}{4} \text{ inches}} = \frac{3 \text{ inches}}{x \text{ inches}}$$

$$5x = 3 \cdot 6\frac{1}{4}$$

$$5x = 3 \cdot \frac{25}{4}$$

$$5x = \frac{75}{4}$$

$$\frac{5x}{5} = \frac{\frac{75}{4}}{5}$$

$$x = \frac{75}{20}$$

$$x = \frac{15}{4} \text{ or } 3\frac{3}{4} \text{ inches}$$

67. **HEIGHT OF A TREE**

$\frac{6}{h} = \frac{4}{26}$

$6 \cdot 26 = 4 \cdot h$

$156 = 4h$

$39 = h$

The tree is 39 feet tall.

69. **WIDTH OF A RIVER**

$\frac{w}{20} = \frac{75}{32}$

$32 \cdot w = 20 \cdot 75$

$w = 46.875$

The width of the river is 46.875 feet or $46\frac{7}{8}$ feet.

71. **FLIGHT PATH**

Let $x$ represent the descent, remember that

1 mile $= 5280$ feet.

$(5\text{ miles} = 5 \cdot 5280 = 26{,}400)$

$\frac{1350}{x} = \frac{5280}{26{,}400}$

$1350 \cdot 26{,}400 = 5280 \cdot x$

$35{,}640{,}000 = 5280x$

$6750 = x$

This plane will lose 6750 feet in altitude.

**Writing**

73. Answers will vary

75. Answers will vary

**Review**

77. $\frac{9}{10} = 0.90$

$0.90 \cdot 100\% = 90\%$

79. 30% of 1600 is

$0.30 \cdot 1600 = 480$

81. $m = \frac{-2-(-8)}{-2-(-12)} = \frac{-2+8}{-2+12} = \frac{6}{10} = \frac{3}{5}$

**Vocabulary**

1. The equation $y = kx$ defines **direct** variation.
3. In $y = kx$, the **constant** of variation is $k$.

**Concepts**

5. This graph represents **direct** variation.
7. This graph represents **inverse** variation.

9. a. The equation defines direct variation.
   b. The equation does not define direct variation.
   c. The equation does not define direct variation.
   d. The equation defines direct variation.

11. $g = kf$

13. $t = ks$
$$21 = k(6)$$
$$\frac{21}{6} = k$$
$$\frac{7}{2} = k$$

15. a. This equation does not define inverse variation.
    b. This equation defines inverse variation.
    c. This equation does not define inverse variation.
    d. This equation defines inverse variation.

17. $n = \frac{k}{p}$

19. $y = \frac{k}{x}$
$$15 = \frac{k}{10}$$
$$150 = k$$

**Notation**

21. $d = 21, k = \frac{7}{5}$
$$d = kf$$
$$21 = \frac{7}{5}f$$
$$\frac{5}{7}(21) = \frac{5}{7} \cdot \frac{7}{5}f$$
$$15 = f$$

**Practice**

23. $10 = 2k$
$$5 = k$$

$$y = 5(7)$$
$$y = 35$$

25. $50 = 200k$
$\frac{1}{4} = k$

$l = 25(\frac{1}{4})$
$l = \frac{25}{4} = 6.25$

27. $2 = 30k$
$\frac{1}{15} = k$

$y = 45(\frac{1}{15})$
$y = 3$

29. $8 = \frac{k}{1}$
$8 = k$

$y = \frac{8}{8}$
$y = 1$

31. $600 = \frac{k}{300}$
$180{,}000 = k$

$a = \frac{180{,}000}{15}$
$a = 12{,}000$

33. $4 = \frac{k}{5}$
$20 = k$

$3\frac{1}{3} = \frac{20}{t_2}$
$t_2 = \frac{20}{3\frac{1}{3}} = \frac{20}{\frac{10}{3}}$
$t_2 = 6$

**Applications**

35. COMMUTING DISTANCE
$360 = 15k$
$24 = k$

$y = 7(24)$
$y = 168$
On seven gallons of fuel the auto can travel 168 miles.

37. DOSAGE
$18 = 30k$
$\frac{9}{15} = k$

$y = 45(\frac{9}{15})$
$y = 27$
The correct dosage would be 27 mg.

39. CIDER

$6 = 8k$

$\frac{3}{4} = k$

$c = \frac{3}{4}(30)$

$c = 22.5$

22.5 inches of cinnamon are needed for 30 servings.

41. COMMUTING TIME

$3 = \frac{k}{50}$

$150 = k$

$t = \frac{150}{60}$

$t = 2.5$

This trip will take 2.5 hours at 60 mph.

43. ELECTRICITY

$30 = \frac{k}{4}$

$120 = k$

$c = \frac{120}{15}$

$c = 8$

The current will be 8 amps.

45. COMPUTING PRESSURES

$V = \frac{k}{p}$

$40 = \frac{k}{8}$

$320 = k$

$V = \frac{320}{6}$

$V = 53\frac{1}{3}$

When the gas is subjected to a pressure of 6 pounds per square inch the volume is $53\frac{1}{3}$ cubic meters.

**Writing**

47. Answers will vary

49. Answers will vary

**Review**

51. $x^2 - 5x - 6 = 0$

$(x - 6)(x + 1) = 0$

$x - 6 = 0$ or $x + 1 = 0$

$x = 6 \qquad x = -1$

53. $(t + 2)(t^2 + 7t + 12) = 0$

$(t + 2)(t + 3)(t + 4) = 0$

$t + 2 = 0$ or $t + 3 = 0$ or $t + 4 = 0$

$t = -2 \qquad t = -3 \qquad t = -4$

55. $y^3 - y^2 = 0$

   $y^2(y - 1) = 0$

   $y^2 = 0$ or $y - 1 = 0$

   $y = 0 \qquad y = 1$

57. $(x^2 - 1)(x^2 - 4) = 0$

   $(x - 1)(x + 1)(x - 2)(x + 2) = 0$

   $x - 1 = 0$ or $x + 1 = 0$

   or $x - 2 = 0$ or $x + 2 = 0$

   $x = 1$ or $x = -1$ or $x = 2$ or $x = -2$

**Chapter 12 Key Concepts**

1.  a. $\frac{2x^2-8x}{x^2-6x+8} = \frac{2x(x-4)}{(x-4)(x-2)} = \frac{2x}{x-2}$

    b. $(x-4)$ was the common factor

3.  a. $\frac{x}{x+1} + \frac{x-1}{x}$

    $= \frac{x^2}{x(x+1)} + \frac{(x+1)(x-1)}{x(x+1)}$

    $= \frac{x^2+x^2-1}{x(x+1)}$

    $= \frac{2x^2-1}{x(x+1)}$

    b. The first fraction was multiplied by $\frac{x}{x}$ and  the second fraction was multiplied by $\frac{x+1}{x+1}$

5.  a. $\frac{11}{b} + \frac{13}{b} = 12$

    $b\left(\frac{11}{b} + \frac{13}{b}\right) = 12b$

    $24 = 12b$

    $2 = b$

    b. Multiplied both sides by $b$.

7.  a. $y + \frac{3}{4} = \frac{3y-50}{4y-24}$

    $y + \frac{3}{4} = \frac{3y-50}{4(y-6)}$

    $4(y-6)\left(y+\frac{3}{4}\right) = 3y - 50$

    $4\left(y^2 - \frac{21}{4}y - \frac{18}{4}\right) = 3y - 50$

    $4y^2 - 21y - 18 = 3y - 50$

    $4y^2 - 24y + 32 = 0$

    $4(y^2 - 6y + 8) = 0$

    $4(y-4)(y-2) = 0$

    $y - 4 = 0 \quad \text{or} \quad y - 2 = 0$

    $y = 4 \qquad\qquad y = 2$

    b. Both sides were multiplied by $4(y-6)$.

**Section 12.1**

1. The expression is undefined when the denominator equals zero.
$$x^2 - 16 = 0$$
$$(x + 4)(x - 4) = 0$$
$$x + 4 = 0 \quad \text{or} \quad x - 4 = 0$$
$$x = -4 \qquad\qquad x = 4$$

3.  a. $\dfrac{3x^2}{6x^3} = \dfrac{3x^2}{3x^2 \cdot 2x} = \dfrac{1}{2x}$

   b. $\dfrac{5xy^2}{2x^2y^2} = \dfrac{5}{2x}$

   c. $\dfrac{x^2}{x^2 + x} = \dfrac{x^2}{x(x+1)} = \dfrac{x}{x+1}$

   d. $\dfrac{a^2 - 4}{a + 2} = \dfrac{(a-2)(a+2)}{(a+2)} = a - 2$

   e. $\dfrac{3p-2}{2-3p} = \dfrac{3p-2}{-1(-2+3p)} = -\dfrac{3p-2}{3p-2} = -1$

   f. $\dfrac{8-x}{x^2 - 5x + 24}$
$$= \dfrac{8-x}{(x-8)(x+3)}$$
$$= \dfrac{8-x}{(-1)(8-x)(x+3)}$$
$$= -\dfrac{1}{x+3}$$

   g. $\dfrac{2x^2 - 16x}{2x^2 - 18x + 16}$
$$= \dfrac{2x(x-8)}{2(x^2 - 9x + 8)}$$
$$= \dfrac{x(x-8)}{(x-8)(x-1)}$$
$$= \dfrac{x}{x-1}$$

   h. $\dfrac{x^2 + x - 2}{x^2 - x - 2} = \dfrac{(x+2)(x-1)}{(x-2)(x+1)}$

   This is in lowest terms

5. It would not be correct to divide out the $x$ in $\dfrac{x+1}{x}$ since $x$ is not a common factor of the denominator and the numerator.

**Section 12.2**

7.  a. $\dfrac{3xy}{2x} \cdot \dfrac{4x}{2y^2} = \dfrac{3y}{2} \cdot \dfrac{2x}{y^2} = \dfrac{3y \cdot 2x}{2 \cdot y^2} = \dfrac{3x}{y}$

   b. $56x\left(\dfrac{12}{7x}\right) = 8 \cdot 7x\left(\dfrac{12}{7x}\right) = 8(12) = 96$

   c. $\dfrac{x^2 - 1}{x^2 + 2x} \cdot \dfrac{x}{x+1}$
$$= \dfrac{(x+1)(x-1)}{x(x+2)} \cdot \dfrac{x}{(x+1)}$$
$$= \dfrac{x(x-1)}{x(x+2)}$$
$$= \dfrac{x-1}{x+2}$$

   d. $\dfrac{x^2 + x}{3x - 15} \cdot \dfrac{6x - 30}{x^2 + 2x + 1}$
$$= \dfrac{x(x+1)}{3(x-5)} \cdot \dfrac{6(x-5)}{(x+1)^2}$$
$$= \dfrac{6x(x+1)(x-5)}{3(x-5)(x+1)^2}$$
$$= \dfrac{2x}{(x+1)}$$

9. $\dfrac{b^2+4b+4}{b^2+b-6}\left(\dfrac{b-2}{b-1} \div \dfrac{b+2}{b^2+2b-3}\right)$

$\quad = \dfrac{(b+2)^2}{(b+3)(b-2)}\left(\dfrac{(b-2)}{(b-1)} \cdot \dfrac{b^2+2b-3}{(b+2)}\right)$

$\quad = \dfrac{(b+2)^2}{(b+3)(b-2)}\left(\dfrac{(b-2)}{(b-1)} \cdot \dfrac{(b+3)(b-1)}{(b+2)}\right)$

$\quad = \dfrac{(b+2)^2}{(b+3)(b-2)}\left(\dfrac{(b-2)(b+3)(b-1)}{(b-1)(b+2)}\right)$

$\quad = \dfrac{(b+2)^2}{(b+3)(b-2)}\left(\dfrac{(b-2)(b+3)}{(b+2)}\right)$

$\quad = \dfrac{(b+2)^2(b-2)(b+3)}{(b+3)(b-2)(b+2)}$

$\quad = b + 2$

## Section 12.3

11. a. LCD is $4x^2$

    b. LCD is $18xy^2$

    c. LCD is $(x+1)(x+2)$

    d. LCD is $(y-5)(y+5)$

13. VIDEO CAMERA

Perimeter

$P = 2l + 2w$

$P = 2\left(\dfrac{4}{x+6}\right) + 2\left(\dfrac{3}{x-1}\right)$

$P = \dfrac{8}{x+6} + \dfrac{6}{x-1}$

$P = \dfrac{8(x-1)}{(x+6)(x-1)} + \dfrac{6(x+6)}{(x-1)(x+6)}$

$P = \dfrac{8(x-1)+6(x+6)}{(x+6)(x-1)}$

$P = \dfrac{8x-8+6x+36}{(x+6)(x-1)}$

$P = \dfrac{14x+28}{(x+6)(x-1)}$

Area

$A = lw$

$A = \left(\dfrac{4}{x+6}\right)\left(\dfrac{3}{x-1}\right)$

$A = \dfrac{12}{(x+6)(x-1)}$

15. a. $\frac{3}{x} = \frac{2}{x-1}$

$x(x-1)\left(\frac{3}{x}\right) = \left(\frac{2}{x-1}\right)x(x-1)$

$3x - 3 = 2x$

$x = 3$

Check

$\frac{3}{3} \stackrel{?}{=} \frac{2}{3-1}$

$1 = \frac{2}{2}$

b. $\frac{a}{a-5} = 3 + \frac{5}{a-5}$

$\frac{a}{a-5} - \frac{5}{a-5} = 3$

$\frac{a-5}{a-5} = 3$

$1 \neq 3$

No solutions; 5 is extraneous.

c. $\frac{2}{3t} + \frac{1}{t} = \frac{5}{9}$

$9t\left(\frac{2}{3t} + \frac{1}{t}\right) = 9t\left(\frac{5}{9}\right)$

$6 + 9 = 5t$

$15 = 5t$

$3 = t$

Check

$\frac{2}{3\cdot3} + \frac{1}{3} \stackrel{?}{=} \frac{5}{9}$

$\frac{2}{9} + \frac{3}{9} \stackrel{?}{=} \frac{5}{9}$

$\frac{5}{9} = \frac{5}{9}$

d. $a = \frac{3a-50}{4a-24} - \frac{3}{4}$

$a + \frac{3}{4} = \frac{3a-50}{4a-24}$

$(4a-24)\left(a + \frac{3}{4}\right) = 3a - 50$

$4a^2 + 3a - 24a - 18 = 3a - 50$

$4a^2 - 24a + 32 = 0$

$4(a^2 - 6a + 8) = 0$

$(a-4)(a-2) = 0$

$a - 4 = 0 \qquad$ or $\quad a - 2 = 0$

$a = 4 \qquad\qquad a = 2$

Check both solutions:

$a = 4$

$4 \stackrel{?}{=} \frac{3(4)-50}{4(4)-24} - \frac{3}{4}$

$4 \stackrel{?}{=} \frac{-38}{-8} - \frac{3}{4}$

$4 \stackrel{?}{=} \frac{38}{8} - \frac{6}{8}$

$4 \stackrel{?}{=} \frac{32}{8}$

$4 = 4$

Check

$a = 2$

$2 \stackrel{?}{=} \frac{3(2)-50}{4(2)-24} - \frac{3}{4}$

$2 \stackrel{?}{=} \frac{-44}{-16} - \frac{3}{4}$

$2 \stackrel{?}{=} \frac{44}{16} - \frac{12}{16}$

$2 \stackrel{?}{=} \frac{32}{16}$

$2 = 2$

e. $\frac{4}{x+2} - \frac{3}{x+3} = \frac{6}{x^2+5x+6}$

$\frac{4}{x+2} - \frac{3}{x+3} = \frac{6}{(x+2)(x+3)}$

$(x+2)(x+3)\left(\frac{4}{x+2} - \frac{3}{x+3}\right)$

$\qquad = (x+2)(x+3)\left(\frac{6}{(x+2)(x+3)}\right)$

$4(x+3) - 3(x+2) = 6$

$4x + 12 - 3x - 6 = 6$

$x = 0$

Check

$\frac{4}{0+2} - \frac{3}{0+3} \overset{?}{=} \frac{6}{0^2+5(0)+6}$

$2 - 1 \overset{?}{=} \frac{6}{6}$

$1 = 1$

17. $\frac{1}{r} = \frac{1}{r_1} + \frac{1}{r_2}$

$r_1 r_2 r\left(\frac{1}{r}\right) = r_1 r_2 r\left(\frac{1}{r_1} + \frac{1}{r_2}\right)$

$r_1 r_2 = r_2 r + r_1 r$

$r_1 r_2 - r_1 r = r_2 r$

$r_1(r_2 - r) = r_2 r$

$r_1 = \frac{r_2 r}{(r_2 - r)}$

19. If it takes a maid 4 hours to clean a house then in one hour $\frac{1}{4}$ of the house is cleaned.

21. INVESTMENTS

|       | Principal | Rate  | Interest |
|-------|-----------|-------|----------|
| S&L   | $P$       | $r$   | 100      |
| CU    | $P$       | $r+1$ | 120      |

$I = Prt$

$\frac{100}{r} = \frac{120}{r+1}$

$100(r+1) = 120r$

$100r + 100 = 120r$

$100 = 20r$

$5 = r$

The savings and loan rate was 5%.

23. WIND SPEED

|          | rate      | time | distance |
|----------|-----------|------|----------|
| downwind | $360 + r$ | $t$  | 400      |
| upwind   | $360 - r$ | $t$  | 320      |

$r$ is the velocity of the wind.

$\frac{400}{360+r} = \frac{320}{360-r}$

$400(360 - r) = 320(360 + r)$

$144000 - 400r = 115200 + 320r$

$28800 = 720r$

$40 = r$

The wind velocity is 40 mph.

25. a. $\frac{3}{x} = \frac{6}{9}$

     $9 \cdot 3 = 6x$

     $\frac{27}{6} = x$

     $\frac{9}{2} = x$

   b. $\frac{x}{3} = \frac{x}{5}$

     $5x = 3x$

     $5x - 3x = 0$

     $2x = 0$

     $x = 0$

   c. $\frac{x-2}{5} = \frac{x}{7}$

     $7(x - 2) = 5x$

     $7x - 14 = 5x$

     $-14 = -2x$

     $7 = x$

   d. $\frac{2x}{x+4} = \frac{3}{x-1}$

     $(x - 1)2x = 3(x + 4)$

     $2x^2 - 2x = 3x + 12$

     $2x^2 - 5x - 12 = 0$

     $(2x + 3)(x - 4) = 0$

     $2x + 3 = 0 \quad \text{or} \quad x - 4 = 0$

     $2x = -3 \qquad\qquad x = 4$

     $x = -\frac{3}{2}$

27. $\frac{t}{6} = \frac{12}{3.6}$

   $3.6t = 12 \cdot 6$

   $3.6t = 72$

   $t = 20$

   The telephone pole is 20 feet tall.

## Section 12.7

29. $l = \frac{k}{w}$

   $30 = \frac{k}{20}$

   $30(20) = k$

   $600 = k$

31. Inverse variation

## Chapter 12 Test

1. $x^2 + x - 6 = 0$
$(x + 3)(x - 2) = 0$
$x + 3 = 0 \quad$ or $\quad x - 2 = 0$
$x = -3 \qquad\qquad x = 2$

3. $\dfrac{2x^2 - x - 3}{4x^2 - 9} = \dfrac{(2x-3)(x+1)}{(2x-3)^2} = \dfrac{(x+1)}{(2x-3)}$

5. $-\dfrac{12x^2 y}{15xy} \cdot \dfrac{25y^2}{16x}$
$= -\dfrac{3 \cdot 4x}{3 \cdot 5} \cdot \dfrac{5 \cdot 5y^2}{4 \cdot 4x}$
$= -\dfrac{3 \cdot 4x \cdot 5 \cdot 5y^2}{3 \cdot 5 \cdot 4 \cdot 4x}$
$= -\dfrac{5y^2}{4}$

7. $\dfrac{8x^2}{25x} \div \dfrac{16x^2}{30x}$
$= \dfrac{8x}{25} \cdot \dfrac{30}{16x}$
$= \dfrac{(8x)(30)}{25 \cdot 16x}$
$= \dfrac{(2^3 x)(3 \cdot 5 \cdot 2)}{5^2 \cdot 2^4 x}$
$= \dfrac{3}{5}$

9. $\dfrac{x^2 + x}{x - 1} \cdot \dfrac{x^2 - 1}{x^2 - 2x} \div \dfrac{x^2 + 2x + 1}{x^2 - 4}$
$= \dfrac{x(x+1)}{(x-1)} \cdot \dfrac{(x-1)(x+1)}{x(x-2)} \div \dfrac{(x+1)^2}{(x-2)(x+2)}$
$= \dfrac{x(x+1)(x-1)(x+1)}{(x-1)x(x-2)} \div \dfrac{(x+1)^2}{(x-2)(x+2)}$
$= \dfrac{(x+1)(x+1)}{(x-2)} \div \dfrac{(x+1)^2}{(x-2)(x+2)}$
$= \dfrac{(x+1)(x+1)}{(x-2)} \cdot \dfrac{(x-2)(x+2)}{(x+1)^2}$
$= \dfrac{(x+1)(x+1)(x-2)(x+2)}{(x-2)(x+1)^2}$
$= x + 2$

11. $\dfrac{3y+7}{2y+3} - \dfrac{3(y-2)}{2y+3}$
$= \dfrac{3y+7-3(y-2)}{2y+3}$
$= \dfrac{3y+7-3y+6}{2y+3}$
$= \dfrac{13}{2y+3}$

13. $\dfrac{a+3}{a-1} - \dfrac{a+4}{1-a}$
$= \dfrac{a+3}{a-1} - \dfrac{a+4}{(-1)(a-1)}$
$= \dfrac{a+3}{a-1} + \dfrac{a+4}{a-1}$
$= \dfrac{2a+7}{a-1}$

15. $\dfrac{1+\frac{y}{x}}{\frac{y}{x}-1} = \dfrac{\frac{x}{x}+\frac{y}{x}}{\frac{y}{x}-\frac{x}{x}} = \dfrac{\frac{x+y}{x}}{\frac{y-x}{x}} = \dfrac{x+y}{x} \cdot \dfrac{x}{y-x} = \dfrac{x+y}{y-x}$

17. $\frac{2}{3} = \frac{2c-12}{3c-9} - c$

$\frac{2}{3} + c = \frac{2c-12}{3c-9}$

$(3c-9)(\frac{2}{3} + c) = 2c - 12$

$2c + 3c^2 - 6 - 9c = 2c - 12$

$3c^2 - 9c + 6 = 0$

$3(c^2 - 3c + 2) = 0$

$c^2 - 3c + 2 = 0$

$(c-1)(c-2) = 0$

$c - 1 = 0 \quad$ or $\quad c - 2 = 0$

$c = 1 \qquad\qquad c = 2$

19. $\frac{3}{5} = \frac{6xt}{10xt}$

$3 \cdot 10xt \overset{?}{=} 6xt \cdot 5$

$30xt = 30xt$

This is a proportion.

21. HEALTH RISKS

Healthy: $\frac{\text{waist}}{\text{hip}} = \frac{19}{20}$

Patient: $\frac{\text{waist}}{\text{hip}} = \frac{114}{120} = \frac{19}{20}$

Since the healthy condition is $\leq \frac{19}{20}$, this patient satisfies the condition.

23. POGO STICK

$130 = 6.5k$

$\frac{130}{6.5} = k$

$20 = k$

$f = 5(20)$

$f = 100 \text{ lb}$

25. CLEANING HIGHWAYS

$\frac{1}{7} + \frac{1}{9} = \frac{1}{h}$

$63h\left(\frac{1}{7} + \frac{1}{9}\right) = 63h\left(\frac{1}{h}\right)$

$9h + 7h = 63$

$16h = 63$

$h = \frac{63}{16}$

$h = 3\frac{15}{16}$

It will take the two of them working together $3\frac{15}{16}$ hours.

27. In $\frac{5x}{5}$, 5 is a common factor in the numerator and denominator. In $\frac{5+x}{5}$, 5 is not a common factor in the numerator and denominator.

1.  $9^2 - 3[45 - 3(6 + 4)]$
    $= 9^2 - 3[45 - 3(10)]$
    $= 81 - 3[15]$
    $= 81 - 45$
    $= 36$

3.  Let $t$ represent test scores, then $\bar{t}$ represents test score average.
    $\bar{t} = \frac{80+73+61+73+98}{5}$
    $\bar{t} = 77$

5.  $\frac{3}{4} = \frac{1}{2} + \frac{x}{5}$
    $20\left(\frac{3}{4} = \frac{1}{2} + \frac{x}{5}\right)$
    $15 = 10 + 4x$
    $5 = 4x$
    $\frac{5}{4} = x$

7.  $V = \frac{1}{3}s^2h$
    $V = \frac{1}{3} \cdot 6^2 \cdot 20$
    $V = \frac{1}{3} \cdot 36 \cdot 20$
    $V = 12 \cdot 20$
    $V = 240\,\text{ft}^3$

9.  $2 - 3(x - 5) = 4(x - 1)$
    $2 - 3x + 15 = 4x - 4$
    $17 - 3x = 4x - 4$
    $21 = 7x$
    $3 = x$

11.  $A - c = 2B + r$
    $A - c - r = 2B$
    $\frac{A-c-r}{2} = B$

13.  $\frac{4}{5}d = -4$
    $4d = -20$
    $d = -5$

15.  SPEED OF A PLANE
    Using $d = rt$ or $t = \frac{d}{r}$, let $r_1$ represent the speed of the first plane and $r_1 + 200$ represent the speed of the second plane.
    $d = r_1t + r_2t$
    $6000 = r_15 + (r_1 + 200)5$
    $1200 = r_1 + r_1 + 200$
    $1000 = 2r_1$
    $500 = r_1$
    $r_2 = 500 + 200$
    $r_2 = 700$
    The slower plane is going 500mph.

17. $y = (x+2)^3$

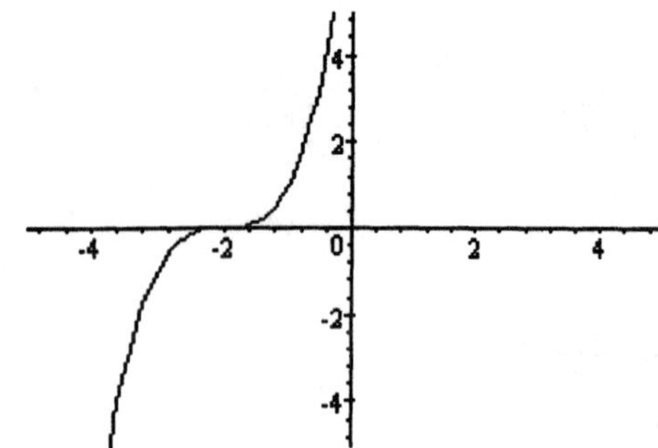

19. The equation of the line passing through $(1, 5)$ with slope $m = 3$ is found using
$$y - y_1 = m(x - x_1)$$
$$y - 5 = 3(x - 1)$$
$$y - 5 = 3x - 3$$
$$y = 3x + 2$$

21. $y = \frac{5}{2}$

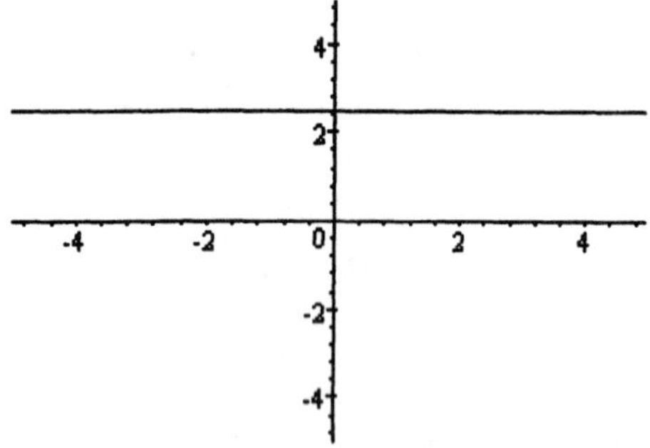

23. The graph of a circle is not a function, it does not pass the vertical line test.

25. $f(x) = \frac{x^2 - 2x}{2}$
$$f(-4) = \frac{(-4)^2 - 2(-4)}{2}$$
$$f(-4) = \frac{16 + 8}{2}$$
$$f(-4) = 12$$

27. $x^4 x^3 = x^{4+3} = x^7$

29. $\left(\frac{y^3 y}{2yy^2}\right)^3 = \left(\frac{y^4}{2y^3}\right)^3 = \left(\frac{y}{2}\right)^3 = \frac{y^3}{8}$

31. $(a^{-2}b^3)^{-4} = a^{(-2)(-4)}b^{3(-4)} = a^8 b^{-12} = \frac{a^8}{b^{12}}$

33. $290{,}000 = 2.9 \times 10^5$

35. $(3x^2 - 3x - 2) + (3x^2 + 4x - 3)$
$$= 6x^2 + x - 5$$

37. $(2y - 5)((3y + 7) = 6y^2 + 14y - 15y - 35 = 6y^2 - y - 35$

39. $\frac{6x+9}{3} = 2x + 3$

41. LICENSE PLATES
$10^3 \cdot 26^3 = 17{,}576{,}000$

43. $k^3 t - 3k^2 t = k^2 t(k - 3)$

45. $2a^2 - 200b^2$
$= 2(a^2 - 100b^2)$
$= 2(a^2 - (10b)^2)$
$= 2(a - 10b)(a + 10b)$

47. $u^2 - 18u + 81 = (u - 9)(u - 9) = (u - 9)^2$

49. $-r^2 + r + 2 = -(r^2 - r - 2) = -(r - 2)(r + 1)$

51. $5x^2 + x = 0$
$x(5x + 1) = 0$
$x = 0 \quad \text{or} \quad 5x + 1 = 0$
$5x = -1$
$x = -\frac{1}{5}$

53. COOKING
$lw = 160$
$w(w + 6) = 160$
$w^2 + 6w = 160$
$w^2 + 6w - 160 = 0$
$(w + 16)(w - 10) = 0$
$w + 16 = 0 \quad \text{or} \quad w - 10 = 0$
$w = -16 \qquad w = 10$
Therefore the width is 10 inches and the length is $10 + 6 = 16$ inches.

55. $\frac{x^2 - 16}{x - 4} \div \frac{3x + 12}{x}$
$= \frac{x^2 - 16}{x - 4} \cdot \frac{x}{3x + 12}$
$= \frac{(x^2 - 16)x}{(x - 4)(3x + 12)}$
$= \frac{(x + 4)(x - 4)x}{3(x - 4)(x + 4)}$
$= \frac{x}{3}$

57. $\frac{2 - \frac{2}{x+1}}{2 + \frac{2}{x}} = \frac{x(x+1)}{x(x+1)} \cdot \left( \frac{2 - \frac{2}{x+1}}{2 + \frac{2}{x}} \right)$
$= \frac{2x(x+1) - 2x}{2x(x+1) + 2(x+1)}$
$= \frac{2x^2 + 2x - 2x}{2x^2 + 2x + 2x + 2}$
$= \frac{2x^2}{2x^2 + 4x + 2}$
$= \frac{2x^2}{2(x^2 + 2x + 1)}$
$= \frac{x^2}{x^2 + 2x + 1}$
$= \frac{x^2}{(x + 1)^2}$

59. $\frac{7}{5x} - \frac{1}{2} = \frac{5}{6x} + \frac{1}{3}$

$\quad 30x\left(\frac{7}{5x} - \frac{1}{2}\right) = 30x\left(\frac{5}{6x} + \frac{1}{3}\right)$

$\quad 6(7) - 15x = 5(5) + 10x$

$\quad 42 - 15x = 25 + 10x$

$\quad 17 = 25x$

$\quad \frac{17}{25} = x$

61. COMPUTING INTEREST

$\quad 700 = k(2)$

$\quad 350 = k$

$\quad i = 350(7)$

$\quad i = \$2{,}450$

63. DRAINING A TANK

Pipe one will drain the pipe at a rate of $\frac{x}{24}$ and pipe two will drain the pipe at a rate of $\frac{x}{36}$, together the pipes will drain the tank

$\quad \frac{x}{24} + \frac{x}{36} = 1$

$\quad \frac{x}{2^3 \cdot 3} + \frac{x}{2^2 \cdot 3^2} = 1$

$\quad \frac{x}{2^3 \cdot 3}\left(\frac{3}{3}\right) + \frac{x}{2^2 \cdot 3^2}\left(\frac{2}{2}\right) = 1$

$\quad \frac{3x + 2x}{72} = 1$

$\quad 5x = 72$

$\quad x = \frac{72}{5} = 14\frac{2}{5}$ hours

The two pipes will drain this tank in $14\frac{2}{5}$ hours.

# Section 13.1 Solving Systems of Equations by Graphing

**Vocabulary**

1.  The pair of equations
    $$x - y = -1$$
    $$2x - y = 1$$
    is called a **system** of equations.

3.  When the graphs of two equations in a system are different lines, the equations are called **independent** equations.

5.  Systems of equations that have no solution are called **inconsistent** systems.

**Concepts**

7.  If the coordinates of point A were substituted into the equation for $l_1$ a **true** statement would result.

9.  If the coordinates of point A were substituted into the equation for $l_2$ a **false** statement would result.

11. If the coordinates of point C were substituted into the equation for $l_1$ a **true** statement would result.

13. a. It will cost \$2000 to make 30 chairs.
    b. The revenue on 30 chairs is \$1200.
    c. If the company makes 30 chairs they will lose \$800.

15. There is one solution to this system of equations. This is a consistent system.

17. Graphing methods tend to be inaccurate with solutions that are fractions.

**Notation**

19. $\frac{1}{6}x - \frac{1}{3}y = \frac{11}{2}$
    $$6\left(\frac{1}{6}x - \frac{1}{3}y\right) = 6\left(\frac{11}{2}\right)$$
    $$6\left(\frac{1}{6}x\right) - 6\left(\frac{1}{3}y\right) = 6\left(\frac{11}{2}\right)$$
    $$x - 2y = 33$$

**Practice**

21. A.  $x + y = 2$
    B.  $2x - y = 1$
    Substitution with $(1, 1)$
    A.  $1 + 1 \overset{?}{=} 2$
    $\qquad 2 = 2$
    B.  $2(1) - 1 \overset{?}{=} 1$
    $\qquad 1 = 1$
    $(1, 1)$ is a solution.

23. A. $2x + y = 4$
   B. $y = 1 - x$
   Substitution with $(3, -2)$
   A. $2(3) - 2 \overset{?}{=} 4$
      $4 = 4$
   B. $-2 \overset{?}{=} 1 - 3$
      $-2 = -2$
   $(3, -2)$ is a solution

25. A. $4x + 5y = -23$
   B. $-3x + 2y = 0$
   Substitution with $(-2, -4)$
   A. $4(-2) + 5(-4) \overset{?}{=} -23$
      $-28 \neq -23$
   B. $-3(-2) + 2(-4) \overset{?}{=} 0$
      $-2 \neq 0$
   $(-2, -4)$ is not a solution

27. A. $2x + y = 4$
   B. $4x - 11 = 3y$
   Substitution with $(\frac{1}{2}, 3)$
   A. $2(\frac{1}{2}) + 3 \overset{?}{=} 4$
      $4 = 4$
   B. $4(\frac{1}{2}) - 11 \overset{?}{=} 3(3)$
      $-9 \neq 9$
   $(\frac{1}{2}, 3)$ is not a solution

29. A. $x - 4y = -6$
   B. $8y = 10x + 12$
   Substitution with $(-\frac{2}{5}, \frac{1}{4})$
   A. $-\frac{2}{5} - 4(\frac{1}{4}) \overset{?}{=} -6$
      $-\frac{7}{5} \neq -6$
   B. $8(\frac{1}{4}) \overset{?}{=} 10(-\frac{2}{5}) + 12$
      $2 \neq 8$
   $(-\frac{2}{5}, \frac{1}{4})$ is not a solution

31. A. $20x + 10y = 7$
   B. $20y = 15x + 3$
   Substitution with $(0.2, 0.3)$
   A. $20(0.2) + 10(0.3) \overset{?}{=} 7$
      $7 = 7$
   B. $20(0.3) \overset{?}{=} 15(0.2) + 3$
      $6 = 6$
   $(0.2, 0.3)$ is a solution

33. $x + y = 2$
$x - y = 0$

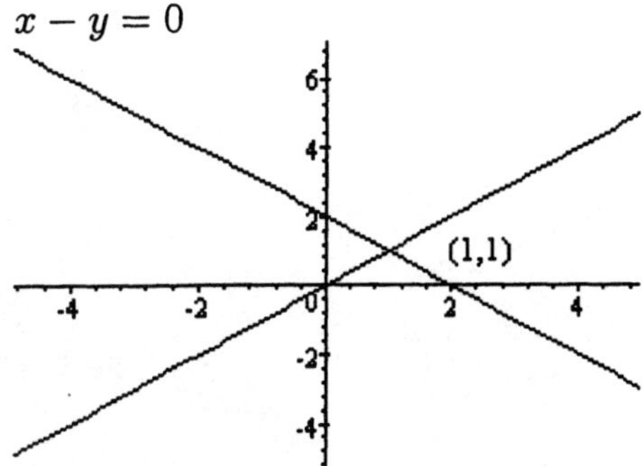



35. $x + y = 2$
$y = x - 4$

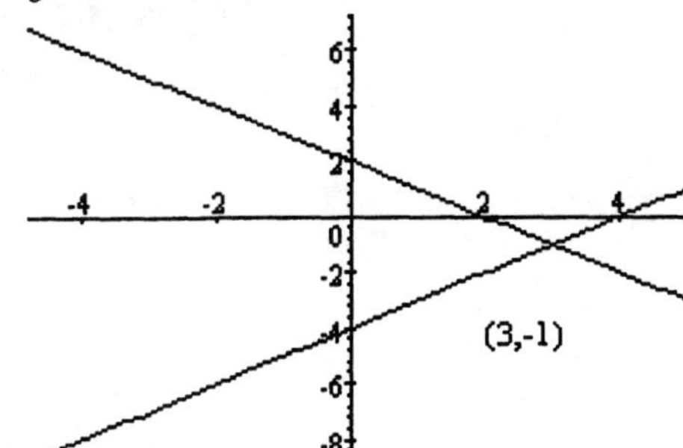



37. $3x + 2y = -8$
$2x - 3y = -1$

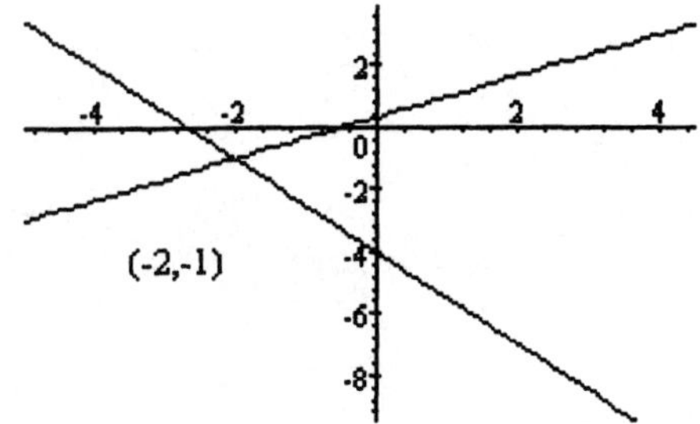



39. $4x - 2y = 8$
$y = 2x - 4$

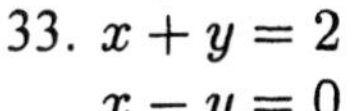

Equations are dependent.

41. $2x - 3y = -18$
$3x + 2y = -1$

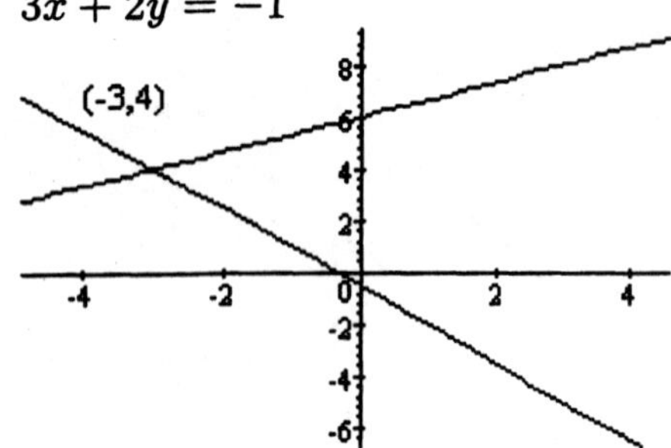


43. $x = 4$
$2y = 12 - 4x$

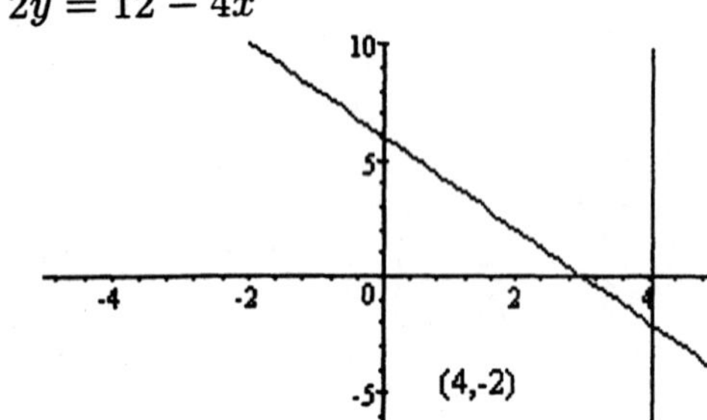


45. $x + 2y = -4$
$x - \frac{1}{2}y = 6$

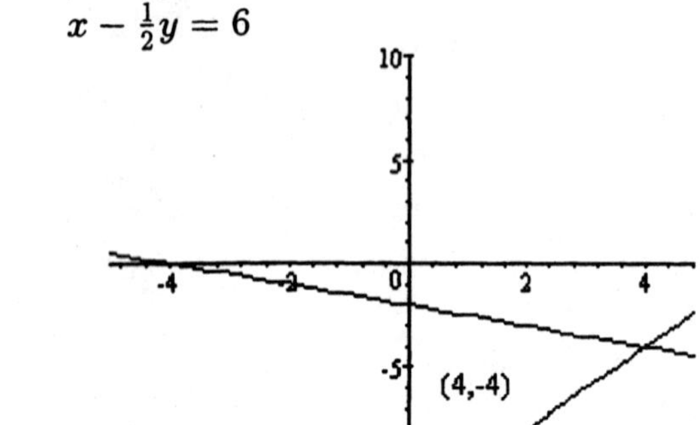


47. $-\frac{3}{4}x + y = 3$

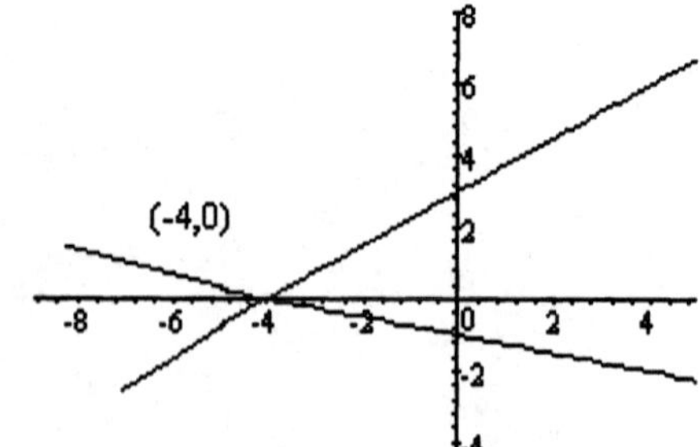


$\frac{1}{4}x + y = -1$

49. $2y = 3x + 2$
    $\frac{3}{2}x - y = 3$

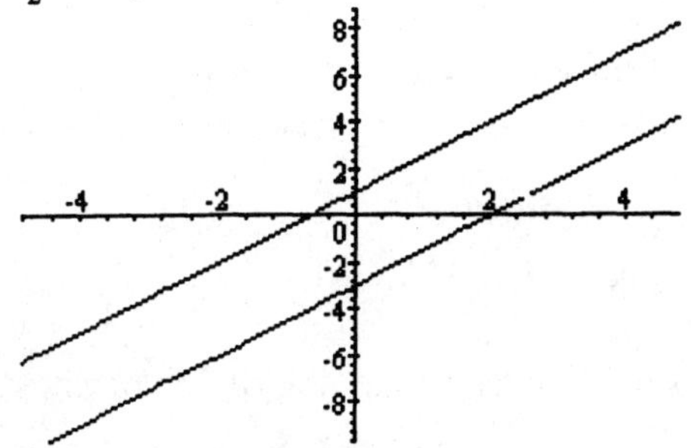

Equations are inconsistent

51. $\frac{1}{3}x - \frac{1}{2}y = \frac{1}{6}$
    $\frac{2x}{5} + \frac{y}{2} = \frac{13}{10}$

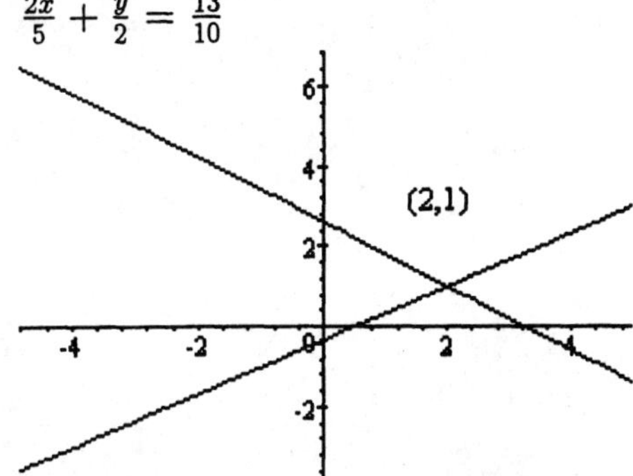



53. $y = 4 - x$
    $y = 2 + x$
    The solution is $(1, 3)$

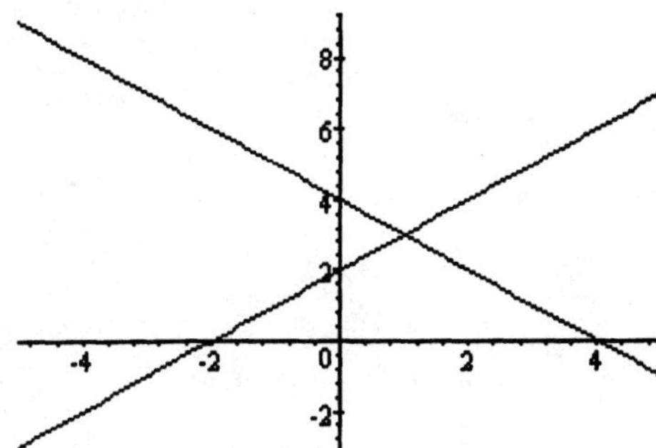

55. $6x - 2y = 5$
    $3x = y + 10$
    The is no solution

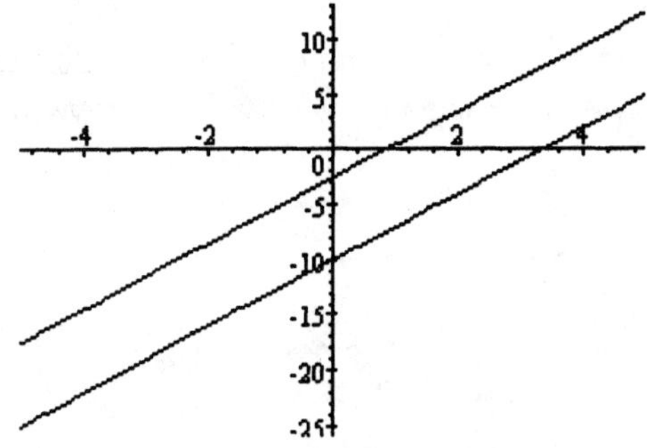

**Applications**

57. TRANSPLANTS
    a. In 1989 the number of Donors exceeded the number of people on the liver transplant list.
    b. In 1994 the number of Donors and the number of people waiting were the same, about 4,100.

59. LATITUDE AND LONGITUDE
    a. St. Augustine, New Orleans and Houston lie on a latitude line of $30^0$ north.
    b. New Orleans, Memphis and St. Louis lie on a longitudinal line of $90^0$ west.
    c. New Orleans lies on on a longitudinal line of $90^0$ west and on a latitude line of $30^0$ north.

61. AIR TRAFFIC CONTROL
$$y = -\tfrac{1}{2}x + 3$$
$$3y = 2x + 2$$

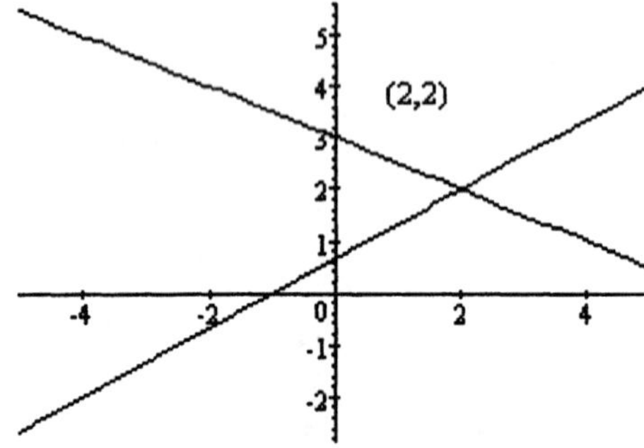

    There is a possibility of a mid-air collision at $(2, 2)$.

**Writing**

63. Answers will vary

**Review**

65. $y = -3x + 4$ has a slope of $-3$ and a $y$-intercept of $(0, 4)$.

67. $f(x) = -4x - x^2$
    $f(3) = -4 \cdot 3 - 3^2$
    $f(3) = -21$

69. The equation for the $y$-axis is $x = 0$.

71. $y - 2 = 7(x - 5)$
    $y - 2 = 7x - 35$
    $y = 7x - 33$
    Answers may vary, but from the initial problem statement a point, this equation passes through is $(5, 2)$. Since it is in this form, $y - y_1 = m(x - x_1)$, the answer can be found without any calculations. Other points are $(0, -33)$, $\left(\tfrac{33}{7}, 0\right)$

### Section 13.2 Solving Systems of Equations by Substitution

**Vocabulary**

1. We say the equation $y = 2x + 4$ is solved for **y** or that $y$ is expressed in **terms** of $x$.
3. When we write $2(x - 6)$ as $2x - 12$, we are applying the **distributive** property.
5. A dependent system has **infinitely** many solutions.

**Concepts**

7. a. Each equation in this system has two variables.
   b. $2x + 3(2x + 4) = 12$
      The resulting equation has one variable.

9. a. $x - 2y = -10$
      $x = 2y - 10$
   b. $x - 2y = -10$
      $x + 10 = 2y$
      $\frac{x+10}{2} = y$
      $y = \frac{x}{2} + 5$
   c. $x$; Solving for $x$ required only one step.

11. a. The error occured when the substitution was made in the second line. The quantity, $(x - 4)$, was not in parentheses to show multiplication by 2. It should read $x + 2(x - 4) = 5$
    b. $x + 2(x - 4) = 5$
       $x + 2x - 8 = 5$
       $3x = 13$
       $x = \frac{13}{3}$

13. a. $x - 2y = 0$
       $6x + 3y = 5$

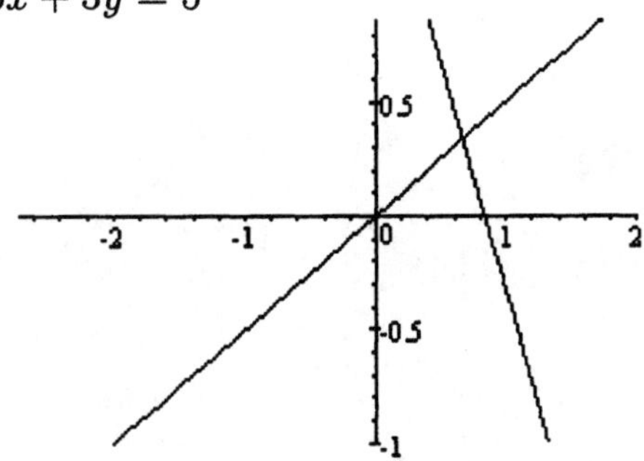

It is difficult to determine the coordinates of the intersection since these coordinates are not integers.

   b. A. $x - 2y = 0$ therefore $x = 2y$
      B. $6x + 3y = 5$
      Substituting A into B yields
      $6(2y) + 3y = 5$
      $12y + 3y = 5$
      $15y = 5$
      $y = \frac{1}{3}$
      Using $x = 2y$, $x = 2(\frac{1}{3}) = \frac{2}{3}$
      The solution $(x, y)$ is $(\frac{2}{3}, \frac{1}{3})$.

**Notation**

15. $y = 3x$
$x - y = 4$
Substitute
$x - (3x) = 4$
$-2x = 4$
$x = -2$

$y = 3(-2)$
$y = -6$
The solution is $(-2, -6)$.

**Practice**

17. A.  $y = 2x$
B.  $x + y = 6$
Substituting A into B yields
$x + 2x = 6$
$3x = 6$
$x = 2$
Substituting this value into A
$y = 2 \cdot 2$
$y = 4$
The solution is $(2, 4)$.

19. A.  $y = 2x - 6$
B.  $2x + y = 6$
Substituting A into B yields
$2x + (2x - 6) = 6$
$4x = 12$
$x = 3$
Substituting this value into A
$y = 2(3) - 6$
$y = 0$
The solution is $(3, 0)$.

21. A.  $y = 2x + 5$
B.  $x + 2y = -5$
Substituting A into B yields
$x + 2(2x + 5) = -5$
$x + 4x + 10 = -5$
$5x = -15$
$x = -3$
Substituting this value into A
$y = 2(-3) + 5$
$y = -1$
The solution is $(-3, -1)$.

23. A. $2a + 4b = -24$
   B. $a = 20 - 2b$
   Substituting B into A yields
   $2(20 - 2b) + 4b = -24$
   $40 - 4b + 4b = -24$
   $40 \neq -24$
   This system is inconsistent.

25. A. $2a = 3b - 13$
   B. $-b = -2a - 7$ therefore $b = 2a + 7$
   Substituting B into A yields
   $2a = 3(2a + 7) - 13$
   $2a = 6a + 21 - 13$
   $-4a = 8$
   $a = -2$
   Substituting this value into B
   $-b = -2(-2) - 7$
   $-b = -3$
   $b = 3$
   The solution is $(a, b) = (-2, 3)$.

27. A. $r + 3s = 9$ therefore $r = 9 - 3s$
   B. $3r + 2s = 13$
   Substituting A into B yields
   $3(9 - 3s) + 2s = 13$
   $27 - 9s + 2s = 13$
   $-7s = -14$
   $s = 2$
   Substituting this value into A
   $r + 3 \cdot 2 = 9$
   $r = 3$
   The solution is $(r, s) = (3, 2)$.

29. A. $0.4x + 0.5y = 0.2$

   B. $3x - y = 11$

   Rewriting equation A

   $10(0.4x + 0.5y) = 10(0.2)$

   $4x + 5y = 2$

   $4x = 2 - 5y$

   $x = \frac{1}{2} - \frac{5}{4}y$

   Substituting this form of equation A into B yields

   $3(\frac{1}{2} - \frac{5}{4}y) - y = 11$

   $\frac{3}{2} - \frac{15}{4}y - y = 11$

   $\frac{-19}{4}y = \frac{19}{2}$

   $y = -\frac{4}{19} \cdot \frac{19}{2}$

   $y = -2$

   Substituting this value into A

   $x = \frac{1}{2} - \frac{5}{4}(-2)$

   $x = \frac{1}{2} + \frac{5}{2}$

   $x = 3$

   The solution is $(3, -2)$.

31. A. $6x - 3y = 5$

   B. $2y + x = 0$ therefore $x = -2y$

   Substituting B into A yields

   $6(-2y) - 3y = 5$

   $-12y - 3y = 5$

   $-15y = 5$

   $y = -\frac{5}{15}$

   $y = -\frac{1}{3}$

   Substituting this value into B

   $-2(-\frac{1}{3}) = x$

   $\frac{2}{3} = x$

   The solution is $(\frac{2}{3}, -\frac{1}{3})$.

33. A. $3x + 4y = -7$
B. $2y - x = -1$ therefore $2y + 1 = x$
Substituting B into A yields
$3(2y + 1) + 4y = -7$
$6y + 3 + 4y = -7$
$10y = -10$
$y = -1$
Substituting this value into B
$2(-1) - x = -1$
$-x = 1$
$x = -1$
The solution is $(-1, -1)$.

35. A. $9x = 3y + 12$
B. $4 = 3x - y$
Rewriting equation A
$\frac{1}{9}(9x) = \frac{1}{9}(3y + 12)$
$x = \frac{1}{3}y + \frac{4}{3}$
Substituting this form of equation A into B yields
$4 = 3(\frac{1}{3}y + \frac{4}{3}) - y$
$4 = y + 4 - y$
$4 = 4$
This is a dependent system.

37. A. $0.02x + 0.05y = -0.02$
B. $-\frac{x}{2} = y$
Rewriting equation A
$100(0.02x + 0.05y) = 100(-0.02)$
$2x + 5y = -2$
Subsituting B into this form of equation A yields
$2x + 5(-\frac{x}{2}) = -2$
$2x - \frac{5}{2}x = -2$
$\frac{4}{2}x - \frac{5}{2}x = -2$
$-\frac{1}{2}x = -2$
$x = -2(-\frac{2}{1})$
$x = 4$
Substituting this value into B
$-\frac{4}{2} = y$
$y = -2$
The solution is $(4, -2)$.

39. A. $b = \frac{2}{3}a$
B. $8a - 3b = 3$
Substituting A into B yields
$8a - 3(\frac{2}{3}a) = 3$
$8a - 2a = 3$
$6a = 3$
$a = \frac{1}{2}$
Substituting this value into A
$b = \frac{2}{3}(\frac{1}{2})$
$b = \frac{1}{3}$
$(a, b) = (\frac{1}{2}, \frac{1}{3})$

41. A. $y - x = 3x$ therefore $y = 4x$
B. $2x + 2y = 14 - y$
Substituting A into B yields
$2x + 2(4x) = 14 - 4x$
$2x + 8x = 14 - 4x$
$10x = 14 - 4x$
$14x = 14$
$x = 1$
Substituting this value into A
$y - 1 = 3(1)$
$y = 4$
The solution is $(1, 4)$.

43. A. $2x - y = x + y$
B. $-2x + 4y = 6$
Rewriting equation A
$2x - x = y + y$
$x = 2y$
Substituting this form of equation A into B yields
$-2(2y) + 4y = 6$
$-4y + 4y = 6$
$0 \neq 6$
This is an inconsistent system.

45. A. $3(x-1)+3 = 8+2y$
B. $2(x+1) = 8+y$
Rewriting equation B
$2x+2 = 8+y$
$2x-6 = y$
Substituting this form of equation B into A yields
$3(x-1)+3 = 8+2(2x-6)$
$3x-3+3 = 8+4x-12$
$3x = 4x-4$
$4 = x$
Substituting this value into B
$2(4+1) = 8+y$
$2(5) = 8+y$
$2 = y$
The solution is $(4, 2)$.

47. A. $\frac{1}{2}x + \frac{1}{2}y = -1$
B. $\frac{1}{3}x - \frac{1}{2}y = -4$
Rewriting equation A
$2(\frac{1}{2}x + \frac{1}{2}y) = 2(-1)$
$x+y = -2$
$x = -2-y$
Substituting this form of equation A into B yields
$\frac{1}{3}(-2-y) - \frac{1}{2}y = -4$
$-\frac{2}{3} - \frac{1}{3}y - \frac{1}{2}y = -4$
$6(-\frac{2}{3} - \frac{1}{3}y - \frac{1}{2}y) = 6(-4)$
$-4 - 2y - 3y = -24$
$-5y = -20$
$y = 4$
Substituting this value back into A
$x = -2-4$
$x = -6$
The solution is $(-6, 4)$.

49. A. $5x = \frac{1}{2}y - 1$

    B. $\frac{1}{4}y = 10x - 1$

Rewriting equation A

$5x = \frac{1}{2}y - 1$

$x = \frac{1}{5}(\frac{1}{2}y - 1)$

$x = \frac{1}{10}y - \frac{1}{5}$

Substituting this form of equation A into B yields

$\frac{1}{4}y = 10(\frac{1}{10}y - \frac{1}{5}) - 1$

$\frac{1}{4}y = y - 2 - 1$

$-\frac{3}{4}y = -3$

$y = -\frac{4}{3}(-3)$

$y = 4$

Substituting this value into A

$x = \frac{1}{10} \cdot 4 - \frac{1}{5}$

$x = \frac{4}{10} - \frac{2}{10}$

$x = \frac{2}{10}$

$x = \frac{1}{5}$

The solution is $(\frac{1}{5}, 4)$.


51. A. $\frac{6x-1}{3} - \frac{5}{3} = \frac{3y+1}{2}$

    B. $\frac{1+5y}{4} + \frac{x+3}{4} = \frac{17}{2}$

Rewriting equation B

$4(\frac{1+5y}{4} + \frac{x+3}{4}) = 4(\frac{17}{2})$

$1 + 5y + x + 3 = 34$

$5y + x = 30$

$x = 30 - 5y$

Substituting this form of equation A into B yields

$\frac{6(30-5y)-1}{3} - \frac{5}{3} = \frac{3y+1}{2}$

$3(\frac{180-30y-1}{3} - \frac{5}{3}) = 3(\frac{3y+1}{2})$

$180 - 30y - 1 - 5 = \frac{9y+3}{2}$

$2(180 - 30y - 6) = 2(\frac{9y+3}{2})$

$360 - 60y - 12 = 9y + 3$

$345 = 69y$

$5 = y$

Substituting this value into B

$x = 30 - 5 \cdot 5$

$x = 5$

The solution is $(5, 5)$.

**Applications**

53. DINING
    If a customer does not want hash browns with their meal the owner will allow substitution of mellon
    since it is the same price.

**Writing**

55. Answers will vary
57. Answers will vary

**Review**

59. The slope of the line represented by
    $y = -\frac{5}{8}x - 12$ is $-\frac{5}{8}$.

61. The $y$-intercept is found when $x = 0$
    $2x - 3y = 18$
    $2 \cdot 0 - 3y = 18$
    $-3y = 18$
    $y = -6$
    $(0, -6)$

63. A circle cannot represent the graph of a function, recall the vertical line test.

65. $s = kt$

## Section 13.3 Solving Systems of Equations by Addition

**Vocabulary**

1. The **coefficient** of the term $-3x$ is $-3$.
3. $Ax + By + C = 0$ is the **general** form of the equation of a line.

**Concepts**

5. The second equation should be written with the $x$ term first, $3x - 2y = 10$.
7. To clear this equation of fractions, multiply both sides by 15.

9.  a. $(2, -3)$

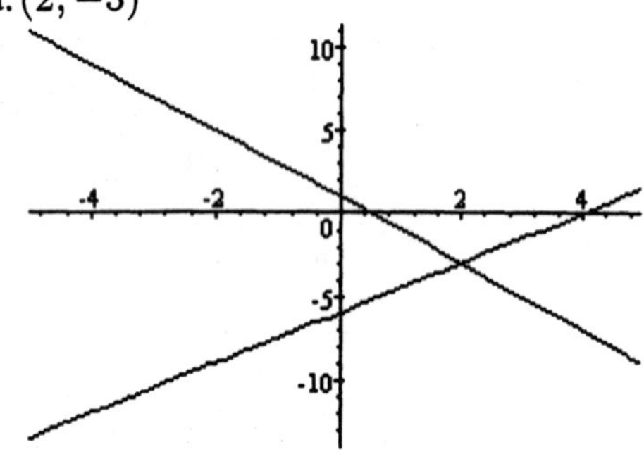

b. $4x + 2y = 2$
$\quad 3x - 2y = 12$
Rewrite the first equation
$4x = 2 - 2y$
$x = \frac{1}{2} - \frac{1}{2}y$
Substitute into the second equation
$3(\frac{1}{2} - \frac{1}{2}y) - 2y = 12$
$\frac{3}{2} - \frac{3}{2}y - 2y = 12$
$-\frac{7}{2}y = \frac{21}{2}$
$y = -\frac{2}{7} \cdot \frac{21}{2}$
$y = -\frac{21}{7}$
$y = -3$
Substitute this value into the first equation
$x = \frac{1}{2} - \frac{1}{2}(-3)$
$x = \frac{4}{2}$
$x = 2$
The solution is $(2, -3)$.

c. $4x + 2y = 2$
$\quad \underline{3x - 2y = 12}$
$\quad 7x = 14$
$\quad x = 2$
Substitute this value into either equation.
$4 \cdot 2 + 2y = 2$
$2y = -6$
$y = -3$
The solution is $(2, -3)$.

**Notation**

11. $x + y = 5$
$\underline{x - y = -3}$
$2x = 2$
$x = 1$
$x + y = 5$
$1 + y = 5$
$y = 4$
The solution is $(1, 4)$.

**Practice**

13. $x - y = -5$
$\underline{x + y = 1}$
$2x = -4$
$x = -2$
$-2 - y = -5$
$3 = y$
The solution is $(-2, 3)$.

15. $2r + s = -1$
$\underline{-2r + s = 3}$
$2s = 2$
$s = 1$
$2r + 1 = -1$
$2r = -2$
$r = -1$
The solution is $(r, s) = (-1, 1)$.

17. $2x + y = -2$
$\underline{-2x - 3y = -6}$
$-2y = -8$
$y = 4$
$2x + 4 = -2$
$2x = -6$
$x = -3$
The solution is $(-3, 4)$.

19. $4x + 3y = 24$
$\underline{4x - 3y = -24}$
$8x = 0$
$x = 0$
$4 \cdot 0 + 3y = 24$
$3y = 24$
$y = 8$
The solution is $(0, 8)$.

21. $x + y = 5$
$x + 2y = 8$

$x + y = 5$
$-1(x + 2y) = -1(8)$

$x + y = 5$
$\underline{-x - 2y = -8}$
$-y = -3$
$y = 3$
$x + 3 = 5$
$x = 2$
The solution is $(2, 3)$.

23. $2x + y = 4$
$2x + 3y = 0$

$2x + y = 4$
$-1(2x + 3y) = -1(0)$

$2x + y = 4$
$\underline{-2x - 3y = 0}$
$-2y = 4$
$y = -2$
$2x - 2 = 4$
$2x = 6$
$x = 3$
The solution is $(3, -2)$.

25. $3x - 5y = -29$
$3x + 4y = 34$

$3x - 5y = -29$
$-1(3x + 4y) = -1(34)$

$3x - 5y = -29$
$\underline{-3x - 4y = -34}$
$-9y = -63$
$y = 7$
$3x - 5(7) = -29$
$3x = 6$
$x = 2$
The solution is $(2, 7)$.

27. $2a - 3b = -6$
$2a - 3b = 8$

$2a - 3b = 6$
$-1(2a - 3b) = -1(8)$

$2a - 3b = -6$
$\underline{-2a + 3b = 8}$
$0 = 2$
This is an inconsistent system.

29. $8x - 4y = 18$
$3x - 2y = 8$

$8x - 4y = 18$
$-2(3x - 2y) = -2(8)$

$8x - 4y = 18$
$\underline{-6x + 4y = -16}$
$2x = 2$
$x = 1$
$8(1) - 4y = 18$
$-4y = 10$
$y = -\frac{10}{4}$
$y = -\frac{5}{2}$
$8x - 4(-\frac{5}{2}) = 18$
$8x + 10 = 18$
$8x = 8$
$x = 1$
The solution is $(1, -\frac{5}{2})$.

31. $2x + y = 10$
$0.1x + 0.2y = 1.0$

$2x + y = 10$
$-20(0.1x + 0.2y) = -20(1.0)$

$2x + y = 10$
$\underline{-2x - 4y = -20}$
$-3y = -10$
$y = \frac{10}{3}$

$2x + \frac{10}{3} = 10$
$2x = 10 - \frac{10}{3}$
$2x = \frac{20}{3}$
$x = \frac{20}{6}$
$x = \frac{10}{3}$
The solution is $(\frac{10}{3}, \frac{10}{3})$.

33. $2x - y = 16$
$0.03x + 0.02y = 0.03$

$2x - y = 16$
$50(0.03x + 0.02y) = 50(0.03)$

$2x - y = 16$
$\underline{1.5x + y = 1.5}$
$3.5x = 17.5$
$\frac{7}{2}x = \frac{35}{2}$
$x = \frac{2}{7} \cdot \frac{35}{2}$
$x = 5$
$2(5) - y = 16$
$-6 = y$
The solution is $(5, -6)$.

35. $6x + 3y = 0$
$5y = 2x + 12$

$6x + 3y = 0$
$-2x + 5y = 12$

$6x + 3y = 0$
$3(-2x + 5y) = 3(12)$

$6x + 3y = 0$
$\underline{-6x + 15y = 36}$
$18y = 36$
$y = 2$
$6x + 3(2) = 0$
$6x = -6$
$x = -1$
The solution is $(-1, 2)$.

37. $-2(x + 1) = 3y - 6$
$3(y + 2) = 10 - 2x$

$-2x - 2 = 3y - 6$
$3y + 6 = 10 - 2x$

$-2x - 3y = -4$
$\underline{2x + 3y = 4}$
$0 + 0 = 0$
$0 = 0$
This is a dependent system.

39. $4(x + 1) = 17 - 3(y - 1)$
$2(x + 2) + 3(y - 1) = 9$

$4x + 4 = 17 - 3y + 3$
$2x + 4 + 3y - 3 = 9$

$4x + 3y = 16$
$2x + 3y = 8$

$4x + 3y = 16$
$-1(2x + 3y) = -1(8)$

$4x + 3y = 16$
$\underline{-2x - 3y = -8}$
$2x = 8$
$x = 4$
$2(4 + 2) + 3(y - 1) = 9$
$12 + 3y - 3 = 9$
$3y = 0$
$y = 0$
The solution is $(4, 0)$.

41. $\frac{3}{5}s + \frac{4}{5}t = 1$
$-\frac{1}{4}s + \frac{3}{8}t = 1$

$10\left(\frac{3}{5}s + \frac{4}{5}t\right) = 10 \cdot 1$
$24\left(-\frac{1}{4}s + \frac{3}{8}t\right) = 24 \cdot 1$

$6s + 8t = 10$
$\underline{-6s + 9t = 24}$
$17t = 34$
$t = 2$
$\frac{3}{5}s + \frac{4}{5} \cdot 2 = 1$
$\frac{3}{5}s = 1 - \frac{8}{5}$
$s = \frac{5}{3}\left(-\frac{3}{5}\right)$
$s = -1$
The solution is $(s, t) = (-1, 2)$.

43. $\frac{3}{5}x + y = 1$
$\frac{4}{5}x - y = -1$

$\frac{3}{5}x + y = 1$
$\underline{\frac{4}{5}x - y = -1}$
$\frac{7}{5}x = 0$
$x = 0$
$\frac{3}{5} \cdot 0 + y = 1$
$y = 1$
The solution is $(0, 1)$.

45. $\frac{x}{2} - \frac{y}{3} = -2$

    $\frac{2x-3}{2} + \frac{6y+1}{3} = \frac{17}{6}$

$$-4\left(\frac{x}{2} - \frac{y}{3}\right) = -4(-2)$$
$$2\left(\frac{2x-3}{2} + \frac{6y+1}{3}\right) = 2\left(\frac{17}{6}\right)$$

$$-2x + \frac{4}{3}y = 8$$
$$2x - 3 + 4y + \frac{2}{3} = \frac{17}{3}$$

$$-2x + \frac{4}{3}y = 8$$
$$\underline{2x + 4y = \frac{15}{3} + 3}$$
$$\frac{16}{3}y = 16$$
$$y = \frac{3}{16} \cdot 16$$
$$y = 3$$
$$\frac{x}{2} - \frac{3}{3} = -2$$
$$\frac{x}{2} = -1$$
$$x = -2$$

The solution is $(-2, 3)$.

47. $\frac{x-3}{2} + \frac{y+5}{3} = \frac{11}{6}$

    $\frac{x+3}{3} - \frac{5}{12} = \frac{y+3}{4}$

$$-24\left(\frac{x-3}{2} + \frac{y+5}{3}\right) = -24\left(\frac{11}{6}\right)$$
$$36\left(\frac{x+3}{3} - \frac{5}{12}\right) = 36\left(\frac{y+3}{4}\right)$$

$$-12x + 36 - 8y - 40 = -44$$
$$12x + 36 - 15 = 9y + 27$$

$$-12x - 8y = -40$$
$$\underline{12x - 9y = 6}$$
$$-17y = -34$$
$$y = 2$$
$$\frac{x-3}{2} + \frac{2+5}{3} = \frac{11}{6}$$
$$6\left(\frac{x-3}{2} + \frac{2+5}{3}\right) = 6\left(\frac{11}{6}\right)$$
$$3x - 9 + 4 + 10 = 11$$
$$3x = 6$$
$$x = 2$$

The solution is $(2, 2)$.

## Writing

49. Answers will vary

51. Answers will vary

53. $8(3x - 5) - 12 = 4(2x + 3)$
    $24x - 40 - 12 = 8x + 12$
    $16x = 64$
    $x = 4$

55. $x - x = 0$

57. $A = \frac{1}{2}bh$
    $A = \frac{1}{2}(4)(3.75)$
    $A = 7.5\,\text{ft}^2$

59. $x - 10$

## Section 13.4 Applications of Systems of Equations

### Vocabulary

1.  A **variable** is a letter that stands for a number.

3.  $a + b = 20$
    $a = 2b + 4$
    is a **system** of linear equations.

### Concepts

5.  Downstream $x + c$, where $c$ represents the water current.
    Upstream $x - c$, where $c$ represents the water current.

7.  a. $\$y$ was invested in the USA Savings account.
    b. The City Bank account earned 5%.
    c.

|            | Principal | Rate | Time  | Interest |
|------------|-----------|------|-------|----------|
| City Bank  | $x$       | 5%   | 1 yr  | $0.05x$  |
| USA        | $y$       | 11%  | 1 yr  | $0.11y$  |

9.  a. Two unknowns require two equations for a solution.
    b. Three methods to solve a system of equations are graphing, substitution and addition.

### Notation

11. $A = lw$
13. $d = rt$
15. $2l + 2w = 90$

### Practice

17. $x = 2y$
    $x + y = 96$
    Substitute
    $2y + y = 96$
    $3y = 96$
    $y = 32$
    then $x = 2 \cdot 32 = 64$
    The integers are $32, 64$.

19. $3x + y = 29$
$x + 2y = 18$
Rewrite
$3x + y = 29$
$3x = 29 - y$
$x = \frac{29}{3} - \frac{1}{3}y$
Substitute
$x + 2y = 18$
$\frac{29}{3} - \frac{1}{3}y + 2y = 18$
$\frac{5}{3}y = 18 - \frac{29}{3}$
$y = \frac{3}{5}\left(18 - \frac{29}{3}\right)$
$y = \frac{3 \cdot 18}{5} - \frac{29}{5}$
$y = 5$
then
$x = \frac{29}{3} - \frac{1}{3} \cdot 5$
$x = 8$
The integers are $8, 5$.

**Applications**

21. TREE TRIMMING
$x$ represents the upper arm
$y$ represents the lower arm

$x + y = 51$
$x = y - 7$
Substitute
$(y - 7) + y = 51$
$2y = 58$
$y = 29$
The lower arm is 29 feet long.
$x = y - 7$
$x = 29 - 7$
$x = 22$
The upper arm is 22 feet long.

23. EXECUTIVE BRANCH
$x$ represents the Presidential salary.
$y$ represents the Vice-Presidential salary.

$x + y = 581,000$
$x = y + 219,000$
Substitute
$(y + 219,000) + y = 581,000$
$2y = 362,000$
$y = 181,000$
The Vice President's salary is \$181,000.
$x = 181,000 + 219,000$
$x = 400,000$
The President's salary is \$400,000.

25. BUYING PAINTING SUPPLIES
   $p$ represents gallons of paint
   $b$ represents brushes

   $8p + 3b = 270$
   $6p + 2b = 200$

   $2(8p + 3b) = 2(270)$
   $-3(6p + 2b) = -3(200)$

   $16p + 6b = 540$
   $\underline{-18p - 6b = -600}$
   $-2p = -60$
   $p = 30$
   The paint costs $30 per gallon
   $8p + 3b = 270$
   $8 \cdot 30 + 3b = 270$
   $3b = 30$
   $b = 10$
   A paintbrush costs $10.

27. BUYING TICKETS
   $a$ represents regular admissions
   $s$ represents senior admissions

   $a + s = 190$
   $4a + 3s = 720$

   $-4(a + s) = -4(190)$
   $4a + 3s = 720$

   $-4a - 4s = -760$
   $\underline{4a + 3s = 720}$
   $-s = -40$
   $s = 40$
   There were 40 senior citizens in attendance.

29. MARINE CORPS
   $a_1$ represents the measure of angle one
   $a_2$ represents the measure of angle two

   $a_1 + a_2 = 180$
   $a_2 = 2a_1 - 15$
   Substitute
   $a_1 + 2a_1 - 15 = 180$
   $3a_1 = 195$
   $a_1 = 65$
   Angle one measures 65 degrees.
   $65 + a_2 = 180$
   $a_2 = 115$
   Angle two measure 115 degrees.

31. THEATER SCREEN

$l$ represents the screen length
$w$ represents the screen width
$P$ represents the screen perimeter
We are told that $P = 332$

$P = 2l + 2w$
$w = l - 26$
Substitute
$332 = 2l + 2(l - 26)$
$332 = 2l + 2l - 52$
$384 = 4l$
$96 = l$
The length of this screen is 96 feet.
$w = 96 - 26$
$w = 70$
The width of this screen is 70 feet.
$A$ represents area
$A = lw$
$A = 96 \cdot 70$
$A = 6,720$
The area of this screen is $6,720$ ft$^2$.

33. MAKING TIRES

|  | setup | per tire cost |
|---|---|---|
| $m_1$, mold 1 | 1,000 | 15 |
| $m_2$, mold 2 | 3,000 | 10 |

$t$ represents the number of manufactured tires

$m_1 = 1,000 + 15t$
$m_2 = 3,000 + 10t$

a) The break point occurs when $m_1 = m_2$
$1,000 + 15t = 3,000 + 10t$
$5t = 2000$
$t = 400$
The break point is 400 tires.

b)

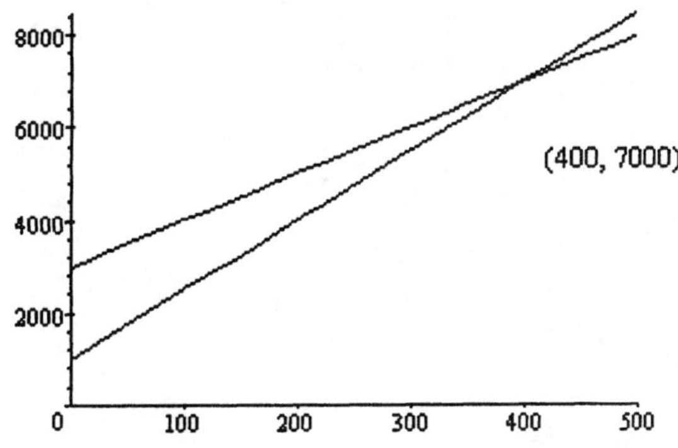

c)A production run of 500 tires would cost:
$$m_1 = 1,000 + 15 \cdot 500$$
$$m_1 = 8,500$$
Using mold one 500 tires would cost
$8,500$.
$$m_2 = 3,000 + 10 \cdot 500$$
$$m_2 = 8,000$$
Using mold two 500 tires would cost $8,000$.
A production run of 500 tires should be done using mold two.

## 35. STUDENT LOANS
$l_1$ represents the loan to the nursing student
$l_2$ represents the loan to the business majore

$$l_1 + l_2 = 5000$$
$$0.05l_1 + 0.07l_2 = 310$$
Rewrite
$$l_1 = 5000 - l_2$$
Substitute
$$0.05(5000 - l_2) + 0.07l_2 = 310$$
$$250 - 0.05l_2 + 0.07l_2 = 310$$
$$0.02l_2 = 60$$
$$l_2 = 3000$$
Substitute
$$l_1 + 3000 = 5000$$
$$l_1 = 2000$$
The nursing student was loaned $2000 and the business major was loaned $3000.

## 37. GULF STREAM
$r_1$ represents the boat speed
$r_2$ represents the water current

|  | rate | time | distance |
|---|---|---|---|
| upstream | $r_1 + r_2$ | 10 hrs | 300 mi |
| downstream | $r_1 - r_2$ | 15 hrs | 300 mi |

$d = rt$, therefore $r = \frac{d}{t}$
$$r_1 + r_2 = \frac{300}{10}$$
$$r_1 - r_2 = \frac{300}{15}$$
$$2r_1 = \frac{300}{10} + \frac{300}{15}$$
$$2r_1 = 30 + 20$$
$$2r_1 = 50$$
$$r_1 = 25$$
The boat is traveling at 25 mph.
Substitute to find the rate of the water.
$$25 + r_2 = \frac{300}{10}$$
$$r_2 = 5$$
The water current is 5 mph.

39. AVIATION

$r_1$ represents the plane speed

$r_2$ represents the wind speed

|  | rate | time | distance |
|---|---|---|---|
| with the wind | $r_1 + r_2$ | 2 hrs | 600 mi |
| against wind | $r_1 - r_2$ | 3 hrs | 600 mi |

$d = rt$, therefore $r = \frac{d}{t}$

$$r_1 + r_2 = \frac{600}{2}$$
$$r_1 - r_2 = \frac{600}{3}$$
$$\overline{2r_1 = 300 + 200}$$
$$2r_1 = 500$$
$$r_1 = 250$$

Substituting to find the wind speed.

$$250 - r_2 = \frac{600}{3}$$
$$r_2 = 50$$

The wind speed is 50 mph.

41. MARINE BIOLOGY

$a_1$ represents aquarium with 6 % solution

$a_2$ represents aquarium with 2 % solution

$a_3$ represents new aquarium being set up

|  | salt | amount |
|---|---|---|
| $a_1$ | 6 % | $x$ |
| $a_2$ | 2 % | $y$ |
| $a_3$ | 3 % | 16 liters |

$$x + y = 16$$
$$0.06x + 0.02y = 0.03(16)$$

$$x + y = 16$$
$$100(0.06x + 0.02y) = 100(0.03)(16)$$

$$x + y = 16$$
$$6x + 2y = 48$$

$$-6(x + y) = -6(16)$$
$$6x + 2y = 48$$

$$-6x - 6y = -96$$
$$\underline{6x + 2y = 48}$$
$$-4y = -48$$
$$y = 12$$

$$x + y = 16$$
$$x + 12 = 16$$
$$x = 4$$

This biologist will need 12 liters of the 2 % solution combined with 4 liters of the 6 % solution to obtain 16 liters of 3 % solution.

43. MIXING NUTS

|          | cost | pounds |
|----------|------|--------|
| peanuts  | $ 3  | $p$    |
| cashews  | $ 6  | $c$    |
| mix      | $ 4  | 48     |

$p + c = 48$

$3p \mid 6c = 4(48)$

$-3(p + c) = -3(48)$
$3p + 6c = 192$

$-3p - 3c = -144$
$\underline{3p + 6c = 192}$
$3c = 48$
$c = 16$

$p + c = 48$
$p + 16 = 48$
$p = 32$

The merchant will need 16 pounds of cashews and 32 pounds of peanuts to make 48 pounds of mixed nuts.

45. MARKDOWN

Retail price $-$ discount $=$ sale price
(1) $r - d = 384$

Discount $=$ Discount rate $\cdot$ retail price
(2) $d = 0.40 \cdot r$

Substitute equation (2) into (1)
$r - 0.4r = 384$
$0.6r = 384$
$r = 640$
The retail price of these golf clubs is $ 640.

**Writing**

47. Answers will vary

**Review**

49. $x < 4$

51. $-1 < x \leq 2$

53. $x^2 - 4 = 0$
$(x - 2)(x + 2) = 0$
$x - 2 = 0 \quad$ or $\quad x + 2 = 0$
$x = 2 \qquad\qquad x = -2$

55. $x^2 - 4x + 4 = 0$
$(x - 2)(x - 2) = 0$
$x - 2 = 0 \quad$ or $\quad x - 2 = 0$
$x = 2 \qquad\qquad x = 2$

**Vocabulary**

1. $2x - y \le 4$ is a linear **inequality** in $x$ and $y$.
3. In the graph in Illustration 1, the line $2x - y = 4$ is the **boundary**.

**Concepts**

5. $5x - 3y \ge 0$
   a. $(1, 1)$ is a solution since $5 \cdot 1 - 3 \cdot 1 = 2 \ge 0$
   b. $(-2, -3)$ is not a solution since $5(-2) - 3(-3) = -1 \le 0$
   c. $(0, 0)$ is a solution since $5 \cdot 0 - 3 \cdot 0 = 0 \ge 0$
   d. $\left(\frac{1}{5}, \frac{4}{3}\right)$ is not a solution since $5 \cdot \frac{1}{5} - 3 \cdot \frac{4}{3} = -3 \le 0$

7. a. $y > -x$ does not include the boundary line.
   b. $5x - 3y \le -2$ does include the boundary line.

9. a. $(1, -3)$ does not satisfy the inequality
   b. $(-2, -1)$ satisfies the inequality
   c. $(2, 3)$ satisfies the inequality
   d. $(3, -4)$ does not satisfy the inequality

11. A test point for an inequality must clearly lie on side or the other of the boundary line. Since $(0, 0)$ lies on the boundary line it cannot be used.

**Practice**

13. $y \le x + 2$
    Test point $(0, 0)$
    $0 \overset{?}{\le} 0 + 2$, yes
    Shade to the right of the graph.

15. $y > 2x - 4$
    Test point $(0, 0)$
    $0 \overset{?}{>} 2 \cdot 0 - 4$, yes
    Shade to the left of the graph.

17. $x - 2y \ge 4$
    Test point $(0, 0)$
    $0 - 2 \cdot 0 \overset{?}{\ge} 4$, no
    Shade to the right of the graph.

19. $y \le 4x$
    Test point $(1, 1)$
    $1 \overset{?}{\le} 4 \cdot 1$, yes
    Shade to the right of the graph.

21. $y \geq 3 - x$

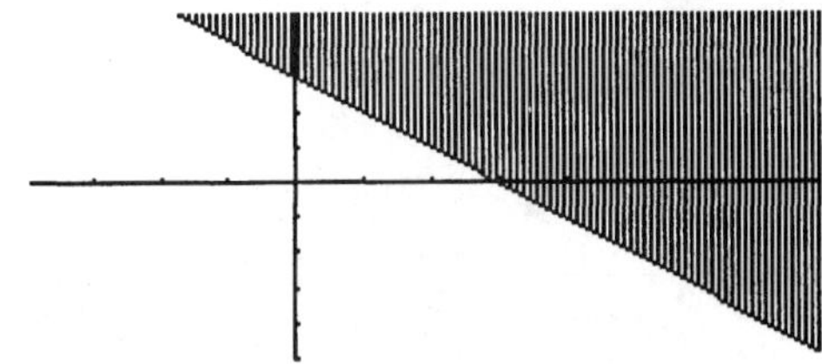

23. $y < 2 - 3x$

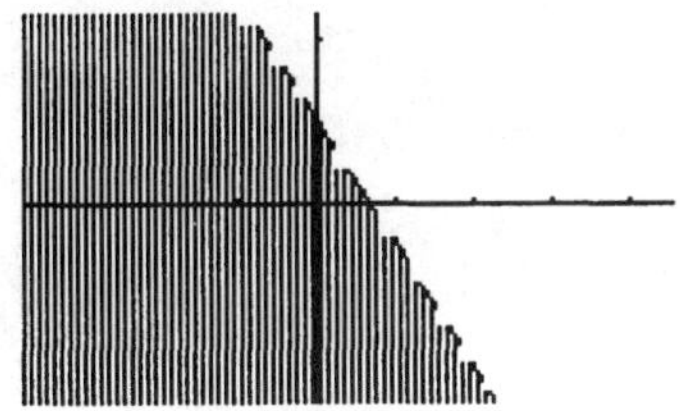

25. $y \geq 2x$

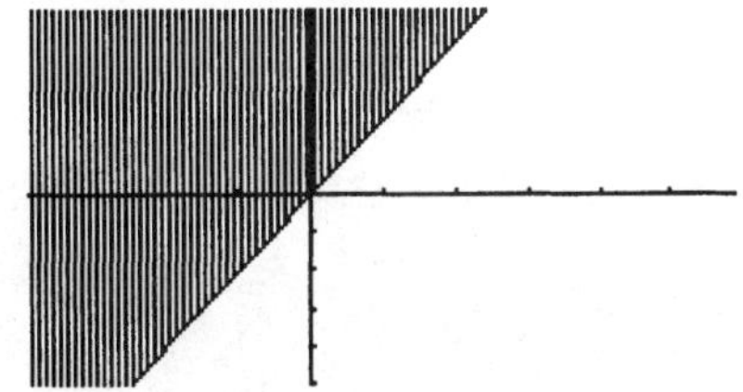

27. $2y - x < 8$

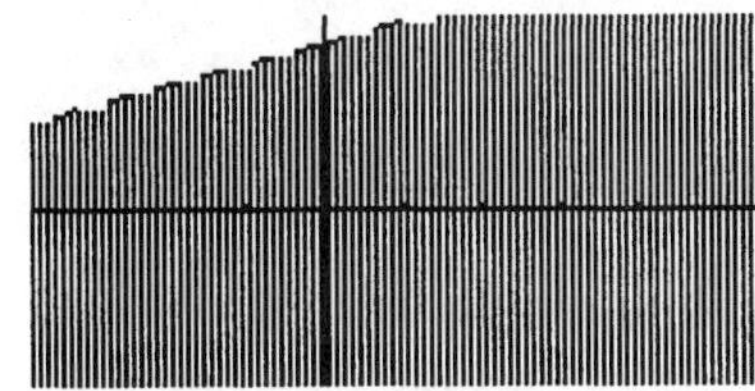

29. $y - x \geq 0$

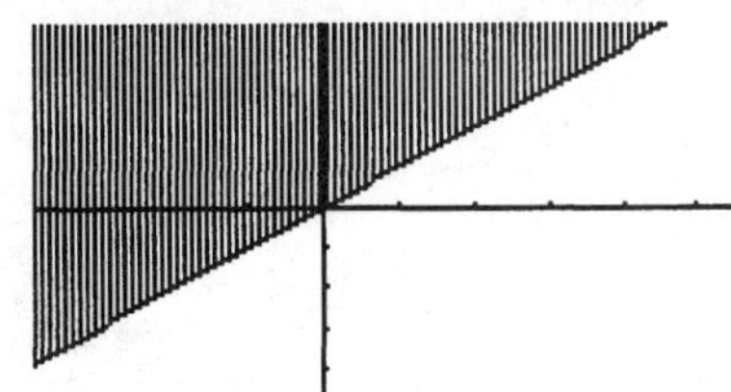

31. $2x + y > 2$

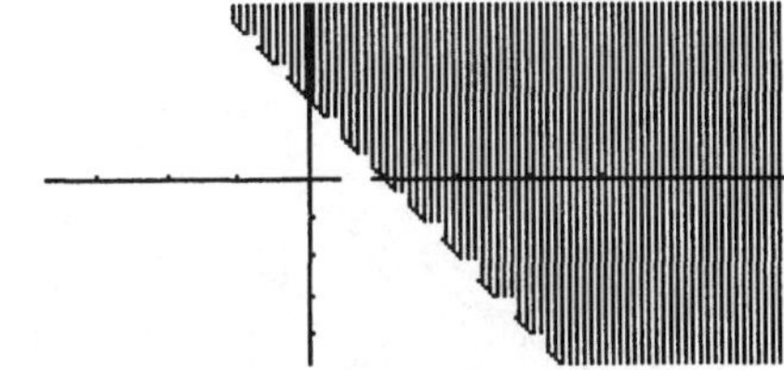

33. $3x - 4y > 12$

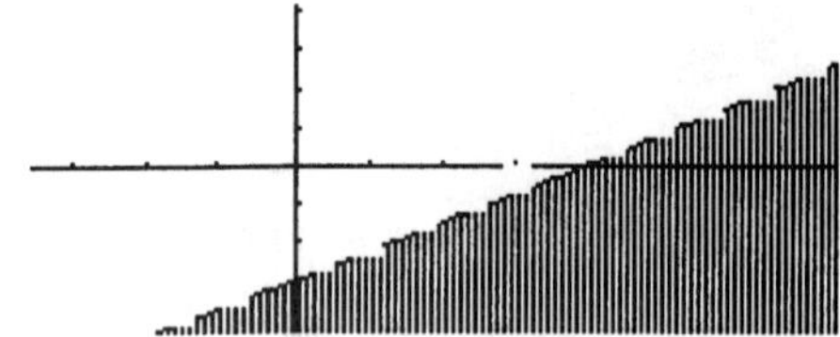

35. $5x + 4y \geq 20$

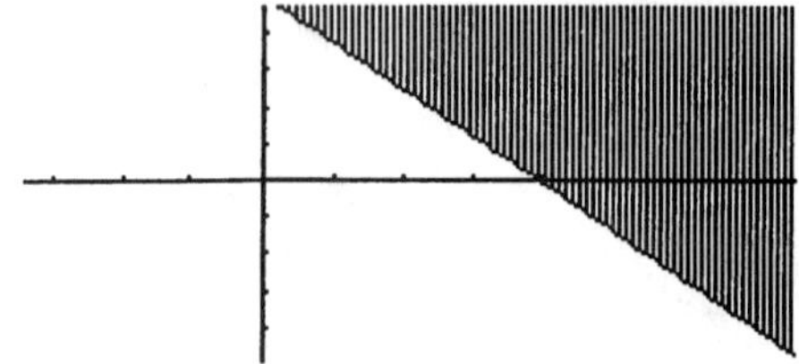

37. $x < 2$

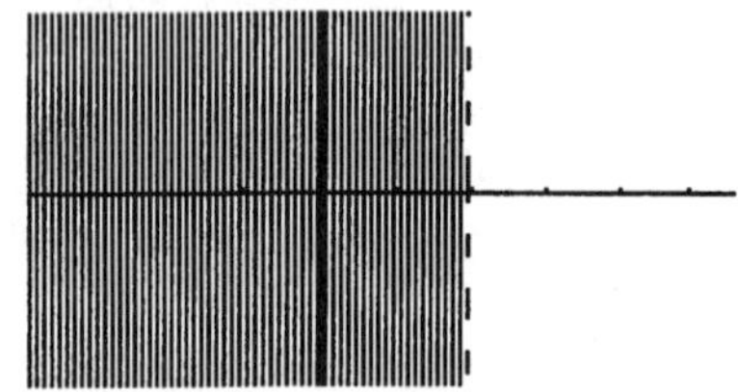

39. $y \leq 1$

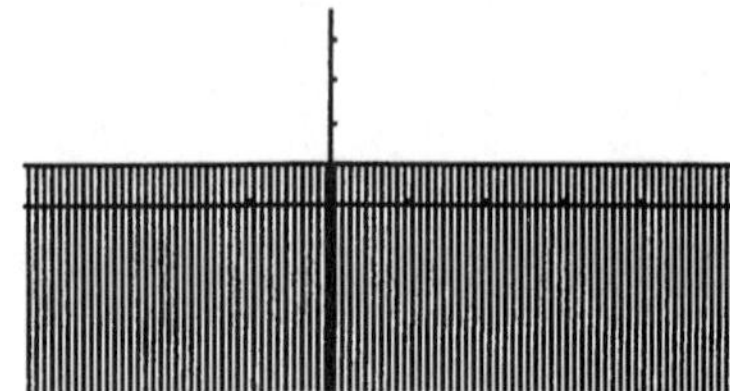

41. $3(x + y) + x < 6$
$3x + 3y + x < 6$
$4x + 3y < 6$

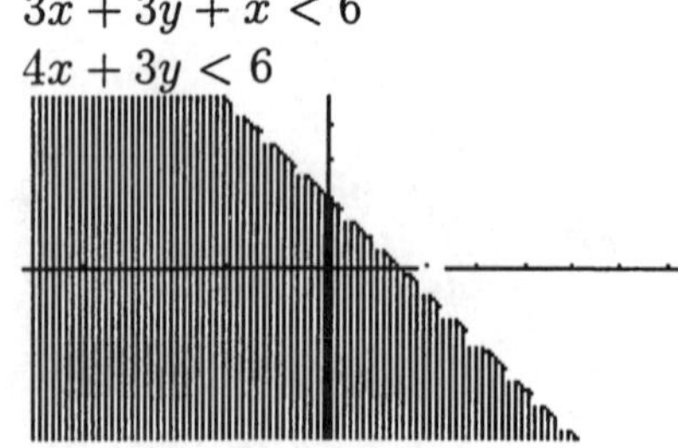

43. $4x - 3(x + 2y) \geq -6y$
$\quad 4x - 3x - 6y \geq -6y$
$\quad x \geq 0$

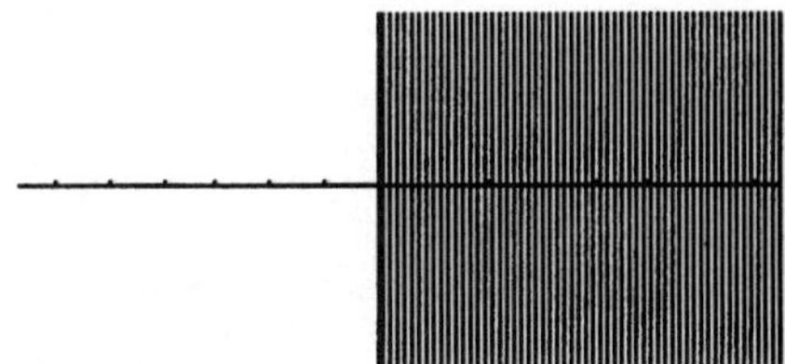

## Applications

45. NATO

The shaded region extends south of the 44th parallel in the countries of Yugoslavia, Montenegro, and Kosovo.

47. PRODUCTION PLANNING

Let $x$ represent cake cost and $y$ represent pie cost then, $3x + 4y \leq 120$

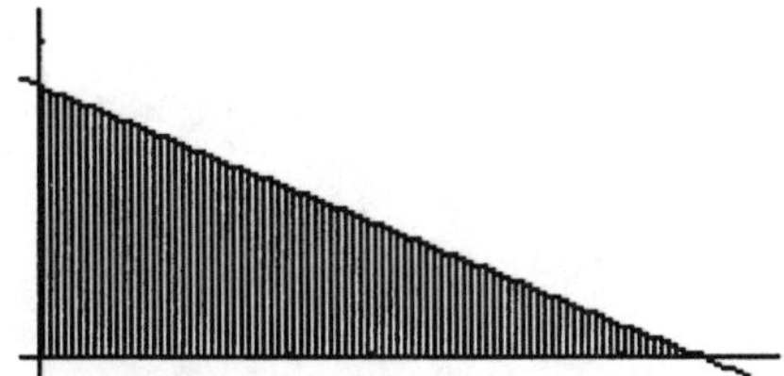

The baker could make any combination that lies in the shaded region or on the boundary line of $y = 30 - \frac{3}{4}x$.
Examples: $(20, 10)$, $(20, 15)$, $(10, 10)$

49. INVENTORY

$100x + 88y \geq 4400$

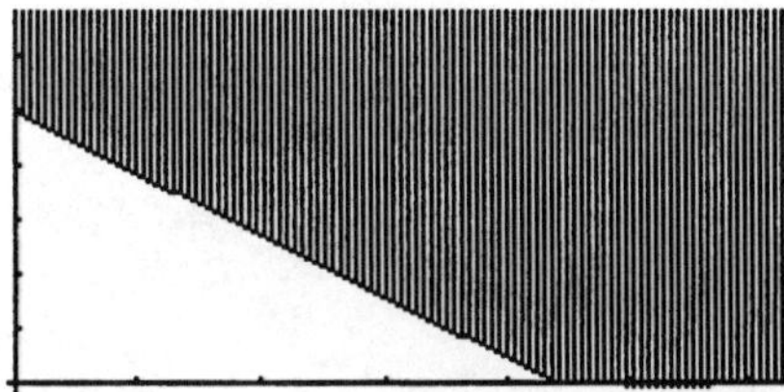

Any combination of leather and nylon jackets in the shaded region, including the boundary line of $y = \frac{4400}{88} - \frac{100}{88}x$ would satisfy the condition of at least $4400 worth of men's jackets.
Examples: $(50, 50)$, $(0, 50)$, $(30, 40)$

51. **INVESTING IN STOCKS**
$40x + 50y \le 8000$

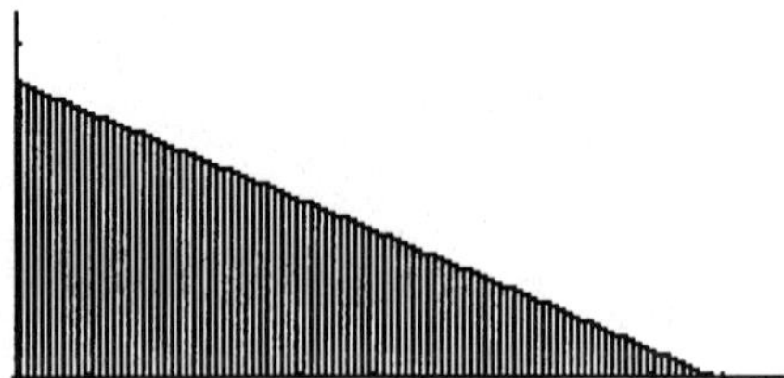

Any combination of stock purchases in the shaded region and on the boundary line $y = \frac{8000}{50} - \frac{40}{50}x$ will satisfy the possible investment options.
Examples: $(80, 80)$, $(100, 60)$, $(0, 160)$

**Writing**

53. Answers will vary

**Review**

55. $g(x) = 3x^2 - 4x + 3$
$g(2) = 3(2)^2 - 4 \cdot 2 + 3$
$g(2) = 7$

57. $x^3 + 27$
Recall: $x^3 + y^3 = (x + y)(x^2 - xy + y^2)$
$x^3 + 3^3 = (x + 3)(x^2 - 3x + 9)$

59. $d = rt$

61. $A = P + Prt$
$A - P = Prt$
$\frac{A-P}{Pr} = \frac{Prt}{Pr}$
$\frac{A-P}{Pr} = t$

**Vocabulary**

1. $x + y > 2$
   $x + y < 4$
   This is a system of linear **inequalities**.

3. Any point in the **doubly shaded** region of the graph of the solution of a system of two linear inequalities has coordinates that satisfy both inequalities of the system.

**Concepts**

5. a. True
   b. False
   c. False
   d. True
   e. True
   f. True

7. a. $(4, -2)$ yes
   b. $(1, 3)$ no
   c. the origin $(0, 0)$ no

**Notation**

9. The graph of the solution of a system of linear inequalities shown in illustration 4 can be described as the triangle **ABC** and the triangular region it encloses.

**Practice**

11. $x + 2y \le 3$
    $2x - y \ge 1$

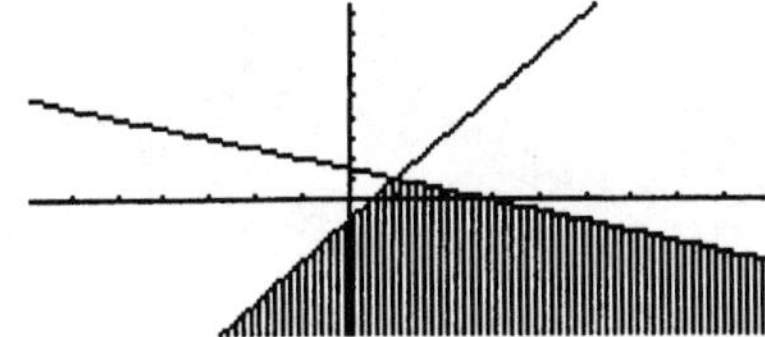

13. $x + y < -1$
    $x - y > -1$

15. $x \geq 2$
    $y \leq 3$

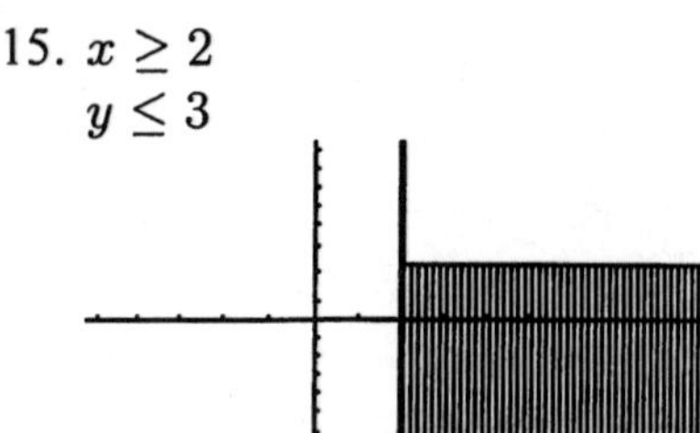

17. $2x - 3y \leq 0$
    $y \geq x - 1$

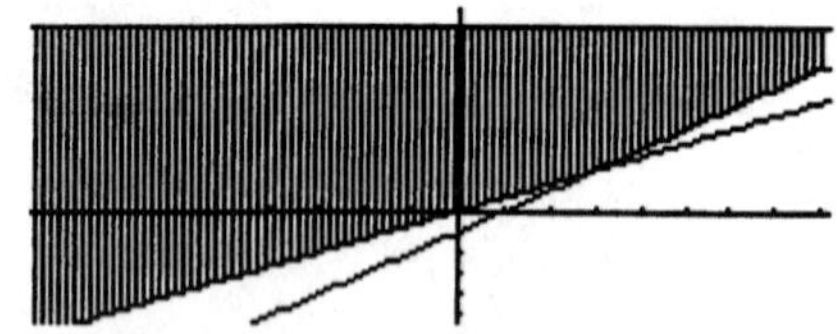

19. $y < -x + 1$
    $y > -x + 3$
    **No Solution**

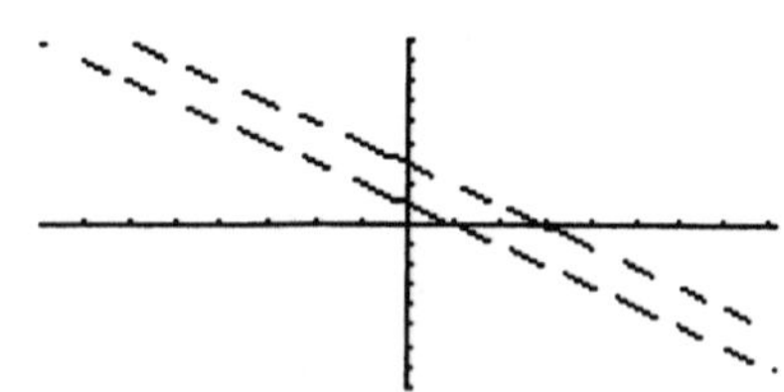

21. $x > 0$
    $y > 0$

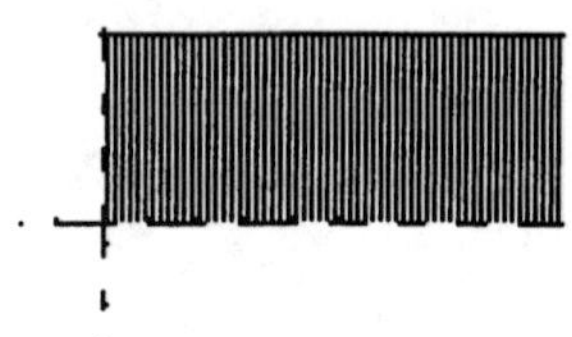

23. $3x + 4y \geq -7$
    $2x - 3y \geq 1$

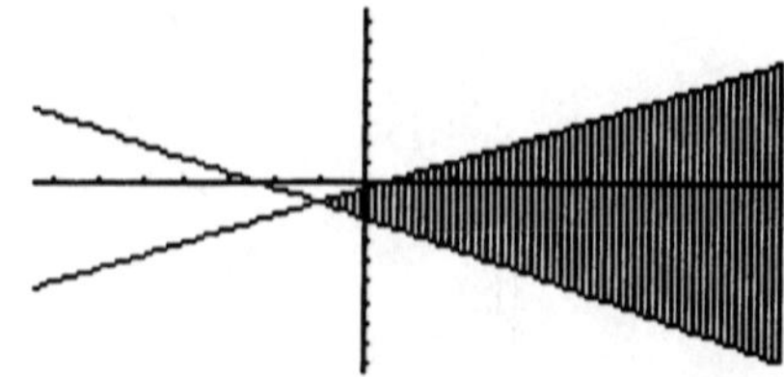

25. $2x + y < 7$
    $y > 2(1 - x)$

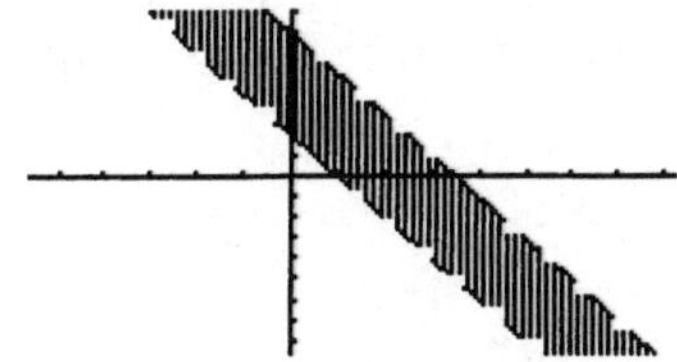

27. $2x - 4y > -6$
    $3x + y \geq 5$

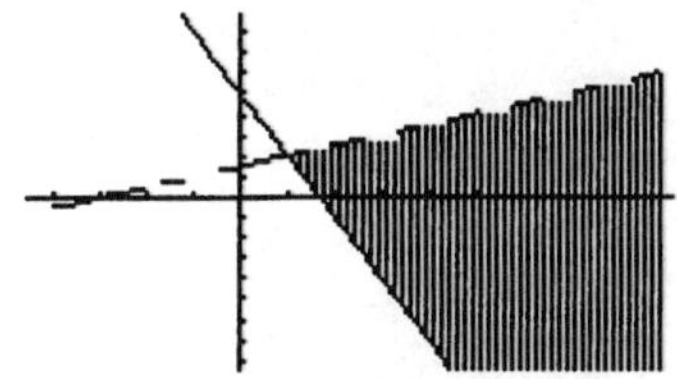

29. $3x - y \leq -4$
    $3y > -2(x + 5)$

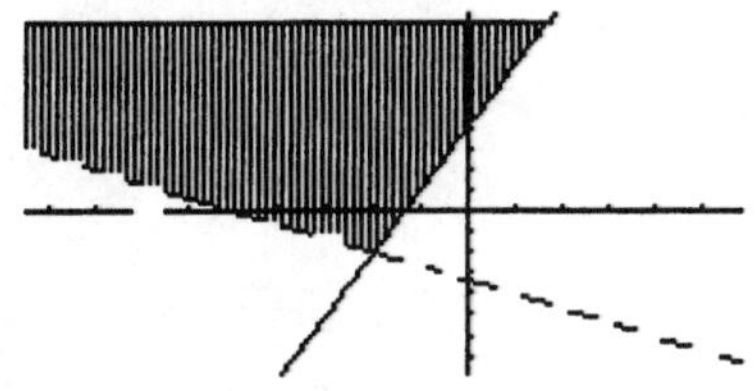

31. $\frac{x}{2} + \frac{y}{3} \geq 2$
    $\frac{x}{2} - \frac{y}{2} < -1$

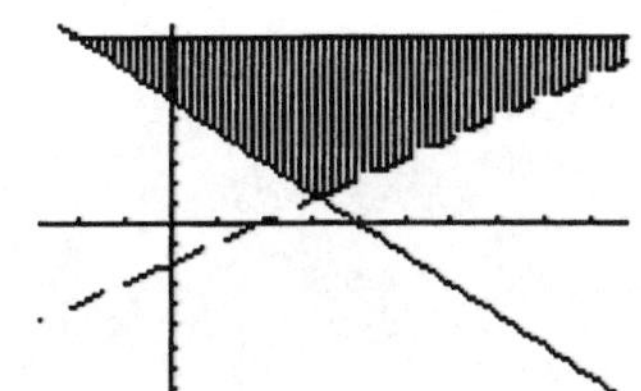

33. $x \geq 0$
    $y \geq 0$
    $x + y \leq 3$

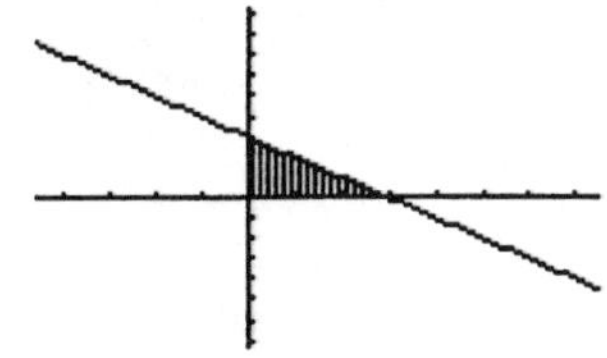

**Applications**

35. BIRDS OF PREY

The area seen by both eyes goes out from the nose at an angle of about $45^0$ in each direction (i.e. the middle section).

37. BUYING COMPACT DISCS

$x$ represents \$10 CD's
$y$ represents \$15 CD's
$10x + 15y \geq 30$
$10x + 15y \leq 60$

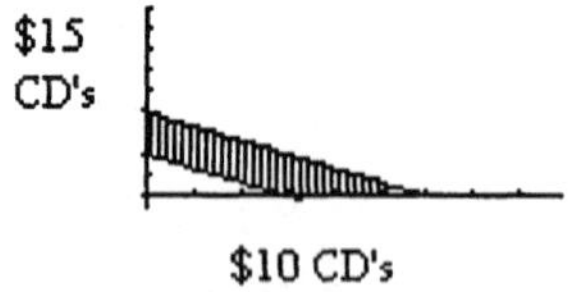


Possible solutions: 1 \$10 CD and 2 \$15 CD's, 4 \$10 CD's and 1 \$15 CD.

39. BUYING FURNITURE

$x$ represents \$150 desk chairs
$y$ represents \$100 side chairs
$150x + 100y \leq 900$
$y > x$

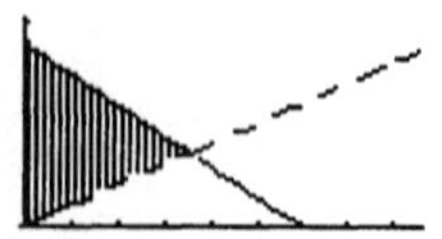


Possible solutions: 2 desk chairs and 4 side chairs, 1 desk chair and 5 side chairs.

41. PESTICIDE

$y \geq -2x + 1$
$y \geq \frac{1}{4}x - 4$

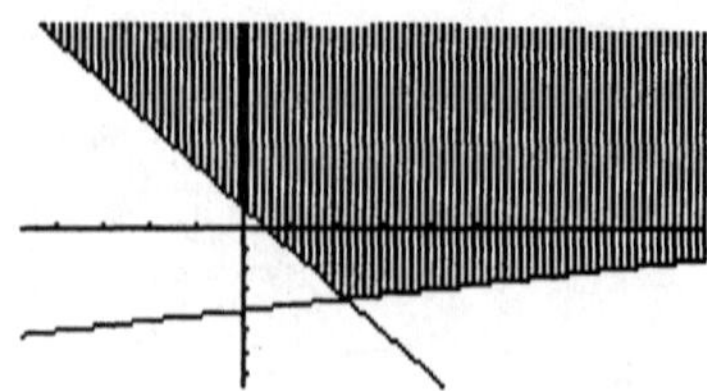

**Writing**

43. Answers will vary
45. Answers will vary

**Review**

47. $y = 2x^2$

| $x$ | $y$ |
| --- | --- |
| 8 | 128 |
| $-2$ | 8 |

49. $f(x) = 4 + x^3$

| Input | Output |
| --- | --- |
| 0 | 4 |
| $-3$ | $-23$ |

1. FOOD SERVICE

| Caterer | Setup | Meal | Equation |
|---|---|---|---|
| Sunshine | $1000 | $4 | $y = 4x + 1000$ |
| Lucy's | $500 | $5 | $y = 5x + 500$ |

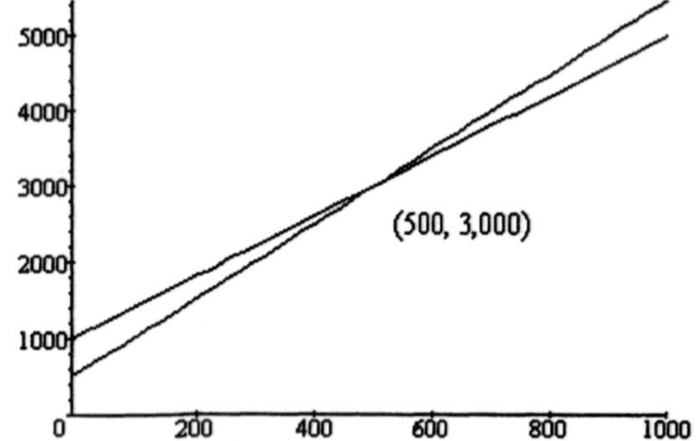



3. $3x + 4y = -7$
$2y - x = -1$
Rewrite
$3x = -7 - 4y$
$x = \frac{-7-4y}{3}$
Substitute
$2y - \frac{-7-4y}{3} = -1$
$3\left(2y - \frac{-7-4y}{3}\right) = 3(-1)$
$6y + 7 + 4y = -3$
$10y = -10$
$y = -1$
Substitute
$3x + 4(-1) = -7$
$3x = -3$
$x = -1$
The solution is $(-1, -1)$.

5. $2x - 3y = -18$
$3x + 2y = -1$

$-3(2x - 3y) = -3(-18)$
$2(3x + 2y) = 2(-1)$

$-6x + 9y = 54$
$\underline{6x + 4y = -2}$
$13y = 52$
$y = 4$

$2x - 3 \cdot 4 = -18$
$2x = -6$
$x = -3$
The solution is $(-3, 4)$.

**Section 13.1**

1.  a. $(2, -3)$ is a solution to
$$3x - 2y = 12$$
$$2x + 3y = -5$$
Since
$$3(2) - 2(-3) \overset{?}{=} 12$$
$$6 + 6 = 12$$
and
$$2(2) + 3(-3) \overset{?}{=} -5$$
$$4 - 9 = -5$$

b. $\left(\frac{7}{2}, -\frac{2}{3}\right)$ is a solution to
$$4x - 6y = 18$$
$$\frac{x}{3} + \frac{y}{2} = \frac{5}{6}$$
Since
$$4\left(\frac{7}{2}\right) - 6\left(-\frac{2}{3}\right) \overset{?}{=} 18$$
$$14 + 4 = 18$$
and
$$\frac{\frac{7}{2}}{3} + \frac{-\frac{2}{3}}{2} \overset{?}{=} \frac{5}{6}$$
$$\frac{7}{6} - \frac{2}{6} = \frac{5}{6}$$

3.  a. $x + y = 7$
$$2x - y = 5$$
$$(4, 3)$$

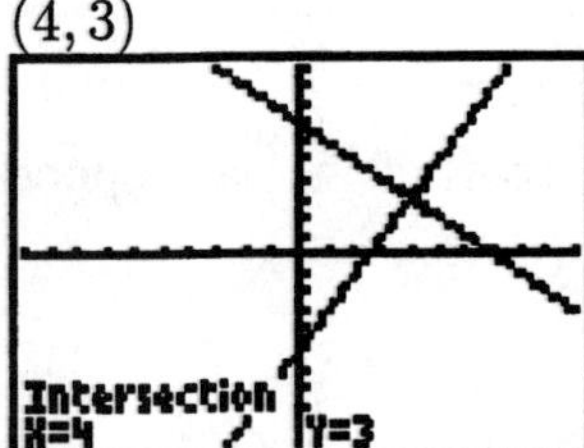


b. $y = -\frac{x}{3}$
$$2x + y = 5$$
$$(3, -1)$$

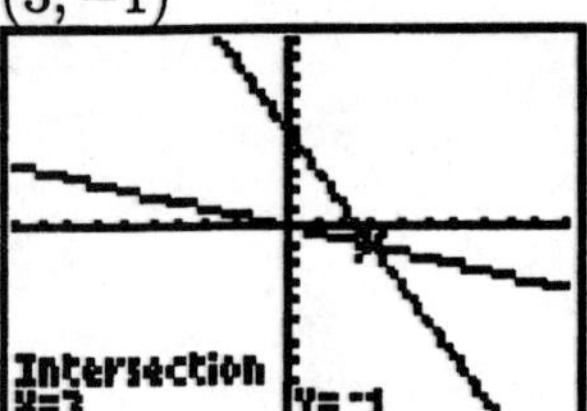

3.  c. $3x + 6y = 6$
$x + 2y = 2$
Dependent System

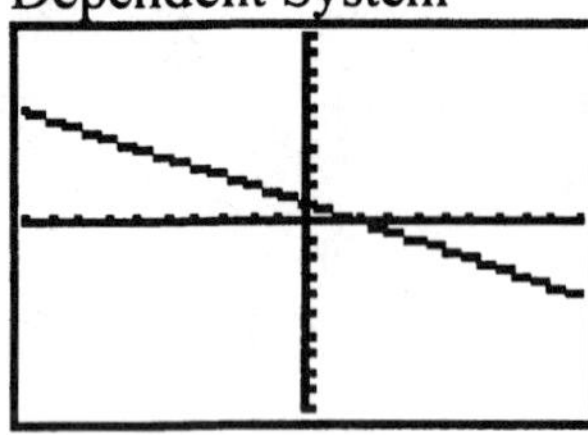

d. $6x + 3y = 12$
$2x + y = 2$
Inconsistent System

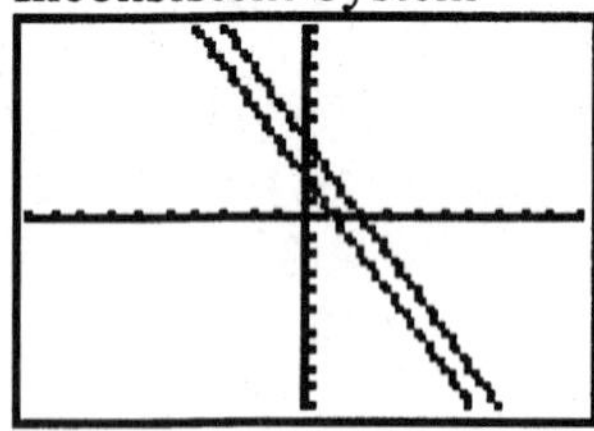

## Section 13.2

5.  a. This system has no solutions.
b. The graph would be two parallel lines
c. This is referred to as an inconsistent system.

## Section 13.3

7.  a. Addition would be a more direct method for solving since the $y$ coefficients have opposite signs and a common multiple of 6.
b. Substitution would be a more direct method since $x$ is given.

## Section 13.4

9.  PAINTING EQUIPMENT
$e$ is the extension
$b$ is the base

$e + b = 35$
$e = b - 7$
Substitute
$(b - 7) + b = 35$
$2b = 42$
$b = 21$
Substitute
$e = 21 - 7$
$e = 14$
The base of the ladder is 21 feet and the extension is 14 feet.

11. **CELEBRITY ENDORSEMENT**

$x$ represents the number of juicers sold

$y_1$ represents the athlete

$y_2$ represents the actor

a. Athlete Earnings: $y_1 = 30,000 + 5x$

Actor Earnings: $y_2 = 20,000 + 10x$

b. When does $y_2 = y_1$?

$30000 + 5x = 20,000 + 10x$

$5x = 10,000$

$x = 2,000$

When 2,000 juicers are sold the actor and athlete would earn the same.

c.

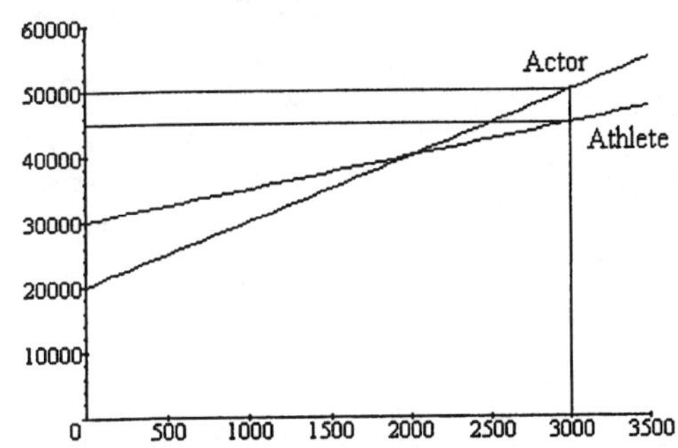


From the graph the athlete would cost less if 3,000 units are sold.

13. **BOATING**

|  | rate | time | distance |
|---|---|---|---|
| upstream | $r = b - c$ | $4 + 3 = 7$ | 56 |
| downstream | $r = b + c$ | 4 | 56 |

$b$ is the boat's rate and $c$ is the current's rate.

$d = rt$ or $r = \frac{d}{t}$

$b + c = \frac{56}{4}$ downstream

$b - c = \frac{56}{7}$ upstream

$b + c = \frac{56}{4}$

$b - c = \frac{56}{7}$

$\overline{\phantom{2b = \frac{56}{4} + \frac{56}{7}}}$

$2b = \frac{56}{4} + \frac{56}{7}$

$2b = 14 + 8$

$b = 11$

$b + c = \frac{56}{4}$

$11 + c = 14$

$c = 3$

The current's rate is 3 miles per hour.

15. INVESTING

|            | Principal  | rate | Interest |
|------------|------------|------|----------|
| certificate | $p$       | 10   | $I_c = 270 - I_p$ |
| passbook   | $3000 - p$ | 6    | $I_p$    |

$I_{\text{total}} = I_{\text{cert}} + I_{\text{pass}}$ where $I_{\text{total}} = 270$

$0.10p = 270 - I_{\text{pass}}$

$0.06(3000 - p) = I_{\text{pass}}$

$0.10p + I_{\text{pass}} = 270$

$180 - 0.06p = I_{\text{pass}}$

$0.10p + I_{\text{pass}} = 270$

$-0.06p - I_{\text{pass}} = -180$

$\overline{0.04p = 90}$

$p = 2250$

Carlos invested \$2250 in the 10% account and $3000 - 2250 = \$750$ in the 6% account.

**Section 13.5**

17. $2x - y \leq -4$
   a. $(0, 5)$ is a solution since $2 \cdot 0 - 5 = -5 \leq -4$
   b. $(2, 8)$ is a solution since $2 \cdot 2 - 8 = -4 \leq -4$
   c. $(-3, -2)$ is a solution since $2 \cdot (-3) - (-2) = 4 \leq -4$
   d. $\left(\frac{1}{2}, -5\right)$ is not a solution since $2 \cdot \frac{1}{2} - (-5) = 7\frac{1}{2} > -4$

19. a. If point A were substituted a true statement would result.
   b. If point B were substituted a false statement would result.
   c. If point C were substituted a false statement would result.

**Section 13.6**

21. a. $5x + 3y < 15$
    $3x - y > 3$

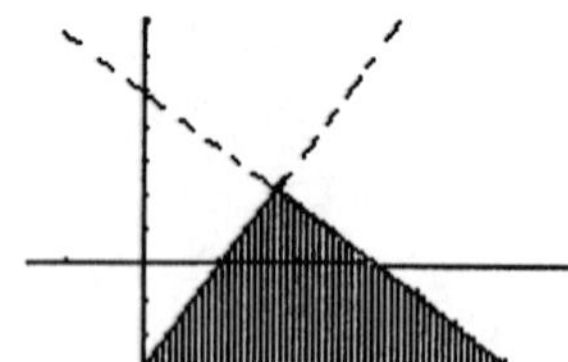

   b. $x \geq 3y$
    $y < 3x$

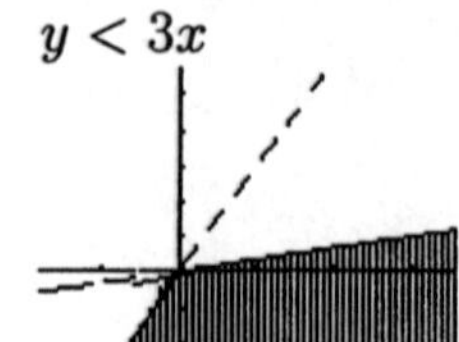

1. $(5, 3)$ is a solution to
$$3x + 2y = 21$$
$$x + y = 8$$
since
$$3 \cdot 5 + 2 \cdot 3 \overset{?}{=} 21$$
$$15 + 6 = 21$$
and
$$5 + 3 = 8$$

3. $3x + y = 7$
$x - 2y = 0$
$(2, 1)$

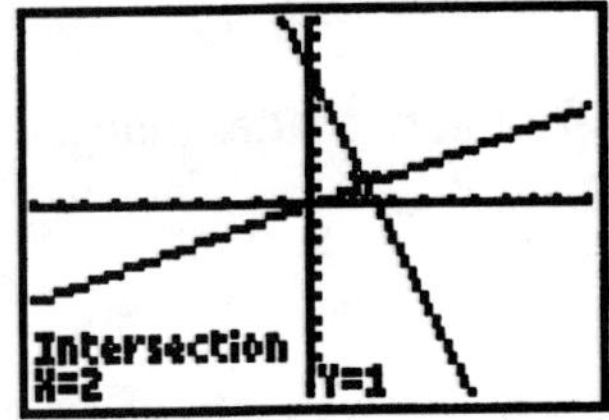


5. $y = x - 1$
$2x + y = -7$

$$2x + (x - 1) = -7$$
$$3x = -6$$
$$x = -2$$

$$y = x - 1$$
$$y = -2 - 1$$
$$y = -3$$
$$(-2, -3)$$

7. $3x - y = 2$
$\underline{2x + y = 8}$
$5x = 10$
$x = 2$

$$3x - y = 2$$
$$3 \cdot 2 - y = 2$$
$$4 = y$$
The solution is $(2, 4)$.

9. $x + y = 4$
$x + y = 6$
This is an inconsistent system since $4 \neq 6$.

11. $5x - 3y = 5$
$3x + 3y = 3$
Addition would be the most efficient method since the $y$ coefficients are opposites and can be eliminated.

13. $2x - 4y > 8$
$(7, 1)$ is a solution since
$2 \cdot 7 - 4 \cdot 1 = 10 > 8$

15. $x - y > -2$

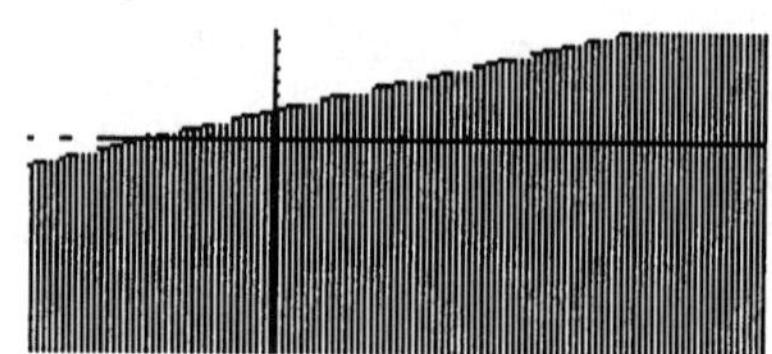

17. The point of intersection is $(30, 3000)$ which is the break even point for the two sales plans. Regardless of which plan, they each pay the same if 30 units are sold.

**Chapters 1 - 13 Cumulative Review**

1. STUDY ABROAD

| Country | 97-98 | 98-99 | % change |
|---------|-------|-------|----------|
| Israel | 1988 | 3302 | $\frac{3302-1988}{1998}*100 \approx 66.1\%$ |
| Mexico | 7574 | 7363 | $\frac{7363-7574}{7574}*100 \approx -2.8\%$ |

3. $3 - 4[-10 - 4(-5)]$
$= 3 - 4[-10 + 20]$
$= 3 - 4(10)$
$= -37$

5. AIR CONDITIONING
$V = \pi r^2 h$ where $r = \frac{1}{2}(6)$ in $= 3(\frac{1}{12})$feet $= \frac{1}{4}$ feet
$V = \pi(\frac{1}{4})^2 6$
$V = \frac{6}{16}\pi$
$V \approx 1.2$ ft$^3$

7. $2 - (4x + 7) = 3 + 2(x + 2)$
$2 - 4x - 7 = 3 + 2x + 4$
$-4x - 5 = 2x + 7$
$-12 = 6x$
$-2 = x$

9. $-4x + 6 > 17$
$-4x > 11$
$x < -\frac{11}{4}$
$(-\infty, -\frac{11}{4})$

11. STOCK MARKET

| % | Amount | Interest |
|---|--------|----------|
| 12 | $x$ | $0.12x$ |
| 6.5 | $45,000 - x$ | $0.065(45,000 - x)$ |

$0.12x + 0.065(45,000 - x) = 4300$
$0.12x + 2925 - 0.065x = 4300$
$0.055x = 1375$
$x = 25,000$
$25,000 was invested in the mutual fund and $20,000 invested in the bonds.

13. $y = -x^3$

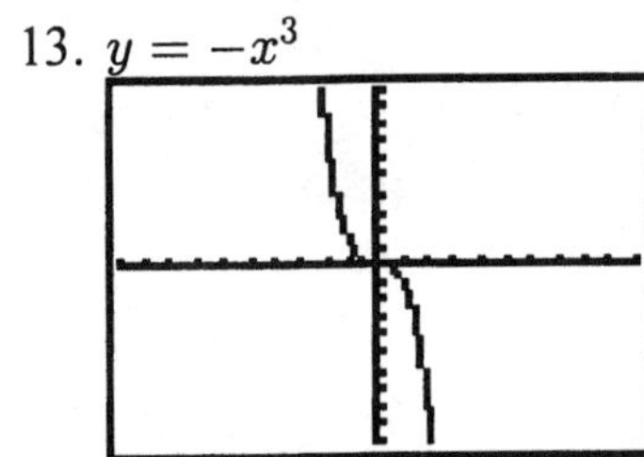

15. $3x + 4y = 8$
$\quad 4y = -3x + 8$
$\quad y = -\frac{3}{4}x + 2$

17. $m = 3,\ (0, -2);\ y = 3x - 2$

19. $f(x) = 4x - 2x^2$
$\quad f(-5) = 4(-5) - 2(-5)^2$
$\quad f(-5) = -20 - 2(25)$
$\quad f(-5) = -20 - 50$
$\quad f(-5) = -70$

21. $(x^5)^2(x^7)^3 = x^{10}x^{21} = x^{31}$

23. $\dfrac{2^{-4}}{3^{-1}} = \dfrac{\frac{1}{16}}{\frac{1}{3}} = \dfrac{1}{16} \cdot \dfrac{3}{1} = \dfrac{3}{16}$

25. $(5x - 8y) - (-2x + 5y)$
$\quad = 5x - 8y + 2x - 5y$
$\quad = 7x - 13y$

27. $(c + 16)^2 = c^2 + 32c + 256$

29. $\dfrac{2x-32}{16x} = \dfrac{2x}{16x} - \dfrac{32}{16x} = \dfrac{1}{8} - \dfrac{2}{x}$

31. $288 = 2^5 \cdot 3^2$

33. $12r^2 - 3rs + 9r^2s^2$
$\quad = 3r(4r - s + 3rs^2)$

35. $2y^2 - 7y + 3$
$\quad = (2y - 1)(y - 3)$

37. $t^3 - v^3$
$\quad = (t - v)(t^2 + tv + v^2)$

39. $8s^2 - 16s = 0$
$\quad 8s(s - 2) = 0$
$\quad 8s = 0 \ \text{ or } \ s - 2 = 0$
$\quad s = 0 \qquad s = 2$

41. $\dfrac{x^2-25}{5x+25} = \dfrac{(x-5)(x+5)}{5(x+5)} = \dfrac{x-5}{5}$

43. $\dfrac{x^2-x-2}{x^2+x} \div \dfrac{2-x}{x}$

$= \dfrac{(x-2)(x+1)}{x(x+1)} \cdot \dfrac{x}{-(x-2)}$

$= -1$

45. $\dfrac{\frac{y}{x}+3y}{y+\frac{2y}{x}}$

$= \dfrac{x}{x}\left(\dfrac{\frac{y}{x}+3y}{y+\frac{2y}{x}}\right)$

$= \dfrac{y+3xy}{xy+2y}$

$= \dfrac{y(1+3x)}{y(x+2)}$

$= \dfrac{3x+1}{x+2}$

47. $\dfrac{12}{a} = \dfrac{15}{20}$

$12(20) = 15a$

$\dfrac{12(20)}{15} = a$

$16 = a$

$\dfrac{b}{6} = \dfrac{20}{15}$

$15b = 6(20)$

$b = \dfrac{6(20)}{15}$

$b = 8$

49. $x + 4y = -2$

$y = -x - 5$

$(-6, 1)$

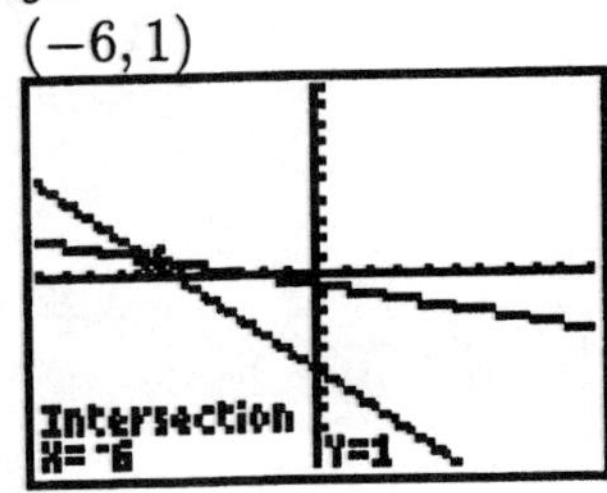

51. $x - 2y = 2$

$2x + 3y = 11$

$x = 2y + 2$

$2x + 3y = 11$

$2(2y + 2) + 3y = 11$

$4y + 4 + 3y = 11$

$7y = 7$

$y = 1$

$x - 2(1) = 2$

$x = 4$

$(x, y) = (4, 1)$

## Section 14.1 Square Roots

### Vocabulary

1. $b$ is a **square** root of $a$ if $b^2 = a$.
3. The principal square root of a positive number is a **positive** number.
5. If a triangle has a right angle, it is called a **right** triangle.

### Concepts

7. The number 25 has two square roots. They are 5 and $-5$.

9. If the length of the hypotenuse of a right triangle is $c$ and the legs are $a$ and $b$, then $c^2 = a^2 + b^2$.

11. If $a$ and $b$ are positive numbers and $a = b$, then $\sqrt{a} = \sqrt{b}$.

13. a. To undo $x$ in $2x = 16$, both sides are divided by 2.
    b. To undo $x$ in $x^2 = 16$, take the positive square root of both sides.

15.

| $x$ | $\sqrt{x}$ |
|---|---|
| 0 | $\sqrt{0} = 0$ |
| $\frac{1}{81}$ | $\sqrt{\frac{1}{81}} = \frac{1}{9}$ |
| 0.16 | $\sqrt{0.16} = 0.4$ |
| 36 | $\sqrt{36} = 6$ |
| 400 | $\sqrt{400} = 20$ |

17. a. The dashed lines help to approximate the $\sqrt{5} \approx 2.2$
    b. $\sqrt{3} \approx 1.7$
    $\sqrt{8} \approx 2.8$

### Notation

19. $c^2 = a^2 + b^2$ where $a = 5, b = 12$
    $c^2 = 5^2 + 12^2$
    $c^2 = 25 + 144$
    $c^2 = 169$
    $c = \sqrt{169}$
    $c = 13$

21. $-\sqrt{9} \neq \sqrt{-9}$ since the left side is a real number, $-\sqrt{9} = -3$, and the right side is not a real number.

### Practice

23. $\sqrt{25} = 5$

25. $-\sqrt{81} = -9$

27. $\sqrt{1.21} = 1.1$

29. $\sqrt{196} = 14$

31. $\sqrt{\frac{9}{256}} = \frac{3}{16}$

33. $-\sqrt{289} = -17$

35. $-\sqrt{2500} = -50$

37. $\sqrt{3600} = 60$

39. $\sqrt{2} \approx 1.414$

41. $\sqrt{11} \approx 3.317$

43. $\sqrt{95} \approx 9.747$

45. $\sqrt{428} \approx 20.688$

47. $-\sqrt{9876} \approx -99.378$

49. $\sqrt{21.35} \approx 4.621$

51. $\sqrt{0.3588} \approx 0.599$

53. $-\sqrt{0.8372} \approx -0.915$

55. $2\sqrt{3} \approx 3.464$

57. $\frac{2+\sqrt{3}}{2} \approx 1.866$

59. $\sqrt{9}$ rational
$\sqrt{17}$ irrational
$\sqrt{49}$ rational
$\sqrt{-49}$ imaginary

61. $f(x) = 1 + \sqrt{x}$

| $x$ | $f(x)$ |
|-----|--------|
| 0 | $1 + \sqrt{0} = 1$ |
| 1 | $1 + \sqrt{1} = 2$ |
| 4 | $1 + \sqrt{4} = 3$ |
| 9 | $1 + \sqrt{9} = 4$ |
| 16 | $1 + \sqrt{16} = 5$ |

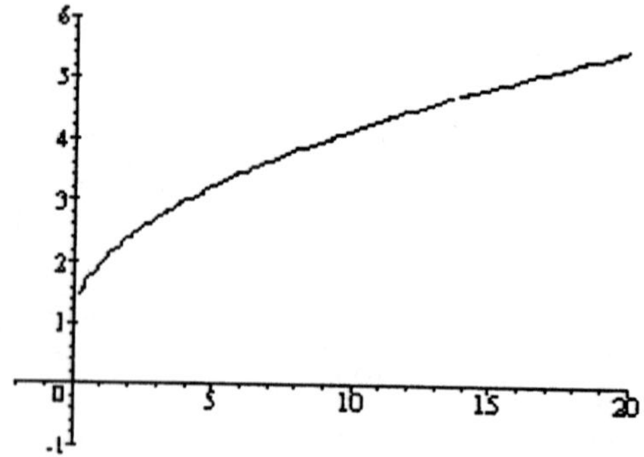

63. $f(x) = -\sqrt{x}$

| $x$ | $f(x)$ |
|-----|--------|
| 0 | $-\sqrt{0} = 0$ |
| 1 | $-\sqrt{1} = -1$ |
| 4 | $-\sqrt{4} = -2$ |
| 9 | $-\sqrt{9} = -3$ |
| 16 | $-\sqrt{16} = -4$ |

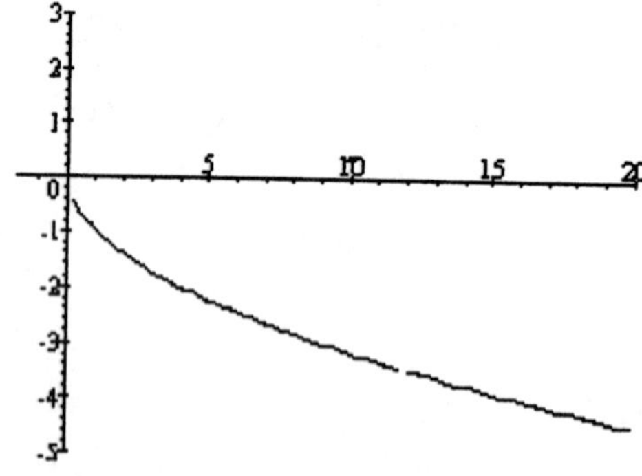

65. $c^2 = a^2 + b^2$

$c^2 = 4^2 + 3^2$

$c^2 = 16 + 9$

$c^2 = 25$

$c = \sqrt{25}$

$c = 5$

67. $c^2 = a^2 + b^2$
$17^2 = 15^2 + b^2$
$17^2 - 15^2 = b^2$
$64 = b^2$
$\sqrt{64} = b$
$8 = b$

69. $c^2 = a^2 + b^2$
$34^2 = a^2 + 16^2$
$34^2 - 16^2 = a^2$
$900 = a^2$
$\sqrt{900} = a$
$30 = a$

71. $c^2 = a^2 + b^2$
$125^2 = 44^2 + b^2$
$125^2 - 44^2 = b^2$
$13{,}689 = b^2$
$\sqrt{13{,}689} = b$
$117 = b$

## Applications

73. ADJUSTING A LADDER
Using $c^2 = a^2 + b^2$ where $c = 20, a = 16$
$20^2 = 16^2 + b^2$
$20^2 - 16^2 = b^2$
$144 = b^2$
$\sqrt{144} = b$
$12 = b$
The base of the ladder is 12 feet from the wall.

75. QUALITY CONTROL
Using the two sides of the square as legs for a right triangle, the carpenter should check to see that the hypotenuse is
$c^2 = 16^2 + 30^2$
$c^2 = 1156$
$c = \sqrt{1156}$
$c = 34$ inches

77. BASEBALL
The distance from home plate to second base is the hypotenuse formed by the right triangle with legs 90 feet long.
$c^2 = 90^2 + 90^2$
$c^2 = 16{,}200$
$c = \sqrt{16{,}200}$
$c \approx 127.28$
The distance from home plate to second base is approximately 127.3 feet.

79. FINDING LOCATION

These archaeologists have formed a right triangle in their exploration,

$c^2 = 4.2^2 + 4.0^2$

$c^2 = 33.64$

$c = \sqrt{33.64}$

$c = 5.8$

As the crow flies, this group is 5.8 miles from their base camp.

81. FOOTBALL

Since ABC represents an isocles triangle we know that the two legs have the same length thus, $a = b$

$c^2 = a^2 + b^2$

$5^2 = 2a^2$

$\frac{25}{2} = a^2$

$3.54 \approx a$

The total yards gained is $6 + 3.54 = 9.54$ yards. The runner did not gain the necessary 10 yards for the first down.

83. PROFESSIONAL WRESTLING

The distance between opposite corners is the hypotenuse of a right triangle with legs 18 feet long,

$c^2 = 18^2 + 18^2$

$c^2 = 2 \cdot (18^2)$

$c^2 = 648$

$c = \sqrt{648}$

$c \approx 25.455844$ feet

The distance between opposite corners is approximately 25.5 feet.

85. HEIGHT OF A TRIANGLE

Area of a triangle is found using the formula $A = \frac{1}{2}bh$. For this triangle the height must first be found using the pythagorean theorem.

$a^2 + b^2 = c^2$

$a^2 + 10^2 = 26^2$

$a^2 = 26^2 - 10^2$

$a^2 = 576$

$a = \sqrt{576}$

$a = 24$

The height of this triangle is 24 feet, now substitute this into the area formula,

$A = \frac{1}{2}(20)(24)$

$A = 240$ in$^2$

87. DRAFTING

a. If the hypotenuse is $\sqrt{2}$ times as long as the leg, and the leg is 6 inches, then the hypotenuse is $c = 6\sqrt{2} \approx 8.48528$ inches

b. If the length of one leg is $\frac{\sqrt{3}}{2}$ times as long as the hypotenuse, then the leg measures,
$b = \frac{\sqrt{3}}{2}(9) \approx 7.794228$ inches

**Writing**

89. Answers will vary
91. Answers will vary

**Review**

93. $(3s^2 - 3s - 2) + (3s^2 + 4s - 3)$
$= 3s^2 - 3s - 2 + 3s^2 + 4s - 3$
$= 6s^2 + s - 5$

95. $(3x - 2)(x + 4)$
$= 3x^2 + 12x - 2x - 8$
$= 3x^2 + 10x - 8$

### Section 14.2 Higher-Order Roots; Radicands That Contain Variables

#### Vocabulary

1. If $p^3 = q$, $p$ is called a **cube** root of $q$.
3. We denote the cube root **function** with the notation $f(x) = \sqrt[3]{x}$.

#### Concepts

5. The **cube** of $-4$ is $-64$, because $(-4)^3 = -64$. The number $-3$ is a cube **root** of $-27$, because $(-3)^3 = -27$.

7. $\sqrt[3]{-216} = -6$, because $(-6)^3 = -216$

9. a. $\sqrt{-125}$ not a real number
   b. $\sqrt[3]{-125} = \sqrt[3]{(-5)^3} = -5$

11. $f(x) = \sqrt[3]{x}$
   a. $f(1) = \sqrt[3]{1} = 1$
   b. $f\left(-\frac{1}{27}\right) = \sqrt[3]{-\frac{1}{27}} = -\frac{1}{3}$
   c. $f(125) = \sqrt[3]{125} = 5$
   d. $f(0.008) = \sqrt[3]{0.008} = 0.2$
   e. $f(1000) = \sqrt[3]{1000} = 10$

#### Notation

13. In the notation $\sqrt[3]{x^6}$, 3 is called the **index** and $x^6$ is called the **radicand**.

15. The "understood" index of the radical expression $\sqrt{55}$ is **2**.

#### Practice

17. $\sqrt[3]{8} = 2$

19. $\sqrt[3]{0} = 0$

21. $\sqrt[3]{-8} = -2$

23. $\sqrt[3]{-64} = -4$

25. $\sqrt[3]{\frac{1}{125}} = \frac{1}{5}$

27. $-\sqrt[3]{-1} = 1$

29. $-\sqrt[3]{64} = -4$

31. $\sqrt[3]{729} = 9$

33. $\sqrt[3]{1,000} = 10$

35. $\sqrt[3]{32,100} \approx 31.78$

37. $\sqrt[3]{-0.11324} \approx -0.48$

39. $f(x) = \sqrt[3]{x} + 1$

| $x$ | $f(x)$ |
|---|---|
| $-8$ | $\sqrt[3]{-8} + 1 = -1$ |
| $-1$ | $\sqrt[3]{-1} + 1 = 0$ |
| $0$ | $\sqrt[3]{0} + 1 = 1$ |
| $1$ | $\sqrt[3]{1} + 1 = 2$ |
| $8$ | $\sqrt[3]{8} + 1 = 3$ |

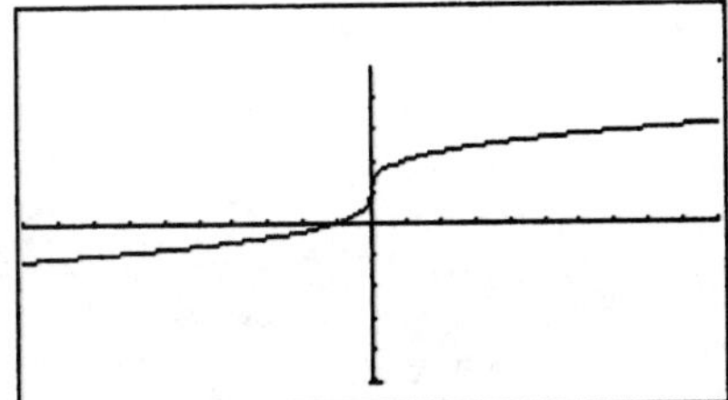

41. $f(x) = -\sqrt[3]{x}$

| $x$ | $f(x)$ |
|---|---|
| $-8$ | $-\sqrt[3]{-8} = 2$ |
| $-1$ | $-\sqrt[3]{-1} = 1$ |
| $0$ | $-\sqrt[3]{0} = 0$ |
| $1$ | $-\sqrt[3]{1} = -1$ |
| $8$ | $-\sqrt[3]{8} = -2$ |

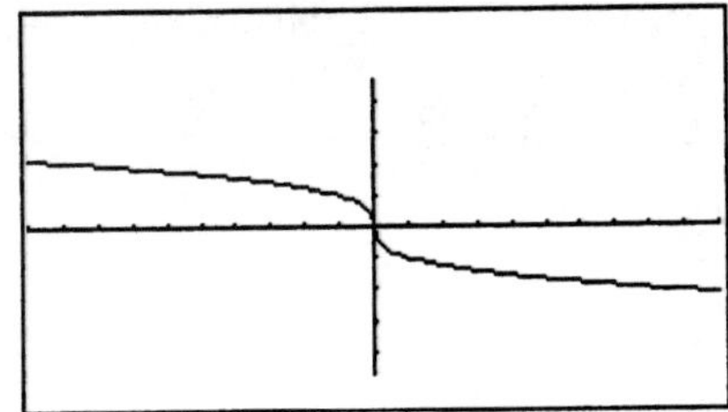

43. $\sqrt[4]{16} = 2$

45. $-\sqrt[5]{32} = -2$

47. $\sqrt[6]{1} = 1$

49. $\sqrt[5]{-32} = -2$

51. $\sqrt[4]{125} \approx 3.34$

53. $\sqrt[5]{-6,000} \approx -5.70$

55. $\sqrt{x^2} = x$

57. $\sqrt{x^6} = x^3$

59. $\sqrt{x^{10}} = x^5$

61. $\sqrt{4z^2} = 2z$

63. $-\sqrt{x^4 y^2} = -x^2 y$

65. $-\sqrt{0.04y^2} = -0.2y$

67. $-\sqrt{25x^4 z^{12}} = -5x^2 z^6$

69. $\sqrt{36z^{36}} = 6z^{18}$

71. $-\sqrt{625z^2} = -25z$

73. $\sqrt[3]{y^6} = y^2$

75. $\sqrt[5]{f^5} = f$

77. $\sqrt[3]{27y^3} = 3y$

79. $\sqrt[3]{-p^6 q^3} = -p^2 q$

81. $\sqrt[4]{x^4} = x$

**Applications**

83. PACKAGING
$$V = s^3$$
$$2 = s^3$$
$$\sqrt[3]{2} = s$$
$$1.2592105 \approx s$$
The length of each side of this box is approximately 1.26 feet.

85. WINDMILLS

$S = \sqrt[3]{\dfrac{P}{0.02}}$

When $P = 400$ watts

$S = \sqrt[3]{\dfrac{400}{0.02}}$

$S = \sqrt[3]{20000}$

$S \approx 27.144176$

When the windmill is producing 400 watts the speed of the wind is approximately 27.14 mph.

87. DEPRECIATION

$r = 1 - \sqrt[n]{\dfrac{S}{C}}$

$S = \$8,000$

$C = \$27,000$

$n = 4$ years

$r = 1 - \sqrt[4]{\dfrac{8,000}{27,000}}$

$r \approx 0.2622121$

The annual depreciation rate for this sound board is approximately 26.22%.

**Writing**

89. Answers will vary

**Review**

91. $m^5 m^2 = m^7$

93. $(3^2)^4 = 3^8$ or $9^4$

95. $(x^2 x^3)^5 = (x^5)^5 = x^{25}$

97. $4x^3(6x^5) = 24x^8$

**Section 14.3 Simplifying Radical Expressions**

**Vocabulary**

1.   Squares of integers such as $4, 9,$ and $16$ are called **perfect** squares.
3.   "To **simplify** $\sqrt{8}$ " means to write it as $2\sqrt{2}$.

**Concepts**

5.   a. The square root of the product of two positive numbers is equal to the **product** of their square roots.
        In symbols $\sqrt{ab} = \sqrt{a}\sqrt{b}$
     b. The square root of the quotient of two positive numbers is equal to the quotient of their square roots.
        In symbols, $\sqrt{\frac{a}{b}} = \frac{\sqrt{a}}{\sqrt{b}}$

7.   Simplifying $\sqrt{20} = \sqrt{16+4}$ is okay, but the error occurs when it is falsely stated that
     $\sqrt{16+4} = \sqrt{16} + \sqrt{4}$

9.   a. If $a = 28$ then from
        $$a = s^2$$
        $$\sqrt{a} = s$$
        $$\sqrt{28} = s$$
        $$\sqrt{4 \cdot 7} = s$$
        $$\sqrt{4}\sqrt{7} = s$$
        $$2\sqrt{7} = s$$
        The length of the side is $2\sqrt{7}$ inches
     b. To the nearest tenth of an inch, the length of the side is $2\sqrt{7} \approx 5.3$ inches.

11.  $\sqrt{b^2 - 4ac}$
     $a = 5, \ b = 10, \ c = 3$
     $\sqrt{10^2 - 4(5)(3)}$
     $= \sqrt{40}$
     $= \sqrt{4 \cdot 10}$
     $= 2\sqrt{10}$

13.  $\sqrt{b^2 - 4ac}$
     $a = -1, b = 6, c = 9$
     $\sqrt{6^2 - 4(-1)(9)}$
     $= \sqrt{72}$
     $= \sqrt{36 \cdot 2}$
     $= 6\sqrt{2}$

**Notation**

15. $\sqrt{80a^3b^2} = \sqrt{16 \cdot 5 \cdot a^2 \cdot a \cdot b^2}$
$\quad = \sqrt{16a^2b^25a}$
$\quad = \sqrt{16a^2b^2}\sqrt{5a}$
$\quad = 4ab\sqrt{5a}$

17. Multiplication is indicated between the two radicals in the expression, $\sqrt{4}\sqrt{3}$.

19. a. $\sqrt{5} \cdot 2 = 2\sqrt{5}$
$\quad$ b. $\sqrt{7a} = a\sqrt{7}$
$\quad$ c. $9\sqrt{x^2}\sqrt{6} = 9x\sqrt{6}$
$\quad$ d. $\sqrt{y}\sqrt{25z^4} = 5z^2\sqrt{y}$

**Practice**

21. $\sqrt{20} = \sqrt{4 \cdot 5} = 2\sqrt{5}$

23. $\sqrt{50} = \sqrt{25 \cdot 2} = 5\sqrt{2}$

25. $\sqrt{45} = \sqrt{9 \cdot 5} = 3\sqrt{5}$

27. $\sqrt{98} = \sqrt{49 \cdot 2} = 7\sqrt{2}$

29. $\sqrt{48} = \sqrt{16 \cdot 3} = 4\sqrt{3}$

31. $-\sqrt{200} = -\sqrt{100 \cdot 2} = -10\sqrt{2}$

33. $\sqrt{192} = \sqrt{64 \cdot 3} = 8\sqrt{3}$

35. $\sqrt{250} = \sqrt{25 \cdot 10} = 5\sqrt{10}$

37. $2\sqrt{24} = 2\sqrt{4 \cdot 6} = 4\sqrt{6}$

39. $-2\sqrt{28} = -2\sqrt{4 \cdot 7} = -4\sqrt{7}$

41. $\sqrt{n^3} = \sqrt{n^2 \cdot n} = n\sqrt{n}$

43. $\sqrt{4k} = \sqrt{4 \cdot k} = 2\sqrt{k}$

45. $\sqrt{12x} = \sqrt{4 \cdot 3x} = 2\sqrt{3x}$

47. $6\sqrt{75t} = 6\sqrt{25 \cdot 3t} = 30\sqrt{3t}$

49. $\sqrt{25x^3} = \sqrt{25x^2 \cdot x} = 5x\sqrt{x}$

51. $\sqrt{a^2b} = \sqrt{a^2 \cdot b} = a\sqrt{b}$

53. $\sqrt{9x^4y} = \sqrt{9x^4 \cdot y} = 3x^2\sqrt{y}$

55. $\frac{1}{5}x^2y\sqrt{50x^2y^2} = \frac{1}{5}x^2y\sqrt{25x^2y^2 \cdot 2}$
$\qquad = \frac{1}{5}x^2y \cdot 5xy\sqrt{2}$
$\qquad = x^3y^2\sqrt{2}$

57. $-12x\sqrt{16x^2y^3} = -12x\sqrt{16x^2y^2 \cdot y}$
$\qquad = -12x \cdot 4xy\sqrt{y}$
$\qquad = -48x^2y\sqrt{y}$

59. $-\frac{2}{5}\sqrt{80mn^4} = -\frac{2}{5}\sqrt{16n^4 \cdot 5m}$
$\qquad = -\frac{2}{5} \cdot 4n^2\sqrt{5m}$
$\qquad = -\frac{8n^2}{5}\sqrt{5m}$

61. $\sqrt{\frac{25}{9}} = \frac{\sqrt{25}}{\sqrt{9}} = \frac{5}{3}$

63. $\sqrt{\frac{81}{64}} = \frac{\sqrt{81}}{\sqrt{64}} = \frac{9}{8}$

65. $\sqrt{\frac{26}{25}} = \frac{\sqrt{26}}{\sqrt{25}} = \frac{\sqrt{26}}{5}$

67. $-\sqrt{\frac{20}{49}} = -\frac{\sqrt{20}}{\sqrt{49}} = -\frac{2\sqrt{5}}{7}$

69. $\sqrt{\frac{48}{81}} = \frac{\sqrt{48}}{\sqrt{81}} = \frac{\sqrt{16 \cdot 3}}{9} = \frac{4\sqrt{3}}{9}$

71. $\sqrt{\frac{32}{25}} = \frac{\sqrt{32}}{\sqrt{25}} = \frac{\sqrt{16 \cdot 2}}{5} = \frac{4\sqrt{2}}{5}$

73. $\sqrt{\frac{72x^3}{y^2}} = \frac{\sqrt{36 \cdot 2 \cdot x^2x}}{\sqrt{y^2}} = \frac{6x\sqrt{2x}}{y}$

75. $\sqrt{\frac{125n^5}{64n}}$
$\qquad = \frac{\sqrt{25 \cdot 5 \cdot n^4n}}{\sqrt{64n}}$
$\qquad = \frac{5n^2\sqrt{5n}}{8\sqrt{n}}$
$\qquad = \frac{5n^2\sqrt{5} \cdot \sqrt{n}}{8\sqrt{n}}$
$\qquad = \frac{5n^2\sqrt{5}}{8}$

77. $\sqrt{\dfrac{128m^3n^5}{81mn^7}}$

$= \dfrac{\sqrt{64m^2n^4 \cdot 2mn}}{\sqrt{81n^6mn}}$

$= \dfrac{8mn^2\sqrt{2mn}}{9n^3\sqrt{mn}}$

$= \dfrac{8mn^2\sqrt{2}\cdot\sqrt{mn}}{9n^3\sqrt{mn}}$

$= \dfrac{8mn^2\sqrt{2}}{9n^3}$

$= \dfrac{8m\sqrt{2}}{9n}$

79. $\sqrt{\dfrac{12r^7s^7}{r^5s^2}}$

$= \dfrac{\sqrt{4r^6s^6 3rs}}{\sqrt{r^4s^2r}}$

$= \dfrac{2r^3s^3\sqrt{3rs}}{r^2s\sqrt{r}}$

$= \dfrac{2r^3s^3\sqrt{3s}\cdot\sqrt{r}}{r^2s\sqrt{r}}$

$= 2rs^2\sqrt{3s}$

81. $\sqrt[3]{24} = \sqrt[3]{8\cdot 3} = 2\sqrt[3]{3}$

83. $\sqrt[3]{-128} = \sqrt[3]{-64\cdot 2} = -4\sqrt[3]{2}$

85. $\sqrt[3]{8x^3} = 2x$

87. $\sqrt[3]{-64x^5} = \sqrt[3]{-64x^3x^2} = -4x\sqrt[3]{x^2}$

89. $\sqrt[3]{54x^3z^6} = \sqrt[3]{27x^3z^6 \cdot 2} = 3xz^2\sqrt[3]{2}$

91. $\sqrt[3]{-81x^2y^3} = \sqrt[3]{-27y^3 \cdot 3x^2} = -3y\sqrt[3]{3x^2}$

93. $\sqrt[3]{\dfrac{27m^3}{8n^6}} = \dfrac{\sqrt[3]{27m^3}}{\sqrt[3]{8n^6}} = \dfrac{3m}{2n^2}$

95. $\sqrt[3]{\dfrac{r^4s^5}{1000t^3}}$

$= \dfrac{\sqrt[3]{r^4s^5}}{\sqrt[3]{1000t^3}}$

$= \dfrac{\sqrt[3]{r^3rs^3s^2}}{\sqrt[3]{1000t^3}}$

$= \dfrac{rs\sqrt[3]{rs^2}}{10t}$

## Applications

97. AMUSEMENT PARK RIDE

$$t = \pi\sqrt{\frac{L}{32}}$$

a. $t = \pi\sqrt{\frac{54}{32}}$

$$t = \pi\frac{\sqrt{9 \cdot 6}}{\sqrt{16 \cdot 2}}$$

$$t = \pi\frac{3\sqrt{6}}{4\sqrt{2}}$$

$$t = \frac{3\pi}{4} \cdot \frac{\sqrt{3}\sqrt{2}}{\sqrt{2}}$$

$$t = \frac{3\pi}{4}\sqrt{3} \text{ seconds}$$

b. To the nearest tenth of a second $t = 4.1$ seconds

99. ARCHAEOLOGY

a. $A(20, 20), B(80, 80)$

$$d = \sqrt{(x_2 - x_1)^2 + (y_2 - y_1)^2}$$

$$d = \sqrt{(80 - 20)^2 + (80 - 20)^2}$$

$$d = \sqrt{(60)^2 + (60)^2}$$

$$d = \sqrt{2(60)^2}$$

$$d = 60\sqrt{2} \text{ centimeters}$$

b. To the nearest tenth the distance is $d \approx 84.9$ centimeters

## Writing

101. Answers will vary

## Review

103. $(-2a^3)(3a^2) = -6a^5$

105. $y - y_1 = m(x - x_1)$
$y - 3 = -2(x - 0)$
$y - 3 = -2x$
$y = -2x + 3$

107. $-x > -5$
$\frac{-x}{-1} < \frac{-5}{-1}$
$x < 5$

**Section 14.4 Adding and Subtracting Radical Expressions**

**Vocabulary**

1.  Like **radicals** have the same index and the same radicand.
3.  Radical expressions such as $\sqrt{8}$ and $\sqrt{18}$ can be **simplified** so that they contain like radicals.

**Concepts**

5.  $5\sqrt{2}$ and $2\sqrt{3}$ do not contain like radicals

7.  $125\sqrt[3]{13a}$ and $-\sqrt[3]{13a}$ contain like radicals

9.  $7\sqrt{5} - 3\sqrt{2} = 4\sqrt{3}$
    These terms cannot be combined since the radicands are different.

11. $7 - 3\sqrt{2} = 4\sqrt{2}$
    Unlike terms cannot be combined.

13.

| $x$ | $\sqrt{x} + \sqrt{3}$ |
|---|---|
| 3 | $\sqrt{3} + \sqrt{3} = 2\sqrt{3}$ |
| 12 | $\sqrt{4 \cdot 3} + \sqrt{3} = 3\sqrt{3}$ |
| 27 | $\sqrt{9 \cdot 3} + \sqrt{3} = 4\sqrt{3}$ |
| 48 | $\sqrt{16 \cdot 3} + \sqrt{3} = 5\sqrt{3}$ |

**Notation**

15. $3\sqrt{80} + 4\sqrt{125}$
$$= 3\sqrt{16 \cdot 5} + 4\sqrt{25 \cdot 5}$$
$$= 3\sqrt{16}\sqrt{5} + 4\sqrt{25}\sqrt{5}$$
$$= 3(4)\sqrt{5} + 4(5)\sqrt{5}$$
$$= 12\sqrt{5} + 20\sqrt{5}$$
$$= 32\sqrt{5}$$

**Practice**

17. $5\sqrt{7} + 4\sqrt{7} = 9\sqrt{7}$

19. $\sqrt{x} - 4\sqrt{x} = -3\sqrt{x}$

21. $5 + 3\sqrt{3} + 3\sqrt{3} = 5 + 6\sqrt{3}$

23. $-1 + 2\sqrt{r} - 3\sqrt{r} = -1 - \sqrt{r}$

25. $\sqrt{12} + \sqrt{27}$
$$= \sqrt{4 \cdot 3} + \sqrt{9 \cdot 3}$$
$$= 2\sqrt{3} + 3\sqrt{3}$$
$$= 5\sqrt{3}$$

27. $\sqrt{18} - \sqrt{8}$
$$= \sqrt{9 \cdot 2} - \sqrt{4 \cdot 2}$$
$$= 3\sqrt{2} - 2\sqrt{2}$$
$$= \sqrt{2}$$

29. $2\sqrt{45} + 2\sqrt{80}$
$$= 2\sqrt{9 \cdot 5} + 2\sqrt{16 \cdot 5}$$
$$= 2(3)\sqrt{5} + 2(4)\sqrt{5}$$
$$= 6\sqrt{5} + 8\sqrt{5}$$
$$= 14\sqrt{5}$$

31. $2\sqrt{80} - 3\sqrt{125}$
$$= 2\sqrt{16 \cdot 5} - 3\sqrt{25 \cdot 5}$$
$$= 2(4)\sqrt{5} - 3(5)\sqrt{5}$$
$$= 8\sqrt{5} - 15\sqrt{5}$$
$$= -7\sqrt{5}$$

33. $\sqrt{20} + \sqrt{180}$
$$= \sqrt{4 \cdot 5} + \sqrt{36 \cdot 5}$$
$$= 2\sqrt{5} + 6\sqrt{5}$$
$$= 8\sqrt{5}$$

35. $\sqrt{12} - \sqrt{48}$
$$= \sqrt{4 \cdot 3} - \sqrt{16 \cdot 3}$$
$$= 2\sqrt{3} - 4\sqrt{3}$$
$$= -2\sqrt{3}$$

37. $\sqrt{288} - 3\sqrt{200}$
$$= \sqrt{144 \cdot 2} - 3\sqrt{100 \cdot 2}$$
$$= 12\sqrt{2} - 30\sqrt{2}$$
$$= -18\sqrt{2}$$

39. $2\sqrt{28} + 2\sqrt{112}$
$$= 2\sqrt{4 \cdot 7} + 2\sqrt{16 \cdot 7}$$
$$= 4\sqrt{7} + 8\sqrt{7}$$
$$= 12\sqrt{7}$$

41. $\sqrt{20} + \sqrt{45} + \sqrt{80}$
$$= \sqrt{4 \cdot 5} + \sqrt{9 \cdot 5} + \sqrt{16 \cdot 5}$$
$$= 2\sqrt{5} + 3\sqrt{5} + 4\sqrt{5}$$
$$= 9\sqrt{5}$$

43. $\sqrt{200} - \sqrt{75} + \sqrt{48}$
$= \sqrt{100 \cdot 2} - \sqrt{25 \cdot 3} + \sqrt{16 \cdot 3}$
$= 10\sqrt{2} - 5\sqrt{3} + 4\sqrt{3}$
$= 10\sqrt{2} - \sqrt{3}$

45. $8\sqrt{6} - 5\sqrt{2} - 3\sqrt{6}$
$= 5\sqrt{6} - 5\sqrt{2}$

47. $\sqrt{24} + \sqrt{150} + \sqrt{240}$
$= \sqrt{4 \cdot 6} + \sqrt{25 \cdot 6} + \sqrt{16 \cdot 15}$
$= 2\sqrt{6} + 5\sqrt{6} + 4\sqrt{15}$
$= 7\sqrt{6} + 4\sqrt{15}$

49. $\sqrt{48} - \sqrt{8} + \sqrt{27} - \sqrt{32}$
$= \sqrt{16 \cdot 3} - \sqrt{4 \cdot 2} + \sqrt{9 \cdot 3} - \sqrt{16 \cdot 2}$
$= 4\sqrt{3} - 2\sqrt{2} + 3\sqrt{3} - 4\sqrt{2}$
$= 7\sqrt{3} - 6\sqrt{2}$

51. $\sqrt{2x^2} + \sqrt{8x^2}$
$= x\sqrt{2} + 2x\sqrt{2}$
$= 3x\sqrt{2}$

53. $\sqrt{2d^3} + \sqrt{8d^3}$
$= d\sqrt{2d} + 2d\sqrt{2d}$
$= 3d\sqrt{2d}$

55. $\sqrt{18x^2 y} - \sqrt{27x^2 y}$
$= 3x\sqrt{2y} - 3x\sqrt{3y}$

57. $\sqrt{32x^5} - \sqrt{18x^5}$
$= \sqrt{16 \cdot 2x^5} - \sqrt{9 \cdot 2x^5}$
$= 4x^2\sqrt{2x} - 3x^2\sqrt{2x}$
$= x^2\sqrt{2x}$

59. $3\sqrt{54b^2} + 5\sqrt{24b^2}$
$= 3\sqrt{9 \cdot 6b^2} + 5\sqrt{4 \cdot 6b^2}$
$= 9b\sqrt{6} + 10b\sqrt{6}$
$= 19b\sqrt{6}$

61. $y\sqrt{490y} - 2\sqrt{360y^3}$
$= y\sqrt{49 \cdot 10y} - 2\sqrt{36 \cdot 10y^3}$
$= 7y\sqrt{10y} - 12y\sqrt{10y}$
$= -5y\sqrt{10y}$

63. $\sqrt{20x^3y} + \sqrt{45x^5y^3} - \sqrt{80x^7y^5}$
$$= \sqrt{4 \cdot 5x^2 \cdot xy} + \sqrt{9 \cdot 5x^4y^2xy} - \sqrt{16 \cdot 5x^6y^4xy}$$
$$= 2x\sqrt{5xy} + 3x^2y\sqrt{5xy} - 4x^3y^2\sqrt{5xy}$$

65. $\sqrt[3]{3} + \sqrt[3]{3} = 2\sqrt[3]{3}$

67. $2\sqrt[3]{x} - 3\sqrt[3]{x} = -\sqrt[3]{x}$

69. $\sqrt[3]{16} + \sqrt[3]{54}$
$$= \sqrt[3]{8 \cdot 2} + \sqrt[3]{27 \cdot 2}$$
$$= 2\sqrt[3]{2} + 3\sqrt[3]{2}$$
$$= 5\sqrt[3]{2}$$

71. $\sqrt[3]{81} - \sqrt[3]{24}$
$$= \sqrt[3]{27 \cdot 3} - \sqrt[3]{8 \cdot 3}$$
$$= 3\sqrt[3]{3} - 2\sqrt[3]{3}$$
$$= \sqrt[3]{3}$$

73. $\sqrt[3]{40} + \sqrt[3]{125}$
$$= \sqrt[3]{8 \cdot 5} + \sqrt[3]{125}$$
$$= 2\sqrt[3]{5} + 5$$

75. $\sqrt[3]{x^4} - \sqrt[3]{x^7}$
$$= \sqrt[3]{x^3x} - \sqrt[3]{x^6x}$$
$$= x\sqrt[3]{x} - x^2\sqrt[3]{x}$$

77. $\sqrt[3]{192x^4y^5} - \sqrt[3]{24x^4y^5}$
$$= xy\sqrt[3]{3 \cdot 64xy^2} - xy\sqrt[3]{8 \cdot 3xy^2}$$
$$= 4xy\sqrt[3]{3xy^2} - 2xy\sqrt[3]{3xy^2}$$
$$= 2xy\sqrt[3]{3xy^2}$$

79. $\sqrt[3]{135x^7y^4} - \sqrt[3]{40x^7y^4}$
$$= x^2y\sqrt[3]{5 \cdot 27xy} - x^2y\sqrt[3]{8 \cdot 5xy}$$
$$= 3x^2y\sqrt[3]{5xy} - 2x^2y\sqrt[3]{5xy}$$
$$= x^2y\sqrt[3]{5xy}$$

**Applications**

81. ANATOMY
The length of the arm will be the sum of the upper arm and the lower arm.
$$2\sqrt{48} + 5\sqrt{12}$$
$$= 2\sqrt{16 \cdot 3} + 5\sqrt{4 \cdot 3}$$
$$= 8\sqrt{3} + 10\sqrt{3}$$
$$= 18\sqrt{3} \text{ inches}$$

83. **READING BLUEPRINTS**

The motor length is $m$

$m = 10\sqrt{50} - (\sqrt{128} + 5\sqrt{18})$

$m = 10\sqrt{25 \cdot 2} - (\sqrt{64 \cdot 2} + 5\sqrt{9 \cdot 2})$

$m = 50\sqrt{2} - (8\sqrt{2} + 15\sqrt{2})$

$m = 50\sqrt{2} - (23\sqrt{2})$

$m = 27\sqrt{2}$ centimeters

85. **FENCING**

The total fencing is the sum of the four sides,

$10\sqrt{150} + 13\sqrt{24} + 9\sqrt{96} + 7\sqrt{54}$

$= 10\sqrt{25 \cdot 6} + 13\sqrt{4 \cdot 6} + 9\sqrt{16 \cdot 6} + 7\sqrt{9 \cdot 6}$

$= 50\sqrt{6} + 26\sqrt{6} + 36\sqrt{6} + 21\sqrt{6}$

$= 133\sqrt{6}$ feet

**Writing**

87. Answers will vary

**Review**

89. $3^{-2} = \frac{1}{3^2} = \frac{1}{9}$

91. $-3^2 = -(3)(3) = -9$

93. $x^{-3} = \frac{1}{x^3}$

95. $3^0 = 1$

**Section 14.5 Multiplying and Dividing Radical Expressions**

**Vocabulary**

1.  The method of changing a radical denominator of a fraction into a rational number is called **rationalizing** the denominator.

3.  $3 + \sqrt{2}$ is the **conjugate** of $3 - \sqrt{2}$.

5.  In the radical expression $3\sqrt{7}$, the number 3 is the **coefficient** of the radical.

**Concepts**

7.  To change $\sqrt{11}$ into a perfect integer square, we multiply it by $\sqrt{11}$.

9.  To rationalize the denominator of $\frac{x}{\sqrt{7}}$ we multiply the numerator and denominator by $\sqrt{7}$.

11. a. $\sqrt{\frac{3}{4}}$ is not in simplified form since the radicand is a fraction.

   b. $\frac{1}{\sqrt{10}}$ is not in simplified form since the denominator contains a radical.

13.

| Rational | Irrational |
|---|---|
| $\frac{\sqrt{5}}{3}$, $\frac{-\sqrt{2}}{8}$, $\frac{1+\sqrt{3}}{4}$ | $\frac{2}{\sqrt{6}}$, $\frac{9}{7-\sqrt{10}}$ |

15. a. $\sqrt{2} + \sqrt{3}$ not possible

   b. $\sqrt{2} \cdot \sqrt{3} = \sqrt{6}$

   c. $\sqrt{2} - \sqrt{3}$ not possible

   d. $\frac{\sqrt{2}}{\sqrt{3}} = \frac{\sqrt{2}}{\sqrt{3}} \cdot \frac{\sqrt{3}}{\sqrt{3}} = \frac{\sqrt{6}}{3}$

   e. $\sqrt{2} + 3\sqrt{2} = 4\sqrt{2}$

   f. $\sqrt{2} \cdot 3\sqrt{2} = 3\sqrt{4} = 3 \cdot 2 = 6$

   g. $\sqrt{2} - 3\sqrt{2} = -2\sqrt{2}$

   h. $\frac{\sqrt{2}}{3\sqrt{2}} = \frac{\sqrt{2}}{3\sqrt{2}} \cdot \frac{\sqrt{2}}{\sqrt{2}} = \frac{2}{6} = \frac{1}{3}$

**Notation**

17. $(\sqrt{x} + \sqrt{2})(\sqrt{x} - 3\sqrt{2})$

$$= \sqrt{x}\sqrt{x} - \sqrt{x} \cdot 3\sqrt{2} + \sqrt{2}\sqrt{x} - \sqrt{2} \cdot 3\sqrt{2}$$
$$= x - 3\sqrt{2x} + \sqrt{2x} - 3 \cdot \sqrt{2}\sqrt{2}$$
$$= x - 2\sqrt{2x} - 3 \cdot 2$$
$$= x - 2\sqrt{2x} - 6$$

**Practice**

19. $\left(\sqrt{5}\right)^2 = 5$

21. $\left(3\sqrt{6}\right)^2 = 9 \cdot 6 = 54$

23. $\sqrt{2}\sqrt{8} = \sqrt{16} = 4$

25. $\sqrt{7}\sqrt{3} = \sqrt{21}$

27. $\sqrt{8}\sqrt{7} = \sqrt{56} = \sqrt{4 \cdot 14} = 2\sqrt{14}$

29. $3\sqrt{2}\sqrt{x} = 3\sqrt{2x}$

31. $\sqrt{x^3}\sqrt{x^5} = \sqrt{x^8} = x^4$

33. $\left(-5\sqrt{6}\right)\left(4\sqrt{3}\right)$
$= -20\sqrt{18}$
$= -20\sqrt{9 \cdot 2}$
$= -60\sqrt{2}$

35. $\left(4\sqrt{x}\right)\left(-2\sqrt{x}\right) = -8\sqrt{x^2} = -8x$

37. $\sqrt{8x}\sqrt{2x^3} = \sqrt{16x^4} = 4x^2$

39. $\sqrt{2}(\sqrt{2} + 1) = \sqrt{2}\sqrt{2} + \sqrt{2} \cdot 1 = 2 + \sqrt{2}$

41. $3\sqrt{3}(\sqrt{27} - 1)$
$= 3\sqrt{81} - 3\sqrt{3}$
$= 3 \cdot 9 - 3\sqrt{3}$
$= 27 - 3\sqrt{3}$

43. $\sqrt{3}(\sqrt{6} + 1)$
$= \sqrt{18} + \sqrt{3}$
$= \sqrt{2 \cdot 9} + \sqrt{3}$
$= 3\sqrt{2} + \sqrt{3}$

45. $\sqrt{x}(\sqrt{3x} - 2)$
$= \sqrt{3x^2} - 2\sqrt{x}$
$= x\sqrt{3} - 2\sqrt{x}$

47. $2\sqrt{x}(\sqrt{9x} + 3)$
$= 2\sqrt{9x^2} + 6\sqrt{x}$
$= 6x + 6\sqrt{x}$

49. $(\sqrt{2} + 1)(\sqrt{2} - 1)$
$= \sqrt{2}\sqrt{2} - 1\sqrt{2} + 1\sqrt{2} - 1 \cdot 1$
$= 2 - 1$
$= 1$

51. $(2\sqrt{7} - x)(3\sqrt{2} + x)$
$$= 2\sqrt{7}\left(3\sqrt{2}\right) + 2\sqrt{7}x - x3\sqrt{2} - x^2$$
$$= 6\sqrt{14} + 2x\sqrt{7} - 3x\sqrt{2} - x^2$$

53. $(\sqrt{6} + 1)^2$
$$= (\sqrt{6} + 1)(\sqrt{6} + 1)$$
$$= \sqrt{6}\sqrt{6} + \sqrt{6} \cdot 1 + 1\sqrt{6} + 1 \cdot 1$$
$$= 6 + 2\sqrt{6} + 1$$
$$= 7 + 2\sqrt{6}$$

55. $(\sqrt{2x} + 3)(\sqrt{8x} - 6)$
$$= \sqrt{2x}\sqrt{8x} - 6\sqrt{2x} + 3\sqrt{8x} - 6 \cdot 3$$
$$= \sqrt{16x^2} - 6\sqrt{2x} + 3\sqrt{8x} - 18$$
$$= 4x - 6\sqrt{2x} + 6\sqrt{2x} - 18$$
$$= 4x - 18$$

57. $\left(-\sqrt[3]{9}\right)^3 = -9$

59. $\left(2\sqrt[3]{4}\right)\left(3\sqrt[3]{3}\right) = 6\sqrt[3]{12}$

61. $\sqrt[3]{7}(\sqrt[3]{49} - 2)$
$$= \sqrt[3]{343} - 2\sqrt[3]{7}$$
$$= 7 - 2\sqrt[3]{7}$$

63. $(\sqrt[3]{2} + 1)(\sqrt[3]{2} + 3)$
$$= \sqrt[3]{2}\sqrt[3]{2} + 3\sqrt[3]{2} + 1\sqrt[3]{2} + 1 \cdot 3$$
$$= \sqrt[3]{4} + 4\sqrt[3]{2} + 3$$

65. $\dfrac{\sqrt{12x^3}}{\sqrt{27x}}$
$$= \dfrac{2x\sqrt{3x}}{3\sqrt{3x}}$$
$$= \dfrac{2x}{3}$$

67. $\dfrac{\sqrt{18x}}{\sqrt{25x}}$
$$= \dfrac{3\sqrt{2x}}{5\sqrt{x}}$$
$$= \dfrac{3\sqrt{2} \cdot \sqrt{x}}{5\sqrt{x}}$$
$$= \dfrac{3\sqrt{2}}{5}$$

69. $\dfrac{\sqrt{196x}}{\sqrt{49x^3}}$
$$= \dfrac{14\sqrt{x}}{7x\sqrt{x}}$$
$$= \dfrac{2}{x}$$

71. $\dfrac{\sqrt[3]{16x^6}}{\sqrt[3]{54x^3}}$

$\quad = \dfrac{x^2\sqrt[3]{16}}{x\sqrt[3]{54}}$

$\quad = \dfrac{2x^2\sqrt[3]{2}}{3x\sqrt[3]{2}}$

$\quad = \dfrac{2x}{3}$

73. $\dfrac{1}{\sqrt{3}} = \dfrac{1}{\sqrt{3}} \cdot \dfrac{\sqrt{3}}{\sqrt{3}} = \dfrac{\sqrt{3}}{3}$

75. $\sqrt{\dfrac{13}{7}} = \dfrac{\sqrt{13}}{\sqrt{7}} \cdot \dfrac{\sqrt{7}}{\sqrt{7}} = \dfrac{\sqrt{91}}{7}$

77. $\dfrac{9}{\sqrt{27}} = \dfrac{9}{3\sqrt{3}} = \dfrac{3}{\sqrt{3}} = \dfrac{3}{\sqrt{3}} \cdot \dfrac{\sqrt{3}}{\sqrt{3}}$

$\quad = \dfrac{3\sqrt{3}}{3} = \sqrt{3}$

79. $\dfrac{3}{\sqrt{32}} = \dfrac{3}{4\sqrt{2}} = \dfrac{3}{4\sqrt{2}} \cdot \dfrac{\sqrt{2}}{\sqrt{2}} = \dfrac{3\sqrt{2}}{8}$

81. $\sqrt{\dfrac{12}{5}} = \dfrac{\sqrt{12}}{\sqrt{5}} = \dfrac{\sqrt{12}}{\sqrt{5}} \cdot \dfrac{\sqrt{5}}{\sqrt{5}}$

$\quad = \dfrac{\sqrt{60}}{5} = \dfrac{2\sqrt{15}}{5}$

83. $\dfrac{10}{\sqrt{x}} = \dfrac{10}{\sqrt{x}} \cdot \dfrac{\sqrt{x}}{\sqrt{x}} = \dfrac{10\sqrt{x}}{x}$

85. $\dfrac{\sqrt{9y}}{\sqrt{2x}} = \dfrac{\sqrt{9y}}{\sqrt{2x}} \cdot \dfrac{\sqrt{2x}}{\sqrt{2x}} = \dfrac{\sqrt{18xy}}{2x} = \dfrac{3\sqrt{2xy}}{2x}$

87. $\dfrac{3}{\sqrt{3}-1}$

$\quad = \dfrac{3}{\sqrt{3}-1} \cdot \dfrac{\sqrt{3}+1}{\sqrt{3}+1}$

$\quad = \dfrac{3\sqrt{3}+3}{3-1}$

$\quad = \dfrac{3\sqrt{3}+3}{2}$

89. $\dfrac{3}{\sqrt{7}+2}$

$\quad = \dfrac{3}{\sqrt{7}+2} \cdot \dfrac{\sqrt{7}-2}{\sqrt{7}-2}$

$\quad = \dfrac{3\sqrt{7}-6}{7-4}$

$\quad = \dfrac{3\sqrt{7}-6}{3}$

$\quad = \sqrt{7} - 2$

91. $\dfrac{12}{3-\sqrt{3}}$

$\quad = \dfrac{12}{3-\sqrt{3}} \cdot \dfrac{3+\sqrt{3}}{3+\sqrt{3}}$

$\quad = \dfrac{36+12\sqrt{3}}{9-3}$

$\quad = \dfrac{36+12\sqrt{3}}{6}$

$\quad = 6 + 2\sqrt{3}$

93. $\dfrac{-\sqrt{3}}{\sqrt{3}+1}$

$= \dfrac{-\sqrt{3}}{\sqrt{3}+1} \cdot \dfrac{\sqrt{3}-1}{\sqrt{3}-1}$

$= \dfrac{-3+\sqrt{3}}{3-1}$

$= \dfrac{-3+\sqrt{3}}{2}$

95. $\dfrac{5}{\sqrt{3}+\sqrt{2}}$

$= \dfrac{5}{\sqrt{3}+\sqrt{2}} \cdot \dfrac{\sqrt{3}-\sqrt{2}}{\sqrt{3}-\sqrt{2}}$

$= \dfrac{5\sqrt{3}-5\sqrt{2}}{3-2}$

$= 5\sqrt{3} - 5\sqrt{2}$

97. $\dfrac{\sqrt{x}+2}{\sqrt{x}-2}$

$= \dfrac{\sqrt{x}+2}{\sqrt{x}-2} \cdot \dfrac{\sqrt{x}+2}{\sqrt{x}+2}$

$= \dfrac{\sqrt{x}\sqrt{x}+2\sqrt{x}+2\sqrt{x}+4}{x-4}$

$= \dfrac{x+4\sqrt{x}+4}{x-4}$

99. $\dfrac{5}{\sqrt[3]{5}} = \dfrac{5}{\sqrt[3]{5}} \cdot \dfrac{\left(\sqrt[3]{5}\right)^2}{\left(\sqrt[3]{5}\right)^2} = \dfrac{5\left(\sqrt[3]{5}\right)^2}{5} = \sqrt[3]{25}$

101. $\dfrac{4}{\sqrt[3]{4}} = \dfrac{4}{\sqrt[3]{4}} \cdot \dfrac{\left(\sqrt[3]{4}\right)^2}{\left(\sqrt[3]{4}\right)^2} = \dfrac{4\left(\sqrt[3]{4}\right)^2}{4} = \sqrt[3]{16} = 2\sqrt[3]{2}$

103. $\dfrac{\sqrt[3]{5}}{\sqrt[3]{2}} = \dfrac{\sqrt[3]{5}}{\sqrt[3]{2}} \cdot \dfrac{\left(\sqrt[3]{2}\right)^2}{\left(\sqrt[3]{2}\right)^2} = \dfrac{\sqrt[3]{5}\left(\sqrt[3]{4}\right)}{2} = \dfrac{\sqrt[3]{20}}{2}$

**Applications**

105. ROTARY LAWNMOWER

$A = \pi r^2$

$r = 6\sqrt{3} \text{ inches}$

$A = \pi\left(6\sqrt{3}\right)^2$

$A = 36 \cdot 3\pi$

$A = 108\pi \text{ in}^2$

107. AIR HOCKEY GAME

$A = lw$

$A = 20\sqrt{3}(30\sqrt{6})$

$A = 600\sqrt{18}$

$A = 600 \cdot 3\sqrt{2}$

$A = 1800\sqrt{2} \text{ in}^2$

The area of this air hockey table is $1800\sqrt{2} \text{ in}^2$.

109. **COSTUME DESIGN**

Using the distance formula $d = \sqrt{(x_2 - x_1)^2 + (y_2 - y_1)^2}$ and labeling the points, starting with the bottom point at $A(0,0)$, moving to the right $B(8,4)$, then moving up, $C(2,16)$, and finally moving to the left, $D(-2,14)$ we have

$$d_{AB} = \sqrt{(8 - 0)^2 + (4 - 0)^2}$$
$$d_{AB} = \sqrt{(8)^2 + (4)^2}$$
$$d_{AB} = \sqrt{80}$$
$$d_{AB} = 4\sqrt{5}$$

$$d_{BC} = \sqrt{(2 - 8)^2 + (16 - 4)^2}$$
$$d_{BC} = \sqrt{(-6)^2 + (12)^2}$$
$$d_{BC} = \sqrt{180}$$
$$d_{BC} = 6\sqrt{5}$$

$$d_{CD} = \sqrt{(-2 - 2)^2 + (14 - 16)^2}$$
$$d_{CD} = \sqrt{(-4)^2 + (-2)^2}$$
$$d_{CD} = \sqrt{20}$$
$$d_{CD} = 2\sqrt{5}$$

$$d_{DA} = \sqrt{(-2 - 0)^2 + (14 - 0)^2}$$
$$d_{DA} = \sqrt{(-2)^2 + (14)^2}$$
$$d_{DA} = \sqrt{200}$$
$$d_{DA} = 10\sqrt{2}$$

Assuming side CD is parallel to side AB, then the area of this trapezoid is found from
$$A = \tfrac{1}{2}d_{BC}(d_{CD} + d_{AB})$$
$$A = \tfrac{1}{2}(6\sqrt{5})(2\sqrt{5} + 4\sqrt{5})$$
$$A = (3\sqrt{5})(6\sqrt{5})$$
$$A = 18 \cdot 5$$
$$A = 90$$
The number of square inches of fabric in the trapezoidal shaped pattern is 90 in$^2$.

**Writing**

111. **Answers will vary**

**Review**

113. $x = -2$ is not a solution to
$$3x - 7 = 5x + 1.$$
$$3(-2) - 7 \overset{?}{=} 5(-2) + 1$$
$$-6 - 7 \overset{?}{=} -10 + 1$$
$$-13 \neq -9$$

115. Addition should be performed first in the expression $2 - (-3 + 4)^2$.

117. $(x-4)(x+4)$
$= x^2 + 4x - 4x - 16$
$= x^2 - 16$

# Section 14.6 Solving Radical Equations; the Distance Formula

## Vocabulary

1. A **radical** equation contains one or more radical expressions with a variable radicand.
3. A false solution that occurs when you square both sides of an equation is called an **extraneous** solution.

## Concepts

5. The squaring property of equality states that if $a = b$ then $a^2 = b^2$.

7. a. To isolate $x$ in $x^2 = 4$ take the positive square root of each side.
   b. To isolate $x$ in $\sqrt{x} = 4$ square both sides of the equation.

9. The error occured when only the left-hand side was squared.

11. On the second line, $\sqrt{a+2}$ was not isolated before squaring both sides.

13. a.

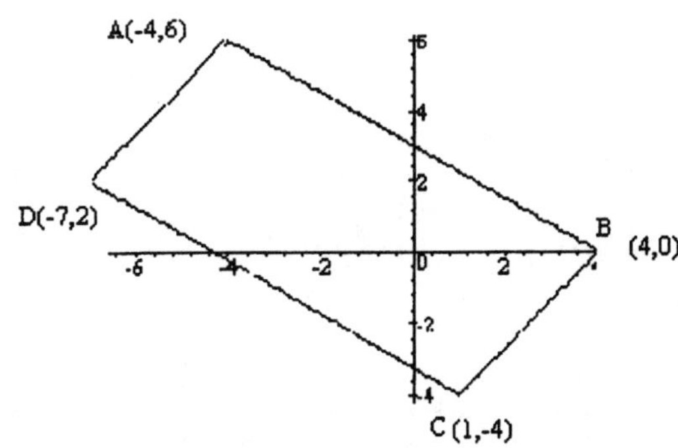


   b. Figure ABCD is a rectangle.

   c.

| Side | Length |
|------|--------|
| AB | $\sqrt{(-4-4)^2 + (6-0)^2} = 10$ |
| BC | $\sqrt{(4-1)^2 + (0-(-4))^2} = 5$ |
| CD | $\sqrt{(1-(-7))^2 + (-4-2)^2} = 10$ |
| DA | $\sqrt{(-7-(-4))^2 + (2-6)^2} = 5$ |

   d. The perimeter is
$$P = 2l + 2w$$
$$P = 2 \cdot 10 + 2 \cdot 5$$
$$P = 20 + 10$$
$$P = 30 \text{ units}$$

**Notation**

15. $\sqrt{x-3} = 5$
$$\left(\sqrt{x-3}\right)^2 = 5^2$$
$$x - 3 = 25$$
$$x = 28$$

**Practice**

17. $\sqrt{x} = 3$
$$(\sqrt{x})^2 = 3^2$$
$$x = 9$$
check
$$\sqrt{9} = 3$$

19. $\sqrt{2a} = 4$
$$(\sqrt{2a})^2 = 4^2$$
$$2a = 16$$
$$a = 8$$
check
$$\sqrt{2 \cdot 8} \overset{?}{=} 4$$
$$\sqrt{16} = 4$$

21. $\sqrt{r} + 4 = 0$
$$\sqrt{r} = -4$$
none

23. $-\sqrt{x} = -5$
$$\frac{-\sqrt{x}}{-1} = \frac{-5}{-1}$$
$$\sqrt{x} = 5$$
$$(\sqrt{x})^2 = 5^2$$
$$x = 25$$
check
$$-\sqrt{25} = -5$$

25. $10 - \sqrt{s} = 7$
$$-\sqrt{s} = -3$$
$$\frac{-\sqrt{s}}{-1} = \frac{-3}{-1}$$
$$\sqrt{s} = 3$$
$$\left(\sqrt{s}\right)^2 = 3^2$$
$$s = 9$$
check
$$10 - \sqrt{9} \overset{?}{=} 7$$
$$10 - 3 \overset{?}{=} 7$$
$$7 = 7$$

27. $\sqrt{x+3} = 2$

$\left(\sqrt{x+3}\right)^2 = 2^2$

$x + 3 = 4$

$x = 1$

check

$\sqrt{1+3} \stackrel{?}{=} 2$

$\sqrt{4} = 2$

29. $\sqrt{3-T} = -2$

$\left(\sqrt{3-T}\right)^2 = (-2)^2$

$3 - T = 4$

$T = -1$

check

$\sqrt{3-(-1)} \stackrel{?}{=} -2$

$\sqrt{4} \neq -2$

none

31. $\sqrt{6+2x} = 4$

$\left(\sqrt{6+2x}\right)^2 = 4^2$

$6 + 2x = 16$

$2x = 10$

$x = 5$

check

$\sqrt{6+2\cdot 5} \stackrel{?}{=} 4$

$\sqrt{16} = 4$

33. $\sqrt{5x-5} - 5 = 0$

$\sqrt{5x-5} = 5$

$\left(\sqrt{5x-5}\right)^2 = 5^2$

$5x - 5 = 25$

$5x = 30$

$x = 6$

check

$\sqrt{5\cdot 6 - 5} - 5 \stackrel{?}{=} 0$

$\sqrt{25} - 5 = 0$

35. $\sqrt{x+3} + 5 = 12$

$\sqrt{x+3} = 7$

$\left(\sqrt{x+3}\right)^2 = 7^2$

$x + 3 = 49$

$x = 46$

check

$\sqrt{46+3} + 5 \stackrel{?}{=} 12$

$\sqrt{49} + 5 = 12$

37. $x - 3 = \sqrt{x^2 - 15}$

$\phantom{37.}\ (x - 3)^2 = \left(\sqrt{x^2 - 15}\right)^2$

$\phantom{37.}\ x^2 - 6x + 9 = x^2 - 15$

$\phantom{37.}\ -6x + 9 = -15$

$\phantom{37.}\ -6x = -24$

$\phantom{37.}\ x = 4$

$\phantom{37.}\ $ check

$\phantom{37.}\ 4 - 3 \overset{?}{=} \sqrt{4^2 - 15}$

$\phantom{37.}\ 1 = \sqrt{1}$

39. $\sqrt{3t - 9} = \sqrt{t + 1}$

$\phantom{39.}\ \left(\sqrt{3t - 9}\right)^2 = \left(\sqrt{t + 1}\right)^2$

$\phantom{39.}\ 3t - 9 = t + 1$

$\phantom{39.}\ 2t - 9 = 1$

$\phantom{39.}\ 2t = 10$

$\phantom{39.}\ t = 5$

$\phantom{39.}\ $ check

$\phantom{39.}\ \sqrt{3 \cdot 5 - 9} \overset{?}{=} \sqrt{5 + 1}$

$\phantom{39.}\ \sqrt{6} = \sqrt{6}$

41. $\sqrt{10 - 3x} = \sqrt{2x + 20}$

$\phantom{41.}\ \left(\sqrt{10 - 3x}\right)^2 = \left(\sqrt{2x + 20}\right)^2$

$\phantom{41.}\ 10 - 3x = 2x + 20$

$\phantom{41.}\ 10 = 5x + 20$

$\phantom{41.}\ -10 = 5x$

$\phantom{41.}\ -2 = x$

$\phantom{41.}\ $ check

$\phantom{41.}\ \sqrt{10 - 3(-2)} \overset{?}{=} \sqrt{2(-2) + 20}$

$\phantom{41.}\ \sqrt{16} = \sqrt{16}$

43. $\sqrt{3c - 8} - \sqrt{c} = 0$

$\phantom{43.}\ \sqrt{3c - 8} = \sqrt{c}$

$\phantom{43.}\ \left(\sqrt{3c - 8}\right)^2 = \left(\sqrt{c}\right)^2$

$\phantom{43.}\ 3c - 8 = c$

$\phantom{43.}\ 2c - 8 = 0$

$\phantom{43.}\ 2c = 8$

$\phantom{43.}\ c = 4$

$\phantom{43.}\ $ check

$\phantom{43.}\ \sqrt{3 \cdot 4 - 8} - \sqrt{4} \overset{?}{=} 0$

$\phantom{43.}\ \sqrt{4} - \sqrt{4} = 0$

45. $x - 1 = \sqrt{x^2 - 4x + 9}$

$(x - 1)^2 = \left(\sqrt{x^2 - 4x + 9}\right)^2$

$x^2 - 2x + 1 = x^2 - 4x + 9$

$-2x + 1 = -4x + 9$

$-2x = -4x + 8$

$2x = 8$

$x = 4$

check

$4 - 1 \overset{?}{=} \sqrt{4^2 - 4 \cdot 4 + 9}$

$3 \overset{?}{=} \sqrt{16 - 16 + 9}$

$3 = \sqrt{9}$

47. $\sqrt{4m^2 + 6m + 6} = -2m$

$\left(\sqrt{4m^2 + 6m + 6}\right)^2 = (-2m)^2$

$4m^2 + 6m + 6 = 4m^2$

$6m + 6 = 0$

$6m = -6$

$m = -1$

check

$\sqrt{4(-1)^2 + 6(-1) + 6} \overset{?}{=} -2(-1)$

$\sqrt{4 - 6 + 6} \overset{?}{=} 2$

$\sqrt{4} = 2$

49. $\sqrt{3x + 3} = 3\sqrt{x - 1}$

$\left(\sqrt{3x + 3}\right)^2 = \left(3\sqrt{x - 1}\right)^2$

$3x + 3 = 9(x - 1)$

$3x + 3 = 9x - 9$

$3 = 6x - 9$

$12 = 6x$

$2 = x$

check

$\sqrt{3 \cdot 2 + 3} \overset{?}{=} 3\sqrt{2 - 1}$

$\sqrt{9} = 3\sqrt{1}$

$3 = 3$

51. $2\sqrt{3x + 4} = \sqrt{5x + 9}$

$\left(2\sqrt{3x + 4}\right)^2 = \left(\sqrt{5x + 9}\right)^2$

$4(3x + 4) = 5x + 9$

$12x + 16 = 5x + 9$

$7x + 16 = 9$

$7x = -7$

$x = -1$

check

$2\sqrt{3(-1) + 4} \overset{?}{=} \sqrt{5(-1) + 9}$

$2\sqrt{1} \overset{?}{=} \sqrt{4}$

$2 = 2$

53. $\sqrt[3]{x} = 7$

$\left(\sqrt[3]{x}\right) = 7^3$

$x = 343$

55. $\sqrt[3]{x - 1} = 4$

$\left(\sqrt[3]{x - 1}\right)^3 = 4^3$

$x - 1 = 64$

$x = 65$

check

$\sqrt[3]{65 - 1} \overset{?}{=} 4$

$\sqrt[3]{64} = 4$

57. $\sqrt[3]{\frac{1}{2}x - 3} = 2$

$\left(\sqrt[3]{\frac{1}{2}x - 3}\right)^3 = 2^3$

$\frac{1}{2}x - 3 = 8$

$\frac{1}{2}x = 11$

$x = 22$

check

$\sqrt[3]{\frac{1}{2}(22) - 3} \overset{?}{=} 2$

$\sqrt[3]{11 - 3} \overset{?}{=} 2$

$\sqrt[3]{8} = 2$

59. $\sqrt[3]{7n - 1} + 1 = 4$

$\sqrt[3]{7n - 1} = 4 - 1$

$\sqrt[3]{7n - 1} = 3$

$\left(\sqrt[3]{7n - 1}\right)^3 = 3^3$

$7n - 1 = 27$

$7n = 28$

$n = 4$

check

$\sqrt[3]{7(4) - 1} + 1 \overset{?}{=} 4$

$\sqrt[3]{27} + 1 \overset{?}{=} 4$

$3 + 1 = 4$

61. $d = \sqrt{(x_2 - x_1)^2 + (y_2 - y_1)^2}$

$P(3, -4) \quad Q(0, 0)$

$d = \sqrt{(3 - 0)^2 + (-4 - 0)^2}$

$d = \sqrt{(3)^2 + (-4)^2}$

$d = \sqrt{9 + 16}$

$d = \sqrt{25}$

$d = 5$

63. $d = \sqrt{(x_2 - x_1)^2 + (y_2 - y_1)^2}$
$P(2, 4) \; Q(5, 9)$
$d = \sqrt{(2 - 5)^2 + (4 - 9)^2}$
$d = \sqrt{(-3)^2 + (-5)^2}$
$d = \sqrt{9 + 25}$
$d = \sqrt{34}$
$d \approx 5.83$

65. $d = \sqrt{(x_2 - x_1)^2 + (y_2 - y_1)^2}$
$P(-2, -8) \; Q(3, 4)$
$d = \sqrt{(-2 - 3)^2 + (-8 - 4)^2}$
$d = \sqrt{(-5)^2 + (-12)^2}$
$d = \sqrt{25 + 144}$
$d = \sqrt{169}$
$d = 13$

67. $d = \sqrt{(x_2 - x_1)^2 + (y_2 - y_1)^2}$
$P(6, 8) \; Q(12, 16)$
$d = \sqrt{(6 - 12)^2 + (8 - 16)^2}$
$d = \sqrt{(-6)^2 + (-8)^2}$
$d = \sqrt{36 + 64}$
$d = \sqrt{100}$
$d = 10$

**Applications**

69. NIAGARA FALLS
$t = \dfrac{\sqrt{s}}{4}$
$3.25 = \dfrac{\sqrt{s}}{4}$
$4(3.25) = \sqrt{s}$
$13 = \sqrt{s}$
$13^2 = \left(\sqrt{s}\right)^2$
$169 = s$
The height of the waterfall is 169 feet.

71. FOUCAULT PENDULUM
$t = 1.11\sqrt{L}$
$8.91 = 1.11\sqrt{L}$
$\dfrac{8.91}{1.11} = \dfrac{1.11\sqrt{L}}{1.11}$
$8.027027 \approx \sqrt{L}$
$8.027027^2 \approx \left(\sqrt{L}\right)^2$
$64.43316 \approx L$
To the nearest tenth of a foot, this pendulum is 64.4 feet long

73. **ROAD SAFETY**

$$s = k\sqrt{d}$$
$$55 = 3.24\sqrt{d}$$
$$\frac{55}{3.24} = \frac{3.24\sqrt{d}}{3.24}$$
$$16.97531 \approx \sqrt{d}$$
$$16.97531^2 \approx \left(\sqrt{d}\right)^2$$
$$288.1611 \approx d$$

With $k = 3.24$ a car traveling 55 mph will skid approximately 288 feet.

75. **SATELLITE ORBITS**

$$\sqrt{r} = \frac{2.029 \times 10^7}{s}$$

With $s = 7 \times 10^3$

$$\sqrt{r} = \frac{2.029 \times 10^7}{7 \times 10^3}$$
$$\sqrt{r} \approx 2898.571429$$
$$\left(\sqrt{r}\right)^2 \approx 2898.571429^2$$
$$r \approx 8,401,716.327$$

Now use the given relationship,

$$r = a + (6.4 \times 10^6)$$
$$a = r - (6.4 \times 10^6)$$
$$a = (8,401,716.327) - (6.4 \times 10^6)$$
$$a = 2,001,716.327$$
$$a \approx 2.001 \times 10^6$$

77. **GEOMETRY**

$$r = \sqrt{\frac{3V}{\pi h}}$$
$$r^2 = \left(\sqrt{\frac{3V}{\pi h}}\right)^2$$
$$r^2 = \frac{3V}{\pi h}$$
$$\pi h r^2 = 3V$$
$$\frac{\pi h r^2}{3} = \frac{3V}{3}$$
$$\frac{\pi h r^2}{3} = V$$

79. NAVIGATION

a. The ordered pair for position one is $(7, 4)$ and the ordered pair on Sicily is $(8, 3)$ so the distance is
$$d = \sqrt{(7 - 8)^2 + (4 - 3)^2}$$
$$d = \sqrt{(-1)^2 + (1)^2}$$
$$d = \sqrt{2}$$
$$d \approx 1.414 \text{ miles}$$

The ordered pair for position one is $(7, 4)$ and the ordered pair on Sardinia is $(6, 5)$ so the distance is
$$d = \sqrt{(7 - 6)^2 + (4 - 5)^2}$$
$$d = \sqrt{(1)^2 + (-1)^2}$$
$$d = \sqrt{2}$$
$$d \approx 1.414 \text{ miles}$$

b. The ordered pair for position two is $(9, 6)$ and the ordered pair on Sicily is $(8, 3)$ so the distance is
$$d = \sqrt{(9 - 8)^2 + (6 - 3)^2}$$
$$d = \sqrt{(1)^2 + (3)^2}$$
$$d = \sqrt{10}$$
$$d \approx 3.2 \text{ miles}$$

The ordered pair for position two is $(9, 6)$ and the ordered pair on Sardinia is $(6, 5)$ so the distance is
$$d = \sqrt{(9 - 6)^2 + (6 - 5)^2}$$
$$d = \sqrt{(3)^2 + (1)^2}$$
$$d = \sqrt{10}$$
$$d \approx 3.2 \text{ miles}$$

## Writing

81. Answers will vary

## Review

83. $(3x^2 + 2x) + (5x^2 - 8x)$
$$= 3x^2 + 2x + 5x^2 - 8x$$
$$= 8x^2 - 6x$$

85. $(x + 3)(x + 3)$
$$= x^2 + 3x + 3x + 9$$
$$= x^2 + 6x + 9$$

87. $(3y - 7)^2 = 9y^2 - 42y + 49$

## Section 14.7 Rational Exponents

### Vocabulary

1. A fractional exponent is also called a **rational** exponent.

### Concepts

3. $x^m x^n = x^{m+n}$

5. $\left(\frac{x}{y}\right)^n = \frac{x^n}{y^n}$

7. $x^{-n} = \frac{1}{x^n}$

9. $x^{1/n} = \sqrt[n]{x}$

11. $\sqrt{5} = 5^{1/2}$

13. $8^{4/3} = \left(\sqrt[3]{8}\right)^4$

15.

| $x$ | $x^{1/2}$ |
|---|---|
| 0 | $(0)^{1/2} = 0$ |
| 1 | $(1)^{1/2} = 1$ |
| 4 | $(4)^{1/2} = 2$ |
| 9 | $(9)^{1/2} = 3$ |

17. $\{8^{1/3}, 17^{1/2}, 2^{3/2}, -5^{2/3}\}$

### Notation

19. $(-216)^{4/3}$
$$= \left(\sqrt[3]{-216}\right)^4$$
$$= (-6)^4$$
$$= 1296$$

### Practice

21. $81^{1/2} = \sqrt{81} = 9$

23. $-144^{1/2} = -\sqrt{144} = -12$

25. $\left(\frac{1}{4}\right)^{1/2} = \sqrt{\frac{1}{4}} = \frac{1}{2}$

27. $\left(\frac{4}{49}\right)^{1/2} = \sqrt{\frac{4}{49}} = \frac{2}{7}$

29. $27^{1/3} = \sqrt[3]{27} = 3$

31. $-125^{1/3} = -\sqrt[3]{125} = -5$

33. $(-8)^{1/3} = \sqrt[3]{(-8)} = -2$

35. $\left(\frac{27}{64}\right)^{1/3} = \frac{\sqrt[3]{27}}{\sqrt[3]{64}} = \frac{3}{4}$

37. $81^{3/2} = \left(\sqrt{81}\right)^3 = 9^3 = 729$

39. $25^{3/2} = \left(\sqrt{25}\right)^3 = 5^3 = 125$

41. $125^{2/3} = \left(\sqrt[3]{125}\right)^2 = 5^2 = 25$

43. $1000^{2/3} = \left(\sqrt[3]{1000}\right)^2 = 10^2 = 100$

45. $(-8)^{2/3} = \left(\sqrt[3]{-8}\right)^2 = (-2)^2 = 4$

47. $\left(\frac{8}{27}\right)^{2/3} = \left(\frac{\sqrt[3]{8}}{\sqrt[3]{27}}\right)^2 = \left(\frac{2}{3}\right)^2 = \frac{4}{9}$

49. $6^{3/5}6^{2/5} = 6^{3/5+2/5} = 6^{5/5} = 6$

51. $5^{2/3}5^{4/3} = 5^{2/3+4/3} = 5^{6/3} = 5^2 = 25$

53. $(7^{2/5})^{5/2} = 7^{\frac{2}{5} \cdot \frac{5}{2}} = 7$

55. $(5^{2/7})^7 = 5^{\frac{2}{7} \cdot 7} = 5^2 = 25$

57. $\frac{8^{3/2}}{8^{1/2}} = 8^{3/2-1/2} = 8^{2/2} = 8$

59. $\frac{5^{11/3}}{5^{2/3}} = 5^{11/3-2/3} = 5^{9/3} = 5^3 = 125$

61. $4^{-1/2} = \frac{1}{4^{1/2}} = \frac{1}{\sqrt{4}} = \frac{1}{2}$

63. $27^{-2/3} = \frac{1}{27^{2/3}} = \frac{1}{\left(\sqrt[3]{27}\right)^2} = \frac{1}{3^2} = \frac{1}{9}$

65. $16^{-3/2} = \frac{1}{16^{3/2}} = \frac{1}{\left(\sqrt{16}\right)^3} = \frac{1}{4^3} = \frac{1}{64}$

67. $(-27)^{-4/3} = \frac{1}{(-27)^{4/3}} = \frac{1}{(\sqrt[3]{-27})^4} = \frac{1}{(-3)^4} = \frac{1}{81}$

69. $(x^{1/2})^2 = x^{\frac{1}{2} \cdot 2} = x$

71. $(x^{12})^{1/6} = x^{12 \cdot \frac{1}{6}} = x^2$

73. $x^{5/6}x^{7/6} = x^{5/6+7/6} = x^{12/6} = x^2$

75. $y^{4/7}y^{10/7} = y^{4/7+1/7} = y^{14/7} = y^2$

77. $\dfrac{x^{3/5}}{x^{1/5}} = x^{3/5-1/5} = x^{2/5}$

79. $\dfrac{x^{1/7}x^{3/7}}{x^{2/7}} = x^{4/7-2/7} = x^{2/7}$

81. $x^{2/3}x^{3/4} = x^{8/12}x^{9/12} = x^{8/12+9/12} = x^{17/12}$

83. $(b^{1/2})^{3/5} = b^{\frac{1}{2} \cdot \frac{3}{5}} = b^{3/10}$

85. $\dfrac{t^{2/3}}{t^{2/5}} = t^{2/3-2/5} = t^{10/15-6/15} = t^{4/15}$

87. $\left(\dfrac{x^{4/5}}{x^{2/15}}\right)^3$

$= \left(x^{4/15-2/15}\right)^3$

$= \left(x^{12/15-2/15}\right)^3$

$= \left(x^{10/15}\right)^3$

$= \left(x^{2/3}\right)^3 = x^{6/3}$

$= x^2$

**Applications**

89. SPEAKERS

$A = V^{2/3}$
$A = (2744)^{2/3}$
$A = (\sqrt[3]{2744})^2$
$A = 14^2$
$A = 196 \text{ in}^2$

Each speaker takes up 196 in$^2$ of floor space, so the pair of speakers take up $2(196) = 392$ in$^2$ of floor space.

91. HOLIDAY DECORATING

$s = (r^2 + h^2)^{1/2}$
$s = \sqrt{10^2 + 24^2}$
$s = \sqrt{100 + 576}$
$s = \sqrt{676}$
$s = 26$

The length $s$ of each string of colored lights is 26 feet.

93. TOY DESIGN

$$r = \left(\frac{3V}{4\pi}\right)^{1/3}$$

$$r = \left(\frac{3(2\pi)}{4\pi}\right)^{1/3}$$

$$r = \left(\frac{6}{4}\right)^{1/3}$$

$$r = \left(\frac{3}{2}\right)^{1/3}$$

$$r \approx 1.1447$$

The radius of this ball is approximately 1.1 inches.

## Writing

95. Answers will vary

## Review

97. $x = 3$

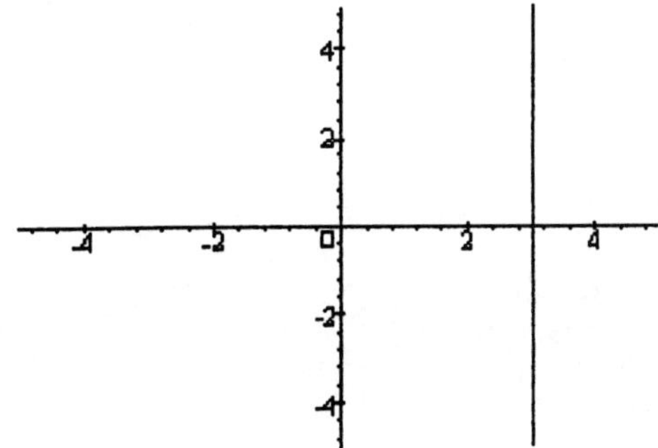

99. $-2x + y = 4$
$\quad y = 2x + 4$

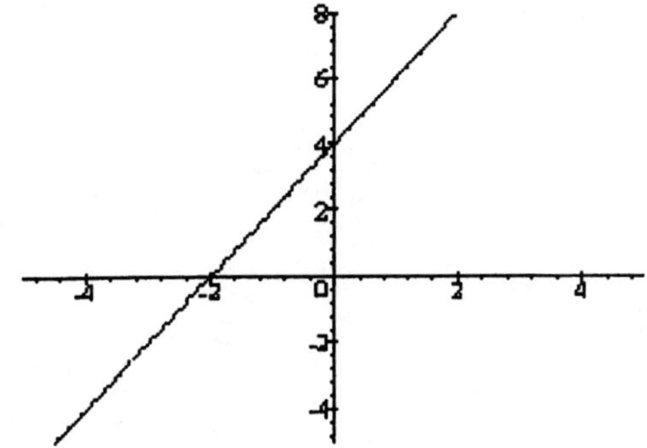

1. $x + 2 = -4$

| operation on variable | inverse operation |
|---|---|
| addition | subtraction |

$x + 2 = -4$
$x + 2 - 2 = -4 - 2$
$x = -6$

3. $-6x = 24$

| operation on variable | inverse operation |
|---|---|
| multiplication | division |

$\frac{-6x}{-6} = \frac{24}{-6}$
$x = -4$

5. $\sqrt{x} = 7$

| operation on variable | inverse operation |
|---|---|
| square root | square |

$\left(\sqrt{x}\right)^2 = 7^2$
$x = 49$

7. $\sqrt[3]{x} = -2$

| operation on variable | inverse operation |
|---|---|
| cube root | cube |

$\left(\sqrt[3]{x}\right)^3 = (-2)^3$
$x = -8$

9. $-2x - 4 = 6$
$-2x = 10$
$x = -5$

11. $\sqrt{x + 1} = 4$
$\left(\sqrt{x + 1}\right)^2 = 4^2$
$x + 1 = 16$
$x = 15$

13. $\sqrt{x} - 3 = 5$
$\sqrt{x} = 8$
$\left(\sqrt{x}\right)^2 = 8^2$
$x = 64$

**Section 14.1**

1.  The **square** of 4 is 16, because $4^2 = 16$; 4 is the **square** root of 16, because $4^2 = 16$.

3.  a. $\sqrt{21} \approx 4.583$
    b. $-\sqrt{15} \approx -3.873$
    c. $2\sqrt{7} \approx 5.292$
    d. $\sqrt{751.9} \approx 27.421$

5.  a. $f(x) = \sqrt{x}$

| $x$ | $\sqrt{x}$ |
|---|---|
| 0 | $\sqrt{0} = 0$ |
| 1 | $\sqrt{1} = 1$ |
| 4 | $\sqrt{4} = 2$ |
| 9 | $\sqrt{9} = 3$ |

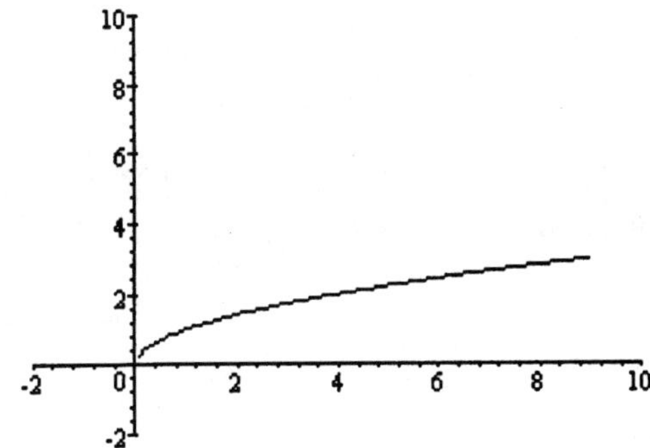

   b. $f(x) = 2 - \sqrt{x}$

| $x$ | $2 - \sqrt{x}$ |
|---|---|
| 0 | $2 - \sqrt{0} = 2$ |
| 1 | $2 - \sqrt{1} = 1$ |
| 4 | $2 - \sqrt{4} = 0$ |
| 9 | $2 - \sqrt{9} = -1$ |

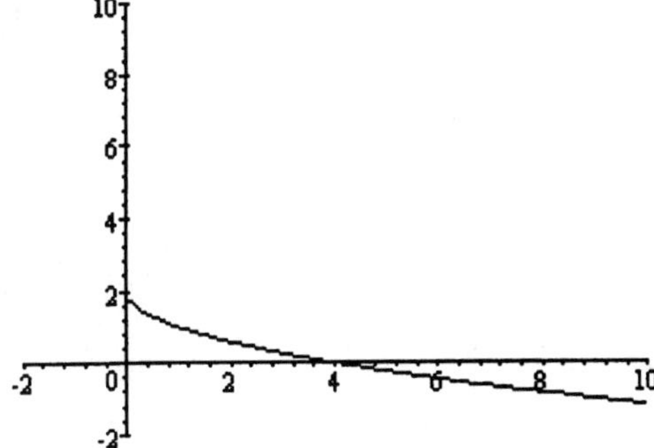

7. THEATER SEATING
$$c^2 = a^2 + b^2$$
$$12.5^2 = 12^2 + b^2$$
$$12.5^2 - 12^2 = b^2$$
$$12.25 = b^2$$
$$\sqrt{12.25} = b$$
$$3.5 = b$$
The seat at the top of the incline is 3.5 feet higher.

**Section 14.2**

9. $\sqrt[3]{125} = 5$, because $5^3 = 125$; 5 is called the **cube** root of 125.

11. a. $\sqrt[3]{16} \approx 2.520$
 b. $\sqrt[3]{-102.35} \approx -4.678$
 c. $\sqrt[4]{6} \approx 1.565$
 d. $\sqrt[5]{34{,}500} \approx 8.083$

13. DICE
$$V = s^3$$
$$1728 = s^3$$
$$\sqrt[3]{1728} = \sqrt[3]{s^3}$$
$$12 = s$$
Dice with a volume of 1728 mm$^3$ have side lengths of 12 mm.

**Section 14.3**

15. a. $\sqrt{\dfrac{16}{25}} = \dfrac{\sqrt{16}}{\sqrt{25}} = \dfrac{4}{5}$

 b. $\sqrt{\dfrac{60}{49}} = \dfrac{\sqrt{60}}{\sqrt{49}} = \dfrac{\sqrt{4 \cdot 15}}{7} = \dfrac{2\sqrt{15}}{7}$

 c. $\sqrt[3]{\dfrac{1000}{27}} = \dfrac{\sqrt[3]{1000}}{\sqrt[3]{27}} = \dfrac{10}{3}$

 d. $\sqrt{\dfrac{242x^4}{169x^2}}$

 $= \sqrt{\dfrac{242x^2}{169}}$

 $= \dfrac{\sqrt{242x^2}}{\sqrt{169}}$

 $= \dfrac{x\sqrt{121 \cdot 2}}{13}$

 $= \dfrac{11x\sqrt{2}}{13}$

17. a. $\sqrt{2} + \sqrt{8} - \sqrt{18}$
    $= \sqrt{2} + 2\sqrt{2} - 3\sqrt{2}$
    $= 3\sqrt{2} - 3\sqrt{2}$
    $= 0$

 b. $\sqrt{3} + 4 + \sqrt{27} - 7$
    $= \sqrt{3} - 3 + 3\sqrt{3}$
    $= 4\sqrt{3} - 3$

 c. $5\sqrt{28} - 3\sqrt{63}$
    $= 5 \cdot 2\sqrt{7} - 3 \cdot 3\sqrt{7}$
    $= 10\sqrt{7} - 9\sqrt{7}$
    $= \sqrt{7}$

 d. $3y\sqrt{5xy^3} - y^2\sqrt{20xy}$
    $= 3y \cdot y\sqrt{5xy} - 2y^2\sqrt{5xy}$
    $= 3y^2\sqrt{5xy} - 2y^2\sqrt{5xy}$
    $= y^2\sqrt{5xy}$

 e. $\sqrt[3]{16} + \sqrt[3]{54}$
    $= \sqrt[3]{8 \cdot 2} + \sqrt[3]{27 \cdot 2}$
    $= 2\sqrt[3]{2} + 3\sqrt[3]{2}$
    $= 5\sqrt[3]{2}$

 f. $\sqrt[3]{2000x^3} - \sqrt[3]{128x^3}$
    $= x\sqrt[3]{2 \cdot 1000} - x\sqrt[3]{64 \cdot 2}$
    $= 10x\sqrt[3]{2} - 4x\sqrt[3]{2}$
    $= 6x\sqrt[3]{2}$

19. GARDENING
    $7\sqrt{45} - 4\sqrt{20}$
    $= 7\sqrt{9 \cdot 5} - 4\sqrt{4 \cdot 5}$
    $= 21\sqrt{5} - 8\sqrt{5}$
    $= 13\sqrt{5}$ inches

21. VACUUM CLEANER NOZZLE
   a. Let $p$ represent perimeter,

$$p = 2l + 2w$$
$$p = 2 \cdot 5\sqrt{3} + 2 \cdot 2\sqrt{6}$$
$$p = \left(10\sqrt{3} + 4\sqrt{6}\right) \text{ inches}$$

Let $a$ represent area

$$a = lw$$
$$a = \left(5\sqrt{3}\right)\left(2\sqrt{6}\right)$$
$$a = 10\sqrt{18}$$
$$a = 10 \cdot 3\sqrt{2}$$
$$a = 30\sqrt{2} \text{ in}^2$$

   b. $p = \left(10\sqrt{3} + 4\sqrt{6}\right)$ inches

$$p \approx 27.1 \text{ inches}$$
$$a = 30\sqrt{2} \text{ in}^2$$
$$a \approx 42.4 \text{ in}^2$$



23. a. $\left(\sqrt{x}\right)^2 = x$
   b. $\left(\sqrt[3]{x}\right)^3 = x$
   c. $\left(2\sqrt{t}\right)^2 = 4t$
   d. $\left(\sqrt{e-1}\right)^2 = e - 1$

25. FERRIS WHEEL

$$t = \sqrt{\frac{d}{16}}$$
$$2 = \sqrt{\frac{d}{16}}$$
$$2^2 = \left(\sqrt{\frac{d}{16}}\right)^2$$
$$4 = \frac{d}{16}$$
$$d = 64$$

If it takes 2 seconds for the coin to hit the ground the ferris wheel is 64 feet high.

27. a. $49^{1/2} = \sqrt{49} = 7$

b. $(-1000)^{1/3} = \sqrt[3]{-1000} = -10$

c. $36^{3/2} = \left(\sqrt{36}\right)^3 = 6^3 = 216$

d. $\left(\frac{8}{27}\right)^{2/3} = \left(\sqrt[3]{\frac{8}{27}}\right)^2 = \left(\frac{2}{3}\right)^2 = \frac{4}{9}$

e. $4^{-3/2} = \frac{1}{4^{3/2}} = \frac{1}{\left(\sqrt{4}\right)^3} = \frac{1}{2^3} = \frac{1}{8}$

f. $8^{2/3} 8^{4/3} = 8^{2/3+4/3} = 8^2 = 64$

g. $\left(3^{2/3}\right)^3 = 3^{\frac{2}{3}\cdot 3} = 3^2 = 9$

h. $\left(a^4 b^8\right)^{-1/2} = \frac{1}{\sqrt{a^4 b^8}} = \frac{1}{a^2 b^4}$

i. $x^{1/3} x^{2/5} = x^{5/15+6/15} = x^{11/15}$

j. $\frac{t^{3/4}}{t^{2/3}} = t^{3/4-2/3} = t^{9/12-8/12} = t^{1/12}$

k. $\frac{x^{2/5} x^{1/5}}{x^{-2/5}} = \frac{x^{3/5}}{x^{-2/5}} = x^{3/5-(-2/5)} = x^{5/5} = x$

l. $\frac{x^{17/7}}{x^{3/7}} = x^{17/7-3/7} = x^{14/7} = x^2$

29. DENTISTRY

16 hours after the original dose there is $\frac{1}{64}$ remaining in the patient's system.

$h^{-3/2} = (16)^{-3/2} = \frac{1}{\left(\sqrt{16}\right)^3} = \frac{1}{4^3} = \frac{1}{64}$

**Chapter 14 Test**

1. $\sqrt{100} = 10$

3. $\sqrt[3]{-27} = -3$

5. $\sqrt{b^2 - 4ac}, a = 2, b = 10, c = 6$
$\sqrt{10^2 - 4(2)(6)}$
$= \sqrt{100 - 48}$
$= \sqrt{52}$
$\approx 7.2$

7. $\sqrt{4x^2} = 2x$

9. $\sqrt{\frac{18x^2y^3}{2xy}} = \sqrt{9xy^2} = 3y\sqrt{x}$

11. a. $A = s^2$, where $A = 24$
$s = \sqrt{A}$
$s = \sqrt{24}$
$s = 2\sqrt{6}$ yards
b. $s = 2\sqrt{6}$ yards
$s \approx 4.9$ yards

13. $\sqrt{8x^3} - x\sqrt{18x}$
$= 2x\sqrt{2x} - 3x\sqrt{2x}$
$= -x\sqrt{2x}$

15. $\sqrt{3}(\sqrt{8} + \sqrt{6})$
$= \sqrt{3}\sqrt{8} + \sqrt{3}\sqrt{6}$
$= \sqrt{24} + \sqrt{18}$
$= 2\sqrt{6} + 3\sqrt{2}$

17. $(2\sqrt{x} + 2)(\sqrt{x} - 3)$
$= 2\sqrt{x}\sqrt{x} - 3(2\sqrt{x}) + 2\sqrt{x} - 3 \cdot 2$
$= 2x - 6\sqrt{x} + 2\sqrt{x} - 6$
$= 2x - 4\sqrt{x} - 6$

19. $\frac{2}{\sqrt{2}} = \frac{2}{\sqrt{2}} \cdot \frac{\sqrt{2}}{\sqrt{2}} = \frac{2\sqrt{2}}{2} = \sqrt{2}$

21. $\sqrt{x} = 15$
$\left(\sqrt{x}\right)^2 = 15^2$
$x = 225$

23. $\sqrt{3x+9} = 2\sqrt{x+1}$
$\left(\sqrt{3x+9}\right)^2 = \left(2\sqrt{x+1}\right)^2$
$3x + 9 = 4(x+1)$
$3x + 9 = 4x + 4$
$x = 5$

25. $d = \sqrt{(x_2 - x_1)^2 + (y_2 - y_1)^2}$
$(-2, -3), (-8, 5)$
$d = \sqrt{(-8 - (-2))^2 + (5 - (-3))^2}$
$d = \sqrt{(-6)^2 + (8)^2}$
$d = \sqrt{36 + 64}$
$d = \sqrt{100}$
$d = 10$

27. $x = 0$ is not a solution to $\sqrt{3x+1} = x - 1$ since $\sqrt{3(0)+1} = (0) - 1$ and $\sqrt{1} \neq -1$

29. CARPENTRY
If the wall is positioned correctly the measurement on the tape should be five feet according to the Pythagorean Theorem.
$3^2 + 4^2 = c^2$
$9 + 16 = c^2$
$25 = c^2$
$\sqrt{25} = \sqrt{c^2}$
$5 = c$

31. $121^{1/2} = \sqrt{121} = 11$

**Chapters 1-14 Cumulative Review**

1. a. All whole numbers are integers is a **true** statement.
   b. $\pi$ is a rational number is **false**, $\pi$ is irrational.
   c. A real number is either rational or irrational is a **true** statement.

3. BACKPACK
   Since the recommendation is for 20% or less, an 85 pound girl should only carry up to $0.20(85) = 17$ pounds in her backpack.

5. $3p - 6(p + z) + p$
   $= 3p - 6p - 6z + p$
   $= -3p - 6z + p$
   $= -2p - 6z$

7. $3 - 3x \geq 6 + x$
   $-3 \geq 4x$
   $-\frac{3}{4} \geq x$ or $x \leq -\frac{3}{4}$
   $\left(-\infty, -\frac{3}{4}\right]$

9. SEARCH AND RESCUE
   Using $d = rt$ where total distance $d = 21$

   |           | rate  | time | distance |
   |-----------|-------|------|----------|
   | Walking   | 2 mph | $t$  | $d_1$    |
   | Horseback | 4 mph | t    | $d_2$    |

   $d_1 + d_2 = 21$
   $2t + 4t = 21$
   $6t = 21$
   $t = \frac{21}{6}$
   $t = 3.5$ hours
   It will take the two teams 3.5 hours to cover 21 miles.

11. SURFACE AREA
    $A = 2lw + 2wh + 2lh$
    $A = 202 \text{ in}^2, l = 9 \text{ in}, w = 5 \text{ in}$
    Solving for $h$ yields,
    $A - 2lw = h(2w + 2l)$
    $\frac{A - 2lw}{2w + 2l} = h$
    Substitution yields,
    $\frac{202 - 2(9)(5)}{2(5) + 2(9)} = h$
    $\frac{202 - 90}{10 + 18} = h$
    $\frac{112}{28} = h$
    $4 = h$
    The height of the box is 4 inches.

13. $y = \frac{1}{2}x$

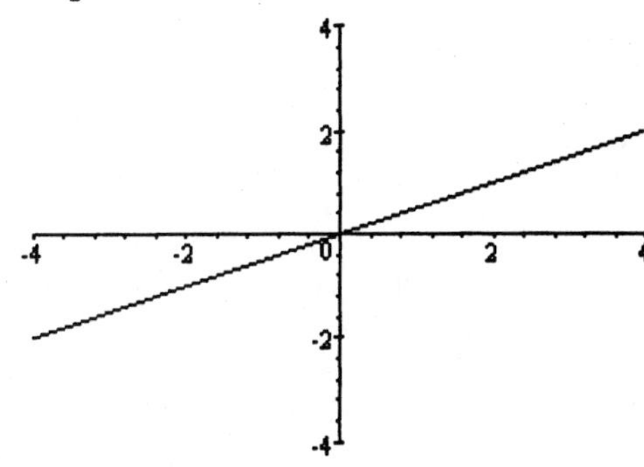

15. $3x + 4y \leq 12$

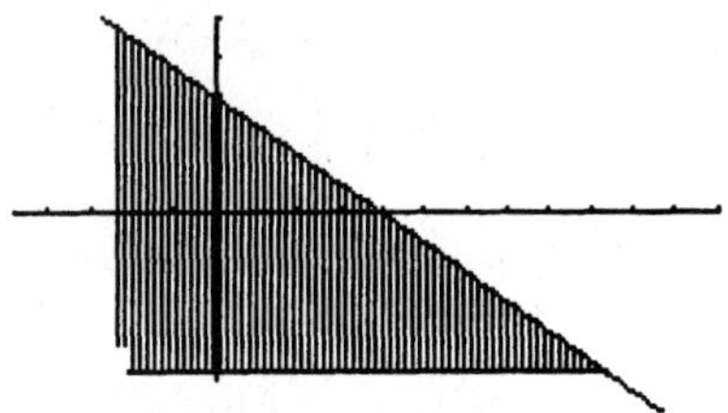

17. a. $y = 3x - 7$ has slope of 3.
    b. $2x + 3y = -10$
       $3y = -2x - 10$
       $y = -\frac{2}{3}x - \frac{10}{3}$
       The slope is $-\frac{2}{3}$.

19. SHOPPING SURGE
The rate of increase over this period of time is the slope of the line between the points $(1991, 724)$ and $(1996, 974)$

$$m = \frac{y_2 - y_1}{x_2 - x_1}$$
$$m = \frac{974 - 724}{1996 - 1991}$$
$$m = \frac{250}{5}$$
$$m = 50$$

The rate of increase of sales over this period is \$50 billion per year.

21. $f(x) = |1 - x|$

| $x$ | $f(x)$ |
|-----|--------|
| 0 | 1 |
| 1 | 0 |
| 2 | 1 |
| 3 | 2 |
| −1 | 2 |
| −2 | 3 |

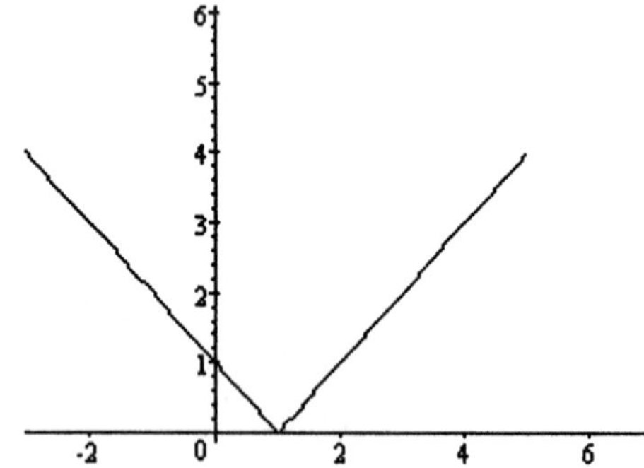

The domain is all real numbers and the range is $y \geq 0$, that is all real numbers greater than or equal to zero.

23. $(x^5)^2(x^7)^3$
$= x^{10}x^{21}$
$= x^{31}$

25. $4^{-3} \cdot 4^{-2} \cdot 4^5$
$= 4^{-5} \cdot 4^5$
$= 4^0$
$= 1$

27. ASTRONOMY

Given that 1 parse $= 3 \times 10^{16}$ meters and the distance to Betelgeuse is $1.6 \times 10^2$ parses, then the distance in meters is $(3 \times 10^{16} \text{ meters/parse})(1.6 \times 10^2 \text{ parse}) = 4.8 \times 10^{18}$ meters

29. $(-r^4 st^2)(2r^2 st)(rst)$
$= 2(-r^4 r^2 r)(sss)(t^2 tt)$
$= -2r^7 s^3 t^4$

31. $(3a^2 - 2a + 4) - (a^2 - 3a + 7)$
$= 3a^2 - 2a + 4 - a^2 + 3a - 7$
$= 2a^2 + a - 3$

33. $\dfrac{4x - 3y + 8z}{4xy}$
$= \dfrac{4x}{4xy} - \dfrac{3y}{4xy} + \dfrac{8z}{4xy}$
$= \dfrac{1}{y} - \dfrac{3}{4x} + \dfrac{2z}{xy}$

35. $3x^2 y - 6xy^2 = 3xy(x - 2y)$

37. $25p^4 - 16q^2$
$$= (5p^2 - 4q)(5p^2 + 4q)$$

39. $x^2 - 11x - 12 = (x - 12)(x + 1)$

41. $6a^2 - 7a - 20 = (2a - 5)(3a + 4)$

43. $x^2 + 3x + 2 = 0$
$$(x + 2)(x + 1) = 0$$
$x + 2 = 0 \quad$ or $\quad x + 1 = 0$
$x = -2 \qquad\qquad x = -1$

45. $6x^2 - x - 2 = 0$
$$(3x - 2)(2x + 1) = 0$$
$3x - 2 = 0 \quad$ or $\quad 2x + 1 = 0$
$3x = 2 \qquad\qquad 2x = -1$
$x = \frac{2}{3} \qquad\qquad x = -\frac{1}{2}$

47. CHILDREN'S STICKER
$A = 20$ cm$^2$
$w = l - 1$
$A = lw$
$A = l(l - 1)$
$20 = l^2 - l$
$l^2 - l - 20 = 0$
$(l - 5)(l + 4) = 0$
$l - 5 = 0 \quad$ or $\quad l + 4 = 0$
$l = 5 \qquad\qquad l = -4$
The sticker is 5 cm long.

49. $\dfrac{x^2 + 2x + 1}{x^2 - 1}$
$$= \frac{(x+1)(x+1)}{(x+1)(x-1)}$$
$$= \frac{x+1}{x-1}$$

51. $\dfrac{p^2 - p - 6}{3p - 9} \div \dfrac{p^2 + 6p + 9}{p^2 - 9}$
$$= \frac{(p-3)(p+2)}{3(p-3)} \div \frac{(p+3)(p+3)}{(p-3)(p+3)}$$
$$= \frac{p+2}{3} \cdot \frac{p-3}{p+3}$$
$$= \frac{(p+2)(p-3)}{3(p+3)}$$

53. $\dfrac{x+2}{x+5} - \dfrac{x-3}{x+7}$
$$= \frac{x+2}{x+5} \cdot \frac{x+7}{x+7} - \frac{x-3}{x+7} \cdot \frac{x+5}{x+5}$$
$$= \frac{x^2 + 9x + 14 - (x^2 + 2x - 15)}{(x+5)(x+7)}$$
$$= \frac{x^2 + 9x + 14 - x^2 - 2x + 15}{(x+5)(x+7)}$$
$$= \frac{7x + 29}{(x+5)(x+7)}$$

55. $\frac{3a}{2b} - \frac{2b}{3a}$

$\quad = \frac{3a}{2b} \cdot \frac{3a}{3a} - \frac{2b}{3a} \cdot \frac{2b}{2b}$

$\quad = \frac{9a^2 - 4b^2}{6ab}$

57. $\frac{4}{a} = \frac{6}{a} - 1$

$\quad -\frac{2}{a} = -1$

$\quad -2 = -a$

$\quad 2 = a$

59. $\frac{1}{r} = \frac{1}{r_1} + \frac{1}{r_2}$

$\quad \frac{1}{r} = \frac{r_2}{r_1 r_2} + \frac{r_1}{r_1 r_2}$

$\quad \frac{1}{r} = \frac{r_2 + r_1}{r_1 r_2}$

$\quad r_1 r_2 = r(r_2 + r_1)$

$\quad \frac{r_1 r_2}{r_2 + r_1} = r$

61. $y = \frac{a}{x}$

$\quad 8 = \frac{a}{2}$

$\quad 16 = a$

When $x = 8$, then

$\quad y = \frac{16}{8}$

$\quad y = 2$

63. A.  $x = y + 4$

B.  $2x + y = 5$

Substitute A into B

$2(y + 4) + y = 5$

$2y + 8 + y = 5$

$3y = -3$

$y = -1$

Substituting $y = -1$ into A

$x = -1 + 4$

$x = 3$

The solution is $(x, y) = (3, -1)$.

65. FINANCIAL PLANNING

|         | Rate | Time | Amount      | Interest            |
|---------|------|------|-------------|---------------------|
| Savings | 6%   | 1 yr | $x$         | $I_s = 540 - I_m$   |
| Mall    | 12%  | 1 yr | $6{,}000 - x$ | $I_m$             |

Total Interest we are told is \$540, which is the sum of the interest from the two accounts.

$$0.06x = 540 - I_m$$
$$0.12(6000 - x) = I_m$$

$$0.06x = 540 - I_m$$
$$720 - 0.12x = I_m$$

$$0.06x + I_m = 540$$
$$\underline{-0.12x - I_m = -720}$$
$$-0.06x = -180$$
$$x = 3000$$
$$6000 - x = 6000 - 3000 = 3000$$

Therefore \$3,000 was invested in the savings account and \$3,000 was invested in the mini-mall project.

67. $\sqrt{\dfrac{49}{225}} = \sqrt{\dfrac{7^2}{15^2}} = \dfrac{7}{15}$

69. $-12x\sqrt{16x^2y^3}$
$= -12x(4xy)\sqrt{y}$
$= -48x^2y\sqrt{y}$

71. $(\sqrt{y} - 4)(\sqrt{y} - 5)$
$= y - 5\sqrt{y} - 4\sqrt{y} + 20$
$= y - 9\sqrt{y} + 20$

73. $\dfrac{4}{\sqrt{20}}$
$= \dfrac{4}{2\sqrt{5}}$
$= \dfrac{2}{\sqrt{5}}$
$= \dfrac{2\sqrt{5}}{5}$

75. $\sqrt{6x + 19} - 5 = 2$
$\sqrt{6x + 19} = 7$
$\left(\sqrt{6x + 19}\right)^2 = 7^2$
$6x + 19 = 49$
$6x = 30$
$x = 5$

**Section 15.1 Completing the Square**

**Vocabulary**

1.  If the polynomial in the equation $ax^2 + bx + c = 0$ doesn't factor, we can solve the equation by **completing** the square.

3.  In the equation $x^2 - 4x + 1 = 0$, the **coefficient** of $x$ is $-4$.

**Concepts**

5.  The equation $x^2 = c$, where $c > 0$, has **two** solutions.

7.  To complete the square on $x^2 + 8x$, we add the **square** of one-half of 8, which is 16.

9.  $x^2 - 2x = 35$
    a. The first step using the factoring method is to subtract 35 from each side.
    b. The first step for completing the square is to add the square of one-half of $-2$ to each side, that is to add 1 to both sides.

11. a. We can't use the factoring method on $x^2 - 2x - 1 = 0$ since it does not factor.
    b. Yes, any quadratic equation can be solved by completing the square.

13. a. $\frac{1}{2}$ of 4 is 2
    b. $\frac{1}{2}$ of $-8$ is $-4$
    c. $\frac{1}{2}$ of 5 is $\frac{5}{2}$
    d. $\frac{1}{2}$ of $-7$ is $-\frac{7}{2}$

15. $2x^2 + 4x - 8 = 0$
    $$\frac{2x^2 + 4x - 8}{2} = \frac{0}{2}$$
    $$x^2 + 2x - 4 = 0$$

**Notation**

17. $(y - 1)^2 = 9$
    $$y - 1 = \sqrt{9} \quad \text{or} \quad y - 1 = -\sqrt{9}$$
    $$y - 1 = 3 \qquad\qquad y - 1 = -3$$
    $$y = 4 \qquad\qquad\quad y = -2$$

19. a. There are two solutions represented by $x = \pm\sqrt{10}$, namely $x = \sqrt{10}$ and $x = -\sqrt{10}$.
    b. The two solutions are $8 + \sqrt{3}$ and $8 - \sqrt{3}$
    $$8 + \sqrt{3} \approx 9.73$$
    $$8 - \sqrt{3} \approx 6.27$$

**Practice**

21. $x^2 = 1$
    $$x = \pm\sqrt{1}$$
    $$x = \pm 1$$

23. $x^2 = 9$
  $x = \pm\sqrt{9}$
  $x = \pm 3$

25. $t^2 = 20$
  $t = \pm\sqrt{20}$
  $t = \pm\sqrt{4 \cdot 5}$
  $t = \pm 2\sqrt{5}$

27. $3m^2 = 27$
  $m^2 = 9$
  $m^2 = \pm\sqrt{9}$
  $m = \pm 3$

29. $4x^2 = 16$
  $x^2 = 4$
  $x^2 = \pm\sqrt{4}$
  $x = \pm 2$

31. $x^2 = \frac{9}{16}$
  $x^2 = \pm\sqrt{\frac{9}{16}}$
  $x = \pm\frac{3}{4}$

33. $(x+1)^2 = 25$
  $\sqrt{(x+1)^2} = \pm\sqrt{25}$
  $x + 1 = 5$ or $x + 1 = -5$
  $x = 4 \qquad x = -6$

35. $(x+2)^2 = 81$
  $\sqrt{(x+2)^2} = \pm\sqrt{81}$
  $x + 2 = 9$ or $x + 2 = -9$
  $x = 7 \qquad x = -11$

37. $(x-2)^2 = 8$
  $\sqrt{(x-2)^2} = \pm\sqrt{8}$
  $x - 2 = 2\sqrt{2}$ or $x - 2 = -2\sqrt{2}$
  $x = 2 + 2\sqrt{2} \qquad x = 2 - 2\sqrt{2}$
  An equivalent answer is $x = 2 \pm 2\sqrt{2}$.

39. $x^2 = 45.82$
  $\sqrt{x^2} = \pm\sqrt{45.82}$
  $x \approx \pm 6.77$

41. $(x+2)^2 = 90.04$
  $\sqrt{(x+2)^2} = \pm\sqrt{90.04}$
  $x + 2 = \pm\sqrt{90.04}$
  $x = -2 \pm\sqrt{90.04}$
  $x \approx 7.49$ or $x \approx -11.49$

43. $y^2 + 4y + 4 = 4$
$(y + 2)^2 = 4$
$\sqrt{(y + 2)^2} = \pm\sqrt{4}$
$y + 2 = \pm 2$
$y + 2 = 2 \qquad \text{or} \qquad y + 2 = -2$
$y = 2 - 2 \qquad \text{or} \qquad y = -2 - 2$
$y = 0 \qquad\qquad\qquad\quad y = -4$

45. $9x^2 - 12x + 4 = 16$
$(3x - 2)^2 = 16$
$\sqrt{(3x - 2)^2} = \pm\sqrt{16}$
$3x - 2 = \pm 4$
$3x = 2 \pm 4$
$x = \frac{2 \pm 4}{3}$
$x = \frac{2}{3} \pm \frac{4}{3}$
$x = \frac{2}{3} + \frac{4}{3} \qquad \text{or} \qquad x = \frac{2}{3} - \frac{4}{3}$
$x = \frac{6}{3} \qquad\qquad\qquad\quad x = -\frac{2}{3}$
$x = 2$

47. $x^2 + 2x$
$x^2 + 2x + \left(\frac{1}{2} \cdot 2\right)^2$
$x^2 + 2x + 1$

49. $x^2 - 4x$
$x^2 - 4x + \left(\frac{1}{2} \cdot -4\right)^2$
$x^2 - 4x + 4$

51. $x^2 + 7x$
$x^2 + 7x + \left(\frac{1}{2} \cdot 7\right)^2$
$x^2 + 7x + \frac{49}{4}$

53. $a^2 - 3a$
$a^2 - 3a + \left(\frac{1}{2} \cdot -3\right)^2$
$a^2 - 3a + \frac{9}{4}$

55. $b^2 + \frac{2}{3}b$
$b^2 + \frac{2}{3}b + \left(\frac{1}{2} \cdot -\frac{2}{3}\right)^2$
$b^2 + \frac{2}{3}b + \frac{4}{36}$
$b^2 + \frac{2}{3}b + \frac{1}{9}$

57. $x^2 + 6x + 8 = 0$
$x^2 + 6x = -8$
$x^2 + 6x + \left(\frac{1}{2} \cdot 6\right)^2 = -8 + \left(\frac{1}{2} \cdot 6\right)^2$
$x^2 + 6x + 9 = -8 + 9$
$(x + 3)^2 = 1$
$x + 3 = \pm\sqrt{1}$
$x = -3 \pm 1$
$x = -3 + 1 \quad \text{or} \quad x = -3 - 1$
$x = -2 \qquad\qquad\quad x = -4$

*Section 15.1 Completing the Square* 546

59. $k^2 - 8k + 12 = 0$
$k^2 - 8k = -12$
$k^2 - 8k + (\frac{1}{2} \cdot -8)^2 = -12 + (\frac{1}{2} \cdot -8)^2$
$k^2 - 8k + 16 = -12 + 16$
$(k - 4)^2 = 4$
$k - 4 = \pm\sqrt{4}$
$k = 4 \pm \sqrt{4}$
$k = 4 + 2$    or    $k = 4 - 2$
$k = 6$          $k = 2$

61. $x^2 - 2x = 15$
$x^2 - 2x + (\frac{1}{2} \cdot -2)^2 = 15 + (\frac{1}{2} \cdot -2)^2$
$x^2 - 2x + 1 = 15 + 1$
$(x - 1)^2 = 16$
$x - 1 = \pm\sqrt{16}$
$x = 1 \pm \sqrt{16}$
$x = 1 + 4$    or    $x = 1 - 4$
$x = 5$          $x = -3$

63. $g^2 + 5g - 6 = 0$
$g^2 + 5g = 6$
$g^2 + 5g + (\frac{1}{2} \cdot 5)^2 = 6 + (\frac{1}{2} \cdot 5)^2$
$g^2 + 5g + \frac{25}{4} = 6 + \frac{25}{4}$
$(g + \frac{5}{2})^2 = \frac{49}{4}$
$g + \frac{5}{2} = \pm\sqrt{\frac{49}{4}}$
$g = -\frac{5}{2} \pm \frac{7}{2}$
$g = -\frac{5}{2} + \frac{7}{2}$   or   $g = -\frac{5}{2} - \frac{7}{2}$
$g = \frac{2}{2}$          $g = -\frac{12}{2}$
$g = 1$          $g = -6$

65. $2x^2 = 4 - 2x$
$2x^2 + 2x = 4$
Dividing both sides by 2 gives $x^2 + x = 2$
$x^2 + x + (\frac{1}{2} \cdot 1)^2 = 2 + (\frac{1}{2} \cdot 1)^2$
$x^2 + x + \frac{1}{4} = 2 + \frac{1}{4}$
$(x + \frac{1}{2})^2 = \frac{9}{4}$
$x + \frac{1}{2} = \pm\sqrt{\frac{9}{4}}$
$x = -\frac{1}{2} \pm \frac{3}{2}$
$x = -\frac{1}{2} + \frac{3}{2}$   or   $x = -\frac{1}{2} - \frac{3}{2}$
$x = \frac{2}{2}$          $x = -\frac{4}{2}$
$x = 1$          $x = -2$

67. $3x^2 + 9x + 6 = 0$

$x^2 + 3x + 2 = 0$

$x^2 + 3x = -2$

$x^2 + 3x + (\frac{1}{2} \cdot 3)^2 = -2 + (\frac{1}{2} \cdot 3)^2$

$x^2 + 3x + \frac{9}{4} = -2 + \frac{9}{4}$

$(x + \frac{3}{2})^2 = \frac{1}{4}$

$x + \frac{3}{2} = \pm\sqrt{\frac{1}{4}}$

$x = -\frac{3}{2} \pm \frac{1}{2}$

$x = -\frac{3}{2} + \frac{1}{2}$ or $\quad x = -\frac{3}{2} - \frac{1}{2}$

$x = -\frac{2}{2} \qquad\qquad x = -\frac{4}{2}$

$x = -1 \qquad\qquad x = -2$

69. $2x^2 = 3x + 2$

$2x^2 - 3x = 2$

$x^2 - \frac{3}{2}x = 1$

$x^2 - \frac{3}{2}x + (\frac{1}{2} \cdot -\frac{3}{2})^2 = 1 + (\frac{1}{2} \cdot -\frac{3}{2})^2$

$x^2 - \frac{3}{2}x + \frac{9}{16} = 1 + \frac{9}{16}$

$(x - \frac{3}{4})^2 = \frac{25}{16}$

$x - \frac{3}{4} = \pm\sqrt{\frac{25}{16}}$

$x = \frac{3}{4} \pm \frac{5}{4}$

$x = \frac{3}{4} + \frac{5}{4}$ or $\quad x = \frac{3}{4} - \frac{5}{4}$

$x = \frac{8}{4} \qquad\qquad x = -\frac{2}{4}$

$x = 2 \qquad\qquad x = -\frac{1}{2}$

71. $4x^2 = 2 - 7x$

$x^2 = \frac{1}{2} - \frac{7}{4}x$

$x^2 + \frac{7}{4}x = \frac{1}{2}$

$x^2 + \frac{7}{4}x + (\frac{1}{2} \cdot \frac{7}{4})^2 = \frac{1}{2} + (\frac{1}{2} \cdot \frac{7}{4})^2$

$x^2 + \frac{7}{4}x + \frac{49}{64} = \frac{1}{2} + \frac{49}{64}$

$(x + \frac{7}{8})^2 = \frac{81}{64}$

$x + \frac{7}{8} = \pm\sqrt{\frac{81}{64}}$

$x = -\frac{7}{8} \pm \frac{9}{8}$

$x = -\frac{7}{8} + \frac{9}{8}$ or $\quad x = -\frac{7}{8} - \frac{9}{8}$

$x = \frac{2}{8} \qquad\qquad x = -\frac{16}{8}$

$x = \frac{1}{4} \qquad\qquad x = -2$

73. $x^2 + 4x + 1 = 0$

$x^2 + 4x = -1$

$x^2 + 4x + (\frac{1}{2} \cdot 4)^2 = -1 + (\frac{1}{2} \cdot 4)^2$

$x^2 + 4x + 4 = -1 + 4$

$(x + 2)^2 = 3$

$x + 2 = \pm\sqrt{3}$

$x = -2 \pm \sqrt{3}$

$x \approx -0.27$ or $\quad x \approx -3.73$

75. $x^2 - 2x - 4 = 0$
$x^2 - 2x = 4$
$x^2 - 2x + (\frac{1}{2} \cdot -2)^2 = 4 + (\frac{1}{2} \cdot -2)^2$
$x^2 - 2x + 1 = 4 + 1$
$(x - 1)^2 = 5$
$x - 1 = \pm\sqrt{5}$
$x = 1 \pm \sqrt{5}$
$x \approx 3.24 \quad$ or $\quad x \approx -1.24$

77. $x^2 = 4x + 3$
$x^2 - 4x = 3$
$x^2 - 4x + (\frac{1}{2} \cdot -4)^2 = 3 + (\frac{1}{2} \cdot -4)^2$
$x^2 - 4x + 4 = 3 + 4$
$(x - 2)^2 = 7$
$x - 2 = \pm\sqrt{7}$
$x = 2 \pm \sqrt{7}$
$x \approx 4.65 \quad$ or $\quad x \approx -0.65$

79. $4x^2 + 4x + 1 = 20$
$4x^2 + 4x = 19$
$x^2 + x = \frac{19}{4}$
$x^2 + x + (\frac{1}{2} \cdot 1)^2 = \frac{19}{4} + (\frac{1}{2} \cdot 1)^2$
$x^2 + x + \frac{1}{4} = \frac{19}{4} + \frac{1}{4}$
$(x + \frac{1}{2})^2 = \frac{20}{4}$
$x + \frac{1}{2} = \pm\sqrt{\frac{20}{4}}$
$x + \frac{1}{2} = \pm\frac{2\sqrt{5}}{2}$
$x = -\frac{1}{2} \pm \frac{2\sqrt{5}}{2} = \frac{-1 \pm 2\sqrt{5}}{2} = -\frac{1}{2} \pm \sqrt{5}$
$x \approx 1.74 \quad$ or $\quad x \approx -2.74$

81. $2x(x + 3) = 8$
$2x^2 + 6x - 8 = 0$
$x^2 + 3x - 4 = 0$
$x^2 + 3x = 4$
$x^2 + 3x + (\frac{1}{2} \cdot 3)^2 = 4 + (\frac{1}{2} \cdot 3)^2$
$x^2 + 3x + \frac{9}{4} = 4 + \frac{9}{4}$
$(x + \frac{3}{2})^2 = \frac{25}{4}$
$x + \frac{3}{2} = \pm\sqrt{\frac{25}{4}}$
$x = -\frac{3}{2} \pm \frac{5}{2}$
$x = -\frac{3}{2} + \frac{5}{2} \quad$ or $\quad x = -\frac{3}{2} - \frac{5}{2}$
$x = \frac{2}{2} \qquad\qquad\qquad x = -\frac{8}{2}$
$x = 1 \qquad\qquad\qquad\quad x = -4$

83. $6(x^2 - 1) = 5x$

$6x^2 - 5x - 6 = 0$

$x^2 - \frac{5}{6}x - 1 = 0$

$x^2 - \frac{5}{6}x = 1$

$x^2 - \frac{5}{6}x + \left(\frac{1}{2} \cdot -\frac{5}{6}\right)^2 = 1 + \left(\frac{1}{2} \cdot -\frac{5}{6}\right)^2$

$x^2 - \frac{5}{6}x + \frac{25}{144} = 1 + \frac{25}{144}$

$\left(x - \frac{5}{12}\right)^2 = \frac{169}{144}$

$x - \frac{5}{12} = \pm\sqrt{\frac{169}{144}}$

$x = \frac{5}{12} \pm \frac{13}{12}$

$x = \frac{5}{12} + \frac{13}{12}$    or    $x = \frac{5}{12} - \frac{13}{12}$

$x = \frac{18}{12}$                 $x = -\frac{8}{12}$

$x = \frac{3}{2}$                  $x = -\frac{2}{3}$

85. $x(x + 3) - \frac{1}{2} = -2$

$x^2 + 3x + \frac{3}{2} = 0$

$x^2 + 3x = -\frac{3}{2}$

$x^2 + 3x + \left(\frac{1}{2} \cdot 3\right)^2 = -\frac{3}{2} + \left(\frac{1}{2} \cdot 3\right)^2$

$x^2 + 3x + \frac{9}{4} = -\frac{3}{2} + \frac{9}{4}$

$\left(x + \frac{3}{2}\right)^2 = \frac{3}{4}$

$x + \frac{3}{2} = \pm\sqrt{\frac{3}{4}}$

$x = -\frac{3}{2} \pm \sqrt{\frac{3}{4}}$

$x = -\frac{3}{2} \pm \frac{\sqrt{3}}{2} = \frac{-3 \pm \sqrt{3}}{2}$

**Applications**

87. CAROUSELS

Using the formula for the area of a circle, $A = \pi r^2$ where we are told that $A = 2376 \text{ ft}^2$ and remembering that diameter is twice the radius, $d = 2r$, we have

$2376 = \pi r^2$

$\frac{2376}{\pi} = r^2$

$\sqrt{\frac{2376}{\pi}} = r$

$27.5 \approx r$

$d = 2 \cdot 27.5$

$d = 55 \text{ feet}$

89. BICYCLE SAFETY

$A = lw$

$800 = (x + 20)x$

$800 = x^2 + 20x$

$800 + \left(\frac{1}{2} \cdot 20\right)^2 = x^2 + 20x + \left(\frac{1}{2} \cdot 20\right)^2$

$900 = x^2 + 20x + 1000$

$900 = (x + 10)^2$

$\sqrt{900} = x + 10$

$\pm 30 = x + 10$

$30 = x + 10$ or $-30 = x + 10$

$30 - 10 = x$ $\qquad -30 - 10 = x$

$20 = x$ $\qquad\quad -40 = x$

The dimensions are 20 feet by $20 + 20 = 40$ feet.

## Writing

91. Answers will vary
93. Answers will vary

## Review

95. $(y - 1)^2 = y^2 - 2y + 1$

97. $(x + y)^2 = x^2 + 2xy + y^2$

99. $(2z)^2 = 4z^2$

## Section 15.2 The Quadratic Formula

### Vocabulary

1. The general **quadratic** equation is $ax^2 + bx + c = 0$.
3. To **solve** a quadratic equation means to find all the values of the variable that make the equation true.

### Concepts

5. In the quadratic equation $ax^2 + bx + c = 0$, $a$ cannot equal 0.

7. In the quadratic equation $3x^2 - 5 = 0$, $a = 3, b = 0,$ and $c = -5$.

9. The formula for the area of a rectangle is $A = lw$ and the formula for the area of a triangle is $A = \frac{1}{2}bh$.

11. In evaluating the numerator of $\frac{-5 \pm \sqrt{5^2 - 4(2)1}}{2(2)}$, $5^2$ should be performed first.

13. $x = \frac{-3 \pm \sqrt{15}}{2}$
    a. The equation has two solutions.
    b. The two exact solutions are $\frac{-3 + \sqrt{15}}{2}, \frac{-3 - \sqrt{15}}{2}$
    c. To the nearest hundredth the solutions are $x = 0.44$ or $x = -3.44$

15. $x = 2 + \sqrt{3}$    or    $x = 2 - \sqrt{3}$
    $x \approx 3.73$         $x \approx 0.268$

### Notation

17. $x^2 - 5x - 6 = 0$
$$x = \frac{-b \pm \sqrt{b^2 - 4ac}}{2a}$$
$$x = \frac{-(-5) \pm \sqrt{(-5)^2 - 4(1)(-6)}}{2(1)}$$
$$x = \frac{5 \pm \sqrt{25 + 24}}{2}$$
$$x = \frac{5 \pm \sqrt{49}}{2}$$
$$x = \frac{5 \pm 7}{2}$$
$$x = \frac{5 + 7}{2} = 6 \quad \text{or} \quad x = \frac{5 - 7}{2} = -1$$

19. In this work the student does not include the $-4\pm$ in the numerator of the solution.
    The solution should be of the form, $x = \frac{-4 \pm \sqrt{16 - 4(1)(-5)}}{2(1)}$

**Practice**

21. $x^2 + 4x + 3 = 0$
$a = 1$
$b = 4$
$c = 3$

23. $3x^2 - 2x + 7 = 0$
$a = 3$
$b = -2$
$c = 7$

25. $4y^2 = 2y - 1$
$4y^2 - 2y + 1 = 0$
$a = 4$
$b = -2$
$c = 1$

27. $x(3x - 5) = 2$
$3x^2 - 5x - 2 = 0$
$a = 3$
$b = -5$
$c = -2$

29. $7(x^2 + 3) = -14x$
$7x^2 + 14x + 21 = 0$
$a = 7$
$b = 14$
$c = 21$

31. $x^2 - 5x + 6 = 0$
$$x = \frac{-b \pm \sqrt{b^2 - 4ac}}{2a}$$
$$x = \frac{-(-5) \pm \sqrt{(-5)^2 - 4(1)(6)}}{2(1)}$$
$$x = \frac{5 \pm \sqrt{25 - 24}}{2}$$
$$x = \frac{5 \pm \sqrt{1}}{2}$$
$$x = \frac{5+1}{2} = 3 \quad \text{or} \quad x = \frac{5-1}{2} = 2$$

33. $x^2 + 7x + 12 = 0$
$$x = \frac{-b \pm \sqrt{b^2 - 4ac}}{2a}$$
$$x = \frac{-7 \pm \sqrt{7^2 - 4(1)(12)}}{2(1)}$$
$$x = \frac{-7 \pm \sqrt{49 - 48}}{2}$$
$$x = \frac{-7 \pm \sqrt{1}}{2}$$
$$x = \frac{-7+1}{2} \quad \text{or} \quad x = \frac{-7-1}{2}$$
$$x = -3 \qquad\qquad x = -4$$

35. $2x^2 - x - 1 = 0$

$$x = \frac{-b \pm \sqrt{b^2 - 4ac}}{2a}$$

$$x = \frac{-(-1) \pm \sqrt{(-1)^2 - 4(2)(-1)}}{2(2)}$$

$$x = \frac{1 \pm \sqrt{1+8}}{4}$$

$$x = \frac{1 \pm \sqrt{9}}{4}$$

$$x = \frac{1+3}{4} \qquad \text{or} \qquad x = \frac{1-3}{4}$$

$$x = 1 \qquad\qquad\qquad x = -\frac{1}{2}$$

37. $3x^2 + 5x + 2 = 0$

$$x = \frac{-b \pm \sqrt{b^2 - 4ac}}{2a}$$

$$x = \frac{-5 \pm \sqrt{5^2 - 4(3)(2)}}{2(3)}$$

$$x = \frac{-5 \pm \sqrt{25-24}}{6}$$

$$x = \frac{-5 \pm \sqrt{1}}{6}$$

$$x = \frac{-5+1}{6} \qquad \text{or} \qquad x = \frac{-5-1}{6}$$

$$x = -\frac{2}{3} \qquad\qquad\qquad x = -1$$

39. $4x^2 + 4x - 3 = 0$

$$x = \frac{-b \pm \sqrt{b^2 - 4ac}}{2a}$$

$$x = \frac{-4 \pm \sqrt{4^2 - 4(4)(-3)}}{2(4)}$$

$$x = \frac{-4 \pm \sqrt{16+48}}{8}$$

$$x = \frac{-4 \pm \sqrt{64}}{8}$$

$$x = \frac{-4+8}{8} \qquad \text{or} \qquad x = \frac{-4-8}{8}$$

$$x = \frac{1}{2} \qquad\qquad\qquad x = -\frac{3}{2}$$

41. $x^2 + 3x + 1 = 0$

$$x = \frac{-b \pm \sqrt{b^2 - 4ac}}{2a}$$

$$x = \frac{-3 \pm \sqrt{3^2 - 4(1)(1)}}{2(1)}$$

$$x = \frac{-3 \pm \sqrt{9-4}}{2}$$

$$x = \frac{-3 \pm \sqrt{5}}{2}$$

43. $3x^2 - x = 3$

$$3x^2 - x - 3 = 0$$

$$x = \frac{-b \pm \sqrt{b^2 - 4ac}}{2a}$$

$$x = \frac{-(-1) \pm \sqrt{(-1)^2 - 4(3)(-3)}}{2(3)}$$

$$x = \frac{1 \pm \sqrt{1+36}}{6}$$

$$x = \frac{1 \pm \sqrt{37}}{6}$$

45. $x^2 + 5 = 2x$
$x^2 - 2x + 5 = 0$
$x = \frac{-b \pm \sqrt{b^2 - 4ac}}{2a}$
$x = \frac{-(-2) \pm \sqrt{(-2)^2 - 4(1)(5)}}{2(1)}$
$x = \frac{2 \pm \sqrt{4 - 20}}{2}$
$x = \frac{2 \pm \sqrt{-16}}{2}$
**No Real Solutions**

47. $x^2 = 1 - 2x$
$x^2 + 2x - 1 = 0$
$x = \frac{-b \pm \sqrt{b^2 - 4ac}}{2a}$
$x = \frac{-2 \pm \sqrt{2^2 - 4(1)(-1)}}{2(1)}$
$x = \frac{-2 \pm \sqrt{4 + 4}}{2}$
$x = \frac{-2 \pm \sqrt{8}}{2}$
$x = \frac{-2 \pm 2\sqrt{2}}{2}$
$x = -1 \pm \sqrt{2}$

49. $3x^2 = 6x + 2$
$3x^2 - 6x - 2 = 0$
$x = \frac{-b \pm \sqrt{b^2 - 4ac}}{2a}$
$x = \frac{-(-6) \pm \sqrt{(-6)^2 - 4(3)(-2)}}{2(3)}$
$x = \frac{6 \pm \sqrt{36 + 24}}{6}$
$x = \frac{6 \pm \sqrt{60}}{6}$
$x = \frac{6 \pm 2\sqrt{15}}{6}$
$x = \frac{3 \pm \sqrt{15}}{3}$

51. $(2y - 1)^2 = 25$
$2y - 1 = \pm\sqrt{25}$
$2y = 1 \pm \sqrt{25}$
$y = \frac{1 \pm \sqrt{25}}{2}$
$y = \frac{1 + 5}{2}$     or     $y = \frac{1 - 5}{2}$
$y = 3$                  $y = -2$

53. $2x^2 + x = 5$
$2x^2 + x - 5 = 0$
$x = \frac{-b \pm \sqrt{b^2 - 4ac}}{2a}$
$x = \frac{-1 \pm \sqrt{1^2 - 4(2)(-5)}}{2(2)}$
$x = \frac{-1 \pm \sqrt{1 + 40}}{4}$
$x = \frac{-1 \pm \sqrt{41}}{4}$
$x = \frac{-1 + \sqrt{41}}{4} \approx 1.35$
or
$x = \frac{-1 - \sqrt{41}}{4} \approx -1.85$

55. $x^2 - 2x - 1 = 0$

$$x = \frac{-b \pm \sqrt{b^2 - 4ac}}{2a}$$

$$x = \frac{-(-2) \pm \sqrt{(-2)^2 - 4(1)(-1)}}{2(1)}$$

$$x = \frac{2 \pm \sqrt{4+4}}{2}$$

$$x = \frac{2 \pm \sqrt{8}}{2}$$

$$x = \frac{2 \pm 2\sqrt{2}}{2}$$

$$x = 1 \pm \sqrt{2}$$

$$x = 1 + \sqrt{2} \approx 2.41$$

or

$$x = 1 - \sqrt{2} \approx -0.41$$

57. $x^2 - 2x - 35 = 0$

$(x - 7)(x + 5) = 0$

$x - 7 = 0 \quad$ or $\quad x + 5 = 0$

$x = 7 \qquad\qquad x = -5$

59. $x^2 + 2x + 7 = 0$

$$x = \frac{-b \pm \sqrt{b^2 - 4ac}}{2a}$$

$$x = \frac{-2 \pm \sqrt{2^2 - 4(1)(7)}}{2(1)}$$

$$x = \frac{-2 \pm \sqrt{4 - 28}}{2}$$

$$x = \frac{-2 \pm \sqrt{-24}}{2}$$

**No Real Solutions**

61. $4c^2 + 16c = 0$

$4c(c + 4) = 0$

$4c = 0 \quad$ or $\quad c + 4 = 0$

$c = 0 \qquad\qquad c = -4$

63. $18 = 3y^2$

$6 = y^2$

$\pm\sqrt{6} = y$

$y = \sqrt{6} \approx 2.45$

or

$y = -\sqrt{6} \approx -2.45$

65. $2.4x^2 - 9.5x + 6.2 = 0$

$$x = \frac{-b \pm \sqrt{b^2 - 4ac}}{2a}$$

$$x = \frac{-(-9.5) \pm \sqrt{(-9.5)^2 - 4(2.4)(6.2)}}{2(2.4)}$$

$$x = \frac{9.5 \pm 5.543}{4.8}$$

$$x = \frac{9.5 + 5.543}{4.8} \quad \text{or} \quad x = \frac{9.5 - 5.543}{4.8}$$

$$x \approx 3.1 \qquad\qquad x \approx 0.8$$

**Applications**

67. HEIGHT OF A TRIANGLE
$A = 30 \text{ in}^2$
$A = \frac{1}{2}bh, \text{ where } b = (x+4), h = x$
$A = \frac{1}{2}(x+4)x$
$30 = \frac{1}{2}(x+4)x$
$60 = (x+4)x$
$60 = x^2 + 4x$
$0 = x^2 + 4x - 60$
$0 = (x-6)(x+10)$
$x - 6 = 0 \quad \text{or} \quad x + 10 = 0$
$x = 6 \qquad\qquad x = -10$
The height of this triangle is 6 inches.

69. FLAGS
$A = 3102 \text{ ft}^2$
$l = w + 19$
$A = lw$
$3102 = (w+19)w$
$3102 = w^2 + 19w$
$0 = w^2 + 19w - 3102$
$0 = (w-47)(w+66)$
$w - 47 = 0 \quad \text{or} \quad w + 66 = 0$
$w = 47 \qquad\qquad w = -66$
The width of this flag is 47 feet and its length is $w + 19 = 47 + 19 = 66$ feet.

71. COMMUNITY GARDENS
$A = 180 \text{ ft}^2$
$A = lw$
$l = 24 - 2x$
$w = 16 - 2x$
$180 = (24 - 2x)(16 - 2x)$
$180 = 384 - 48x - 32x + 4x^2$
$180 = 384 - 80x + 4x^2$
$0 = 384 - 80x + 4x^2 - 180$
$0 = 4x^2 - 80x + 204$
$0 = 4(x^2 - 20x + 51)$
$0 = 4(x-17)(x-3)$
$0 = (x-17)(x-3)$
$x - 17 = 0 \quad \text{or} \quad x - 3 = 0$
$x = 17 \qquad\qquad x = 3$
Substitute these values into the equations for length and width.
Using $x = 17$, gives $l = 24 - 2(17) = -10$ and $w = 16 - 2(17) = -18$.
Using $x = 3$, gives $l = 24 - 2(3) = 18$ and $w = 16 - 2(3) = 10$.
The largest possible garden is 10 feet wide and 18 feet long.

73. DAREDEVILS

Using $s = 16t^2$, where $s = 200$

$200 = 16t^2$

$\frac{200}{16} = t^2$

$\sqrt{\frac{200}{16}} = t$

$\frac{10}{4}\sqrt{2} = t$

$3.536 \approx t$

To the nearest tenth it took about 3.5 seconds to hit the water.

75. ABACUS

Using the pythagorean theorem,

$h^2 + (h + 21)^2 = 39^2$

$h^2 + h^2 + 42h + 441 = 1521$

$2h^2 + 42h - 1080 = 0$

$2(h^2 + 21h - 540) = 0$

$2(h - 15)(h + 36) = 0$

$(h - 15)(h + 36) = 0$

$h - 15 = 0 \quad$ or $\quad h + 36 = 0$

$h = 15 \qquad\qquad h = -36$

This abacus is 15 cm high and $15 + 21 = 36$ cm long.

77. NAVIGATION

Since these boats are traveling south and east, their positions form a right triangle,

$x^2 + (x + 10)^2 = 50^2$

$x^2 + x^2 + 20x + 100 = 2500$

$2x^2 + 20x - 2400 = 0$

$2(x^2 + 10x - 1200) = 0$

$2(x - 30)(x + 40) = 0$

$(x - 30)(x + 40) = 0$

$x - 30 = 0 \quad$ or $\quad x + 40 = 0$

$x = 30 \qquad\qquad x = -40$

The boats travel 30 nautical miles and 40 nautical miles.

79. INVESTING

$A = P(1 + r)^2$

$5724.50 = 5000(1 + r)^2$

$\frac{5724.50}{5000} = \frac{5000(1+r)^2}{5000}$

$1.1449 = (1 + r)^2$

$\pm\sqrt{1.1449} = 1 + r$

$-1 \pm \sqrt{1.1449} = r$

$r = -1 + \sqrt{1.1449} \quad$ or $\quad r = -1 - \sqrt{1.1449}$

$r = 0.07 \quad$ or $\quad r = -2.07$

An interest rate of 7% will allow \$5000 to grow to \$ 5724.50 in  2 years.

81. MANUFACTURING

$R = -\frac{1}{6}x^2 + 450x$

$R = -\frac{1}{6}(600)^2 + 450(600)$

$R = 210,000$

If 600 televisions are manufactured this firm's revenue, if the televisions are sold, would be \$210,000.

83. METAL FABRICATION

$A = s^2$

$64 = (12 - 2x)^2$

$\sqrt{64} = 12 - 2x$

$8 = 12 - 2x$

$2x = 12 - 8$

$2x = 4$

$x = 2$

A 2 inch square removed from each side will result in a box with a base area of 64 in$^2$. The length of the square, 2 inches, is also the depth of the box.

**Writing**

85. Answers will vary

87. Answers will vary

**Review**

89. $A = p + prt$

$A - p = prt$

$\frac{A-p}{pt} = \frac{prt}{pt}$

$\frac{A-p}{pt} = r$

91. $m = \frac{3}{5}, (0, 12)$

$y - 12 = \frac{3}{5}(x - 0)$

$y - 12 = \frac{3}{5}x$

$y - \frac{3}{5}x = 12$

$5y - 5(\frac{3}{5}x) = 5(12)$

$5y - 3x = 60$

$-1(5y - 3x) = -1(60)$

$3x - 5y = -60$

93. $\sqrt{80} = \sqrt{16 \cdot 5} = 4\sqrt{5}$

95. $\frac{x}{\sqrt{7x}} = \frac{x}{\sqrt{7x}} \cdot \frac{\sqrt{7x}}{\sqrt{7x}} = \frac{x\sqrt{7x}}{7x} = \frac{\sqrt{7x}}{7}$

## Section 15.3 Graphing Quadratic Functions

### Vocabulary

1. A function defined by the equation $y = ax^2 + bx + c \, (a \neq 0)$ is called a **quadratic** function.
3. The point where a parabola intersects the $y$-axis is called the ***y*-intercept**.
5. For a parabola that opens upward or downward, the vertical line that passes through its vertex and splits the graph into two identical pieces is called the axis of **symmetry**.

### Concepts

7. The graph of $y = ax^2 + bx + c$, where $a \neq 0$, opens upward when $a > 0$.

9. The $y$-intercept of the graph of $f(x) = ax^2 + bx + c$ is the point $(0, c)$.

11. a. This curve is called a parabola.
    b. The $x$-intercepts are $(1, 0)$ and $(3, 0)$.
    c. The $y$-intercept is $(0, -3)$.
    d. The vertex is $(2, 1)$.

13. Answers will vary.
    zero $x$-intercepts

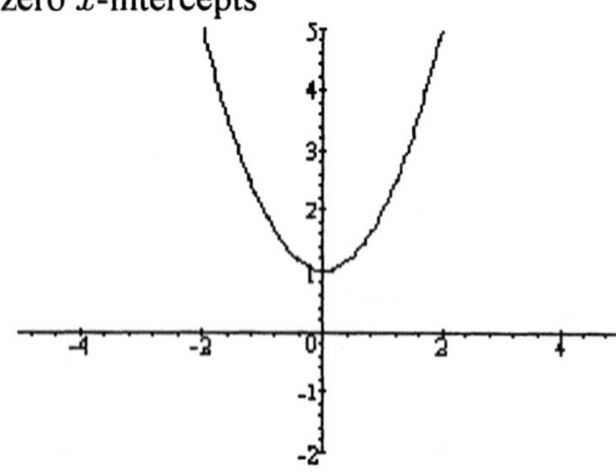

one $x$-intercept

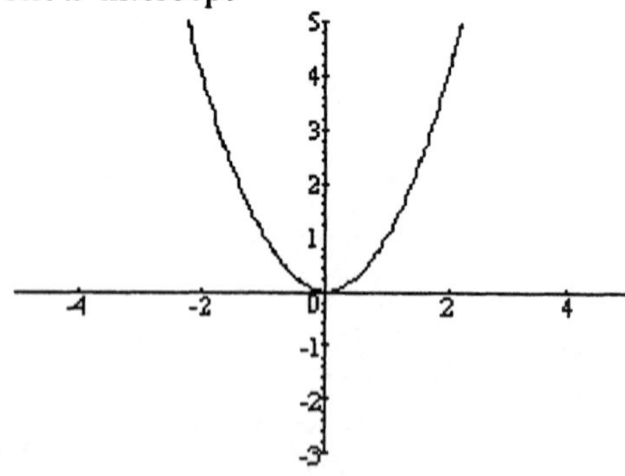

(see following page also)

two $x$-intercepts

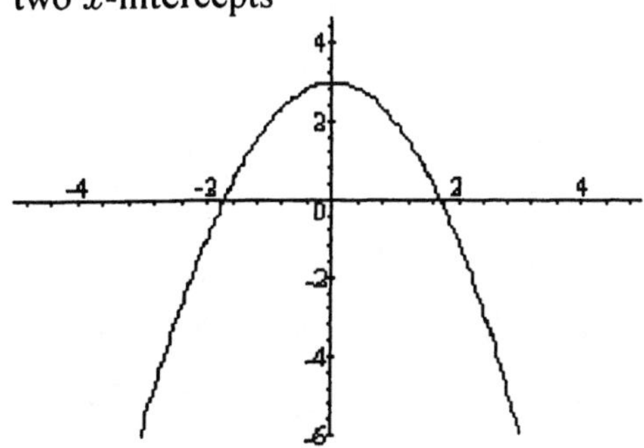

15. HEALTH DEPARTMENT
    This flu outbreak lasted for 10-months with the greatest number of cases occuring in week 5.

**Notation**

17. The equations $y = 2x^2 - x - 2$ and $f(x) = 2x^2 - x - 2$ are the same is a true statement.

19. $-\frac{b}{2a} = -\frac{3}{2(3)} = -\frac{3}{6} = -\frac{1}{2}$

**Practice**

21. $y = x^2 + 1$

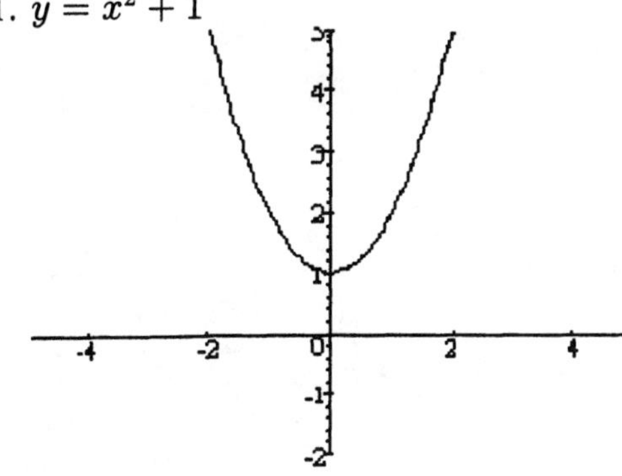

This function is moved up one unit compared to $y = x^2$.

23. $f(x) = -x^2$

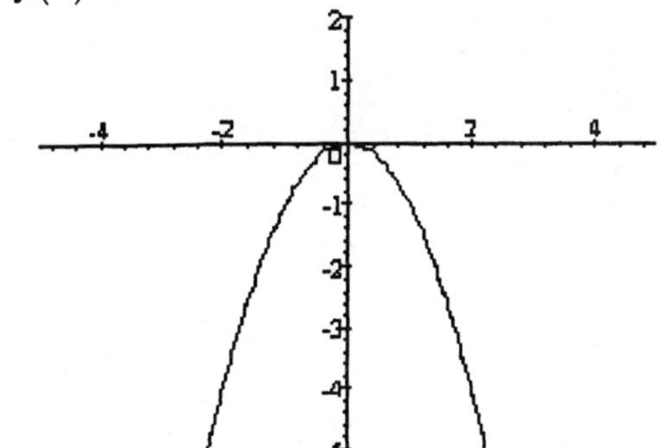

This function opens the opposite direction as $y = x^2$.

25. $y = -x^2 + 6x - 8$
$x = -\frac{b}{2a} = -\frac{6}{2(-1)} = 3$ so
$y = -3^2 + 6 \cdot 3 - 8 = 1$,
The vertex is $(3, 1)$.

27. $f(x) = 2x^2 - 4x + 1$
$x = -\frac{b}{2a} = -\frac{-4}{2(2)} = 1$ so
$y = 2(1)^2 - 4(1) + 1 = -1$
The vertex is $(1, -1)$.

29. $f(x) = x^2 - 2x + 1$
$f(0) = 0^2 - 2 \cdot 0 + 1 = 1$
The $y$-intercept is $(0, 1)$.
$x^2 - 2x + 1 = 0$
$(x - 1)^2 = 0$
$x - 1 = 0$
$x = 1$
The $x$-intercept is $(1, 0)$

31. $y = -x^2 - 10x - 21$
If $x = 0, y = -0^2 - 10 \cdot 0 - 21 = -21$
The $y$-intercept is $(0, -21)$.
If $y = 0$,
$0 = -x^2 - 10x - 21$
$0 = -1(x^2 + 10x + 21)$
$0 = x^2 + 10x + 21$
$0 = (x + 7)(x + 3)$
$x + 7 = 0$ or $\quad x + 3 = 0$
$x = -7 \qquad\qquad x = -3$
The $x$-intercepts are $(-7, 0)$ and $(-3, 0)$.

33. $y = x^2 - 2x$

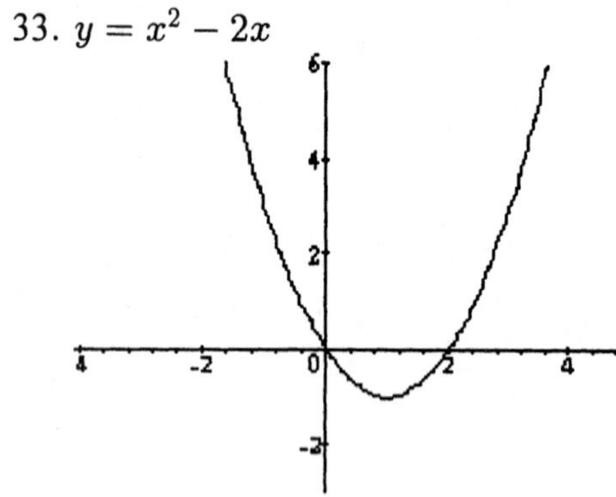

35. $f(x) = -x^2 + 2x$

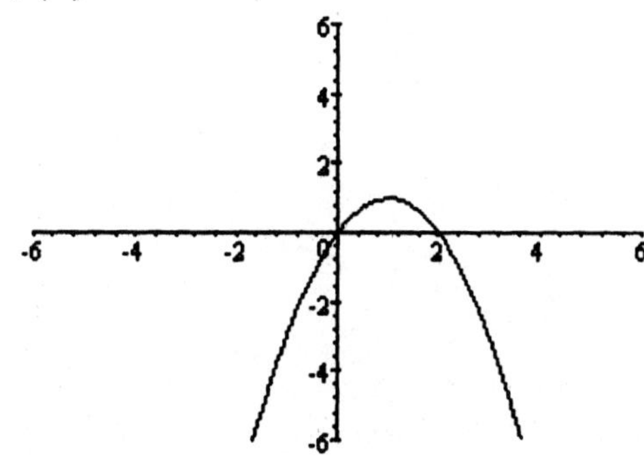

37. $f(x) = x^2 + 4x + 4$

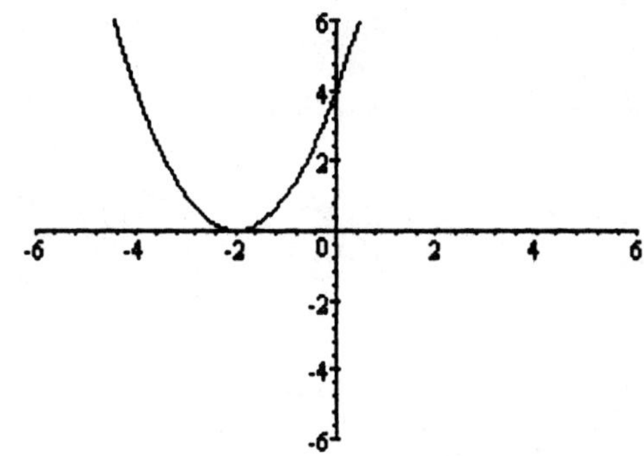

39. $y = -x^2 - 2x - 1$

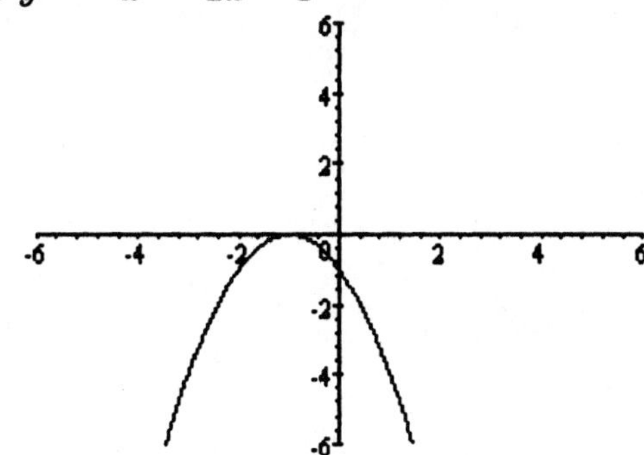

41. $y = x^2 + 2x - 3$

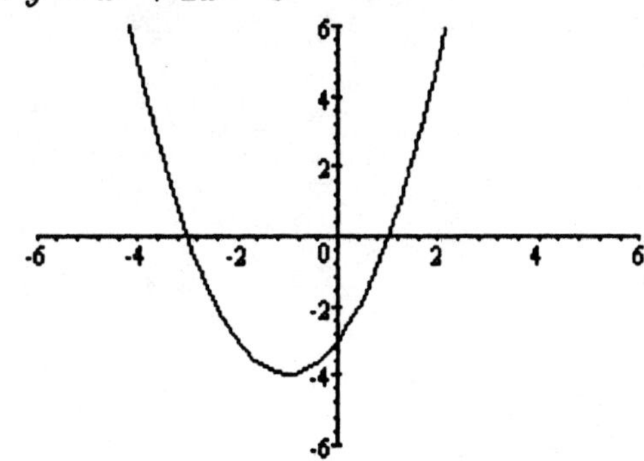

43. $f(x) = 2x^2 + 8x + 6$

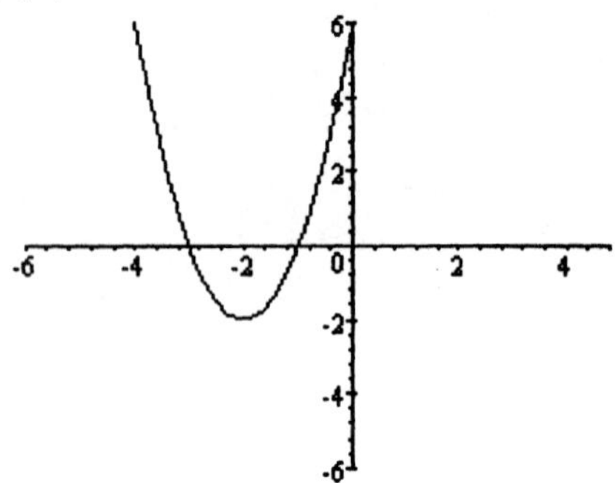

45. $y = x^2 - 2x - 8$

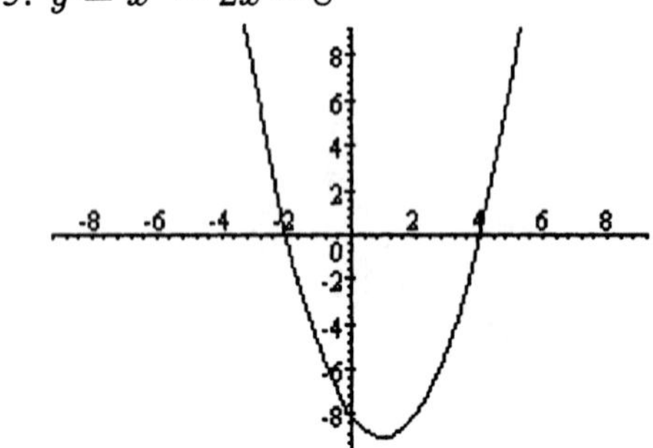

47. $y = x^2 - x - 2$

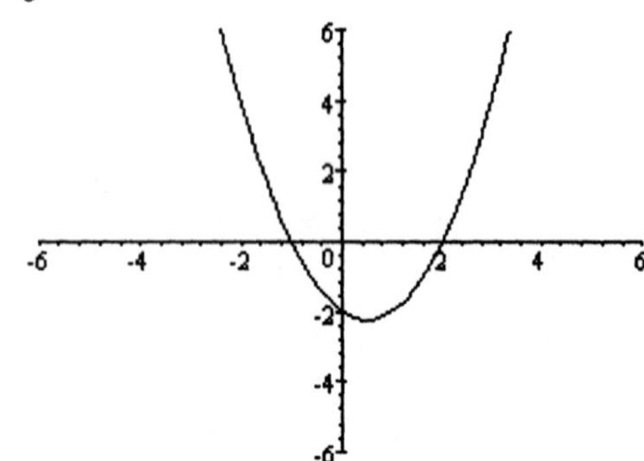

49. $f(x) = 2x^2 + 3x - 2$

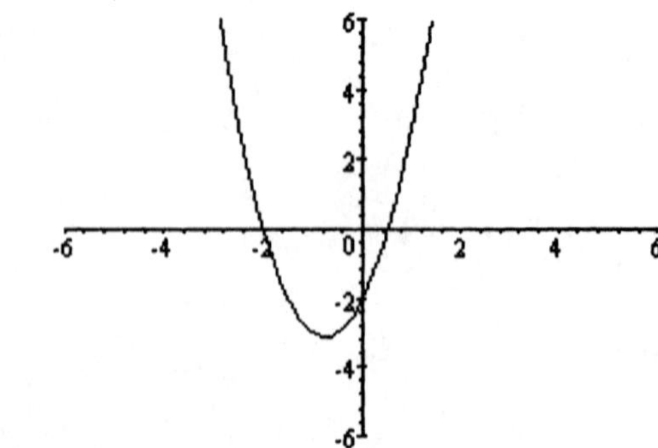

**Applications**

51. TRAMPOLINE
    a. She is 14 feet above the ground after $\frac{1}{2}$ second.
    b. She is 9 feet above the ground at 0.25 second and at 1.75 seconds.
    c. The maximum number of feet above ground is 18 feet which occurs at 1.0 second.

53. BRIDGE
    $y = 0.005x^2$

| $x$ | $-80$ | $-60$ | $-40$ | $-20$ | 0 | 20 | 40 | 60 | 80 |
|---|---|---|---|---|---|---|---|---|---|
| $y$ | 32 | 18 | 8 | 2 | 0 | 2 | 8 | 18 | 32 |

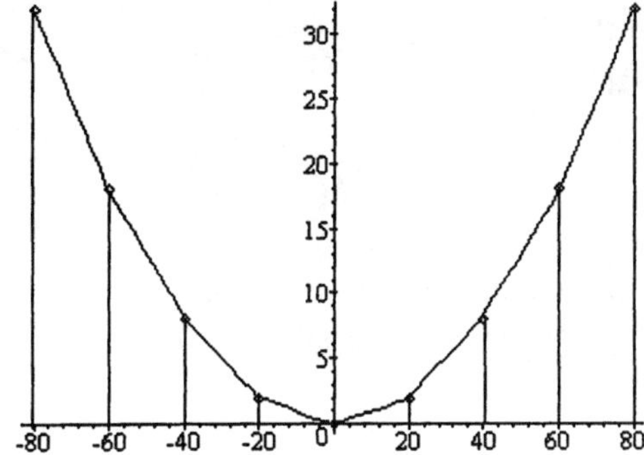

55. SELLING CD PLAYERS
    a. $(150 - \frac{1}{10}n)n = -\frac{1}{10}n^2 + 150n$

    The maximum revenue would be the $x$-coordinate of the vertex of the above function.
    $x = -\frac{150}{2(-\frac{1}{10})} = 750$ units

    The retailer would have to buy 750 units for the wholesaler to maximize revenue.
    b. The maximum revenue is
    revenue $= (150 - \frac{1}{10} \cdot 750)750 = \$ 56{,}250$

**Writing**

57. Answers will vary
59. Answers will vary

**Review**

61. $\sqrt{12} + \sqrt{27} = 2\sqrt{3} + 3\sqrt{3} = 5\sqrt{3}$

63. $(\sqrt{3} + 1)(\sqrt{3} - 1)$
    $= 3 - 1$
    $= 2$

65. $\sqrt{6 + 2x} = 4$
    $\left(\sqrt{6 + 2x}\right)^2 = 4^2$
    $6 + 2x = 16$
    $2x = 10$
    $x = 5$

**Chapter 15 Key Concepts**

1. $y = 3x + 7$ is not a quadratic equation

3. $2x^2 - 3x + 4 = 0$ is a quadratic equation

5. $a^2 + 7a - 1 > 0$ is not a quadratic equation

7. $5 = y - y^2$ is a quadratic equation

9. $\sqrt{x + 7} = 4$ is not a quadratic equation

11. $\frac{m}{2} - \frac{1}{3} = \frac{1}{4}$ is not a quadratic equation

13. $x^2 = 6$
$$\frac{x^2}{2} = \frac{6}{2}$$
$$x = 3$$
The student's error occured when she divided by the exponent on each side instead of taking the square root of each side.

15. $a^2 = 20$
$$a = \pm\sqrt{20}$$
$$a = \pm 2\sqrt{5}$$
The student forgot to use the $\pm$ when taking the square root.

17. $4x^2 - x = 0$
$$x(4 - x) = 0$$
$$x = 0 \quad \text{or} \quad 4 - x = 0$$
$$x = 4$$

19. $x^2 = 36$
$$x = \pm\sqrt{36}$$
$$x = \pm 6$$

21. $x^2 + 4x + 1 = 0$
$$x^2 + 4x = -1$$
$$x^2 + 4x + (\tfrac{1}{2} \cdot 4)^2 = -1 + (\tfrac{1}{2} \cdot 4)^2$$
$$x^2 + 4x + 4 = -1 + 4$$
$$(x + 2)^2 = 3$$
$$x + 2 = \pm\sqrt{3}$$
$$x = -2 \pm \sqrt{3}$$

23. $R = 4x - x^2$

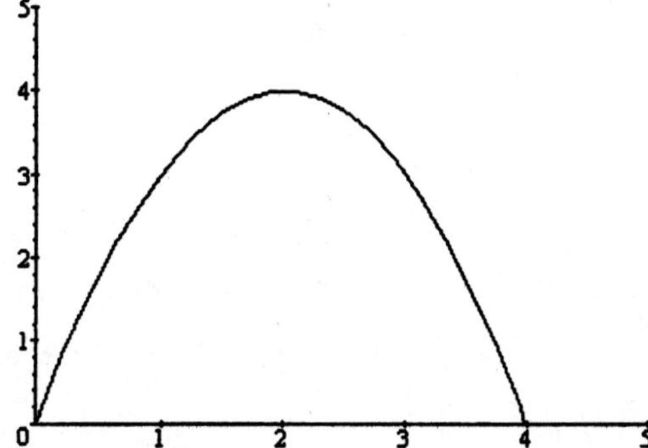

**Section 15.1**

1. a. $x^2 = 25$
    $x = \pm\sqrt{25}$
    $x = \pm 5$
    b. $x^2 = 400$
    $x = \pm\sqrt{400}$
    $x = \pm 20$
    c. $2x^2 = 18$
    $x^2 = 9$
    $x = \pm\sqrt{9}$
    $x = \pm 3$
    d. $4y^2 = 9$
    $y^2 = \frac{9}{4}$
    $y = \pm\sqrt{\frac{9}{4}}$
    $y = \pm\frac{3}{2}$
    e. $t^2 = 8$
    $t = \pm\sqrt{8}$
    $t = \pm 2\sqrt{2}$
    f. $2x^2 - 1 = 149$
    $2x^2 = 150$
    $x^2 = 75$
    $x = \pm\sqrt{75}$
    $x = \pm 5\sqrt{3}$

3. a. $x^2 = 12$
    $x = \pm 2\sqrt{3}$
    $x \approx \pm 3.46$
    b. $(x - 1)^2 = 55$
    $x - 1 = \pm\sqrt{55}$
    $x = 1 \pm\sqrt{55}$
    $x \approx 8.42$  or  $x \approx -6.42$

5. $x^2 + 4x + 1 = 0$ does not factor

7. $x^2 + 4x + 1 = 0$
    $x^2 + 4x = -1$
    $x^2 + 4x + (\frac{1}{2} \cdot 4)^2 = -1 + (\frac{1}{2} \cdot 4)^2$
    $x^2 + 4x + 4 = -1 + 4$
    $(x + 2)^2 = 3$
    $x + 2 = \pm\sqrt{3}$
    $x = -2 \pm\sqrt{3}$
    $x \approx -0.27$  or    $x \approx -3.73$

9. $\dfrac{-b \pm \sqrt{b^2 - 4ac}}{2a}$

    a. $x^2 - 2x - 15 = 0$

$$x = \frac{-(-2) \pm \sqrt{(-2)^2 - 4(1)(-15)}}{2(1)}$$

$$x = \frac{2 \pm \sqrt{4 + 60}}{2}$$

$$x = \frac{2 \pm \sqrt{64}}{2}$$

$$x = \frac{2 \pm 8}{2}$$

$$x = 1 \pm 4$$

$$x = 1 + 4 \quad \text{or} \quad x = 1 - 4$$

$$x = 5 \qquad\qquad x = -3$$

    b. $x^2 - 6x - 7 = 0$

$$x = \frac{-(-6) \pm \sqrt{(-6)^2 - 4(1)(-7)}}{2(1)}$$

$$x = \frac{6 \pm \sqrt{36 + 28}}{2}$$

$$x = \frac{6 \pm \sqrt{64}}{2}$$

$$x = \frac{6 \pm 8}{2}$$

$$x = 3 \pm 4$$

$$x = 3 - 4 \quad \text{or} \quad x = 3 + 4$$

$$x = -1 \qquad\qquad x = 7$$

    c. $6x^2 - 7x - 3 = 0$

$$x = \frac{-(-7) \pm \sqrt{(-7)^2 - 4(6)(-3)}}{2 \cdot 6}$$

$$x = \frac{7 \pm \sqrt{49 + 72}}{12}$$

$$x = \frac{7 \pm \sqrt{121}}{12}$$

$$x = \frac{7 \pm 11}{12}$$

$$x = \frac{7 - 11}{12} \quad \text{or} \quad x = \frac{7 + 11}{12}$$

$$x = -\frac{4}{12} \qquad\qquad x = \frac{18}{12}$$

$$x = -\frac{1}{3} \qquad\qquad x = \frac{3}{2}$$

    d. $x^2 - 6x + 7 = 0$

$$x = \frac{-(-6) \pm \sqrt{(-6)^2 - 4(1)(7)}}{2(1)}$$

$$x = \frac{6 \pm \sqrt{36 - 28}}{2}$$

$$x = \frac{6 \pm \sqrt{8}}{2}$$

$$x = \frac{6 \pm 2\sqrt{2}}{2}$$

$$x = 3 \pm \sqrt{2}$$

11. $10x^2 + 2x + 1 = 0$

$$x = \frac{-2 \pm \sqrt{2^2 - 4(10)(1)}}{2 \cdot 10}$$

$$x = \frac{-2 \pm \sqrt{4 - 40}}{20}$$

$$x = \frac{-2 \pm \sqrt{-36}}{20}$$

There are no real solutions.

13. MILITARY

$h = 3000 + 40t - 16t^2$

When the bomb strikes the ground $h = 0$.

$0 = 3000 + 40t - 16t^2$

$-16t^2 + 40t + 3000 = 0$

$t = \dfrac{-40 \pm \sqrt{40^2 - 4(-16)(3000)}}{2 \cdot (-16)}$

$t = \dfrac{-40 \pm \sqrt{1600 + 192{,}000}}{-32}$

$t = \dfrac{-40 \pm \sqrt{193{,}600}}{-32}$

$t = \dfrac{-40 \pm 440}{-32}$

$t = \dfrac{-40 + 440}{-32}$    or    $t = \dfrac{-40 - 440}{-32}$

$t = -\dfrac{25}{2}$         $t = 15$

$t = -12.5$

It will take this bomb 15 seconds to strike the target.

## Section 15.3

15. a. The $x$-intercepts are $(-3, 0)$ and $(1, 0)$.
     b. The $y$-intercept is $(0, -3)$.
     c. The vertex is $(-1, -4)$.
     d. The axis of symmetry is $x = -1$.

17. a. $y = 2x^2 - 4x + 7$ opens upward

$x = -\dfrac{b}{2a}$

$x = -\dfrac{-4}{2(2)}$

$x = 1$ then

$y = 2(1)^2 - 4 \cdot 1 + 7$

$y = 5$

The vertex is located at $(1, 5)$.

     b. $f(x) = -3x^2 + 18x - 11$ opens downward

$x = -\dfrac{b}{2a}$

$x = -\dfrac{18}{2(-3)}$

$x = 3$ then

$y = -3(3)^2 + 18(3) - 11$

$y = 16$

The vertex is located at $(3, 16)$.

19. a. $y = x^2 + 2x - 3$

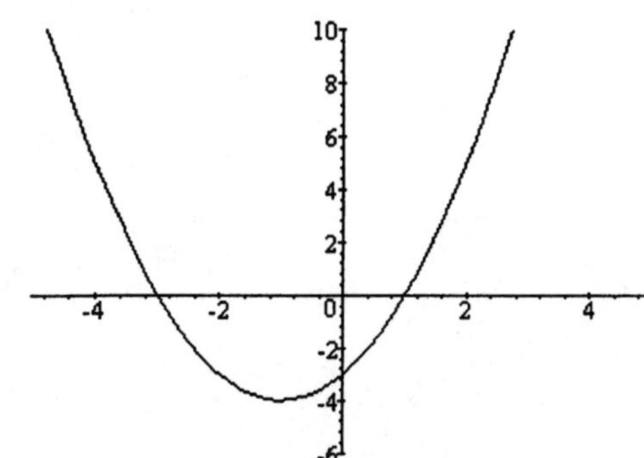

b. $f(x) = -2x^2 + 4x - 2$

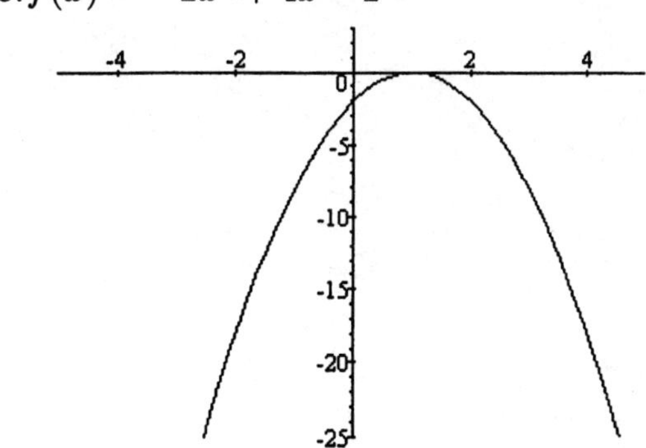

**Chapter 15 Test**

1. $x^2 = 16$
   $x = \pm\sqrt{16}$
   $x = \pm 4$

3. $4y^2 = 25$
   $y^2 = \frac{25}{4}$
   $y = \pm\sqrt{\frac{25}{4}}$
   $y = \pm\frac{5}{2}$

5. ARCHERY
   $A = \pi r^2$
   $A = 5{,}026.5$
   $5{,}026.5 = \pi r^2$
   $r^2 = \frac{5{,}026.5}{\pi}$
   $r = \pm\sqrt{\frac{5{,}026.5}{\pi}}$
   $r \approx 39.9998$
   $r \approx 40$ cm

7. $a^2 + 2a - 4 = 0$
   $a^2 + 2a = 4$
   $a^2 + 2a + \left(\frac{1}{2} \cdot 2\right)^2 = 4 + \left(\frac{1}{2} \cdot 2\right)^2$
   $a^2 + 2a + 1 = 4 + 1$
   $(a + 1)^2 = 5$
   $a + 1 = \pm\sqrt{5}$
   $a = -1 \pm \sqrt{5}$
   $a \approx 1.24$     or     $a \approx -3.24$

9. $x^2 + 3x - 10 = 0$
   $x = \frac{-b \pm \sqrt{b^2 - 4ac}}{2a}$
   $x = \frac{-3 \pm \sqrt{3^2 - 4(1)(-10)}}{2(1)}$
   $x = \frac{-3 \pm \sqrt{9 + 40}}{2}$
   $x = \frac{-3 \pm \sqrt{49}}{2}$
   $x = \frac{-3 \pm 7}{2}$
   $x = \frac{-3 + 7}{2}$     or     $x = \frac{-3 - 7}{2}$
   $x = 2$                 $x = -5$

11. $x^2 + 5 = 2x$
    $x^2 - 2x + 5 = 0$
    $x = \frac{-b \pm \sqrt{b^2 - 4ac}}{2a}$
    $x = \frac{-(-2) \pm \sqrt{(-2)^2 - 4(1)(5)}}{2 \cdot 1}$
    $x = \frac{2 \pm \sqrt{4 - 20}}{2}$
    $x = \frac{2 \pm \sqrt{-16}}{2}$
    No real solutions

13. $3x^2 - x - 1 = 0$

$$x = \frac{-b \pm \sqrt{b^2 - 4ac}}{2a}$$

$$x = \frac{-(-1) \pm \sqrt{(-1)^2 - 4(3)(-1)}}{2 \cdot (3)}$$

$$x = \frac{1 \pm \sqrt{1 + 12}}{6}$$

$$x = \frac{1 \pm \sqrt{13}}{6}$$

$$x = \frac{1 + \sqrt{13}}{6} \quad \text{or} \quad x = \frac{1 - \sqrt{13}}{6}$$

$$x \approx 0.77 \qquad\qquad x \approx -0.43$$

15. ADVERTISING

The vertex tells us that 3 television advertisements run during the week resulted in the maximum number of units sold, 18.

17. According to the given graph, $-x^2 - 2x - 1 = 0$ when $x = -1$. To check this, substitute $x = -1$ into the equation

$$f(-1) = -(-1)^2 - 2(-1) - 1$$
$$f(-1) = -1 + 2 - 1$$
$$f(-1) = 0$$

1.  a.Every rational number can be written as a ratio of two integers is a true statement.
    b.The set of real numbers corresponds to all points on the number line is a true statement.
    c.The whole numbers and their opposites form the set of integers is a true statement.

3.  DRIVING SAFETY
    Two hours before the salt is spread on the roads the accident rate is the worst.

5.  $3p - 6(p + z) + p$
    $= 3p - 6p - 6z + p$
    $= -2p - 6z$

7.  $-(3a + 1) + a = 2$
    $-3a - 1 + a = 2$
    $-2a = 3$
    $a = -\frac{3}{2}$

9.  ENTREPRENEURS

| % | Amount | Interest |
|---|---|---|
| 7 | $x$ | $0.07x$ |
| 10 | $28000 - x$ | $0.10(28000 - x)$ |
| | 28000 | 2560 |

$0.07x + 0.10(28000 - x) = 2560$
$0.07x + 2800 - 0.10x = 2560$
$-0.03x = -240$
$x = 8000$
There was $8,000 loaned at the 7% rate and $20,000 loaned at the 10% rate.

11.  $\left|\frac{4}{5} \cdot 10 - 12\right|$
    $= |8 - 12|$
    $= |-4|$
    $= 4$

13.  $2y - 2x = 6$
    $2y = 2x + 6$
    $y = x + 3$

15. $y = -x + 2$

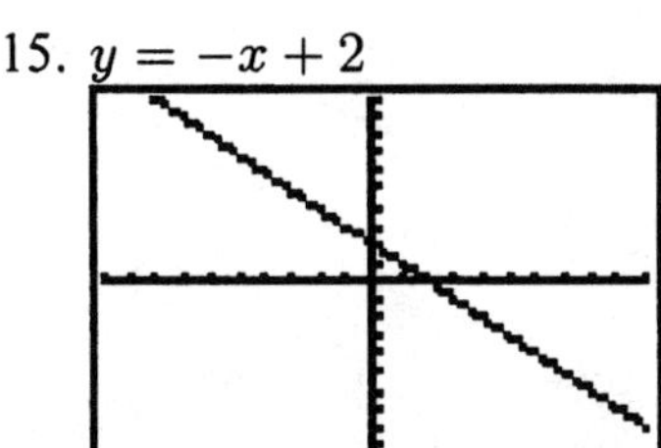

17. $(-2, -1), m = \frac{4}{3}$
$$y - (-1) = \frac{4}{3}(x - (-2))$$
$$y + 1 = \frac{4}{3}(x + 2)$$
$$y + 1 = \frac{4}{3}x + \frac{8}{3}$$
$$y = \frac{4}{3}x + \frac{5}{3}$$

19. $m = -2, (0, 1)$
$$y = -2x + 1$$

21. $4x + 5y = 6$
$$5y = -4x + 6$$
$$y = -\frac{4}{5}x + \frac{6}{5}$$
$$m = -\frac{4}{5}$$

23. Domain is associated with the input of a function.

25. $y^3(y^2 y^4) = y^9$

27. $\frac{10a^4 a^{-2}}{5a^2 a^0} = \frac{10a^2}{5a^2} = 2$

29. FIVE-CARD POKER
$$2.6 \times 10^6 = 2,600,000$$
$2,600,000$ to $1$

31. $(-2a^3)(3a^2) = -6a^5$

33. $(2x + 5y)^2$
$$= 4x^2 + 10xy + 10xy + 25y^2$$
$$= 4x^2 + 20xy + 25y^2$$

35. $6a^2 - 12a^3 b + 36ab = 6a(a - 2a^2 b + 6b)$

37. $b^3 + 125 = (b + 5)(b^2 - 5b + 25)$

39. $3x^2 + 8x = 0$

$\quad x(3x + 8) = 0$

$\quad x = 0 \ \text{ or } \ 3x + 8 = 0$

$\qquad\qquad\qquad 3x = -8$

$\qquad\qquad\qquad\ x = -\frac{8}{3}$

41. $P = 2(2x^3 - x) + 2(x^3 + 3x)$

$\quad P = 4x^3 - 2x + 2x^3 + 6x$

$\quad P = 6x^3 + 4x$

43. When $x = -8$ this rational expression is undefined.

45. $\dfrac{x^2-x-6}{2x^2+9x+10} \div \dfrac{x^2-25}{2x^2+15x+25}$

$\quad = \dfrac{(x-3)(x+2)}{(2x+5)(x+2)} \cdot \dfrac{(2x+5)(x+5)}{(x-5)(x+5)}$

$\quad = \dfrac{x-3}{x-5}$

47. $\dfrac{x}{x-2} + \dfrac{3x}{x^2-4}$

$\quad = \dfrac{x}{x-2} + \dfrac{3x}{(x-2)(x+2)}$

$\quad = \dfrac{x(x+2)}{(x-2)(x+2)} + \dfrac{3x}{(x-2)(x+2)}$

$\quad = \dfrac{x^2+2x+3x}{(x-2)(x+2)}$

$\quad = \dfrac{x^2+5x}{(x-2)(x+2)}$

49. $\dfrac{3r}{2} - \dfrac{3}{r} = \dfrac{3r}{2} + 3$

$\quad 2r\left(\dfrac{3r}{2} - \dfrac{3}{r}\right) = 2r\left(\dfrac{3r}{2} + 3\right)$

$\quad 3r^2 - 6 = 3r^2 + 6r$

$\quad -6 = 6r$

$\quad -1 = r$

51. $\dfrac{1}{a} + \dfrac{1}{b} = 1$

$\quad \dfrac{1}{a} = 1 - \dfrac{1}{b}$

$\quad 1 = a\left(1 - \dfrac{1}{b}\right)$

$\quad \dfrac{1}{1-\frac{1}{b}} = a$

$\quad$ or

$\quad \dfrac{b}{b}\left(\dfrac{1}{1-\frac{1}{b}}\right) = a$

$\quad \dfrac{b}{b-1} = a$

53. LOSING WEIGHT

If 350 days at 100 calories per days is 10 pounds, then 700 days will equate to 20 pounds and then $\frac{1}{2}(350)$ for the remaining 5 pounds or 175 days. To lose 25 pounds this person must cut down their caloric intake by 100 calories for $350 + 350 + 175 = 875$ days.

55. $x + y = 1$
$y = x + 5$
$(-2, 3)$

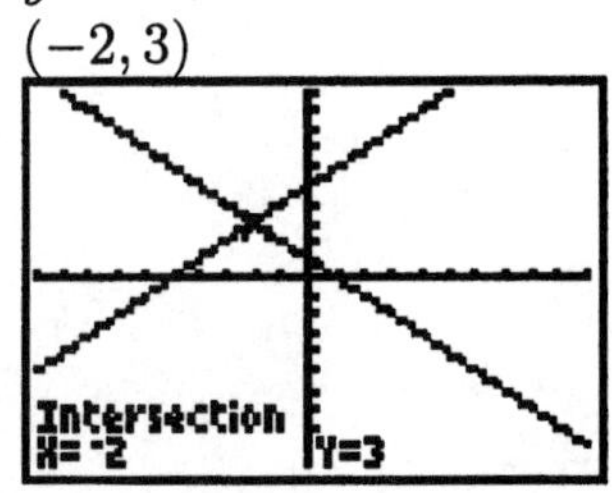

57. $\frac{3}{5}s + \frac{4}{5}t = 1$
$-\frac{1}{4}s + \frac{3}{8}t = 1$

$5\left(\frac{3}{5}s + \frac{4}{5}t\right) = 5(1)$
$8\left(-\frac{1}{4}s + \frac{3}{8}t\right) = 8(1)$

$3s + 4t = 5$
$-2s + 3t = 8$

$2(3s + 4t) = 5(2)$
$3(-2s + 3t) = 8(3)$

$6s + 8t = 10$
$\underline{-6s + 9t = 24}$
$17t = 34$
$t = 2$

Substitute this value into one of the original equations

$\frac{3}{5}s + \frac{4}{5}(2) = 1$
$\frac{3}{5}s + \frac{8}{5} = 1$
$\frac{3}{5}s = -\frac{3}{5}$
$s = -1$

$(s, t) = (-1, 2)$

59. $3x + 4y \geq -7$
$2x - 3y \geq 1$

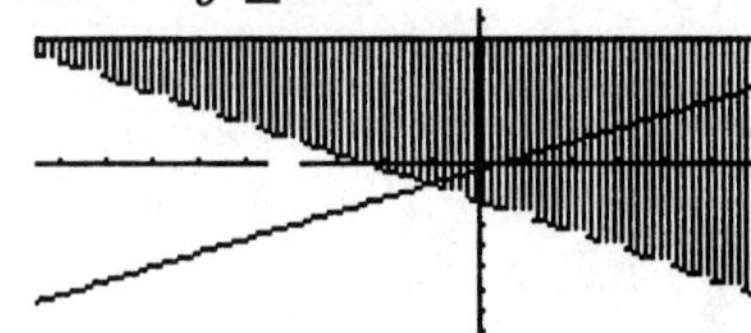

61. $\sqrt{50x^2} = 5x\sqrt{2}$

63. $3\sqrt{24} + \sqrt{54}$
$= 3(2)\sqrt{6} + 3\sqrt{6}$
$= 9\sqrt{6}$

65. $\sqrt{\dfrac{72x^3}{y^2}}$

$= \dfrac{\sqrt{72x^3}}{\sqrt{y^2}}$

$= \dfrac{6x\sqrt{2x}}{y}$

67. $\sqrt{6x+1} + 2 = 7$

$\sqrt{6x+1} = 5$

$6x + 1 = 25$

$6x = 24$

$x = 4$

69. $t^2 = 75$

$t = \pm\sqrt{25 \cdot 3}$

$t = \pm 5\sqrt{3}$

71. $3x^2 - x - 1 = 0$

$x = \dfrac{-b \pm \sqrt{b^2 - 4ac}}{2a}$

$x = \dfrac{-(-1) \pm \sqrt{(-1)^2 - 4(3)(-1)}}{2(3)}$

$x = \dfrac{1 \pm \sqrt{1 + 12}}{6}$

$x = \dfrac{1 \pm \sqrt{13}}{6}$

$x \approx -0.43$

$x \approx 0.77$

73. $y = x^2 + 6x + 5$

The vertex $x$ coordinate is $\dfrac{-6}{2(1)} = -3$. $y = (-3)^2 + 6(-3) + 5$

$y = 9 - 18 + 5$

$y = -4$

The vertex is located at $(-3, -4)$.

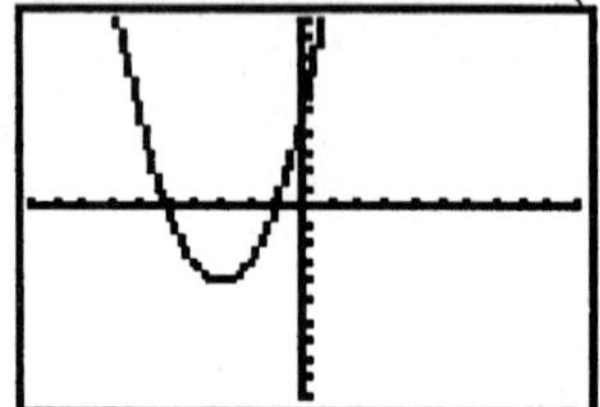

**Appendix 1 Inductive and Deductive Reasoning**

**Vocabulary**

1.  **Inductive** reasoning draws general conclusions from specific observations.

**Concepts**

3.  The pattern is circular.
5.  The pattern is alternating.
7.  The pattern is alternating.

9.  ROOM SCHEDULING
    The room is available on Wednesday at 10 a.m.

**Practice**

11. $1, 5, 9, 13, \underline{17}, \ldots$
13. $-3, -5, -8, -12, \underline{-17}, \ldots$
15. $-7, 9, -6, 8, -5, 7, -4, \underline{6}, \ldots$
17. $9, 5, 7, 3, 5, 1, \underline{3}, \ldots$
19. $-2, -3, -5, -6, -8, -9, \underline{-11}, \ldots$
21. $6, 8, 9, 7, 9, 10, 8, 10, 11, \underline{9}, \ldots$
23.

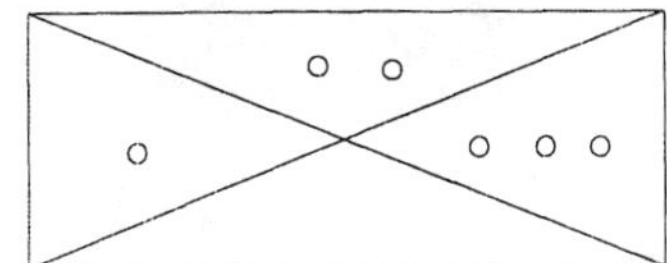

25.

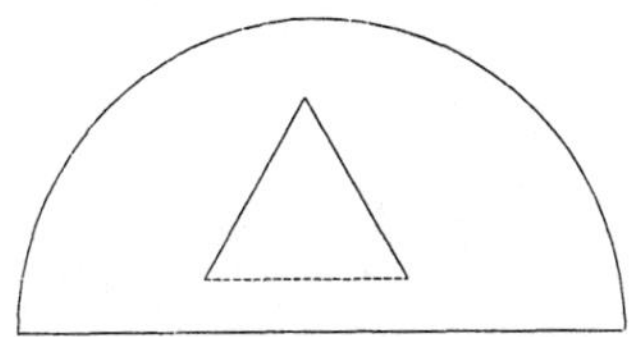

27. $A, c, E, g, \underline{I}, \ldots$
29. $d, h, g, k, j, n, \underline{m}, \ldots$
31. Maria must be the teacher since we know the baker is Luis and that Maria is unmarried.
33. Think of the four vehicles in spaces one through four, left to right. The Buick is in space 1, for the Ford to be between the Dodge and Mercedes, the Mercedes must be in space 4, the right end.
35. From top to bottom the flag colors are green, blue, yellow, then red.

## 37. JURY DUTY

There are 18,935 respondents who have served on neither a criminal court nor a civil jury, see the Venn diagram.

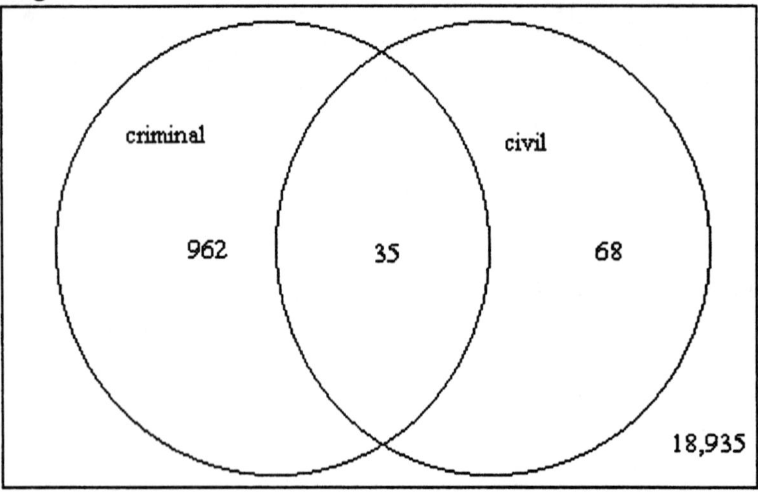


## 39. THE SOLAR SYSTEM

There are zero planets that are neither rocky nor have moons.

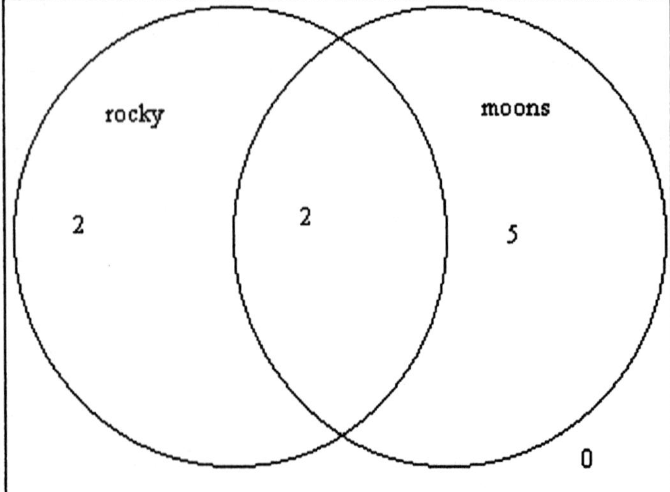

# Appendix II Measurement

## II.1 American Units of Measurement

### Vocabulary

1.  Inches, feet and miles are examples of American units of **length**.
3.  The value of any unit conversion factor is **1**.
5.  Some examples of American units of **capacity** are cups, pints, quarts, and gallons.

### Concepts

7.  12 in = 1 ft
9.  1 mi = 5,280 ft
11. 16 ounces = 1 pound
13. 1 cup = 8 fluid ounces
15. 2 pints = 1 quart
17. 1 day = 24 hours

19. From left to right the arrows point to $\frac{5}{8}$ inch, $1\frac{3}{4}$ inch and $2\frac{5}{16}$ inches.

21. a. $\dfrac{1 \text{ ton}}{2,000 \text{ lb}}$

    b. $\dfrac{2 \text{ pints}}{1 \text{ qt}}$

23. a. Length of the U.S. coastline matches with iv) 12,383 mi
    b. Height of a Barbie doll matches with i) 11 ½ inches
    c. Span of the Golden Gate Bridge matches with ii) 4,200 feet
    d. Width of a football field matches with iii) 53.5 yd

25. a. Amount of blood in an adult matches with iii) 5 qt
    b. Size of the Exxon Valdez oil spill in 1989 matches with iv) 10,080,000 gal
    c. Amount of nail polish in a bottle matches with i) ½ fluid oz
    d. Amount of flour to make 3-dozen cookies matches with ii) 2 cups

### Notation

27. $12 \text{ yd} = 12 \text{ yd} \cdot \dfrac{36 \text{ in}}{1 \text{ yd}}$

      $= 12 \cdot 36 \text{ in}$

      $= 432 \text{ in}$

29. $12 \text{ pt} = 12 \text{ pt} \cdot \dfrac{1 \text{ qt}}{2 \text{ pt}} \cdot \dfrac{1 \text{ gallon}}{4 \text{ qt}}$

      $= \dfrac{12 \cdot 1 \cdot 1}{2 \cdot 4} \text{ gallons}$

      $= 1.5 \text{ gallons}$

**Practice**

31. The width of dollar bill is approximately 2 5/8 inches.
33. The length of this page is approximately 10 ¾ inches.
35. 4 feet is 48 inches

37. $3\frac{1}{2}$ feet is 42 inches

39. 24 inches is 2 feet
41. 8 yards is 288 inches

43. 90 inches is $2\frac{1}{2}$ yards

45. 56 inches is $4\frac{2}{3}$ feet

47. 5 yards is 15 feet

49. 7 feet is $2\frac{1}{3}$ yards

51. 15,840 feet is 3 miles
53. ½ mile is 2,640 feet
55. 80 ounces is 5 pounds
57. 7,000 pounds is 3 ½ tons
59. 12.4 tons is 24,800 pounds
61. 3 quarts is 6 pints
63. 16 pints is 2 gallons
65. 32 fluid ounces is 2 pints
67. 240 minutes is 4 hours
69. 7,200 minutes is 5 days

**Applications**

71. THE GREAT PYRAMID
    450 feet is 150 yards
    $$\frac{450}{3} = 150$$

73. THE GREAT SPHINX
    240 feet is 2,880 inches
    $$240(12) = 2,880$$

75. THE SEARS TOWER
    1,454 feet is approximately 0.28 miles
    $$\frac{1,454}{5,280} \approx 0.27538$$

77. NFL RECORDS
    35 miles is 61,600 yards
    $$\frac{35(5,280)}{3} = 61,600$$

79. WEIGHT OF WATER
    1 gallon is $16(8) = 128$ ounces

81. HIPPOS

9,900 pounds is 4.95 tons

$$\frac{9,900}{2,000} = 4.95$$

83. BUYING PAINT

17 gallons is 68 quarts

$$17(4) = 68$$

85. SCHOOL LUNCHES

575 pints is $71\frac{7}{8}$ gallons

$$\frac{575}{8} = 71.875$$

87. CAMPING

2 ½ gallons is 320 ounces

$$\frac{5}{2}(128) = 320$$

89. SPACE TRAVEL

147 hours is $6\frac{1}{8}$ days

$$\frac{147}{24} = 6.125$$

## II.2 Metric Units of Measurement

### Vocabulary

1. Deka means **tens.**
3. Kilo means **thousands.**
5. Centi means **hundredths.**
7. Meters, grams, and liters are units of measurement in the **metric** system.

### Concepts

9. From left to right the measurement arrows point to 1 cm, 3 cm, and 6 cm.

11. a. $\dfrac{1\text{ km}}{1,000\text{ m}}$

    b. $\dfrac{100\,\text{cg}}{1\text{ g}}$

    c. $\dfrac{1,000\text{ ml}}{1\text{ liter}}$

13. a. Thickness of a phone book matches with iii) 6 cm
   b. Length of the Amazon River matches with i) 6,275 km
   c. Height of a soccer goal matches with ii) 2 m

15. a. Amount of blood in an adult matches with ii) 6 L
   b. Cola in an aluminum can matches with iii) 355 mL
   c. Kuwait's daily production of crude oil matches with i) 290,000 kL

17. 1 dekameter = 10 meters

19. 1 centimeter = $\dfrac{1}{100}$ meter

21. 1 millimeter = $\dfrac{1}{1,000}$ meters

23. 1 gram = 1,000 milligrams
25. 1 kilogram = 1,000 grams
27. 1 liter = 1,000 cubic centimeters

29. 1 centiliter = $\dfrac{1}{100}$ liter

31. 100 liters = 1 hectoliter

### Notation

33. $20\text{ cm} = 20\text{ cm} \cdot \dfrac{1m}{100\text{ cm}}$

          $= \dfrac{20}{100}\text{ m}$

          $= 0.2\text{ m}$

35. $2 \text{ km} = 2 \text{ km} \cdot \dfrac{1{,}000 \text{ m}}{1 \text{ km}} \cdot \dfrac{10 \text{ dm}}{1 \text{ m}}$

$\quad = 2 \cdot 1{,}000 \cdot 10 \text{ dm}$

$\quad = 20{,}000 \text{ dm}$

## Practice

37. The length of a dollar bill is approximately 156 mm.
39. The length of this page is approximately 28 cm.
41. 3 m = 300 cm
43. 5.7 m = 570 cm
45. 0.31 dm = 3.1 cm
47. 76.8 hm = 7,680,000 mm
49. 4.72 cm = 0.472 dm
51. 453.2 cm = 4.532 m
53. 0.325 dm = 0.0325 m
55. 3.75 cm = 37.5 mm
57. 0.125 m = 125 mm
59. 675 dam = 675,000 cm
61. 638.3 m = 6.383 hm
63. 6.3 mm = 0.63 cm
65. 695 dm = 69.5 m
67. 5,689 m = 5.689 km
69. 576.2 mm = 5.762 dm
71. 6.45 dm = 0.000645 km
73. 658.23 m = 0.65823 km
75. 3 g = 3,000 mg
77. 2 kg = 2,000 g
79. 1,000 kg = 1,000,000 g
81. 500 mg = 0.5 g
83. 3 kL = 3,000 L
85. 500 cL = 5,000 mL
87. 10 mL = 10 cc

## Applications

89. SPEED SKATING
    500 m = 0.5 km
    1,000 m = 1 km
    1,500 m = 1.5 km
    5,000 m = 5 km
    10,000 m = 10 km

91. HEALTH CARE
    120 mm = 12 cm
    80 mm = 8 cm

93. WEIGHT OF A BABY
    4 kg = 400,000 cg

95. CONTAINERS
    2(2L) = 4L = 40 dL

## 97. BUYING OLIVES

1,000 g = 1 kg so 4 bottles of olives is 1,136 grams or 1.136 kg.

## 99. MEDICINE

In the bottle there are $60(50) = 3,000$ mg $= 3$ g of active ingredient.

## Appendix II.3 Converting between American and Metric Units

### Vocabulary

1.  In the American system, temperatures are measured in degrees **Fahrenheit**.

### Concepts

3.  a. A meter is longer than a yard.
    b. A meter is longer than a foot.
    c. An inch is longer than a centimeter.
    d. A mile is longer than a kilometer.

5.  a. A liter is greater than a pint.
    b. A liter is greater than a quart.
    c. A gallon is greater than a liter.

### Notation

7.  $4,500 \text{ ft} = 4,500 \text{ ft} (0.3048 \text{ m/ft})$
    $$= 1371.6 \text{ m}$$
    $$= 1.3716 \text{ km}$$

9.  $8 \text{ L} = 8 \text{ L} (0.264 \text{ gal/L})$
    $$= 2.112 \text{ gal}$$

### Practice

11. $3 \text{ ft} = 91.4 \text{ cm}$
13. $3.75 \text{ m} = 147.6 \text{ in}$
15. $12 \text{ km} = 39,372 \text{ ft}$
17. $5,000 \text{ in} = 127 \text{ m}$
19. $37 \text{ oz} = 1 \text{ kg}$
21. $25 \text{ lb} = 11,350 \text{ g}$
23. $0.5 \text{ kg} = 17.6 \text{ oz}$
25. $17 \text{ g} = 0.6 \text{ oz}$
27. $3 \text{ fl oz} = 0.1 \text{ L}$
29. $7.2 \text{ L} = 243.4 \text{ fl oz}$
31. $0.75 \text{ qt} = 710 \text{ mL}$
33. $500 \text{ mL} = 0.5 \text{ qt}$
35. $50^0 \text{ F} = 10^\circ \text{ C}$
37. $50^0 \text{ C} = 122^\circ \text{ F}$
39. $-10^0 \text{ C} = 14^\circ \text{ F}$
41. $-5^0 \text{ F} = -20.6^\circ \text{ C}$

**Applications**

43. THE MIDDLE EAST
To the nearest mile, 8 km is 8(0.6214) = 5 miles.

45. CHEETAH
112 km/hr = 112(0.6214) = 170 mph

47. MOUNT WASHINGTON
6,288 ft = 6,288(0.3048/1000) = 1.9 km

49. HAIR GROWTH
¾ inch per month is approximately ¾(2.54) = 1.9 cm per month

51. WEIGHTLIFTING
Ms. Steenrod's weight class was 82.5/0.454 = 181.7 pounds and her bench press was 132.5/0.454 = 291.9 pounds. Mr. Magruder's weight class was 110/0.454 = 242.3 pounds and his bench press was 270/0.454 = 594.7 pounds.

53. OUNCES AND FLUID OUNCES
a. 8 oz = 8(28.35) = 226.8 g
b. 8 fluid oz = 8/33.8 = 0.24 L

55. POSTAL REGULATIONS
32 kg = 32(2.2) = 70.4 pounds therefore it cannot be sent by priority mail

57. COMPARISON SHOPPING
Using 1 qt = 0.946 L, then 3 qt = 2.838 L, the price per liter is $\dfrac{\$\,4.50}{2.838\text{ L}} = \$\,1.59$ per liter . The price per liter for the two liter root beer is $\dfrac{\$\,3.60}{2\text{ L}} = \$1.80$ per liter .

The 3 quarts is a better buy.

59. HOT SPRINGS
$$C = \frac{5}{9}(F - 32)$$
$$C = \frac{5}{9}(143 - 32)$$
$$C = 61.7^0$$

61. TAKING A SHOWER
The temperature to choose is $28^0$ C.

$$F = \frac{9}{5}C + 32 \qquad F = \frac{9}{5}C + 32 \qquad F = \frac{9}{5}C + 32$$
$$F = \frac{9}{5}(28) + 32 \qquad F = \frac{9}{5}(15) + 32 \qquad F = \frac{9}{5}(50) + 32$$
$$F = 82.4^0 \qquad\qquad F = 59^0 \qquad\qquad F = 122^0$$

63. SNOWY WEATHER

It might snow at $-5^0$ C and $0^0$ C.

$$F = \frac{9}{5}C + 32 \qquad F = \frac{9}{5}C + 32 \qquad F = \frac{9}{5}C + 32$$

$$F = \frac{9}{5}(0) + 32 \qquad F = \frac{9}{5}(-5) + 32 \qquad F = \frac{9}{5}(10) + 32$$

$$F = 32^0 \qquad F = 23^0 \qquad F = 50^0$$

**Appendix III Solving Absolute Values and Inequalities**

**Vocabulary**

1. $|2x - 1| = 10$ is an absolute value **equation**.

3. To **isolate** the absolute value in $|3 - x| - 4 = 5$, we add 4 to both sides.

**Concepts**

5. $|x| \geq 0$ for all real numbers $x$.

7. To solve $|x| > 5$, we must find the coordinates of all points on a number line that are **more than** 5 units from the origin.

9. To solve $|x| = 5$, we must find the coordinates of all points on a number line that are 5 units from the origin.

11. a. $x = -3$ is a solution for $|x - 1| = 4$
    b. $x = -3$ is not a solution for $|x - 1| > 4$
    c. $x = -3$ is a solution for $|x - 1| \leq 4$
    d. $x = -3$ is not a solution for
       $|5 - x| = |x + 12|$

**Notation**

13. a. ii
    b. iii
    c. i

15. $-4 < x < 4, |x| < 4$

17. $x + 3 < -6$ or $x + 3 > 6, |x + 3| > 6$

**Practice**

19. $|8| = 8$

21. $-|0.02| = -0.02$

23. $-\left|-\frac{31}{16}\right| = -\frac{31}{16}$

25. $|\pi| = \pi$

27. $|x| = 23, x = 23$ or $x = -23$

29. $|x - 3.1| = 6$
    $x - 3.1 = 6$ or $x - 3.1 = -6$
    $x = 9.1$ or $x = -2.9$

31. $|3x + 2| = 16$
    $3x + 2 = 16$ or $3x + 2 = -16$
    $3x = 14$ or $3x = -18$
    $x = \frac{14}{3}$ or $x = -6$

33. $|\frac{7}{2}x + 3| = -5$
No Solution

35. $|3 - 4x| = 5$
    $3 - 4x = 5$ or $3 - 4x = -5$
    $-4x = 2$ or $-4x = -8$
    $x = -\frac{1}{2}$ or $x = 2$

37. $2|3x + 24| = 0$
    $3x + 24 = 0$
    $3x = -24$
    $x = -8$

39. $\left|\frac{3x+48}{3}\right| = 12$
    $\frac{3x+48}{3} = 12$ or $\frac{3x+48}{3} = -12$
    $3x + 48 = 36$ or $3x + 48 = -36$
    $3x = -12$ or $3x = -84$
    $x = -4$ or $x = -28$

41. $|x + 3| + 7 = 10$
    $|x + 3| = 3$
    $x + 3 = 3$ or $x + 3 = -3$
    $x = 0$ or $x = -6$

43. $|2x + 1| = |3x + 3|$
    $2x + 1 = 3x + 3$ or $2x + 1 = -(3x + 3)$
    $-2 = x$ or $2x + 1 = -3x - 3$
    $-2 = x$ or $5x = -4$
    $-2 = x$ or $x = -\frac{4}{5}$

45. $|2 - x| = |3x + 2|$
    $2 - x = 3x + 2$ or $2 - x = -(3x + 2)$
    $0 = 4x$ or $2 - x = -3x - 2$
    $0 = x$ or $2x = -4$
    $0 = x$ or $x = -2$

47. $|\frac{x}{2} + 2| = |\frac{x}{2} - 2|$
    $\frac{x}{2} + 2 = \frac{x}{2} - 2$ or $\frac{x}{2} + 2 = -\left(\frac{x}{2} - 2\right)$
    $0 \neq -4$ or $\frac{x}{2} + 2 = -\frac{x}{2} + 2$
    $0 \neq -4$ or $x = 0$

49. $|x + \frac{1}{3}| = |x - 3|$

    $x + \frac{1}{3} = x - 3$ or $x + \frac{1}{3} = -(x - 3)$

    $0 \neq -\frac{10}{3}$ or $x + \frac{1}{3} = -x + 3$

    $0 \neq -\frac{10}{3}$ or $2x = \frac{8}{3}$

    $0 \neq -\frac{10}{3}$ or $x = \frac{4}{3}$

51. $|x| < 4, (-4, 4),$   $\longleftarrow(\underline{\qquad})\longrightarrow$
                                      $-4$      $4$

53. $|x + 9| \leq 12$

    $x + 9 \leq 12$ or $x + 9 \geq -12$

    $x \leq 3$ or $x \geq -21$

    $[-21, 3]$

    $\longleftarrow[\underline{\qquad}]\longrightarrow$
      $-21$      $3$

55. $|3x - 2| \leq 10$

    $-10 \leq 3x - 2 \leq 10$

    $-8 \leq 3x \leq 12$

    $-\frac{8}{3} \leq x \leq 4$

    $\left[-\frac{8}{3}, 4\right]$

    $\longleftarrow[\underline{\qquad}]\longrightarrow$
      $-\frac{8}{3}$      $4$

57. $|3x + 2| \leq -3$

    No Solution

59. $|x| > 3$

    $x > 3$ or $x < -3$

    $(-\infty, -3) \cup (3, \infty)$

    $\longleftarrow\underline{\qquad})\underline{\quad}(\underline{\qquad}\longrightarrow$
        $-3$    $3$

61. $|x - 12| > 24$

    $x - 12 > 24$ or $x - 12 < -24$

    $x > 36$ or $x < -12$

    $(-\infty, -12) \cup (36, \infty)$

    $\longleftarrow\underline{\qquad})\underline{\quad}(\underline{\qquad}\longrightarrow$
       $-12$    $36$

63. $|3x + 2| > 14$

    $3x + 2 > 14$ or $3x + 2 < -14$

    $3x > 12$ or $3x < -16$

    $x > 4$ or $x < -\frac{16}{3}$

    $\left(-\infty, -\frac{16}{3}\right) \cup (4, \infty)$

    $\longleftarrow\underline{\qquad})\underline{\quad}(\underline{\qquad}\longrightarrow$
      $-\frac{16}{3}$    $4$

65. $|4x + 3| > -5$

Absolute value is always positive or zero, therefore it will always be greater than $-5$.

$(-\infty, \infty)$

67. $|2 - 3x| \geq 8$

$2 - 3x \geq 8 \text{ or } 2 - 3x \leq -8$

$-3x \geq 6 \text{ or } -3x \leq -10$

$x \leq -2 \text{ or } x \geq \frac{10}{3}$

$\left(-\infty, -2\right] \cup \left[\frac{10}{3}, \infty\right)$

69. $-|2x - 3| < -7$

$|2x - 3| > 7$

$2x - 3 > 7 \text{ or } 2x - 3 < -7$

$2x > 10 \text{ or } 2x < -4$

$x > 5 \text{ or } x < -2$

$(-\infty, -2) \cup (5, \infty)$

71. $\left|\frac{x-2}{3}\right| \leq 4$

$-4 \leq \frac{x-2}{3} \leq 4$

$-12 \leq x - 2 \leq 12$

$-10 \leq x \leq 14$

$[-10, 14]$

73. $|3x + 1| + 2 < 6$

$|3x + 1| < 4$

$-4 < 3x + 1 < 4$

$-5 < 3x < 3$

$-\frac{5}{3} < x < 1$

$\left(-\frac{5}{3}, 1\right)$

75. $\left|\frac{1}{3}x + 7\right| + 5 > 6$

$\left|\frac{1}{3}x + 7\right| > 1$

$\frac{1}{3}x + 7 > 1 \text{ or } \frac{1}{3}x + 7 < -1$

$\frac{1}{3}x > -6 \text{ or } \frac{1}{3}x < -8$

$x > -18 \text{ or } x < -24$

$(-\infty, -24) \cup (-18, \infty)$

**Applications**

77. TEMPERATURE RANGE
$$|t - 78| \leq 8$$
$$-8 \leq t - 78 \leq 8$$
$$70 \leq t \leq 86$$

79. AUTO MECHANICS
a. Let $c$ represent the camber angle, then
$$|c - 0.6| \leq 0.5$$
b. $|c - 0.6| \leq 0.5$
$$c - 0.6 \leq 0.5 \ \text{ or } \ c - 0.6 \geq -0.5$$
$$c \leq 1.1 \ \text{ or } \ c \geq 0.1$$
$$[0.1, 1.1]$$

81. ERROR ANALYSIS
a. $[p - 25.46| \leq 1.00$
$$p - 25.46 \leq 1.00 \ \text{ or } \ p - 25.46 \geq -1.00$$
$$p \leq 26.46 \ \text{ or } \ p \geq 24.46$$
$$[24.46, 26.46]$$
From the illustration 26.45% and 24.76% satisfy the absolute value inequality.
b. The amount of error in part a is less than or equal to one percent.

**Appendix IV Probability**

**Vocabulary**

1. An experiment is any process for which the outcome is uncertain.

**Concepts**

3. If an event E can occur in n ways out of $s$ equally likely ways, then $P(E) = \frac{n}{s}$.

5. If an event cannot happen, its probability is 0.

**Practice**

7. $\{(1, H), (2, H), (3, H), (4, H), (5, H), (6, H), (1, T), (2, T), (3, T), (4, T), (5, T), (6, T)\}$

9. $\{a, e, i, o, u\}$

11. $\frac{1}{6}$

13. $\frac{2}{3}$

15. $\frac{5}{36}$

17. 0

19. $\frac{19}{42}$

21. $\frac{13}{42}$

23. $\frac{1}{4}$

25. $\frac{1}{4}$

27. $\frac{1}{2}$

29. $\frac{3}{13}$

31. $\frac{5}{12}$

**Applications**

33. Let $f$ represent failure and $s$ represent non-failure, then the sample space is:
$$\{(ffff), (fffs), (ffsf), (fsff), (sfff),$$
$$(ffss), (fsfs), (fssf), (ssff), (sfsf), (sffs)$$
$$(ssss), (sssf), (ssfs), (sfss), (fsss)\}$$

35. $\frac{4}{16} = \frac{1}{4}$

37. $\frac{4}{16} = \frac{1}{4}$

39. $\frac{1}{16} + \frac{1}{4} + \frac{3}{8} + \frac{1}{4} + \frac{1}{16}$
    $= \frac{1}{16} + \frac{4}{16} + \frac{6}{16} + \frac{4}{16} + \frac{1}{16}$
    $= \frac{16}{16}$
    $= 1$

41. $\frac{32}{119}$